Multi-Sensor Data Fusion with MATLAB®

Multi-Sensor Data Fusion with MATLAB®

Jitendra R. Raol

CRC Press is an imprint of the
Taylor & Francis Group, an **informa** business

CRC Press
Taylor & Francis Group
6000 Broken Sound Parkway NW, Suite 300
Boca Raton, FL 33487-2742

Printed in the United States of America on acid-free paper
10 9 8 7 6 5 4 3 2 1

International Standard Book Number: 978-1-4398-0003-4 (Hardback)

Library of Congress Cataloging-in-Publication Data

Raol, J. R. (Jitendra R.), 1947-
Multi-sensor data fusion with MATLAB / Jitendra R. Raol.
p. cm.
"A CRC title."
Includes bibliographical references and index.
ISBN 978-1-4398-0003-4 (hardcover : alk. paper)
1. Multisensor data fusion—Data processing. 2. MATLAB. 3. Detectors. I. Title.

TA331.R36 2010
681'.2—dc22 2009041607

Visit the Taylor & Francis Web site at
http://www.taylorandfrancis.com

and the CRC Press Web site at
http://www.crcpress.com

The book is dedicated

in loving memory to

Professor P. N. Thakre

(M. S. University of Baroda, Vadodara),

Professor Vimal K. Dubey

(Nanyang Technological University, Singapore),

and

Professor Vinod J. Modi

(University of British Columbia, Canada)

Contents

Part II: Fuzzy Logic and Decision Fusion

(J. R. Raol and S. K. Kashyap)

Part III: Pixel- and Feature-Level Image Fusion

(J. R. Raol and V. P. S. Naidu)

Preface

The human brain routinely and almost spontaneously carries out a lot of information processing and fusion. This is possible due to the biological neural networks in our brains, which are actually parallel processing systems of adaptive switching (chemical circuits) units. The main objectives are to collect measurements and simple observations from various similar or dissimilar sources and sensors, extract the required information, draw logical inferences, and then combine or fuse these with a view toward obtaining an enhanced status and identity of the perceived or observed object, scene, or phenomenon. These acts of information processing and decision making are very crucial for the survival and growth of human beings, as well as many other living creatures, and can be termed *multisource multisensor information fusion* (MUSSIF), more popularly known as *sensor data fusion* (DF).

MUSSIF is very rapidly emerging as an independent discipline to be reckoned with and finds ever-increasing applications in many biomedical, automation industry, aerospace, robotics, and environmental engineering processes and systems, in addition to typical defense applications. MUSSIF offers one or more of the following benefits: more spatial coverage of the object under observation, redundancy of measurements, robustness of the system's performance and higher accuracy (basically reduced uncertainty of prediction) of inferences, and an overall assured performance of the sensor-integrated systems. The complete process of MUSSIF involves the study of several related disciplines: (1) signal and image processing; (2) numerical computational algorithms; (3) statistical and probabilistic approaches and methods; (4) sensor modeling, management, control, and optimization; (5) neural networks, fuzzy logic systems, and genetic algorithms; (6) system identification and state or parameter estimation; and (7) database management. Many principles and techniques from these fields strengthen the definition of tasks, analysis, and performance evaluations of multisensor DF (MSDF) systems. Several of these aspects are briefly discussed in this book.

In this book, theories, concepts, and applications of MSDF are treated in three parts: (1) kinematic-level fusion (including theory of DF); (2) fuzzy logic and decision fusion; and (3) pixel- and feature-level image fusion. The development elucidates aspects and concepts of DF strategies, algorithms, and performance evaluations, mainly for aerospace applications. However, the concepts and methods discussed are equally applicable to other systems. Where possible, this is illustrated with examples via numerical simulations coded in MATLAB®. (MATLAB® is the trade mark of The MathWorks Inc. For product information, please contact: The MathWorks, Inc., 3 Apple Hill Drive, Natick, MA 01760-2098 USA, Tel: 508 647 7000, Fax: 508-647-7001, E-mail: info@mathworks.com, Web: www.mathworks.com.) The user should have access to PC-based MATLAB software and other toolboxes such as signal processing, control systems, system identification, neural networks, fuzzy logic, and image processing.

There are other books on sensor DF; however, the treatment of the aspects outlined above is somewhat limited or highly specialized. The treatment of these topics in the present book is comprehensive and also briefly covers several related disciplines that will help in understating sensor DF concepts and methods from a practical point of view. An engineering approach is employed, rather than a purely mathematical approach, without loosing sight of the necessary mathematics. Where appropriate, some novel methods, approaches, techniques, and algorithms are presented.

The end users of this integrated technology of MSDF will be systems, aerocontrol, mechanical, and civil educational institutions; several research and development laboratories; aerospace and other industries; medical diagnostic and biomedical units; civil–military transportation; the automation and mining industries; robotics; and mobile intelligent autonomous systems.

Acknowledgments

Researchers all over the world have been making important contributions to this specialized field for the last three to four decades. This field is rapidly emerging as an enabling technology to be reckoned with, especially in aerospace science and technology applications, robotics, and some industrial spin-offs.

More than a decade ago, DF research and development was also initiated in defense laboratories in India, with a major focus on target tracking and building expert systems to aid range safety officers at flight testing agencies. The authors are very grateful to Air Commodores P. Banerjee and R. Appavu Raj (of Integrated Test Range [ITR], Defence Research and Development Organisation [DRDO]) for conceiving of the idea of the RTFLEX expert system, which gave an incentive to a chain of developmental projects from other organizations in the country in the area of MSDF. Some theoretical and software development work was also initiated in certain educational institutions. The authors are very grateful to Dr. T. S. Prahlad and Dr. S. Srinathkumar of the National Aerospace Laboratories (NAL) for supporting DF activities for more than a decade. The authors are also very grateful to Professor T. K. Ghoshal of Jadavapur University in Kolkata for his support of DF research in various forms and for various applications. The authors are also grateful to Dr. V. V. Murthy and Dr. R. N. Bhattacharjee of the Defence Research and Development Laboratory (DRDL) in Hyderabad, and to Professor Ananthasayanam at The Indian Institute of Science (IISc) in Bangalore for their technical and moral support of DF activities at NAL. We are also grateful to the previous and present directors of NAL for supporting DF research and project activities.

Dr. Raol is very grateful to Dr. S. Balakrishna of NAL; Professor N. K. Sinha (professor emeritus at McMaster University in Canada) Professor R. C. Desai (M.S. University of Baroda), Dr. Gangan Prathap (VC, CUSAT, Kerala), Professor M. R. Kaimal (University of Kerala), Dr. M. R. Nayak (Advisor, M and A, NAL), Professor P. R. Viswanath (NAL and IISc, Bangalore), Dr. U. N. Sinha, Dr. R. M. V. G. K. Rao, and Dr. T. G. Ramesh for their encouragement and moral support for several years. The author is very grateful to Professor Florin Ionescu, the Director of Mechatronics at the University of Applied Sciences in Konstanz, Germany for supporting him during his DFG fellowship in 2001. The author is also very grateful to Andre Nepgen, Dr. Motodi M., Johan Strydom, and Dr. Ajith Gopal for providing him the opportunity for sabbatical at the CSIR (SA). The constant technical support from several colleagues of the Flight Mechanics and

Control Division of NAL (FMCD), ITR, and DRDL is gratefully appreciated. Certain interactions with the DLR Institute of Flight Systems for a number of years have been very useful to us. The author is also very grateful to Mrs. Sarala Udayakanth, Mrs. Prasanna Mukundan, V. Kodantharaman, R. Giridhar, K. Nagaraj, T. Vijeesh, and R. Rajesh of the FMCD office staff, who helped him for several years in managing the administrative tasks and office affairs of the division. We are also grateful to CRC Press (USA) and especially to Mr. Jonathan Plant for their full support of this book project. As ever, we are very grateful to our spouses and children for their endurance, care, affection, patience, and love.

Author

Jitendra R. Raol received BE and ME degrees in electrical engineering from the M.S. University of Baroda, Vadodara, in 1971 and 1973, respectively, and a PhD (in electrical and computer engineering) from McMaster University, Hamilton, Canada, in 1986. He taught for two years at the M.S. University of Baroda before joining the National Aerospace Laboratories (NAL) in 1975. At NAL, he was involved in activities on human pilot modeling in fixed and motion-based research simulators. He rejoined NAL in 1986 and retired on July 31, 2007 as Scientist-G (and head of the flight mechanics and control division at NAL). He has visited Syria, Germany, the United Kingdom, Canada, China, the United States, and South Africa on deputation and fellowships to work on research problems on system identification, neural networks, parameter estimation, MSDF, and robotics, and to present several technical papers at several international conferences. He became a fellow of the IEE in the United Kingdom and a senior member of the IEEE in the United States. He is a life fellow of the Aeronautical Society of India and a life member of the System Society of India. In 1976, he won the K.F. Antia Memorial Prize of the Institution of Engineers in India for his research paper on nonlinear filtering. He was awarded a certificate of merit by the Institution of Engineers in India for his paper on parameter estimation for unstable systems. He has received a best poster paper award and was one of the recipients of the Council of Scientific and Industrial Research's (CSIR, India) prestigious technology shield for the year 2003 for his contributions to integrated flight mechanics and control technology for aerospace vehicles.

Dr. Raol has published 100 research papers and several reports. He has guest-edited two special issues of *Sadhana*, an engineering journal published by the Indian Academy of Sciences in Bangalore, covering topics such as advances in modeling, system identification, parameter estimation, and multi-source multi-sensor information fusion (MUSSIF). He has guided several doctoral and master students and has coauthored an IEE (UK) control series book, *Modeling and Parameter Estimation of Dynamic Systems* and a CRC Press (USA) book, *Flight Mechanics Modeling and Analysis*. He has served as a member and chairman of several advisory, technical project review, and doctoral examination committees.

Dr. Raol is now professor emeritus at M.S. Ramaiah Institute of Technology, where his main research interests have been DF, state and parameter estimation, flight mechanics/flight data analysis, filtering, artificial neural networks, fuzzy systems, genetic algorithms, and robotics. He has also authored a book called *Poetry of Life*, a collection of his poems.

Contributors

Dr. G. Girija, Scientist
Head of Data Fusion Group, and Deputy Head, Flight Mechanics and Control Division (FMCD), National Aerospace Laboratories (NAL)
Bangalore, India

Dr. A. Gopal, Senior Design Engineer
British Aerospace Systems (BAeS)
Johannesburg, South Africa

Dr. S. K. Kashyap, Scientist
Data Fusion Group, FMCD, NAL
Bangalore, India

Dr. V. P. S. Naidu, Scientist
Data Fusion Group, FMCD, NAL
Bangalore, India

Dr. N. Shanthakumar, Scientist
Data Fusion Group, FMCD, NAL
Bangalore, India

Dr. S. Utete, Acting Competency Area Manager
MIAS (Mobile Intelligent Autonomous Systems), Modeling and Digital Science (MDS) Unit, Council for Scientific and Industrial Research (CSIR)
Pretoria, South Africa

Introduction

Humans accept input from five sense organs and senses: touch, smell, taste, sound, and sight in different physical formats (and even the sixth sense as mystics tell us) [1]. By some incredible process, not yet fully understood, humans transform input from these organs within the brain into the sensation of being in some "reality." We need to feel or be assured that we are somewhere, in some coordinates, in some place, and at some time. Thus, we obtain a more complete picture of an observed scene than would have been possible otherwise (i.e., using only one sense organ or sensor). The human activities of planning, acting, investigating, market analysis, military intelligence, complex art work, complex dance sequences, creation of music, and journalism are good examples of activities that use advanced data fusion (DF) aspects and concepts [1] that we do not yet fully understand. Perhaps, the human brain combines such data or information without using any automatic aids, because it has a powerful associative reasoning ability, evolved over thousands of years.

Ours is the information technology (IT) age, and in this context multi-source multi-sensor information fusion (MUSSIF) encompasses the theory, methods, and tools conceived and used for exploiting synergy in information acquired from multiple sources (sensors, databases, information gathered by human senses, etc.). The resulting final understanding of the object (or a process or scene), decision, or action is in some sense better, qualitatively or quantitatively, in terms of accuracy, robustness, etc., and more intelligent than would be possible if any of these sources were used individually without such synergy exploitation. The above seems to be an accepted definition of information fusion [2].

In simple terms, the main objective in sensor DF is to collect measurements and sample observations from various similar or dissimilar sources and sensors, extract the required information, draw inferences, and make decisions. These derived or assessed information and deductions can be combined and fused, with an intent of obtaining an enhanced status and identity of the perceived or observed object or phenomenon. This process is crucial for the survival and growth of humans and many other living creatures, and can be termed MUSSIF, which is popularly called *sensor DF* or even simply *DF* (of course, with derived and simpler meanings and lower level of information processing). Multi-sensor data fusion (MSDF) would primarily involve: (1) hierarchical transformations between observed parameters to generate decisions regarding the location (kinematics and even dynamics), characteristics (features and structures), and the identity of an entity; and (2) inference and interpretation (decision fusion) based on a detailed analysis of the observed scene or entity in the context of a surrounding environment and relationships to other entities.

In a target-tracking application, observations of angular direction, range, and range rate (a basic measurement level fusion of various data) are used for estimating a target's positions, velocities, and accelerations in one or more axes. This is achieved using state-estimation techniques like a Kalman filter [3]. The observations of the target's attributes, such as radar cross section (RCS), infrared spectra, and visual image probably with fusion in mind, may be used to classify the target and assign a label for its identity. Pattern recognition techniques (for feature identification and identity determination) based on clustering algorithms and artificial neural networks (ANNs) or decision-based methods can be used for this purpose, to have more feature-based tracking. Understanding the direction and speed of the target's motion may help us to determine the intent of the target, which may also require automated reasoning or artificial intelligence using implicit and explicit information. For this purpose, knowledge-based methods leading to decision fusion can be used.

DF in a military context has the following meaning [4]: "a multilevel, multifaceted process dealing with the detection, association, correlation, estimation, and combination of data and information from multiple sources to achieve a refined state and identity estimation, and complete and timely assessments of situation and threat."

DF is very rapidly emerging as an independent discipline to be reckoned with and is finding ever-increasing applications in many biomedical, industrial automation, aerospace, and environmental engineering processes and systems, in addition to military defense applications [4–15]. The benefits of DF are more spatial coverage of the object under observation, redundancy of measurements so that they are always available for analysis, robustness of the system's performance, more accurate inferences (meaning better prediction with less uncertainty), and overall assured performance of the multisensor integrated systems. The complete process of DF involves the study of several allied disciplines: (1) signal or image processing; (2) computational and numerical techniques and algorithms; (3) information theoretical, statistical, and probabilistic methods; (4) sensor modeling, sensor management, control, and optimization; (5) neural networks, approximate reasoning, fuzzy logic systems, and genetic algorithms; (6) system identification and state-parameter estimation (least square methods including Kalman filtering); and (7) computational data base management. Several principles and techniques from these fields strengthen the analytical treatment and performance evaluation of DF fusion systems. Some of the foregoing aspects and methods are discussed in the present volume, spread over three parts.

The importance of DF stems from the following considerations, aspects, and feasible applications:

1. Several types of sensors are used in satellite launch vehicles and in satellites.
2. The integrated navigation, guidance, and control–propulsion control (NGC–PC) systems and sensors, sensor data processing and performance requirements, sensor redundancy management, and performance evaluation of NGC–PC system are important in missile, aircraft, rotorcrafts, and spacecraft applications.
3. Use of fused data from gyro and horizon sensors for precision pointing of imaging spacecraft.
4. Importance of interactive multiple modeling (IMM) approach for tracking of maneuvering targets [3] and subsequent use for DF in the multisensor multitarget scenarios.
5. Fusion of information from low-cost inertial platform and sensor systems with global positioning system (GPS)/differential DGPS data in Kalman filters for autonomous vehicles (e.g., unmanned aerial vehicles [UAVs] and robots).
6. Performance evaluation of MSDF application to target tracking in a range safety environment and in general for any DF system.
7. For situation assessment, fuzzy logic–based interfaces for decision fusion can be developed; essentially the output of kinematic fusion and situation assessment can be taken into consideration using a fuzzy logic approach to make a final decision or action in the surveillance volume at any instant of time (in combination with Bayesian networks).
8. Static and dynamic Bayesian networks can be used in situation assessment and DF, leading to more accurate decisions.
9. Integration of identity of targets with the kinematic information.
10. Aircraft cockpit DF for pilot aid and autonomous navigation of vehicles.
11. DF for Mars Rovers and space shuttles.
12. DF in mobile intelligent autonomous systems and robotics (field, service and medical robotics).

Military applications of DF include (1) antimarine warfare; (2) tactical air warfare; and (3) land battle. More specifically, these include [4] (1) command and control (C2) nodes and operations for military forces; (2) intelligence collection data systems; (3) indication and warning (I&W) systems to assess threats and hostile intent; (4) fire control systems—multiple sensors for acquisition, tracking, and command guidance; (5) broad area surveillance—distributed sensor networks (NW); (6) command, control, and communications (C3) operations—use of multiple sensors leading to autonomous weaponry; (7) more recent command, control, communications, and computer intelligence or information (C4-I2) operations with network-centric decision-making capabilities.

Basic algorithms and constituents for multisensor multitarget tracking and fusion (MSMT/F) should incorporate the following aspects or processes: (1) track clustering; (2) data association (often called correlation); (3) track scoring and updating (to help pruning of tracks if need be); (4) track initiation; (5) track management; and (6) track fusion [3]. The central problem in MSMT/F-state estimation is data association—the problem of determining from which target, if any, a particular measurement might have originated. The techniques generally used for MSMT/F applications are (1) simple gating techniques in sparse scenarios; (2) recursive deterministic and PDA techniques in medium-density scenarios; and (3) multiple hypotheses tracking for dense scenarios [4]. Maneuvering target tracking using IMM involves the selection of appropriate models for different regimes of the target flight. Because of the switching between models, there is an exchange of information between the mode filters. 
During each sampling interval, all of the filters are or might be in operation. The overall state estimate is a combination of the estimates from the individual filters.

The radio frequency seeker mounted on a missile gives measurements of range, range rate, the line of sight rate (LOSR), and gimbal angles in a gimbal frame. The range, range rate, and LOSR measurements are corrupted by receiver noise (thermal or shot noise), glint noise, and even RCS fluctuations. Glint noise exhibits a non-Gaussian distribution due to random wandering of the apparent measured position of a target. This is due to the interference of reflections from different elements of the target and also to RCS variation. This noise could be modeled as a combination of Gaussian and Lapalacian noise with appropriate mixing probabilities. Two methods can be used to handle Glint noise: (1) a Masreliez filter based on a nonlinear score function evaluation; which is used as a corrective term in the state estimation equation; and (2) a fuzzy logic–based adaptive filter [9]. Use of models of such noise processes would enhance the accuracy of the DF process.

All of the current generation transport aircraft are equipped with a highly integrated suite of sensors. The data collection devices are used to deliver to the pilot the level of detail of the outside world that she or he requires to fulfill the overall mission. It is essential that a timely, single unified picture of the environment, generated by the most appropriate sensors, is presented to the pilot in a format that can be readily understood and interpreted so that appropriate action can be taken as required. Increasing air traffic control requirements dictate a need for more mental computation on the part of the pilot, which can distract him or her from safe airplane operation. Therefore, there is a need for an on-board aid such as a computational hardware and software (HW/SW) system that increases the pilot's situational awareness, decreases diversions to routine tasks and computations, and anticipates upcoming needs. The cockpit and mission management systems in a transport aircraft should have fully integrated flight and navigation displays (such as multifunction displays) as well as network-centric capabilities, including tactical data link integration, correlation or data association, and DF. The fusion of data from the on-board, off-board, and ground-based sensors to present a single unified picture is at the heart of the cockpit environment for any transport aircraft. The United States has space-based satellites like GPS/DGPS data; Europe, India, and so on, have similar systems. Such systems would also be very useful for military aircraft and UAVs.

Situation assessment models or software for assisting the pilot in decision making during combat missions requires the use of modern classification methods such as fuzzy models for event detection and Bayesian networks to handle uncertainty in modeling situations. Bayesian networks can include many kinds of situations or systems because of their high modularity and expandability.

In this book, the theory, concepts, and applications of DF are treated in four parts: Part I—kinematic-level DF (including theory and concepts of DF), Part II—fuzzy logic and decision fusion, Part III—pixel and feature level image fusion, and Part IV—a brief on DF in other systems. The development elucidates aspects of DF strategies, algorithms (and SW, where appropriate), and performance evaluation, mainly for aerospace applications. However, the concepts and methods discussed are also equally applicable to many other systems. This book will also illustrate certain concepts and algorithms using numerical simulation examples with SW in MATLAB® and will also use exercises, the solution manual for which will be made available to instructors. The SW is not part of the book due to its proprietary nature, so the reader will need access to MATLAB's main SW and related toolboxes. The user should have access to PC-based MATLAB software and other toolboxes: signal and image processing, control systems, system identification, neural networks, and fuzzy logic. There are some good books on sensor DF; however, the treatment of the aspects outlined above is either somewhat limited or highly specialized [3–15]. The treatment in the present book is comprehensive and covers several related disciplines to help in understating sensor DF concepts and methods from a practical point of view. A brief preview of the chapters is given next.

Chapters 1–4 of Part I deal with the theory of DF and kinematic-level DF. Chapter 2 briefly covers the basic concepts and theory of DF including fusion models, fusion methods, sensor modeling, and management. Chapter 3 presents several practical and useful strategies and algorithms for DF and tracking. Where appropriate, some new methods, algorithms, and results are presented. Kalman and information filters are discussed for tracking and decentralized problems, and various data association strategies and algorithms are also discussed. Chapter 4 deals with performance evaluations of some DF algorithms, software, and systems.

Chapters 5–8 of Part II discuss theory of fuzzy logic, fuzzy implication functions, and their application to decision fusion. Chapter 6 discusses the fundamentals of FL concepts, fuzzy sets and their properties, FL operators, fuzzy propositions/rule-based systems, inference engines, and defuzzification methods. A new MATLAB graphical user interface (GUI) tool is developed, explained, and used for evaluating fuzzy implication functions. Chapter 7 discusses using fuzzy logic to estimate the unknown states of a dynamic system by processing sensor data. A fuzzy logic–Bayesian combine is used for situation assessment problems. Chapter 8 highlights some applications and presents some performance evaluation studies.

Chapters 9–11 of Part III cover various aspects of pixel- and image-level registration and fusion. Chapter 10 uses the principal component analysis, spatial frequency, and wavelet-based image fusion algorithms for the fusion of image data from sensors. It also explores the preprocessing of image sequences, spatial domain algorithms, medians, and mean filters. Image segmentation and clustering techniques are employed to detect the position of the target in the background image. Data association techniques such as nearest neighbor Kalman filters (NNKFs) and PDAF are used to track the target in the presence of clutter. Chapter 11 presents procedures for combing tracks obtained from imaging sensor and ground-based radar, as well as some methods and algorithms and their performance evaluation studies.

Chapters 12 and 13 of Part IV briefly discuss some DF aspects as applicable to other systems; however, the importance of the field need not be underestimated.

Appropriate performance evaluation measures and metrics are discussed in the relevant chapters and novel approaches (methods and algorithms) are presented where appropriate. The material of this book could benefit many applications: the integration of the identity of targets with the kinematic information, air and road traffic control and management, cockpit DF, autonomous navigation of vehicles, some security systems, and mobile intelligent autonomous systems (including field, medical, and mining robotics). The end users of this integrated sensor DF technology will be systems, aerocontrol, mechanical, and civil educational institutions; several research and development laboratories; aerospace and other industries; and the transportation and automation industries. Although enough care has been taken in working out the solutions of examples, exercises, and in the presentation of various concepts, theories, algorithms, and case study and numerical results in the book, any practical applications of these should be made with proper care and caution. Any such endeavors would be the readers' own responsibility. Certain MATLAB programs developed for illustrating various concepts via examples in the book are accessible to the readers through CRC.

References

1. Wilfried, E. 2002. *An Introduction to Sensor Fusion*. Research Report 47/2001. Institute fur Technische Informatik, Vienna University of Technology, Austria.
2. Dasarathy, B. V. 2001. Information fusion—What, where, why, when, and how? (Editorial) *Inf Fusion* 2: 75–6.
3. Bar-Shalom, Y., and X.-R. Li. 1995. *Multi-target Multi-sensor Tracking (Principles and Techniques)*, Storrs, CT: YBS.
4. Waltz E., and J. Llinas. 1990. *Multisensor data fusion*. Boston: Artech House Publishers.
5. Hall, D. L. 1992. *Mathematical techniques in multisensor data fusion*. Norwood, MA: Artech House.
6. Abidi, M. A., and R. C. Gonzalez, eds. 1992. *Data fusion in robotics and machine intelligence*. USA: Academic Press.
7. Hager, G. D. 1990. *Task-directed sensor fusion and planning (A Computational Approach)*. Norwell, MA: Kluwer Academic Publishers.
8. Klein, L. A. 2004. *Sensor and data fusion: A tool for information assessment and decision making*. Bellingham, WA: SPIE Press.
9. Harris, C., X. Hong, and Q. Gan. 2002. *Adaptive modelling, estimation and fusion from data: A neurofuzzy approach*. Berlin: Springer.
10. Manyika, J., and H. Durrant-White. 1994. *Data fusion and sensor management: A decentralized information—Theoretic approach*. Ellis Horwood Series.
11. Dasarathy, B. V. 1994. *Decision fusion*. USA: Computer Society Press.
12. Mitchell, H. B. 2007. *Multi-sensor data fusion*. Berlin: Springer.
13. Varshney, P. K. 1997. *Distributed detection and data fusion*. Berlin: Springer.
14. Clark, J. J. 1990. *Data fusion for sensory information processing systems*. Norwell, MA: Kluwer.
15. Goodman, I. R., R. P. S. Mahler, and H. T. Nguyen. 1997. *Mathematics of data fusion*. Norwell, MA: Kluwer.

Part I

Theory of Data Fusion and Kinematic-Level Fusion

J. R. Raol, G. Girija, and N. Shanthakumar

1

Introduction

Animals recognize their changing environment by processing and evaluation of signals, such as data and some crude information, from multiple directions and multiple sensors—sight, sound, touch, smell, and taste [1]. Nature, through the process of natural selection in very long cycles, has evolved a way to integrate the information from multiple sources and sensing organs to form a reliable recognition of an object, entity, scene, or phenomenon. Even in the case of signal or data loss from a sensor, certain systems are able to compensate for or tolerate the lack of information and continue to function at a lower, degraded level. They can do this because they reuse the data obtained from other similar or dissimilar sensors or sensing organs. Humans combine the signals and data from the body senses—sight, sound, touch, smell, and taste—with sometimes vague knowledge of the environment to create and update a dynamic model of the world in which they live and function. It is often said that humans even use the sixth sense! Based on this integrated information within the human brain, an individual interacts with the surrounding environment and makes decisions about her or his immediate present and near future actions [1]. This ability to fuse multisensory data has evolved due to a process popularly known as natural selection, which occurs to a high degree in many animal species, including humans, and has been happening for millions of years. Charles Darwin, some 150 years ago, and a few others have played a prominent role in understanding evolution. The use and application of sensor data fusion concepts in technical areas has led to new disciplines that span and influence many fields of science, engineering, and technology and that have been researched for the last few decades. A broad idea of sensor data fusion is depicted for target tracking in Figure 1.1.

The art of ventriloquism involves a process of human multisensory integration and fusion [2]. The visual information comes from the dummy's lips, which are moved and operated by the ventriloquist, and the auditory information comes from the ventriloquist herself. Although her mouth is closed, she transmits the auditory signal. This dual information is fused by the viewer and he or she feels as if the dummy is speaking, which particularly amuses the children. If the angle between the dummy's lips and the ventriloquist's face is more than 30°, then the coordination is very weak [2]. A rattlesnake is responsive to both visual and infrared (IR) information being represented on the surface of the optic tectum in a similar

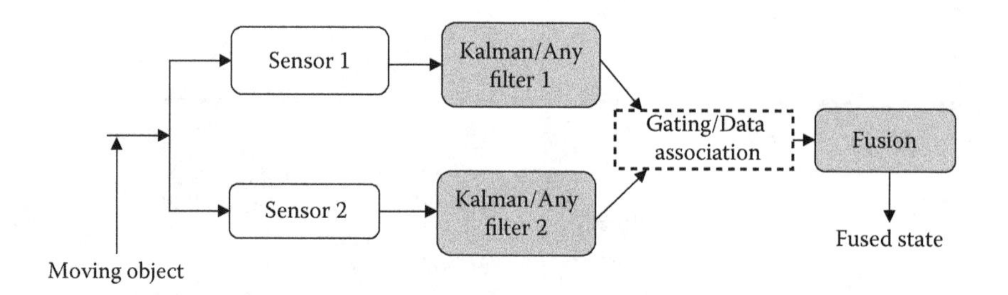

FIGURE 1.1
Concept of sensor data fusion for target tracking (more sensors may be used).

spatial orientation [2]. Visual information about the existence of prey, for example, a rat, is received by the snake's eyes and its IR information is received by the pit organ, from the prey. This information is fused in the snake's brain, and the snake then recognizes the presence of the prey.

The term information fusion (IF) is used for the fusion of any kind of data and data sources [1,3] and is also applicable in the context of data mining and database integration. This term covers all aspects of the fusion field, except nuclear fusion or fusion of different types of music, which may be discordant. However, in the latter kind of system, some fusion might be happening in a sense similar to data fusion, since musical signals can be thought of as data and fusion can be performed, at least for the synthetic musical systems. The term *sensor fusion* means the combination of sensory data or data derived from sensory data, such that the resulting information is better than it would be if these sensors were used individually; in other words, data fusion from multiple sensors, which could be of the same type or of different types. Data fusion means combining information from several sources, in a sensible way, in order to estimate or predict some aspect of an observed scene, leading to the building of a world model of the environment. This is useful for robot motion planning or an airplane pilot's situational assessment and awareness, because it provides a better world model.

Direct fusion means the fusion of sensor data from a set of heterogeneous or homogeneous sensors, soft sensors, and history values of sensor data [1]. Indirect fusion uses information from *a priori* knowledge about the environment, as well as human input. Sensor fusion can be described as a direct fusion system. Information fusion encompasses indirect fusion processes. In general, a sensor measurement system will have one or more of the following problems [1]:

1. A breakdown of a sensor or sensor channel, causing a loss of data from the viewed object or scene.
2. An individual sensor covers only a restricted region of the environment and provides measurement data of only local events, aspects, or attributes.

3. Many sensors need a finite time to perform some basic operations and transmit measurement data to the required place, thereby limiting the frequency of measurements.
4. Measurements from sensors depend on the accuracy and precision of the basic sensing element used in the sensor and the data gathered would thus have limited accuracy.
5. Uncertainty depends on the object being observed; if it is in a noisy environment, the measurement data noise effects would be present.

Uncertainty would arise if (1) some features were missing, such as obstacles in the path, eclipsing of radar cross section (RCS), Glint noise, or background clutter; (2) the sensor cannot measure all relevant attributes and sensor limitations; or (3) the observation is ambiguous or doubtful. Hence, a single sensor cannot reduce uncertainty in its perception of the observed scene, and one of the effective solutions to many of the ongoing problems is multisensor data fusion.

By the fusion of sensors' data we can expect one or more of the following potential advantages [1,4,5]:

1. Multiple sensors would provide redundancy which, in turn, would enable the system to provide information in case of partial failure, data loss from one sensor—i.e., fault tolerance capability—robust functional and operational performance.
2. One sensor can look where other sensors cannot look and provide observations—enhanced spatial or geometrical coverage, and complementary information is made available.
3. Measurements of one sensor are confirmed by the measurements of the other sensors, obtaining cooperative arrangement and enhancing the confidence—increased confidence in the inferred results.
4. Joint information would tend to reduce ambiguous interpretations and hence less uncertainty.
5. Increased dimensionality of the measurement space, say measuring the desired quantity with optical sensors and ultrasonic sensors, the system is less vulnerable to interferences providing a sustained availability of the data.
6. Multiple independent measurements when fused would improve the resolution—enhanced spatial resolution.
7. Extended temporal coverage—the information is continually available.
8. Improved detection of the objects because of less uncertainty provided by the fusion process.

For a given data fusion system, one or more of the above benefits could occur. One can design and use the task-oriented multisensor data fusion (MSDF) system in order to obtain one or more of the above benefits.

When building an MSDF system, the following aspects pertaining to an actual application are of great importance: (1) use of optimal techniques and numerically stable and reliable algorithms for estimation, prediction, and image processing; (2) choice of data fusion architectures such as sensor nodes and the decision-making unit's connectivity and data transmission aspects; (3) accuracy that can be realistically achieved by a data fusion process or system and conditions under which data fusion provides improved performance; and (4) keeping track of the data collection environment in a database management system. The main point is that one makes a synergistic utilization of the data from multiple sensors to extract the greatest amount of information possible about the sensed object, scene, or environment. Humans use data from sight, sound, scent, touch, and *a priori* knowledge, then assess the world and events around them. This involves the conversion of these data (from images, sounds, smell, shapes, textures, etc.) into meaningful perceptions and features involving or leading to a large number of distinctly intelligent decision processes.

However, sensor or data fusion should not be considered as a universal method. Data fusion is a good idea, provided the data are of reasonably good quality. Just manipulating many bad data would not produce any great results—it might produce some results, but at a very high cost [1]. Many fused very poor quality sensors would not make up for a few good ones, and it may not be easy to fix the errors in the initial processing of the data at a later stage, after the data have been processed and inferences are being drawn.

It is often said that the fusion of sensor data enhances accuracy, but of what? What is actually meant is that the overall accuracy of the prediction of the state of the system is increased. However, the accuracy of any sensor cannot increase beyond what is specified by the manufacturer—that is the intrinsic accuracy. Ultimately, the idea is that the overall uncertainty level, after the fusion of data, is reduced, since information from various sensors is utilized in the fusion process. The point is that since uncertainty is reduced (and certainty is enhanced) the impression is that the accuracy has improved. It just makes sense to believe that the accuracy is increased, because the prediction is more certain than it would have been with only a single sensor. This is true even though the accuracy of the sensor cannot simply increase. Hence, the meaning of the term "increased accuracy" in the data fusion process, interpreted in the context of reduction of prediction uncertainty, should not create any fundamental confusion.

There are certain positive outlooks on the use and applications of MSDF [1]: (1) In reference [6] the authors present a proof that the use of additional sensors improves performance in the specific cases of majority vote and decision fusion; and (2) Dasarathy describes the benefits in increasing the number of inputs to the fusion process [7], and discusses what, where, why, when, and how the information fusion occurs in reference [3]. In the analysis limited to the augmentation of two-sensor systems by one additional sensor, increasing the number of sensors may lead to a performance gain depending on the fusion algorithm employed.

There are several other applications where some benefits of MSDF have been realized; these can be found in the various references cited in this book. However, a slight skepticism on "perfect" or "optimal" fusion methods is appropriate [1]. All DF systems may not be optimal, but some merit can be derived from these systems, which can be conditionally optimal or pareto-optimal. This should not prevent sensor data fusion activity, and should give impetus to data fusion activities to search for better and better data fusion solutions and systems with enhanced performance.

Fusion processes are often categorized in three levels of modes: low-, intermediate-, and high-level fusion, as follows:

1. Low-level fusion or raw-data fusion combines several sources of essentially the same type of raw preprocessed (RPP) data to produce a new raw data set that is expected to be more informative and useful than the inputs. Of course, the data should be available in the required engineering units, and could even be RPP data. This can be expanded to include the fusion of estimated states of a system or object by using data from several, often dissimilar, sensors and is called state vector fusion. The entire process is called kinematic data fusion, which can include kinematic states as well as dynamic states. This is dealt with in Part I, along with some fundamental concepts of data fusion.
2. Intermediate-level, mid-level fusion, or feature-level fusion combines various features such as edges, lines, corners, textures, or positions into a feature map. This map is used for segmentation of images, detection of objects, etc. This process of fusion is called pixel-, feature-, or image-level fusion and is dealt with in Part III.
3. High-level fusion, or decision fusion, combines decisions from several experts. Methods of decision fusion are voting, fuzzy logic, and statistical methods. Some aspects, mainly fuzzy logic-based decision fusion, are dealt with in Part II.

In the following systems and applications, data fusion aspects can be incorporated, or data fusion can be utilized [2,8,9]:

1. Modular robotics: Such a system is constructed from a small number of standard units, with each module having its own

hardware/software (HW/SW), driven and steered or controlled units, sensors, data communication links, power units, path-planning algorithms, obstacle avoidance provisions, sensor fusion, and control systems. There is no central processor on the vehicle. The vehicle employs multiple sensors to measure its body position, orientation, wheel position and velocities, obstacle locations, and changes in the terrain. The sensor data from the module are fused in a decentralized manner and used to generate local control for each module. Such a modular system reduces cost and has flexibility, increased system reliability, scalability and good survivability (and modularity).

2. The NASA Mars Pathfinder Mission's Sojourner Rover (1997): The autonomous operations are: terrain navigation, rock inspections, terrain mapping, and response to contingencies. It has five laser stripe projectors and two charge-coupled device (CCD) cameras to detect and avoid rocks. It also has bumpers, articulation sensors, and accelerometers. It is an example of a multisensor or multi-actuator system. The fusion algorithms in the Rover are based on the state-space methods (extended Kalman filter [EKF] is used) and are centralized.
3. The Russian Mir Space Station and the Columbia Space Shuttle could use decentralized data fusion and control strategies.
4. A brief summary of data fusion applications in manufacturing and related processes is given in [9]:
   a. Online prediction of surface finish and tool wear using a neural network-based sensor fusion scheme to estimate, predict, and control the online surface finish.
   b. A tool failure detection system using multisensor: A normalized mean cutting force derived from a tool dynamometer, maximum acceleration (obtained from an accelerometer), and an amplitude of displacement acquired from a gap sensor are fused to successfully detect impending tool breakage online.
   c. Nondestructive testing (NDT) data fusion at pixel level: The fusion of data obtained from NDT of a carbon fiber reinforced composite (CFC) material inspected using eddy current and infrared (IR) thermograph.
   d. Data from three different types of sensors—acoustic emission sensor, vibration sensor, and motor current sensor—are combined to determine the tool wear. The wear data is categorized into four classes—no wear, little wear, medium wear, and high wear—and are defined in terms of fuzzy membership functions. Four different classification tools—nearest neighbor,

two types of artificial neural networks (ANN), and a fuzzy inference system (FIS)—are used to classify the incoming ware descriptive data from the different sensors. Then the decisions from these classifiers are combined to derive the final decision.

e. Nondestructive flaw detection in concrete structures: The data from ultrasonic and impact echo tests are combined using Dempster–Shafer evidential reasoning models to accomplish decision-level identity fusion.

f. NDT: The information from heterogeneous data sources such as x-ray, ultrasonic, and eddy current is exploited in a synergy among complementary information.

Other possible applications of data fusion include [9]: (1) human identity verification, (2) gas turbine power plants, (3) wind-tunnel flow measurements, and (4) mine detection. A survey of many systems that use data fusion aspects in military applications, as well as other nontraditional MSDF applications, is presented in [4].

Chapters 2 to 4 cover some important data fusion process models and propose a small modification in one data fusion process model. They also cover estimation fusion models and rules, data fusion methods, and strategies and algorithms for kinematic-level data fusion (for tracking of moving objects), and will present some performance studies.

2

Concepts and Theory of Data Fusion

Data fusion (DF) or multisensor data fusion (MSDF) is the process of combining or integrating measured or preprocessed data or information originating from different active or passive sensors or sources to produce a more specific, comprehensive, and unified dataset or world model about an entity or event of interest that has been observed [1–8]. A conceptual DF chain is depicted in Figure 2.1, wherein the fusion symbol is indicative of the fusion process: addition, multiplication (through operations involving probabilities, e.g. Bayesian or Dempster–Shafer [DS] fusion rule), or logical derivation; in addition, the hierarchy is indicated by two circles. This implies that fusion is not just an additive process. If successful, fusion should achieve improved accuracy (reduce the uncertainty of predicting the state or declaring the identity of the observed object) and more specific inferences than could be achieved using a single sensor alone. Multiple sensors can be arranged and configured in a certain manner to obtain the desired results in terms of sensor nodes or decision connectivity. Occasionally, the arrangement is dictated by the geometrical or geographical disposition of the available sensors, such as radars.

2.1 Models of the Data Fusion Process and Architectures

Sensor-fusion networks (NWs) are organized topologies that have specified architectures and are categorized according to the type of sensor configuration, as described below (see also Figure 2.2) [4]:

1. Complementary type: In this type of sensor configuration, the sensors do not depend on each other directly. One sensor views one part of the region, and another views a different part of the region, thereby giving a complete picture of the entire region. Because they are complementary, they can be combined to establish a more complete picture of the phenomenon being observed and hence the sensor datasets would be complete. For example, the use of multiple cameras, each observing different parts of a room, can provide a complete view of the room. The four radars

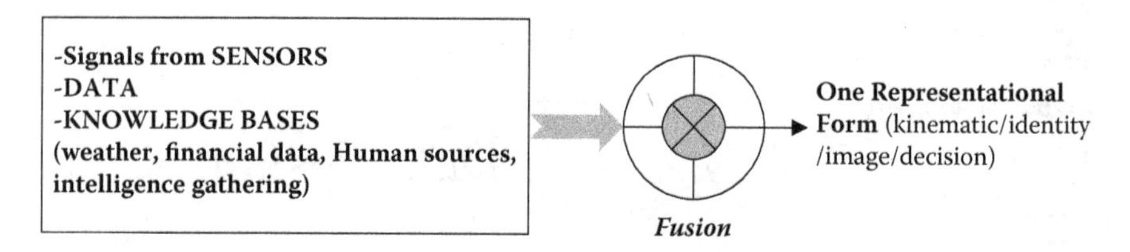

FIGURE 2.1
The conceptual chain of data fusion from data to the fusion result.

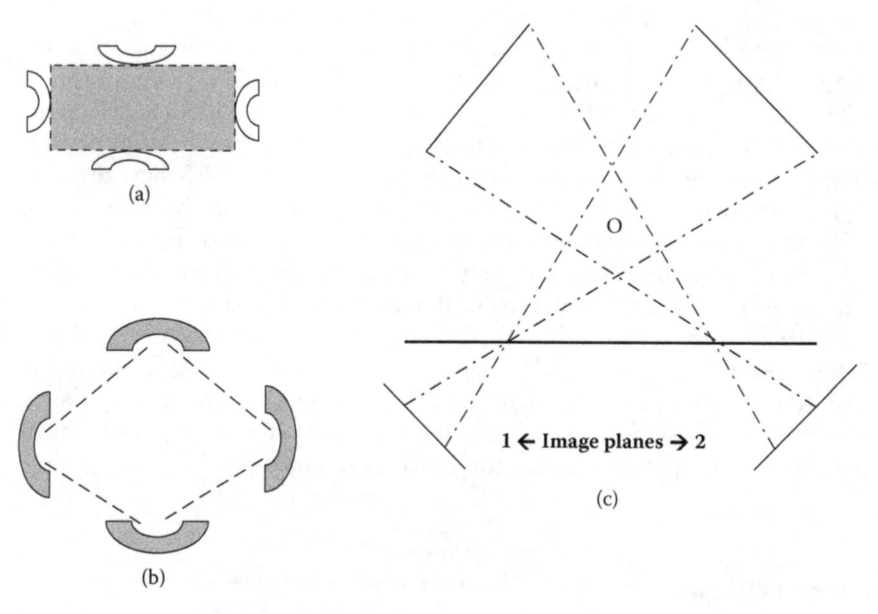

FIGURE 2.2
Data-fusion sensor networks: (a) Complementary sensor network: viewing different regions. (b) Competitive sensor network: viewing the same area. (c) Cooperative sensor network: two vision cameras producing a single 3D image of "O."

around a geographical region would provide a complete picture of the area surrounding the region (Figure 2.2a). Fusion of complementary data is relatively easy because the data from independent sensors can be appended to each other.

2. Competitive type: In this type of configuration (Figure 2.2b), each sensor delivers independent measurements of the same attribute or feature. Fusion of the same type of data from different sensors or the fusion of measurements from a single sensor obtained at different instants is possible. This configuration would provide robustness and fault-tolerance because comparison with another

competitive sensor can be used to detect faults. Such robust systems can provide a degraded level of service in the presence of faults; moreover, the competing sensors in this system do not necessarily have to be identical.

3. Cooperative type: In this type of configuration, data provided by two independent sensors are used to derive information that would not be available from a single sensor (see Figure 2.2c), as in a stereoscopic vision system. By combining the two-dimensional (2D) images from two cameras located at slightly different angles of incidence (viewed from two image planes), a three-dimensional (3D) image of the observed scene, marked "O," is determined. Cooperative sensor fusion is difficult to design, and the resulting data will be sensitive to the inaccuracies in all the individual sensors.

In terms of usage, the three categories of sensor configurations are not mutually exclusive, because more than one of the three types of configurations can be used in most cases. In a hybrid configuration, multiple cameras that monitor a specific area can be used. In certain regions monitored by two or more cameras, the sensor configuration could be either competitive or cooperative.

2.1.1 Data Fusion Models

MSDF is a system-theoretic process (a synergy of sensing, signal and data processing, estimation, control, and decision making) that is very involved, and hence an overall model that interconnects the various aspects and tasks of the DF activities is very much needed. This would also lend a systems approach to the DF process. We discuss several such models here.

2.1.1.1 Joint Directors of Laboratories Model

One very popular model, originating from the US Joint Directors of Laboratories (JDL), Department of Defense, is the JDL fusion model (DoD, USA, 1985/1992) [1,4]. The JDL model has five levels of data processing and a database. These levels are interconnected by a common bus and need not be processed in a sequential order; nevertheless, they can be executed concurrently. There are three main levels—levels 1, 2, and 3, as shown in Figure 2.3 [1,4]. There are many sublevels and auxiliary levels for processing data, combining the available information, and evaluating the performance of the DF system. Several such aspects are briefly reviewed here. Although the terms used here have been borrowed from defense applications, similar terms are quite appropriate for other non-defense civilian DF applications. For such civilian applications, some

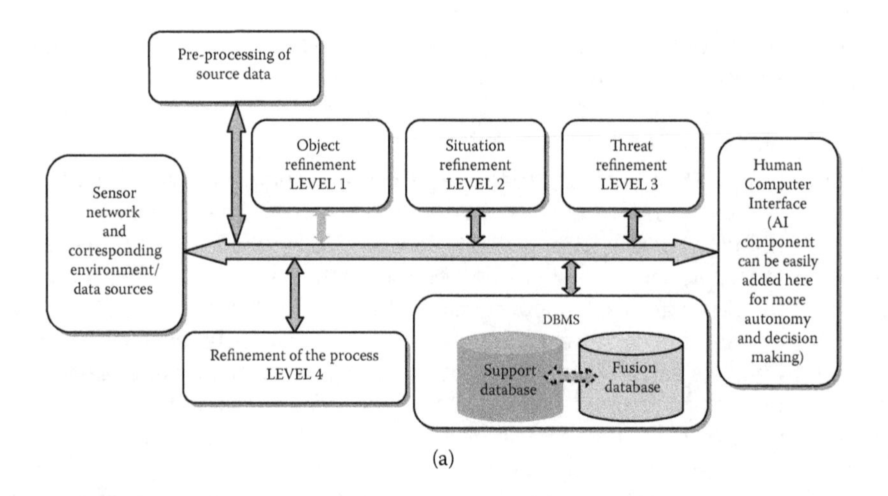

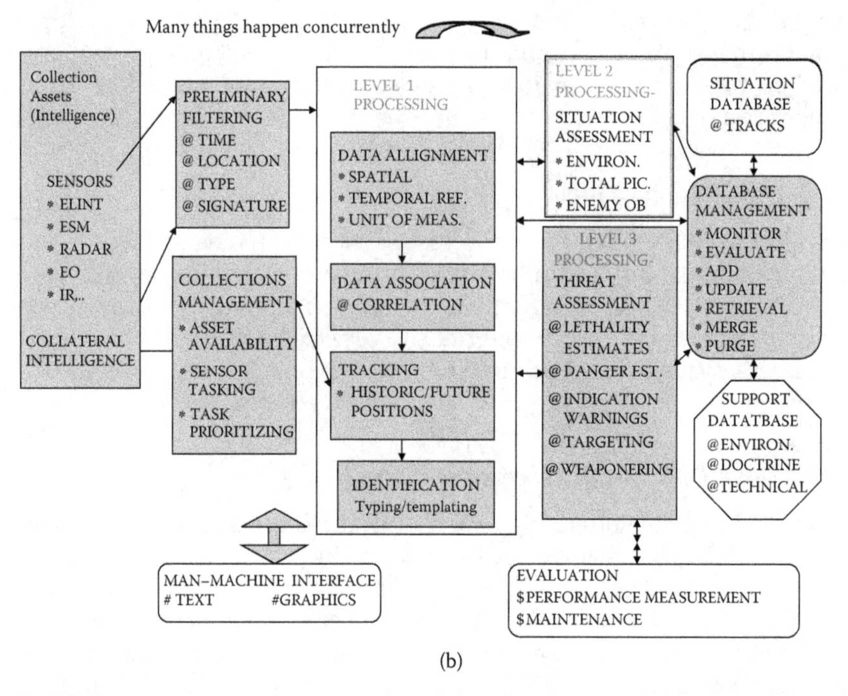

FIGURE 2.3
JDL data-fusion process models (USA): (a) top-level process model (adapted from Hall, D. L. 1992. *Mathematical techniques in multi-sensor data fusion*. Norwood, MA: Artech House. With permission); (b) the detailed process model (adapted from Waltz, E. and J. L. Llinas. 1990. *Multi-sensor data fusion*. Boston: Artech House. With permission.)

parts of the JDL model would be more appropriate, and interested users can determine the appropriate terms.

The JDL fusion-process model is a functionality-oriented DF model and is intended to be very general and useful across multiple application areas—from a sensor NW to the human-computer interface (HCI), with sub-buses having up to four levels of signal-fusion processing and performance evaluation. The output from the DF process in the model—the identification and characterization of the individual entities or objects—is supposed to be minimally ambiguous [1,4]. It is also expected to facilitate a higher-level interpretation of certain entities within the context of the intended application environment. The JDL-data fusion process (DFP) model is a conceptual model that identifies the processes, functions, categories of techniques, and specific techniques applicable for DF. DFP is defined and conceptualized by sources of information, human-computer interaction, preprocessing at the source, DF levels, and finally, the data-management system, including performance evaluation. The JDL model is an information-centered, abstract model, with specific characteristics as explained below [1,2,4].

Information from a variety of data sources, such as sensors, *a priori* information, databases, human input (called humint, for human intelligence), and electronic intelligence, is normally required and collected for DF. This could, in an overall sense, be called intelligence-collection assets, regardless of whether a particular sensor is intelligent or not. The idea is that some intelligence could have been used somewhere in the sensor system or NW. The sensor system could be called intelligent if it is supported by mechanisms of artificial intelligence (AI), which utilize artificial neural networks (ANNs) for learning, fuzzy logic (rule-based fuzzy approximate reasoning for logical decision making), and/or evolutionary algorithms for detection, acquisition, or preprocessing of measurements. The sources of information include [1,4] the following: (1) local sensors associated with the DF system, (2) distributed sensors linked electrically or through radio waves to a fusion system, and (3) reference information, geographical information, and so on. The collected data are preprocessed, and a preliminary filtering is attempted. The time and location and the type and signatures of the collected data are defined or determined at this level, often called level 0 (this is not explicitly indicated in Figure 2.3, but implied as a part of the second column). The data can be grouped and classified, and some signatures can be added or integrated along with the type of the data. Priorities for the further processing of certain data can be assigned, and then level-1 processing can begin, as shown in Figure 2.3.

Object refinement (level 1): At this level, several tasks are carried out, including data alignment (transformation of data into a consistent reference frame and units), association (using correlation methods; better known as data association), tracking actual and future positions of objects, and identification using classification methods. This level consists of numerical procedures, such as estimation, target tracking, and pattern recognition. Object refinement aids in object assessments by combining the location, parametric, and identity information to achieve refined representations of individual objects (such as emitters, platforms, and weapons) regarding their type, identity, position, velocity, acceleration, and so on. Level 1 performs the following four functions: (1) transforms data into a consistent set of units and coordinates; (2) refines and extends to a future time the estimates of an object's position, kinematics, dynamics or attributes; (3) assigns data to the objects to allow the application of statistical estimation techniques; and (4) refines the estimation of an object's identity or classification. Level-1 fusion can be categorically divided into two parts: (1) kinematic fusion, which involves fusion of local information to determine the position, velocity, and acceleration of moving objects, such as missile, aircraft, ships, people, and ground vehicles; and (2) identity fusion, which involves fusion of parametric data to determine the identity of an observed object, e.g., deciding whether a moving object is a missile or an aircraft. Identity estimation can be augmented by rule-based expert systems, wherein various types of factual or procedural information can be exploited to aid the identity estimation.

Situation refinement (level 2): In this level, an attempt is made to find a contextual description of the relationship between the objects and observed events. After processing the data at level 1, the situation is ascertained and further analysis is carried out to refine the situation, if needed. The main goal is to obtain a total picture of the enemy's objective for purposes such as defense application. This is a very complex process.

Threat refinement (level 3): On the basis of the *a priori* knowledge and predictions about the future situation, inferences about vulnerabilities and opportunities for operation are constructed. During threat assessment, several aspects are considered, such as (1) estimation of danger, (2) indication of warning, and (3) targeting. The ultimate aim is to obtain a refined assessment of the threat (and its perception), on which important decisions and actions can be based.

Process refinement, often added as level 4, is a metaprocess that monitors the system performance, e.g., real-time constraints, and reallocates sensors and sources to achieve particular goals or mission objectives. At this level, we are not concerned with the data (processing). However, sensor management is an appropriate aspect to study and employ at this level for optimal use of sensor suites.

Some DF theorists and practitioners also include the cognitive-refinement level [4] as level 5, which is between level 3 and the HCI, thereby introducing the concept of AI at this stage in a limited manner.

The JDL model is supported by a database-management system, which monitors, evaluates, adds, updates, and provides information for the fusion processes, because a large amount of data is involved. Man–machine interaction (in particular, HCI) is very crucial and provides an interface for human input and communication of the fusion results to operators and users. HCI allows human inputs, such as commands, information requests, human assessments of inferences, and reports from human operators. HCI is the mechanism by which a fusion system communicates results via alerts, displays, and dynamic overlays of both positional and identity information on geographical displays. The JDL model is more appropriate for the application of MSDF in defense and is more suited for the C4I2 systems, i.e. command, communication, control (without any feedback), computer, intelligence, and information. However, of the total technical tasks of the C4I2/JDL [4], the DF share is about 25–30%.

2.1.1.2 Modified Waterfall Fusion Model

The waterfall fusion process (WFFP) model emphasizes the processing functions of the lower levels, as depicted in Figure 2.4 [1,10]. The processing stages of the waterfall model have some similarities with the JDL model. They are as follows: (1) Sensing and signal processing relating to source preprocessing (level 0 of JDL model); (2) feature extraction and pattern processing corresponding to object refinement (level 1 of JDL model);

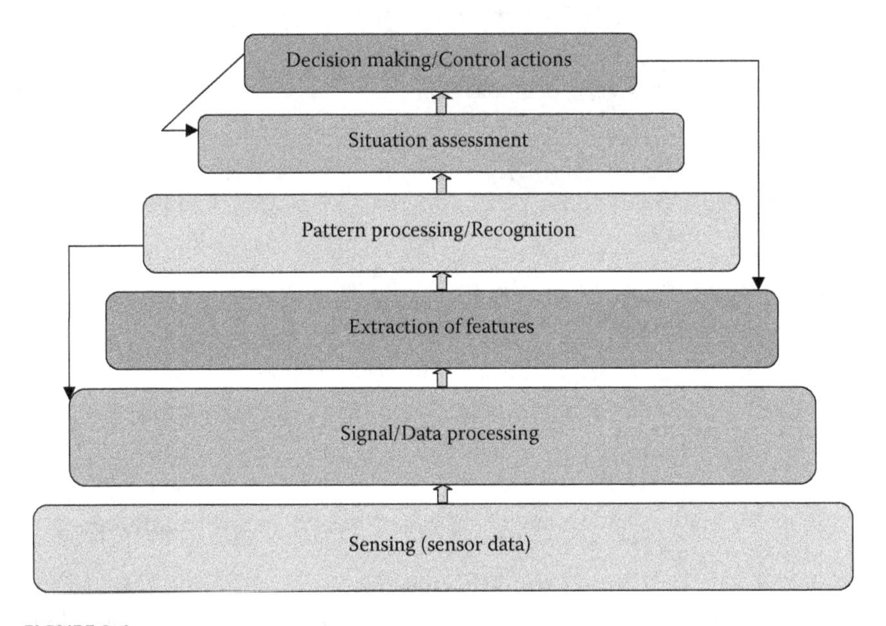

FIGURE 2.4
Modified waterfall fusion-process model: The shadings indicate the increasing task complexity in each three-stage processing; it includes at least one open-loop control action.

(3) situation assessment (SA) relating to situation refinement (level 2 of JDL model); and (4) decision making relating to threat refinement (level 3 of JDL model). The WFFP model is similar in some aspects to the JDL model; however, the WFFP model is very easy to apply. There is no feedback of data flow in the WFFP model because this is an acyclic model. In Figure 2.4, a modified WFFP model is proposed and presented, where (1) the increasing complexity of the subtask is indicated by darker shades within the first (from sensing to extraction of features) and second (from pattern processing to decision making and control actions) stages of the three-stage process; and (2) there are some feedback loops. The model represented in Figure 2.4 is a modified version of the original model [1,10] and can include at least one open-loop control action or some feedback-loop control actions. Thus, the proposed modified WFFP model is an action-oriented model. It has been modified using local feedback loops, as follows (Figure 2.4): (1) From the decision making and control actions to the situation-assessment block, to reflect the situation refinement and its use as new control action; (2) from pattern processing to signal processing, to reflect the enhanced pattern recognition and its use in refining the SA; and (3) from the decision making and control actions to the feature-extraction block, to reflect the improved decision making process and new control action based on the enhanced feature extraction. Thus, the modified WFFP model is more than an abstract model; it is action-oriented, when compared with the JDL or the conventional WFFP models.

2.1.1.3 Intelligence Cycle–Based Model

Because the DF process has some inherent cyclic processing behavior that is not captured in the JDL model, the intelligence cycle–based (IC) model tries to capture these cyclic characteristics [1,11], comprised of the following five stages:

1. In the planning and direction stage, the intelligence requirements are determined.
2. In the collection stage, the appropriate information is gathered.
3. In the collation stage, the collected information is streamlined.
4. In the evaluation stage, the available information is used and the actual fusion is carried out.
5. In the dissemination stage, the fused intelligence and inferences are distributed.

This model is a macrolevel data-processing DF model and looks more like a top-level model than the WFFP model. The sublevel actions and processing tasks are not defined or indicated in the IC model, although these tasks can be implicitly presumed to be present. It would be appropriate to regard the IC model as a superset model, and it appears to be more abstract than the JDL and the modified WFFP models.

2.1.1.4 Boyd Model

The Boyd control cyclic (or observe, orient, decide, and act [OODA]) loop (BCL) model, shown in Figure 2.5 [1,12], represents the classic decision-support mechanism in military-information operations and has been widely used for sensor fusion; it has a feedback loop. It uses the OODA cycle, described below: (1) the *observation* stage is mainly comparable to the source preprocessing step of the JDL model and as a part of the collection phase of the IC model; (2) the *orientation* phase contains, perhaps implicitly, the functions of levels 1, 2, and 3 of the JDL model, whereas in the IC model, this stage corresponds to the structured elements of the collection and the collation phases; (3) the *decision* stage compares with level 4 of the JDL model and the process refinement and dissemination phases of the IC model; and (4) the *action* phase is a comparable to the dissemination phase in the IC model, and in the JDL model, there is no apparent closed loop. The BCL model is more comparable to the modified WFFP model, and it is abstract in nature, specifying only an O-O-D-A loop; being more abstract than the IC model, it is a top-level model. An appropriate combination of the IC and BCL loop models would yield a better fusion-process model, making the new combination model less abstract and more action-oriented. One such combination has yielded the omnibus (OB) model.

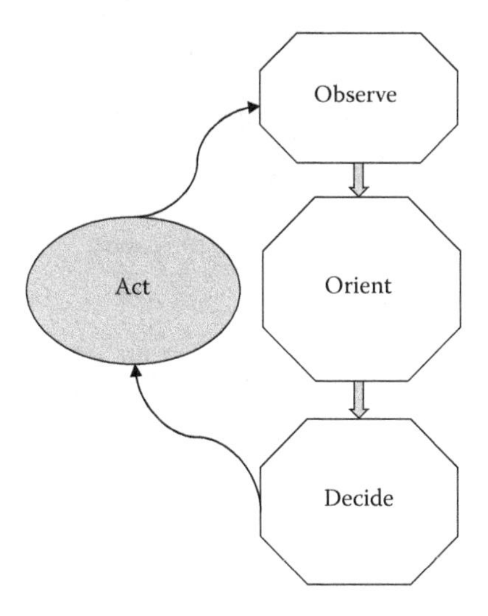

FIGURE 2.5
The Boyd observe, orient, decide, and act cyclic-loop model.

2.1.1.5 Omnibus Model

The OB model integrates most of the beneficial features of other approaches, as shown in Figure 2.6 [1,13]. The OB model defines the order of the processes involved to make the cyclic nature more explicit and additionally, it uses a general terminology. Its cyclic nature is comparable to the BCL model. It provides reasonably detailed processing levels compared to the BCL model. The various levels are as follows: (1) The sensing and signal processing steps are conducted by the observation phase of the BCL model; (2) the feature extraction and pattern processing comprise the orientation phase; (3) the context processing and decision making are included in the decision phase; and (4) the control and resource tasking are conducted by the action phase. The sensor DF is the route from the observation to the orientation phases. The path from orientation to decision is through soft-decision fusion. The route from decision to action is through hard-decision fusion, and from action back to observation involves the sensor-management process. Some aspects of the WFFP model are also found in the OB model. The OB fusion-process model is more complete than the WFFP (and modified WFFP), IC, and BCL fusion models because it encompasses many important features and functional aspects of these models. The

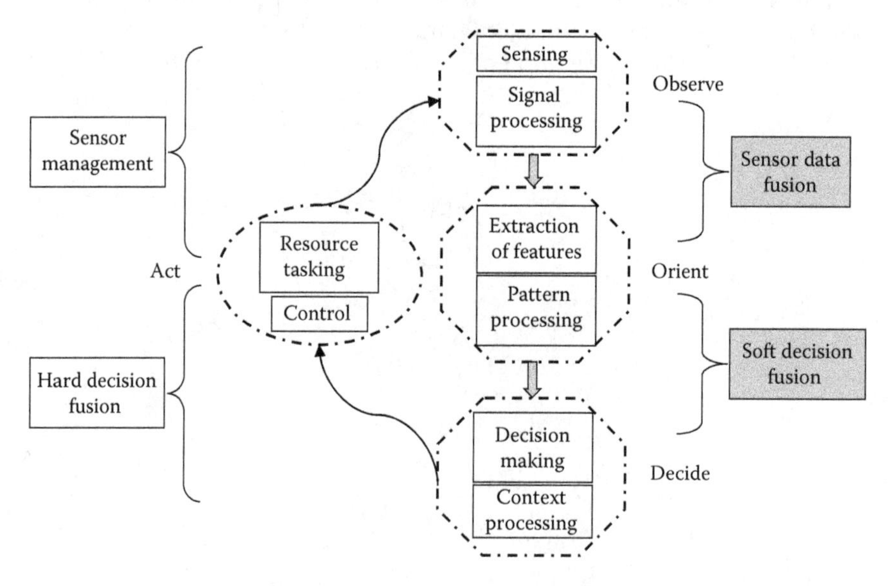

FIGURE 2.6
The omnibus cyclic data-fusion model: The rectangles define the omnibus model in more detail compared to the Boyd observe, orient, decide, and act model (BCL in dotted lines).

OB model is simple and easy to apply and follow for many non-defense DF applications. In addition, the OB model is more generalized than the previous three models and has a cyclic and closed-loop action element, making it more attractive than the JDL fusion model. Thus, the OB model can be regarded as the standard fusion-process model for non-defense DF processing and applications.

2.1.2 Fusion Architectures

The MSDF process, as described above, involves integration of sensors, data preprocessing, estimation, and a further higher level of processing and decision making. It demands a definite arrangement of sensors and sensor-data acquisition systems and signal-processing aspects, which are, in turn, dictated by the fusion architectures. Mainly, there are three types of architecture, as discussed in Sections 2.1.2.1 to 2.1.2.3 [4,8].

2.1.2.1 Centralized Fusion

Centralized fusion architecture, as shown in Figure 2.7a [4], used mainly for similar sensors, involves time-synchronization and bias correction of sensor data, transformation of the sensor data from sensor-based units and coordinates into convenient coordinates and units for central processing, e.g., polar to earth–centered, earth-fixed coordinates, gating and association in case of multiple targets, and measurement fusion. Decisions are based on the maximum possible information gathered from the system of sensors, also called measurement fusion. Thus, centralized fusion is a conventional estimation problem with distributed data.

2.1.2.2 Distributed Fusion

Distributed fusion (Figure 2.7b) [4,8] is mainly used for dissimilar sensors (sensors with different observation frames) i.e. infrared and radar; however, it can still be used for similar types of sensors. In this architecture, observation data from each sensor is processed by an individual Kalman filter (KF), extended KF, or square-root information filter (SRIF) at each node. The local track, consisting of the estimated state vector and covariance matrix from each filter, is used as input to the state-vector fusion process, and the output is the fused state vector and its (fused) covariance matrix. The information is processed locally at each node; there is no central fusion here. This architecture is useful for large flexible and smart structures, monitoring of aircraft or spacecraft health, huge automated plants, large sensor NWs, and chemical industries. This is also referred to as estimation fusion (EF) in target-tracking applications.

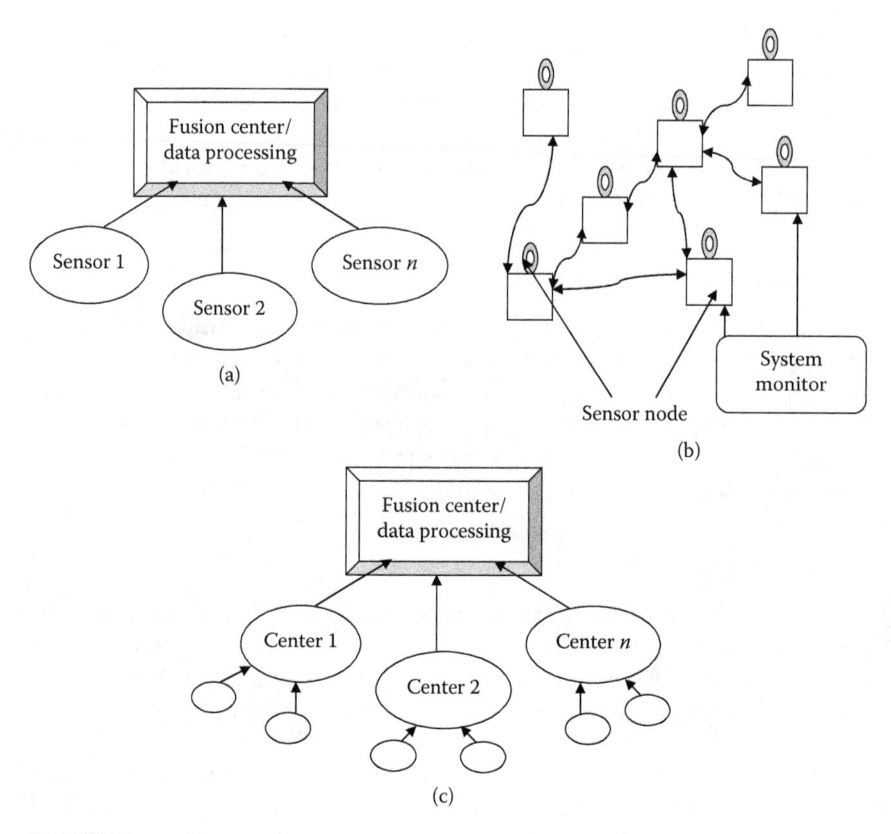

FIGURE 2.7
Three sensor data-fusion architectures: (a) centralized processing in the fusion center, (b) decentralized processing at each node-processing unit, and (c) hierarchical processing.

2.1.2.3 Hybrid Fusion

Hybrid fusion, as shown in Figure 2.7c [4], involves both centralized- and distributed-fusion schemes, based on the disposition of the required sensor configurations. During ordinary operations, distributed fusion is used to reduce the computational workload and communication demands; however, under specific circumstances, such as when more accuracy is desirable or the tracking environment is dense, centralized fusion can be used. Alternatively, based on the sensors available, a combination of both schemes may be used to obtain the fused state of a particular target of interest. This architecture is very suitable for the data processing and fusion system of a flight-test range (in Sections 4.1 and 4.2, a realistic hierarchical MSDF scheme is shown in Figure 4.1).

2.2 Unified Estimation Fusion Models and Other Methods

Fusion methods (actual fusion of data, derived data or information, and inferences) can be based on: (1) probabilistic and statistical models such as Bayesian reasoning, evidence theory, robust statistics, and recursive operators; (2) least-square (LS) and mean square methods such as KF, optimization, regularization, and uncertainty ellipsoids; or (3) other heuristic methods such as ANNs, fuzzy logic, approximate reasoning, and computer vision [2]. The basic process and taxonomy of DF are illustrated in Figure 2.8 [4]. This is called the positional concept; however, it is valid for other state variables of the object being tracked, such as velocity, acceleration, jerk (change in accelerations, and even surge, if need be), and bearing (angle and orientation data). The parametric association involves the measure of association and the association strategies. The following tools can be used for measuring the data association: (1) correlation coefficients to correlate shapes, scatter, and elevation; (2) distance measures, such as the Euclidean norm and the Mahalanobis distance; and (3) probabilistic similarities. The association strategies involve (1) gating techniques (e.g., elliptical or rectangular gates and kinematic models); and (2) assignment strategy (nearest neighbor, probability data association, and so on). There are several parameter- and state-estimation methods that may be useful [14]. Two methods are briefly described in the Appendix.

In this section, the theory and concept of unified fusion models (UM) and fusion rules are discussed as a general and systematic method to estimate and track the fusion approach [15–19]. The approach is called DF, for estimation, or EF, for estimation fusion, e.g., target tracking using data

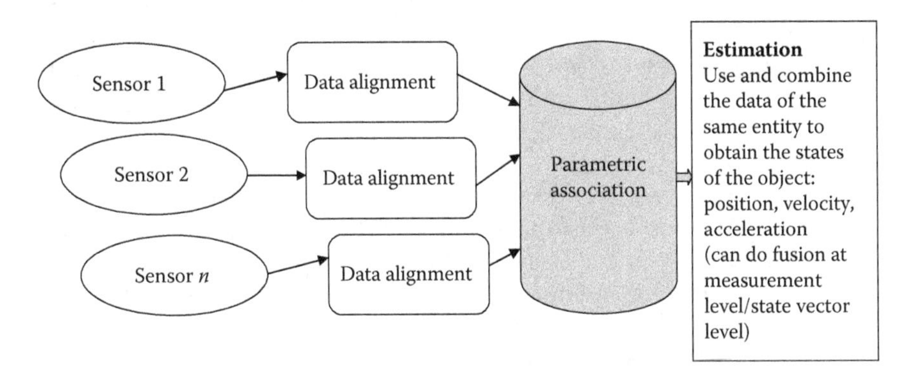

FIGURE 2.8
Basic sensor data-fusion process and taxonomy: implicitly involves certain aspects from the JDL–DFP model.

from multiple sensors (also referred to as track fusion or track-to-track fusion) [15–19]. It addresses the issue of how to utilize the information available from the multiple sets of data and multiple sources and sensors with the aim of estimating an unknown quantity or a parameter of the process [14]. In addition, there is the possibility of using multiple data collected from a single source over a long period. The problems of filtering, prediction, and smoothing can be regarded as special EF problems [14–19]. We consider, from the EF viewpoint, two basic fusion architectures: (1) centralized measurement fusion in target tracking, and (2) decentralized (or distributed) EF in target tracking. Some of the classical limitations of conventional fusion methods are as follows [15–19]: (1) Sensor measurement errors are regarded as uncorrelated. However, in practice, many such errors are correlated, because a quantity or entity is observed by multiple sensors in a common noisy environment. (2) The individual nodes in the dynamic models are assumed to be identical; this is not true in practice, because different models might become necessary, mainly because of the differences in the sensor models. (3) The NW configuration and information-flow patterns of the distributed systems are assumed to be simple. For a complicated NW, it is very difficult to arrive at a simple fusion rule that would be equivalent to the corresponding central-fusion rule. The fusion rules discussed are uniform and are optimal in the linear class for centralized, distributed, and hybrid architectures [15–19]. For the development of EF-UMs and rules, it is hypothesized that the local estimates are "observations" of the estimates. The quantity (say, x) to be estimated is called an "estimatee" [15–19].

2.2.1 Definition of the Estimation Fusion Process

A distributed sensor system with n sensors and a fusion center is considered here, wherein each local sensor is connected to the fusion center. Let the following be specified at each sensor (local node): Z_i observations with R_i as the covariance matrix of the corresponding noises, x as the quantity to be estimated (an estimatee), $\hat{x}_i$ as the local estimate of x, with its covariance matrix $P_i = \text{cov}(\tilde{x}_i)$ {of the estimation error: $\tilde{x}_i = x - \hat{x}_i$}. At the fusion center, we have $\hat{x}$ and $P = \text{cov}(\tilde{x})$ as the estimate and its covariance (for error $\tilde{x} = x - \hat{x}$). We need estimates of x and its covariance matrix, which can be derived by using all the available information $Y = \{y_1, ..., y_n\}$ at the fusion center. The EF is categorized in the following ways: (1) centralized fusion (central-level fusion or measurement fusion), in which $Y = Z$, and all unprocessed data or measurements are available at the fusion center; (2) decentralized fusion (distributed fusion, sensor-level fusion, or autonomous fusion), in which $Y = D = \{g_1(Z_1), ..., g_n(Z_n)\}$, implying that only data-processed measurements are available at the fusion center; (3) standard distributed fusion (known as EF in target tracking), in which $Y = \{(\hat{x}_1, P_1), \ldots, (\hat{x}_n, P_n)\}$; and (4) hybrid fusion, in which both processed and unprocessed data are presented to the fusion center.

2.2.2 Unified Fusion Models Methodology

Let us assume the following:

$$z_i = h_i x + \eta_i \tag{2.1}$$

Here, z_i is the measurement of the ith sensor, and η_i is the measurement noise. Next, a local estimate is viewed as an observation of the estimate [15–19], as shown below:

$$\hat{x}_i = x + (\hat{x}_i - x) = x + (-\tilde{x}_i) \tag{2.2}$$

Equation 2.2 is similar to an observation or measurement equation, where the "new observation" $\hat{x}_i$ is actually the estimate of x and the additive term is regarded as an error (noise). The above model is referred to as the data model for standard distributed fusion. Also consider the following equation:

$$\begin{aligned}\widehat{y}_i = g_i(z_i) = a_i + B_i z_i &= B_i(h_i x + \eta_i) + a_i \\ &= B_i h_i x + (B_i \eta_i + a_i) = B_i h_i x + \widehat{\eta}_i\end{aligned} \tag{2.3}$$

In Equation 2.3, z_i is processed linearly and sent to the fusion center. In this case, a_i and B_i are known. This model is referred to as the linearly-processed data model for distributed fusion.

2.2.2.1 Special Cases of the Unified Fusion Models

If $B_i = I$ and $\widehat{\eta}_i = \eta_i$, then we have $\widehat{y}_i = h_i x + \eta_i$, which is a centralized fusion model of Equation 2.1. If $B_i h_i = I$ and $\widehat{\eta}_i = -\tilde{x}_i$, then we have $\widehat{y}_i = x + (-\tilde{x}_i)$, which is Equation 2.2 in the standard distributed-fusion model. Next, the unified model of the data available to the fusion center is defined as

$$y_i = H_i x + v_i \tag{2.4}$$

In the batch-processing mode, it is represented as follows:

$$y^n = Hx + v^n \tag{2.5}$$

The above model is valid for centralized, distributed, and hybrid DF architectures. From Equations 2.4 and 2.5 in the above model, we get the following definitions [15–19]:

$$y_i = \begin{Bmatrix} z_i -> CL \\ x_i -> SD \\ \widehat{y}_i -> DL \end{Bmatrix}; \quad v_i = \begin{Bmatrix} \eta_i \quad -> CL \\ -\tilde{x}_i -> SD \\ \widehat{\eta}_i \quad -> DL \end{Bmatrix}$$

Here, CL⇒centralized with linear observations, SD⇒standard distributed, and DL⇒distributed with linearly processed data. We further have:

$$H_i = \begin{Bmatrix} h_i \quad ->CL \\ I \quad -> SD \\ B_i h_i -> DL \end{Bmatrix}; \quad y^n = \begin{Bmatrix} y_1 \\ . \\ . \\ . \\ y_n \end{Bmatrix}; \quad H = \begin{Bmatrix} H_1 \\ . \\ . \\ . \\ H_n \end{Bmatrix}; \quad v^n = \begin{Bmatrix} v_1 \\ . \\ . \\ . \\ v_n \end{Bmatrix} \tag{2.6}$$

2.2.2.2 Correlation in the Unified Fusion Models

Let $C = \text{cov}(v^n) = \begin{Bmatrix} R \quad -> CL \\ .\Re -> SD \\ \widehat{R} \quad -> DL \end{Bmatrix}$ be the general covariance matrix, with $R = \text{cov}(\eta_1,\ldots,\eta_n)$ (measurement-noise covariance matrix); $\Re = \text{cov}(\tilde{x}_1,\ldots,\tilde{x}_n)$ (the joint error covariance of the local estimates); and $\widehat{R} = \text{cov}(\widehat{\eta}_1,\ldots,\widehat{\eta}_n)$ (the covariance of equivalent observation noise). The noise components $(v_1,\ldots,v_n)$ of the UM could be correlated, and then "C" is not necessarily a block-diagonal matrix. R is usually assumed to be block-diagonal; however, this is not always true, because the measurement noises of a discrete-time multisensor system obtained by sampling the continuous time system would be correlated. Moreover, these components would be correlated if x is observed in a common noisy environment and if the sensors are in the same platform. In addition, measurement-noise statistics would depend on the distance (x, y, z) of the target from the measurement suite. The matrix $\Re$ is seldom block-diagonal, and the noise v and the estimate x could be correlated as follows:

$$\text{Cov}(x, v^n) = A \quad \text{and} \quad \text{cov}(x, v_i) = A_i$$

Thus, $A_i = \text{cov}(x, v_i) = \text{cov}\{(\tilde{x}_i + \hat{x}_i - \bar{x}), (-\tilde{x}_i)\}$

From Equation 2.1, we have $x = \hat{x} + \tilde{x}_i$, and after expansion and simplification, we have $\text{cov}(x, v_i) = -P_i$, according to the orthogonality principle.

It follows that Equations 2.4 and 2.5 are the unified data model for the three linear types of fusion architectures described earlier; however, the measurement noises would be correlated with each other and with x. An alternative distributed fusion model is provided next.

Let $\hat{x}_i$ be represented as follows:

$$\hat{x}_i = \bar{x}_i + K_i(z_i - h_i\bar{x}_i) = [I - K_i h]\bar{x}_i + K_i z_i \quad (2.7)$$

using the linear measurement model of Equation 2.1 for z_i. Then, we have y_i as the linearly processed measurement form [15–19], as shown below:

$$\hat{x}_i = [I - K_i h]\bar{x}_i + K_i(h_i x + \eta_i) \quad (2.8)$$

$$y_i = \hat{x}_i - [I - K_i h]\bar{x}_i = K_i h_i x + K_i \eta_i = B_i h_i x + v_i \quad (2.9)$$

The covariance of v_i is $C = \text{cov}(v) = [C_{ij}] = [K_i R_{ij} K_j'] = KRK'$; $K = diag[K_1 \ldots K_n]$. We can see that C would be block-diagonal if R is also block-diagonal. We need to send data y_i to and gain K_i to the fusion center. This architecture is referred to as simple nonstandard distributed fusion.

2.2.3 Unified Optimal Fusion Rules

In Sections 2.2.3.1 through 2.2.3.5, some unified optimal fusion rules and their associated mathematical derivations are given.

2.2.3.1 Best Linear Unbiased Estimation Fusion Rules with Complete Prior Knowledge

Let us consider the UM of Equation 2.5. The best linear unbiased estimation (BLUE) fusion rule (FR) with complete prior knowledge is given when the prior mean $E\{x\} = \bar{x}$ and the prior covariance $P_0 = \text{cov}(x)$ (and A) are specified:

$\hat{x}^B = \arg\min E\{(x - \hat{x})(x - \hat{x})' \mid Y \mid\}$; the minimum is taken over $\hat{x} = a + By^n$.

The details are as follows [15–19]:

$$\begin{aligned}
\hat{x} &= \bar{x} + K(y^n - H\bar{x}) \\
P &= \text{cov}(x + \hat{x} \mid Y \mid) = P_0 - KSK' \\
S &= HP_0H' + C + HA + A'H' \\
U &= I - KH \\
P &= UP_0U' + KCK' - UAK' - (UAK')' \\
\hat{x} &= \bar{x} + K(y^n - H\bar{x})
\end{aligned} \quad (2.10)$$

2.2.3.2 Best Linear Unbiased Estimation Fusion Rules without Prior Knowledge

If there is no prior knowledge about the estimate (i.e., $E\{x\}$ is not known) or the prior covariance matrix either is not known or does not exist, then the BLUE FR for the model Equation 2.5 is represented as follows [15–19]:

$$\begin{aligned} \hat{x} &= Ky^n = \tilde{K}y^n \\ P &= KCK' = \tilde{K}C\tilde{K}' \\ \tilde{K} &= H^+(I - C\{TCT\}^+) \\ K &= \tilde{K} = H^+(I - CT^{1/2}(T^{1/2\prime}CT^{1/2})^{-1}T^{1/2\prime})) \end{aligned} \tag{2.11}$$

with $(H, C^{1/2})$ having full-row rank; $T^{1/2}$ is the full-rank square root of $T = I - HH^+$ *and* $C^{1/2}$ *is the* $SQRT(C)$. If C is nonsingular, then $\tilde{K} = PH'C^{-1}$.

2.2.3.3 Best Linear Unbiased Estimation Fusion Rules with Incomplete Prior Knowledge

If prior information about some components of $E\{x\}$ is not available, then it is proper to assume that P_0 does not exist (i.e., P_0^{-1} is singular). We can then set certain elements (or the eigenvalues) of P_0 as infinity. We can assume that $\bar{x} = E\{x\}$, $Cov(x, v^n) = A$, and a positive semidefinite symmetric but singular matrix P_0^{-1} is given. Then, for the UM of Equation 2.5, the optimal BLUE FR is generated by the following equations [15–19]:

$$\begin{aligned} \hat{x} &= VK[(V_1'\bar{x})', y^{n\prime}] \\ P &= VK\tilde{C}K'V' \\ K &= \tilde{H}^+(I - \tilde{C}\{T\tilde{C}T\}^+) \end{aligned} \tag{2.12}$$

Here, $\tilde{H} = \begin{bmatrix} [I_{rxr}, 0] \\ HV \end{bmatrix}$; $\tilde{C} = \begin{bmatrix} \Lambda_1 & -V_1'A \\ -(V_1'A)' & C \end{bmatrix}$; $V = [V_1, V_2]$

The matrix V diagonalizes P_0^{-1} as $P_0^{-1} = V\,\text{diag}(\Lambda_1^{-1}, 0, \ldots, 0)V'$, with $\Lambda_1 = \text{diag}(\lambda_1, \ldots, \lambda_r) > 0$; $r = \text{rank}(P_0^{-1})$.

2.2.3.4 Optimal-Weighted Least Squares Fusion Rule

For the UM of Equation 2.5, we have the following weighted least-squares (WLS) fusion equations [15–19]:

$$\begin{aligned} \hat{x} &= Ky^n \\ P &= KCK' = [H'C^{-1}H]^+ \\ K &= PH'C^{-1} \end{aligned} \tag{2.13}$$

2.2.3.5 Optimal Generalized Weighted Least Squares Fusion Rule

For similar conditions as in the BLUE fusion rule with complete prior knowledge, we have the following WLS fusion rule formulae.

Let the model be written as:

$$y_o = x + v_o = \bar{x} \tag{2.14}$$

with $Cov(v_o) = \text{cov}(\bar{x} - x) = P_o$. In addition, let

$$\hat{y}^n = \begin{bmatrix} \bar{x} \\ y_n \end{bmatrix} \quad \hat{H} = \begin{bmatrix} I \\ H \end{bmatrix} \quad \hat{v}^n = \begin{bmatrix} v_o \\ v_n \end{bmatrix} \quad \hat{C} = \begin{bmatrix} P_o & -A \\ -A' & C \end{bmatrix} \quad \hat{y}^n = \hat{H}x + \hat{v}^n \tag{2.15}$$

Then, the estimator is represented as [15–19]:

$$\begin{aligned} \hat{x} &= \arg\min (\hat{y}^n - \hat{H}\hat{x})' \hat{C}^{-1} (\hat{y}^n - \hat{H}\hat{x}) = K\hat{y}^n \\ P &= K\hat{C}K = [\hat{H}'\hat{C}^{-1}\hat{H}]^{+} \\ K &= P\hat{H}'\hat{C}^{-1} \end{aligned} \tag{2.16}$$

The derivations, many comparative aspects, and detailed analyses of these UM-based optimal, linear fusion rules are treated extensively in papers by Rong Li et al. [15], Rong Li and Wang [16], Rong Li and Zhang [17,18], and Rong Li et al. [19].

These EF rules and algorithms are very useful for target tracking and fusion applications, in addition to image tracking. A basic LS method for target-position EF is described in Section 11.3.

2.2.4 Kalman Filter Technique as a Data Fuser

The KF technique is very useful in the kinematic fusion process. The three widely used methods to perform fusion at the kinematic level are (1) fusion of the raw observational and measurement data (the data converted to engineering units), called centralized fusion; (2) fusion of the estimated state vectors or state-vector fusion; and (3) the hybrid approach, which allows fusion of raw data and the processed state vector, as desired. Kalman filtering has evolved to become a very high-level state-of-the-art method for estimation of the states of dynamic systems [14]. The main reason for its success is that it has a very intuitively appealing state-space formulation and a predictor-corrector estimation and recursive-filtering structure; furthermore, it can be easily implemented on digital computers and digital signal processing units. It is a numerical data processing algorithm, which has tremendous real-time and online application potential.

This is due to its recursive formulation: new estimate = previous estimate + gain times the residuals of the estimation. This is a very powerful and yet simple estimation data-processing structure. Most of the real-time and online estimation and filtering algorithms have similar data-processing structures or algorithms. However, KF is a mathematical model–based approach.

We describe a dynamic system as follows:

$$x(k+1) = \phi x(k) + Bu(k) + Gw(k) \tag{2.17}$$

$$z(k) = Hx(k) + Du(k) + v(k) \tag{2.18}$$

Here, x is the $n \times 1$ state vector; u is the $px1$ control input vector to the dynamic system; z is the $m \times 1$ measurement vector; w is a white Gaussian process-noise sequence, with zero mean and covariance matrix Q; v is a white Gaussian measurement-noise sequence, with zero mean and covariance matrix R; ϕ is the $n \times n$ transition matrix that propagates the state (x) from k to $k + 1$; B is the input gain or magnitude vector or matrix; H is the $m \times n$ measurement model or sensor-dynamic matrix; and D is the $m \times p$ feed forward or direct-control input matrix, which is often excluded from the KF development. In addition, B is often omitted if there is no explicit control input playing a role. Modification of the KF with inclusion of B and D is relatively straightforward. Although most dynamic systems are continuous in time, the KF technique is the best discussed and is mostly used in the discrete-time form. The problem of state estimation using KF is formulated as follows: given the model of the dynamic system, statistics regarding the noise (Q, R) processes, the noisy measurement data (z), and the input (u), determine the optimal estimate of the state, x, of the system. We presume that the state estimate at k has evolved to $k + 1$. At this stage, a new measurement is made available, and it hopefully contains new information regarding the state, as per Equation 2.18. Hence, the idea is to incorporate the measurement into the data-fusion (i.e., update or filtering) process and obtain an improved and refined estimate of the state.

2.2.4.1 Data Update Algorithm

We have the measurement z, know H, and have assumed R; we further have/assume $\tilde{x}(k) \rightarrow$ the "a priori" estimate of state at time k, i.e., before the measurement data is incorporated, and $\tilde{P} \rightarrow$ the "a priori" covariance matrix of the state-estimation error (the time-index k is omitted for simplifying the equations). Then, the measurement-update algorithm (essentially, the filtering of the state vector by considering the measurement data) to obtain $\hat{x}(k) \rightarrow$ the updated estimate of state at time k, i.e., after the measurement data is incorporated, is given as:

Residual equation:

$$r(k) = z(k) - H\,\tilde{x}(k) \tag{2.19}$$

Kalman gain:

$$K = \tilde{P}H^T(H\tilde{P}H^T + R)^{-1} \tag{2.20}$$

Filtered state estimate:

$$\hat{x}(k) = \tilde{x}(k) + K\,r(k) \tag{2.21}$$

Covariance matrix (posteriori):

$$\hat{P} = (I - KH)\tilde{P} \tag{2.22}$$

2.2.4.2 State-Propagation Algorithm

This part of the KF method, which applies the previous estimates of x and P, is represented as:

State estimate:

$$\tilde{x}(k+1) = \phi\,\hat{x}(k) \tag{2.23}$$

Covariance matrix:

$$\tilde{P}(k+1) = \phi\hat{P}(k)\phi^T + GQG^T \tag{2.24}$$

In the KF method, $K = \tilde{P}H^TS^{-1}$ and $S = H\tilde{P}H^T + R$, and matrix S is the covariance matrix of residuals (also called innovations). The actual residuals can be computed and compared with the standard deviations obtained by calculating the square root of the diagonal elements of S. The process of tuning the filter to bring the computed residuals within the bounds of at least two standard deviations is an important filter-tuning exercise for obtaining the correct solution to the problem [14].

We can clearly see that, through the inclusion of z (if it is a vector of several observables, such as position and angles), the KF in the form given above is itself a measurement data–level fusion algorithm. It combines the measurements of these observables directly at the data level and produces an optimal estimate of the state x. For each measurement type, one should choose appropriate H vectors or matrices and their corresponding R matrices. Therefore, the KF fundamentally accomplishes a DF task. The other fusion process, state-vector fusion, is illustrated in Section 3.1.1 (also, refer Exercise I.26).

2.2.5 Inference Methods

An inference (based on some initial observations) [2,4,20,21] is defined as an act of passing from one proposition, statement, or judgment considered or believed to be true to another, whose truth is believed to follow, by some logic, formulae, or process, from that of the former proposition or statement. Inference methods (IM) are used for decision making or fusion to arrive at a decision from the available knowledge. The decision concerning whether the road in front of a vehicle is blocked or free, given the measurements of multiple distance sensors, can be regarded as an inference problem. The classical IM conducts tests on an assumed hypothesis to an alternative hypothesis, and it yields the probability of the actually observed data being present if the chosen hypothesis were true. The classical IM does not support the usage of *a priori* information about the probability of a proposed hypothesis, whereas the *a priori* probability is taken into account in the Bayesian inference method (BIM). The Bayesian theorem quantifies the probability of hypothesis H, given that an event E has occurred. Using multiple hypotheses, the BIM can be used for solving classification problems; the Bayes' rule will then produce a probability for each hypothesis. Due to certain limitations of the BIM, Dempster [20] generalized the Bayesian theory of subjective probability. Dempster's rule of combination, which operates on belief or mass functions as Bayes' rule does on probabilities, was more advanced. Shafer extended Dempster's work and developed a mathematical theory of evidence [21], which can be applied for representation of incomplete knowledge, updating of beliefs, and for combination of evidence [22]. BIM is studied in detail in Section 2.3.

2.2.6 Perception, Sensing, and Fusion

An intelligent agent (say, a robot) should reason about its environment in order to successfully plan and execute actions [23]. A description of the agent's environment is provided by fusing certain perceptions from different sensing organs, different interpretation procedures, or from any conceivable sensors at different times. As such, "perception" denotes an internal description of the external environment around the robot. This

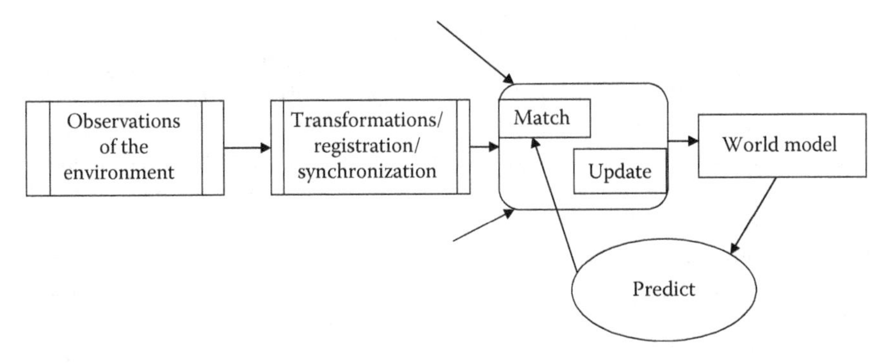

FIGURE 2.9
Dynamic world-model framework.

process is called dynamic world modeling (DWM). Thus, a robot uses a "model," the internal description, to reason about the external environment. The methods of estimation theory provide a theoretical foundation for fusion in the case of numerical data. Other approaches are based on ANNs. Inference techniques can be used as computational mechanisms in cases of symbolic information. Fusion of such symbolic information would require reasoning and inference in the presence of uncertainty. The AI community uses rule-based inference engines. This could be based on forward- or backward-chaining processes, such as fuzzy rule-based fuzzy implication functions, which have to satisfy some criteria, as discussed in Section 6.3. Perceptual fusion is fundamental to the process of DWM. DWM is an iterative process of integrating (or fusing) the observations into an internal description. The framework of the DWM process is shown in Figure 2.9 [23]. This cyclic process consists of predict, match, and update phases.

2.3 Bayesian and Dempster–Shafer Fusion Methods

KF can be viewed as a prediction-corrector (state propagation or evolution and data updating) filtering algorithm, which is widely used for tracking moving object and targets. In addition, the KF is in itself a data–level fusion algorithm. It is possible to consider KF as a Bayesian fusion algorithm [24]. The Bayesian approach involves the definition of priors (*a priori* probabilities), their specifications, and computations of the posteriors. Primarily, the probability theory is based on crisp logic, comprising "zero" or "one" (yes or no, on or off, –1 or +1). It does not consider any third possibility, because the probability definition is based on set theory, which is in turn based on crisp logic; it considers only the probability of occurrence or nonoccurrence of an event. The Dempster–Shafer philosophy incorporates a third aspect—the "unknown." The idea of "mass," or measure of confidence in each alternative, is also introduced. This mass can be considered a probability, although it is not a probability in itself. Sensor DF can also be used to draw inferences about the joint distribution of two sets of random variables, keeping in mind that the measurements being corrupted with noise are themselves random variables. However, in general, this process may not be easy, because each dataset has some information that is also common to both sets. Let us assume that we have two datasets: Set 1 = (x, t) and Set 2 = (y, T), such that x and y have some common variables and t and T are the target variables. The idea is to form inferences about the joint distribution of (t, T) using the information in the datasets (Sets 1 and 2). Subsequently, this estimate of the joint distribution can be used to solve decision problems.

2.3.1 Bayesian Method

Let us define $p(A)$ as the probability of occurrence of an event A, and $p(A, B)$ as the probability of occurrence of two events A and B. Then, the conditional probability of occurrence of A, given that the event B has already occurred, can be related as follows:

$$p(A, B) = p(A \mid B)\, p(B)$$

We will also notice that because $p(A, B) = p(B, A)$,

$$p(B, A) = p(B \mid A)\, p(A)$$

$$p(A \mid B) = p(B \mid A)\, p(A)/p(B)$$

The above relation can also be written as follows:

$$p(A \mid B) = \frac{p(B \mid A)\, p(A)}{\sum_i p(B \mid A_i)\, p(A_i)} \tag{2.25}$$

The denominator acts as a normalization factor if there are several events of A that can be distinguished from B in a few ways. The above equation is known as Bayes' rule. Replacing A with x and B with z, we obtain the following relation:

$$p(x \mid z) = \frac{p(z \mid x)\, p(x)}{\sum_i p(z \mid x_i)\, p(x_i)} \tag{2.26}$$

The items x and z are regarded as random variables, and x is a state or parameter of the system and z is the sensor measurements. Thus, the Bayes' theorem is interpreted as the computation of the posterior probability, given the prior probability of the state or parameter ($p(x)$), and the observation probability ($p(z|x)$): the value of x that maximizes the term $(x|\text{data})$. The maximum likelihood is related to this term if $p(x)$ is considered a uniform distribution: the value of x that maximizes the term $(\text{data}|x)$. The term $p(z|x)$ assumes the role of a sensor model in the following way: (1) First build a sensor model: fix x, and then find the probability-density function (pdf) on z. (2) Use the sensor model: observe z, and then find the pdf on x. (3) For each fixed value of x, a distribution in z is defined. (4) As x varies, a family of distributions in z is formulated. For the observation z of a target-tracking problem with state x, the Gaussian-observation model is given as a function of both z and x as shown below:

$$p(z\,|\,x) = \frac{1}{\sqrt{2\pi\;\sigma_z^2}}\exp\left(-\frac{(z-x)^2}{2\sigma_z^2}\right) \tag{2.27}$$

When the model is built, the state is fixed, and the distribution is then a function of z. When the observations are made, the distribution is a function of x. The prior $p(x)$ is given as

$$p(x) = \frac{1}{\sqrt{2\pi\;\sigma_x^2}}\exp\left(-\frac{(x-x_p)^2}{2\sigma_x^2}\right) \tag{2.28}$$

Then, using the Bayes' rule, the posterior is given as follows after noting the observation:

$$\begin{aligned} P(x/z) &= \text{Const.}\frac{1}{\sqrt{2\pi\sigma_z^2}}\exp\left(-\frac{(z-x)^2}{2\sigma_z^2}\right)\frac{1}{\sqrt{2\pi\sigma_z^2}}\exp\left(-\frac{(x-x_p)^2}{2\sigma_x^2}\right) \\ &= \frac{1}{\sqrt{2\pi\;\sigma^2}}\exp\left(-\frac{(x-\bar{x})^2}{2\sigma^2}\right) \end{aligned} \tag{2.29}$$

where

$$\bar{x} = \frac{\sigma_x^2}{\sigma_x^2+\sigma_z^2}z + \frac{\sigma_z^2}{\sigma_x^2+\sigma_z^2}x_p \tag{2.30}$$

and

$$\sigma^2 = \frac{\sigma_z^2\sigma_x^2}{\sigma_x^2+\sigma_z^2} = \left(\frac{1}{\sigma_z^2}+\frac{1}{\sigma_x^2}\right)^{-1} \tag{2.31}$$

From Bayes' rule, we obtain the following general application rule for the independent likelihood pool:

$$p(x \mid Z^n) = \{p(Z^n)\}^{-1} p(x) \prod_{i=1}^{n} p(z_i \mid x) \tag{2.32}$$

The conditional probabilities of $p(z/x)$ are stored "a priori" as functions of z and x. For fusion, it is assumed that the information obtained from different sources and sensors is independent. Only the underlying state is common between the sources. For a set of observations, we thus obtain:

$$\begin{aligned} p(x \mid Z^n) &= \{p(Z^n)\}^{-1} p(Z^n \mid x) p(x) \\ p(x \mid Z^n) &= \frac{p(z_1, \ldots, z_n \mid x) p(x)}{p(z_1, \ldots, z_n)} \end{aligned} \tag{2.33}$$

The joint distribution of Z should be known completely. From the foregoing discussion, it is clear that Bayes' formula can be used for sensor DF.

2.3.1.1 Bayesian Method for Fusion of Data from Two Sensors

Each sensor is assumed to have made an observation and processed these data to estimate the type of the aircraft (or any moving body) using some tracking algorithm based on the current measurement and the previous measurements [24]. Hence, some prior probabilities are listed below. We assume that for sensor 1, the new set $Z_1{}^1$ is obtained from the current measurement z_1^1 and the old dataset Z_0^1, and for sensor 2, the new set $Z_1{}^2$ is obtained from the current measurement z_1^2 and the old dataset Z_0^2. Essentially, at the fusion node, the probability of x (a particular aircraft) being one of the three aircraft types is to be computed based on the latest set of data: $p(x \mid Z_1{}^1 Z_1{}^2)$. Using Bayes' rule, we obtain the following relationship [24]:

$$\begin{aligned} p(x \mid Z_1^1 Z_1^2) &= p(x \mid z_1^1 z_1^2 Z_0^1 Z_0^2) \\ &= \frac{p(z_1^1 z_1^2 \mid x, Z_0^1 Z_0^2) p(x \mid Z_0^1 Z_0^2)}{p(z_1^1 z_1^2 \mid Z_0^1 Z_0^2)} \end{aligned} \tag{2.34}$$

Because the sensor measurements are assumed independent, we obtain

$$p(x \mid Z_1^1 Z_1^2) = \frac{p(z_1^1 \mid x, Z_0^1) p(z_1^2 \mid x, Z_0^2) p(x \mid Z_0^1 Z_0^2)}{p(z_1^1 z_1^2 \mid Z_0^1 Z_0^2)} \tag{2.35}$$

On the basis of the Equation 2.35, we derive the equation

$$p(x \mid Z_1^1 Z_1^2) = \frac{p(x \mid Z_1^1)p(z_1^1 \mid Z_0^1)p(x \mid Z_1^2)p(z_1^2 \mid Z_0^2)p(x \mid Z_0^1 Z_0^2)}{p(x \mid Z_0^1)p(x \mid Z_0^2)p(z_1^1 z_1^2 \mid Z_0^1 Z_0^2)} \quad (2.36)$$

Finally, at the fusion node, the posterior probability for the aircraft x is given as follows [24]:

$$p(x \mid Z_1^1 Z_1^2) = \frac{p(x \mid Z_1^1)p(x \mid Z_1^2)p(x \mid Z_0^1 Z_0^2)}{p(x \mid Z_0^1)p(x \mid Z_0^2)} \quad (2.37)$$

times the normalization factor.

Thus, Equation 2.37 yields the required fusion solution using the Bayesian approach.

EXAMPLE 2.1

Let us assume that two sensors are observing an aircraft [24]. From the signature of the aircraft, it could be either one of the three aircraft: (1) Learjet (LJ) (the jet-powered Bombardier), (2) Dassault Falcon (DF), or (3) Cessna (CC; the propeller-driven Cessna Caravan). The probability values are given in Table 2.1 [24]. The next step is to calculate the fused probabilities for the latest data. The computed values of 0.5, 0.4, and 0.1 (at the fusion node, in Table 2.1) could be the prior estimates of what the aircraft could reasonably be (or they could be based on a previous iteration using the old data). Now, using the values from Table 2.1, we can compute the fused probabilities at the fusion center as follows:

1) $p(x = \text{LJ} \mid Z_1^1 Z_1^2) \approx \frac{0.7 * 0.80 * 0.5}{0.4 * 0.6} = 1.1667$

2) $p(x = \text{DF} \mid Z_1^1 Z_1^2) \approx \frac{0.29 * 0.15 * 0.4}{0.4 * 0.3} = 0.145$

3) $p(x = \text{CC} \mid Z_1^1 Z_1^2) \approx \frac{0.01 * 0.05 * 0.1}{0.2 * 0.1} = 0.0025$

After the fusion of probabilities using Bayes' rule (and after including the normalization factor of 0.7611), we finally obtain the following results (based on the old probabilities and the new data):

The probability that (a) the aircraft is a Learjet, $p(x = Lj)$, is 0.888, (b) the aircraft is a Falcon is 0.1104, and (c) the aircraft is a Cessna is 0.0019. From this result, we infer that the aircraft is most likely the Learjet (88.8%). The probability that it would be a Falcon is much less (11%) and, almost certainly, it would not be a Cessna (0.2%) because it has negligible posterior probability.

TABLE 2.1

Dataset from Two Sensors (Example 2.1)

Sensor 1	Sensor 2
Posterior probabilities (from the previous iteration)	
$p(x = \text{LJ} \mid Z_0^1) = 0.4$	$p(x = \text{LJ} \mid Z_0^2) = 0.6$
$p(x = \text{DF} \mid Z_0^1) = 0.4$	$p(x = \text{DF} \mid Z_0^2) = 0.3$
$p(x = \text{CC} \mid Z_0^1) = 0.2$	$p(x = \text{CC} \mid Z_0^2) = 0.1$
The computed values at the fusion node	
$p(x = \text{LJ} \mid Z_0^1 Z_0^2) = 0.5$	
$p(x = \text{DF} \mid Z_0^1 Z_0^2) = 0.4$	
$p(x = \text{CC} \mid Z_0^1 Z_0^2) = 0.1$	
Updated posterior probabilities	
$p(x = \text{LJ} \mid Z_1^1) = 0.7$	$p(x = \text{LJ} \mid Z_1^2) = 0.8$
$p(x = \text{DF} \mid Z_1^1) = 0.29$	$p(x = \text{DF} \mid Z_1^2) = 0.15$
$p(x = \text{CC} \mid Z_1^1) = 0.01$	$p(x = \text{CC} \mid Z_1^2) = 0.05$

Source: Challa, S., and D. Koks. 2004. *Sadhana* 29(2):145–76.

2.3.2 Dempster–Shafer Method

In addition to the third state, this method deals with the assigned measures of "belief" (in any of these three states) in terms of "mass," in contrast to the probabilities. The DS method and the Bayes' approach both assign some weighting: either masses or probabilities [24]. These masses could more or less be regarded as probabilities, but not as true probabilities in the usual sense. The DS method provides a rule for determining the confidence measures for the three states of knowledge, based on data from old and new evidence. Reliable estimates of the identities of the sensor measures are required. The results from the application of the DS method depend on the choice of parameters that determine these measures. The DS method might fail to yield a good solution if it is used to fuse two irreconcilable datasets. In a situation wherein the two sensors contribute strongly differing opinions on the identity of an emitter, but agree very weakly on the third state, the DS model will favor the third alternative. This problem needs to be resolved. In the DS method, we have to assign masses to all the subsets of the entities of a system. For a system with n members, there are 2^n (that is, the power set) subsets possible when masses are assigned, say, "1" to each element within the subset and "0" to each one that is not. A measure of the confidence in each of the states has to be assigned—occupied, empty, and unknown; and the total must add up to 100%, although these are not probabilities throughout the entire power set. The masses can be fused using the DS rule of combination. It also introduces the concepts of "support" and "plausibility." The DS method provides a rule [22,24] for computing the confidence measure of each state, based on the data from the old and new evidences, as shown below:

$$m^{1,2}(C) \propto \sum_{A \cap B = C} m^1(A) m^2(B) \tag{2.38}$$

Here, the superscript denotes sensors 1 and 2. Compared to the Bayesian fusion rule, we have two extra states for the available knowledge (Example 2.2): (1) the "unknown" state that represents the decision state regarding the possibility of the aircraft not being a certain type; and (2) the "fast" state, where we are not able to distinguish between the two (relatively fast) aircraft. First, for each sensor (1 and 2), the masses for each element of the power set should be allocated. For example, in the case of a set of three aircraft, there are eight possible states—each aircraft: three states, two aircraft at a time: three states, all three aircraft at a time (one state), and no aircraft (one state). Only the useful states need be assigned masses, and these should add up to 1. Subsequently, these masses are fused based on the new information (say, from the second source of information) using the DS fusion rule, as shown in Equation 2.38. The DS rule with the normalization is written as

$$m^{1,2}(C) = \left\{ \sum_{A \cap B = C} m^1(A) m^2(B) \right\} \Big/ \left\{ 1 - \sum_{A \cap B = \varnothing} m^1(A) m^2(B) \right\} \tag{2.39}$$

EXAMPLE 2.2

Consider Example 2.1. Here, instead of the prior probabilities, we use the masses shown in Table 2.2 [24]. Computing the fused mass for the possible states using the DS fusion rule, the values of the masses obtained are presented in Table 2.2. The computation is illustrated for the state that the aircraft Learjet (LJ) occurs, meaning that, in Equation 2.39, the variable C = Learjet aircraft. All other results are shown in Table 2.2. Here, it is assumed that there is a possible

TABLE 2.2
Masses Assigned for Each Sensor for the Computation of DS Rule

State in the Power Set	Sensor 1	Sensor 2	DS-Fused Final Masses for Each State (after the Application of the DS-Fusion Rule and the Normalization Factor)
	Mass = m^1	Mass = m^2	
Learjet (LJ)	0.30	0.40	0.55
Falcon (DF)	0.15	0.10	0.16
Caravan (CC)	0.03	0.02	0.004
Fast (LJ, DF)	0.42	0.45	0.29
Unknown (LJ, DK, CC)	0.10	0.03	0.003
Total mass	1.00	1.00	1.007

Source: Challa, S., and D. Koks. 2004. *Sadhana* 29(2):145–76.

state of "fast," being the aircraft LL and DF, and the unknown state, being the aircraft LJ, DF, and CC. Hence, we obtain the following result [24]:

$$\begin{aligned} m^{1,2}(\text{LJ}) &\propto m^1(\text{LJ})m^2(\text{LJ}) + m^1(\text{LJ})m^2(\text{LJ},\text{DF}) + m^1(\text{LJ})m^2(\text{LJ},\text{DF},\text{CC}) \\ &\quad + m^1(\text{LJ},\text{DF})m^2(\text{LJ}) + m^1(\text{LJ},\text{DF},\text{CC})m^2(\text{LJ}) \\ &= 0.3 * 0.4 + 0.3 * 0.45 + 0.3 * 0.03 + 0.42 * 0.4 + 0.1 * 0.4 \\ &= 0.47 \end{aligned}$$

From the results, it can be inferred that the aircraft is possibly a Learjet, as its final mass is 0.55 (55%). The "fast" state's fused mass is 0.29 (29%), which shows that the aircraft would certainly not be the Cessna (CC), because it has a fused mass of only 0.4%.

2.3.3 Comparison of the Bayesian Inference Method and the Dempster–Shafer Method

The Bayesian inference method (BIM) uses probability theory and hence does not have a third state called "unknown." The probability itself is based on the occurrence of an event when numerous experiments are carried out. It is based on only two states, and there is an element of chance involved in BIM. In contrast, the DS theory considers a space of elements that reflect the state of our knowledge after making a measurement. In the BIM method, there is no "unknown" state; either an event has occurred or not, or either an event A or an alternative event B has occurred. In the DS model, the state "unknown" could be the state of our knowledge at any time about an emitter, but we are not sure. DS requires the masses to be assigned in a meaningful way to all the states, whereas BIM requires the priors (probabilities) to be assigned. However, a preliminary assignment of masses could be required to reflect the initial knowledge of the system. The DS model allows computation of support and plausibility, in addition to involving more computations compared to the BIM model.

2.4 Entropy-Based Sensor Data Fusion Approach

It is important for a sensor NW to perform efficiently in certain difficult environments. The NW should process available information efficiently and share it such that decision accuracies are enhanced. One interesting approach is to measure the value of information (VOI) obtained from the various sensors, and to then fuse the information if the value (a gain involving significant importance and appreciation) is added in terms of the decision accuracy (DA) [25,26]. The concept is based on the information-theoretic (metric entropy) measure and related concepts. The DF here is conditioned upon whether the VOI has improved the DA. In certain other situations, we have two aspects to consider in a single image, such as the spatial resolution and the spectral resolution (of the imaging sensor or system). A high spatial resolution might be accompanied by low spectral information, and high spectral resolution might be accompanied by low spatial resolution. To obtain high spatial and spectral resolutions, specific methods of image DF need to be applied. This is a useful process for merging similar-sensor and multisensor images to enhance the image information. For this purpose, a metric system based on entropy can be useful. Entropy perceives information as a frequency of change in images (the digital numbers). It is unfortunate that, in some literature, entropy is used directly as information, but the context clears the ambiguity in presentation. In fact, when a new set of data is added and used for analysis and inferences, then the new entropy (uncertainty) will be reduced compared to the old entropy (uncertainty), and the difference will be the gain in the information. The use of entropy must be viewed in this context.

2.4.1 Definition of Information

In 1948, Claude Shannon applied the probabilistic concept in modeling message communications, with the proposal that a particular message is one element (possibility one) from a set of all possible messages. Given the finite set of the number of messages, any monotonic function of this number can be used as a measure of the information when one message is chosen from this set. The information is modeled as a probabilistic process. It is essential to know the probability of each message occurring, the intention being to isolate one message from all of the possible messages (in a set). Thus, the occurrence of this random event x is the probability $p(x)$ of the message. The $I(x)$ is the self-information of x, and this is the inverse of the probability, because if the event x always occurs, then $p(x) = 1$, and no new information can be transferred.

$$I(x) = \log \frac{1}{p(x)} = -\log\{p(x)\} \tag{2.40}$$

The above definition of information is intuitively appealing from an engineering viewpoint. The average self-information in the set of messages with N outputs will be [25]

$$I(x) = -N\,p(x_1)\log\{p(x_1)\} - N\,p(x_2)\log\{p(x_2)\}, \ldots, N\,p(x_n)\log(\{p(x_n)\}) \tag{2.41}$$

Then, the average information per source output is represented by the following equation:

$$H = -N\sum_{i=1}^{n} p(x_i)\log\{p(x_i)\} \tag{2.42}$$

Here, H is also known as Shannon's entropy. However, in general, the value of N is set to 1, and hence we obtain the following relation using the natural logarithm:

$$H = -\sum_{i=1}^{n} p(x_i)\ln_2\{p(x_i)\} \tag{2.43}$$

Shannon introduced the mathematical concept of information and established a technical value and a meaning of information. From the above development, we can see that the entropy (somewhat directly related to the covariance) or uncertainty of a random variable X having a probability-density function $p(x)$ is defined as

$$H(x) = -E_x\{\log p(x)\} \tag{2.44}$$

It is the –ve expected value of the logarithm of the pdf of the random variable X. Entropy can be roughly thought of as a measure of disorder or lack of information. Now, let $H(\beta) = -E_\beta[\log p(\beta)]$, the entropy before collecting data "z," and $p(\beta)$ the prior density function of β, a specific parameter of interest. When the data z is collected, we have the following relation:

$$H(\beta \mid z) = -E_{\beta/z}\{\log p(\beta \mid z)\} \tag{2.45}$$

The measure of the average amount of information provided or gained by the experiment with data z on the parameter β is provided by the following relationship:

$$I = H(\beta) - E_z\{H(\beta, z)\} \tag{2.46}$$

This is the "mean information" in z about β. We note that entropy implies the dispersion or covariance of the density function and, hence, the uncertainty. Thus, the information I is perceived as the difference between the prior uncertainty (which is generally large) and the "expected" posterior uncertainty (which is now reduced due to the new data adding some information about the parameter or variable of interest). This indicates that due to experimentation, collection, and the use of data z, the (posterior) uncertainty (which is expected to reduce) is reduced and, information is gained. Thus, the information is a nonnegative measure, and it is zero if $p(z,\beta) = p(z) \cdot p(\beta)$; i.e., if the data are independent of the parameters, which implies that the data does not contain any information regarding that specific parameter.

2.4.2 Mutual Information

The information could be in the form of features for classification or data for detection of an object. Let $H(x)$ be the entropy of the observed event, and let z be a new event with its uncertainty (entropy) as $H(z)$. Then, we can evaluate the uncertainty of x after the event z has occurred and incorporate it to compute the new entropy:

$$H(x \mid z) = H(x, z) - H(z) \tag{2.47}$$

This conditional entropy $H(x|z)$ signifies the amount of uncertainty remaining about x after z has been observed or accounted for. Thus, if the uncertainty is reduced, information has been gained by observing and incorporating z, the new information. The mutual information $I(x,z)$ is a measure of the uncertainty after observing and incorporating z and can be represented as follows:

$$I(x, z) = H(x) - H(x \mid z) \tag{2.48}$$

Thus, the VOI is useful when we want to assess the information available from multiple sensors on a single node or from different sensors from neighboring nodes.

2.4.3 Entropy in the Context of an Image

Information-content quantification can be similarly applied to the information regarding an image (as digital numbers and related attributes such as the pixels of an image). Shannon's formula is modified as follows [25,26]:

$$H = -\sum_{i=1}^{N_g} d(i) \ln_2 \{d(i)\} \quad (2.49)$$

Here, N_g is the number of the gray level of a histogram of the image. This histogram range, for an 8-bit image, is between 0 and 255. The variable $d(i)$ is the normalized frequency of occurrence of each gray level, and hence the average information content is estimated in units of bit per pixel. It should be noted here that the average information content of the image is not affected by the size of the image or by the pattern of gray levels. However, it is affected by the frequency of each gray level.

2.4.4 Image-Noise Index

For image fusion, the concept of entropy can be used to estimate the change in quantity of information between the original images (called the parent images) and the new images (called the child images) obtained by fusion of two parent images (from two image-sensing devices). The procedure for computation of the image-noise index (INI) is explained next [26]. In the forward pass, the images are fused by a chosen method, and the fused image is obtained. This fused image is restored in the reverse pass to its original condition (i.e., the parent image) by using the reverse process, and the entropy of the restored image is estimated. To bring the fused image to its original condition, the original spatial-resolution value of the multispectral image is substituted in the reverse process. Because we know the three entropy values (a, b, and c) of the original image, the fused image, and the restored image, respectively, we can easily determine the INI. The quantity $|a-c|$ represents the unwanted information and, hence, the noise. The difference $(b-a)$ represents the increase in information, which could be useful information, the noise, or both. Thus, the INI is defined as follows [26]:

$$INI = \frac{(b-a)-(|a-c|)}{|a-c|} = \frac{(b-c)}{|a-c|} - 1 \quad (2.50)$$

The numerator of the first term is the amount of useful information added (the signal). Thus, the INI gives the incremental signal-to-noise ratio (SNR) of the fusion process.

2.5 Sensor Modeling, Sensor Management, and Information Pooling

Recent advances in electronics, computer science and engineering, and optics have resulted in a massive increase in the types of sensors available for use in autonomous systems [27–29]. Automated systems must be able to react to changes in their environment to be effective, and these systems must therefore rely on sensors for information about their environment. The sensors provide dependable interfaces between an often ill-defined and chaotic reality and a computer with a data-acquisition system. Sensors can handle data only when they are within specified bounds and are presented in a precise format for ease of understanding and interpretation. Often sensors are complex and might have limited resolution and reliability. In complex systems, the sensors do not necessarily correspond to a single concrete sensor. A sensor might have some preprocessing built-in unit or some logic or intelligence-processing unit built within the sensor system. The readings from the sensors can be fused to create one abstract sensor.

Sensors that correspond to only one concrete sensor are called "simple sensors," and sensors made up of several concrete sensors are called "sensor NWs," and are created from many simple sensors. After fusion, the data from a sensor NW can be treated as if the data originated from a simple sensor. Sensor NWs can have many different levels of complexity, architectures, and configurations. These architectures have different requirements for data communication and have varying computational complexities. The performance of sensors is measured in terms of accuracy, repeatability, linearity, sensitivity, resolution, reliability, and the range it can handle. The sensor output could be a continuous signal, digital signal, frequency output, amplitude, pulse modulated, and so on.

2.5.1 Sensor Types and Classification

Active sensors inject varying types of signals into their environments and then measure the interaction of the signal with the environment, e.g., radar, sonar, active far (AF) infrared (IR) sensor, an active optical-seam tracker, and an active computer-vision system. These sensors require external power to operate. The signal being measured is thus modified by the sensor to produce an output signal. These sensors are also called parametric sensors, e.g., in a thermistor, a change of resistance occurs due to a change in the temperature. The properties of the sensor can change in response to an external signal. Passive sensors simply record the energy or signals already present in the system's environment, e.g., thermometers, video cameras, and IR cameras. The energy of the input stimulus is converted by the sensor or transducer into output energy without the need for an additional power source or injection of energy into the sensor, e.g., a thermocouple, a piezoelectric sensor, a pyroelectric detector, and passive IR sensor.

2.5.1.1 Sensor Technology

There are mainly two types of sensor technologies [29]: (1) internal state sensors (ISSs), and (2) external navigational state sensors (ENSS).

ISSs [29] measure the internal states of the system; for example, in the case of a robot, the velocity, attitude and bearings, accelerations, current, voltage, temperature, pressure, balance, and so on are recorded so that the static and dynamic stability of the robot can be monitored and maintained, and potential failure situations can be detected and avoided. Some examples of such sensors are potentiometers, tachometers, accelerometers, and optical encoders. These interoceptive sensors inform the body of its own state. They are mainly related to the problems of internal diagnosis and fault detection. Two types of ISSs are described in the following two paragraphs.

Contact-state sensors (CSSs) involve direct physical contact with the objects of interest; e.g., microswitches and touch, force, and tactile potentiometers [29]. Contact sensors are used to handle objects, reach a specific location, and protect a robot from colliding with obstacles. These are relatively inexpensive, have a quick response, and are easy to construct and operate. The tactile sensors are magnetic, capacitive, and include piezoelectric technology. A binary tactile sensor can be used for detecting the presence or absence of touch, and a 2D tactile sensor can provide information regarding the size, shape, and position of an object. CSSs have limited life span because of frequent contacts with objects.

Non-contact state sensors (NCSSs) [29] are synchros, resolvers, compasses, accelerometers, gyroscopes, and optic encoders. Synchros and resolvers are rotating electromechanical devices that measure the angular position information; these devices are normally large and heavy. Gyros are mechanical and optical devices and are used to maintain the stability and attitude of robotic systems (useful in unmanned flying robots, underwater vehicles, space robots that navigate in a 3D environment, unmanned aerial vehicles [UAVs], and so on). Gyroscopes can be used to measure the movement of a system relative to its frame of reference. Compasses, gyroscopes, and encoders are generally combined into an inertial navigation sensor (INS), which is used to estimate the trajectory followed by an object. Mechanical gyros operate on the basis of conservation of momentum, whereas optical gyros have no moving parts. The absolute optic encoders are used to measure and control the angle of the steering wheel during the path control of a wheeled robot. The incremental optic encoders are used for measuring and controlling motor speed and acceleration in a mobile robot.

ENSSs monitor the system's geometry and dynamics in relation to its tasks and environment, e.g., proximity devices, strain gauges, sonar (ultrasonic range sensor), and pressure and electromagnetic (EM) sensors [29]. External navigational sensors measure environmental features such as range, color, gap, road width, room size, and object shape. These are used for correcting the errors in the world model, detecting environmental changes, and avoiding unexpected obstacles. The exteroceptive sensors deal with the external world. The sense of "balance" can be monitored mechanically using gyroscopes. The system's joint positions can be sensed by the mechanical system using encoders. Both balance and joint positions can be handled by the INS to some extent.

Nonvision-based navigation sensors (NVBENSS) [29] are force and magnetic sensors, which are passive navigation sensors. They do not generate EM waves, are reliable, and produce little noise. The force sensor can be used to monitor whether there is an object in contact with the robot for avoiding collision, e.g., a full cover with a force-detecting bumper can be implemented on a cylindrical robot for safe operation. An actuated whisker can be used to measure the profiles of the contacted objects for object identification and obstacle avoidance. Sound (energy), smell, IR radiation, optics, laser, radio frequency (RF), satellites, and radars are other NVBENSS sensors. The data measured by the active sensors are the reflection of either the changes in emitted-energy properties, such as the frequency and phase, or simple time-of-flight calculation. These sensors provide one-dimensional data at a high rate and are useful for following the path, reaching a goal, avoiding obstacles, and mapping an environment. Sonar (ultrasonic) sensors have long wavelengths.

Global positioning systems (GPS) provide a robust alternative to many other methods of navigation. Satellite beacons are used to determine an object's position anywhere on the earth. Other beacon systems are used in transportation NWs, especially in maritime and aviation systems. GPS and RF systems are widely used in robot localization and tracking of objects and humans, using the services of 24 satellites at a height of 6900 nautical miles. The absolute 3D position of any GPS receiver can be determined using a triangulation method. Time-of-flight radio signals, which are uniquely coded and transmitted from the satellites, are routinely used worldwide. Problems with GPS include incorrect time synchronization between the satellite and receiver, precise real-time location of the satellites, EM noise, and other interferences. Differential GPS (DGPS) provides much more precise data than the GPS alone regarding the positions of the objects. The integration of data from the GPS and other INSs leads to fusion of data that reduces the effects of uncertainties. An odor-detection system allows a robot to follow trails of volatile chemicals laid on the floor. The sensor uses controlled flows of air to draw odor-laden air over the sensor crystal to increase the speed of its response.

Vision-based ENSSs mimic human eyes. The visual information obtained from a vision sensor proceeds through three processing stages [29]: image registration and transformation, image segmentation and analysis, and image interpretation (and image fusion, if included in this process)—a difficult task to achieve in real-time. Charge-coupled devices (CCDs), both active and passive, can be used. Active CCDs use structured lighting to illuminate the scene and enhance the area of interest to make image-processing more efficient; the data in the enhanced area only is processed, projecting a pattern of light strips onto the scene, whereas the depth information can be obtained by looking for discontinuities or deformation in the resulting line image. Passive vision-based sensors work under normal ambient illumination and are used in wheeled, legged, tracked, underwater, and flying robots. They are mainly used for object and landmark recognition, line-following, and goal-seeking tasks.

2.5.1.2 Other Sensors and their Important Features and Usages

Many other sensors are equally important [4,29]. Smell and taste involve measuring the concentration of specific chemicals in the environment. Odor sensing detects chemical compounds and estimates their density within an area. Odors from different materials have different particle compositions, and hence have different locations in the spectra. Automated systems can sense odors by using EM radiation in the process of spectroscopy. Recent advances in fiber optics have produced fiber-optic probes that are capable of detecting biochemical agents.

Touch involves measuring items through physical contact, and in automated systems, this is usually carried out using pneumatics or whiskers. Tactile sensing can be accomplished by numerous methods: the simplest sensors detect the presence or lack of an object; e.g., whiskers that are long, light, and relatively stiff protrude from a system or a body, similar to the whiskers in our nostrils and inner ears. When a whisker touches an object, the whisker is displaced. A sensor at the base of the whisker senses this movement and signals the presence of an object. The ambient EM radiation that has been reflected from the object can be detected and used for measurements.

IR cameras capture IR radiation, whose wavelength is longer than visible light. Forward-looking IR (FLIR) sensors sense differences in the heat of objects and construct images from that information. These sensors are useful in night-vision applications. Radar is an EM system for the detection and location of objects, which consists of transmitting antennae emitting EM radiation generated by an oscillator, receiving antennae, and an energy-detecting device or receiver. A portion of the transmitted signal is reflected by a target and reradiated in all directions. This reflected EM radiation (the returned energy) is collected by the receiving antennae and is delivered to a receiver, where it is processed. From the processing of this data, the object is detected and its location and relative velocity are measured by ranging (which is proportional to the speed of light times the total fly-around time/2). The direction or angular position is determined from the direction of the arrival (Section 11.3) of the returned wave front (the acoustic wave for sonar systems). If there is a relative motion between the detected target and the radar, the shift in the carrier frequency of the reflected wave—the Doppler effect—gives the measure of the relative velocity of the object. These measurements are very useful in target tracking.

Microwave radars use wavelengths longer than infrared radiation, and are insensitive to interference from clouds and inclement weather. They are better in terms of detection range and immunity to attenuation caused by atmosphere. However, their capabilities for classification and identification are not very good. Laser ranging systems (laser radar or LADAR) provide images of very accurate depth from a scene. The "identify friend-or-foe" function is used to identify the origin of an aircraft. Apart from these, sound navigation and ranging is an established technology that is approximately equivalent to radar. A sound wave is transmitted in space, and its return signal is measured. By measuring the time needed for the signal to return, the distance of an object from the sensor can be estimated. The velocity of the object relative to the sensor can be estimated using the frequency shift of the returned signal in comparison to the original signal (Doppler shift). These sensors have been developed for both underwater and terrestrial applications, such as detection of submarines, manufacture of self-focusing cameras, and navigation systems for autonomous robotics.

Bumpers are large protuberances from the system, which, when touched, signal contact with an object through an electrical switch. Light-emitting diodes can be used as reflecting sensors to detect objects in the immediate neighborhood. Proprioceptive sensors indicate a system's absolute position or the positions of the system's joints. The positions of a system's joints can be determined by using the direct-drive mechanism, as in a stepper motor, or with an optical encoder attached to the joint. By observing the motion of the encoder, the relative position of a wheel, joint, or track in relation to the rest of the system can be calculated.

An EM-induction (EMI) sensor is used for metal detection. The idea is to detect the secondary magnetic field produced by the eddy currents induced in the to-be-detected metal (say, hidden underground) by a time-varying primary magnetic field, in the frequency range of a few tens of kHz [30]. EMI sensors can be used to detect both ferrous and nonferrous metallic objects. If multiple targets (for example, many metal pieces or devices present underground and considered target objects) must be detected, then the method of data association can be used. The intensity of the received field depends on the depth of the target hidden underground. Thus, the attenuation of the received field would be much higher if the target is deep down. EMI sensors can be used in a variety of modes, such as (1) high-sensitivity search, to detect even small quantities of metal; (2) low-sensitivity search, which has a low probability of detection amidst metallic clutter; and (3) shape search, where the features measured are the metal content and the shape [30].

Ground-penetrating radars (GPRs) detect buried objects by emitting radio waves into the ground. The reflected waves are analyzed. The GPR system has antennae to emit a signal and receive the return signal and a computerized signal-processing system [30]. The frequency of the radio wave is the major control parameter, because the scale or level of detection is proportional to the wavelength of the input signal, i.e., if the wavelength decreases (and consequently, the frequency increases), the quality of the image is enhanced. Signal-processing algorithms also play an important role because they are used for filtering out the clutter signals. The GPR system can build a 3D image of the ground zone for detection. GPRs can be used in three modes as follows: (1) depth search—the operating frequency is maintained low to penetrate the ground at the greatest possible depth; (2) resolution search—the information from the first mode is used to tune the operating frequency for obtaining the best possible resolution for the image (and shape and size of target); and (3) anti-ground bounce-effect search—the frequency is set to a very high value to overcome the masking effect of the ground-air discontinuity, and this mode is suitable for both surface and shallow targets. Many other features of GPR are detailed in a study by Vaghi [30].

In contrast to the thermal-detection methods that exploit the variations in temperatures, IR sensors detect anomalous variations in the EM radiation reflected by the object under detection. IR sensors are also lightweight. The quality of the IR images depends on the time of the day, weather, vegetation cover, and so on. The airborne IR sensor can procure images of the ground from a high standoff distance. The IR sensor can be used in three modes, including (1) searching for shallow buried mines; (2) searching for surface mines; and (3) achieving maximum depth of detection [30].

2.5.1.3 Features of Sensors

Different sensors have different characteristics, and it is important for the DF community to know the specific features of the sensors that are needed for primary sensing purposes and subsequent DF performance requirements. We describe some important aspects of the sensor types and their relations to the type of measurement and state determination [4,29].

From radars, we obtain two basic measurements, radar cross-section (RCS) frequency and time, and from these we derive the range, azimuth, elevation, and range-rate (Doppler effect) information. These derived data are in turn used in a state-estimation algorithm to finally determine the position, velocity, and acceleration time histories of a moving object accurately. We can determine of a limited amount of target ID information (identification mark or some similar feature) from the RCS, size, and shape of the target in limited cases. From the radar homing and warning system, we initially detect the RF intensity, and then we derive the direction, which in turn is used to determine the identity of the target. For example, we get a warning by illumination from the missile-guidance radar. From a synthetic aperture radar, the coherent RCS, shape, size, and direction of the target, and ultimately, the identity of the target are determined. This methodology gives enhanced radar resolution through coherent processing using stable or moving antenna. The signature format is a 2D range and cross-range reflectivity image. The useful features are size, aspect ratio, and number and locations of the scatterers.

From the IR recordings, we obtain the pixel intensity and color of the target and spectral data. The derived information is location, size, shape, and temperature of the object, which finally yield the position and identity of the object. This is dependable for ID based on spectral data but not efficient for determination of the position and velocity of the target. From the IR warning system, we primarily detect the IR intensity and subsequently, the direction and identity of the object. This generates a warning that the heat source is approaching the observer. A laser-warning system records the intensity, which is used to determine the identity of the object. The signature formats for the laser radar are the 3D reflectivity image, Doppler modulation (vibration), and the 2D velocity image. The useful features are size, 3D shape, scatter locations, propulsion, structural features, skin frequencies, and spatial distribution of the moving components.

Using microwave radar, we obtain Doppler modulation 2D (range-cross range) reflectivity images. The useful features are velocity, propulsion frequencies, line widths, size, aspect ratio, and number and location of scatterers. A millimeter wave radar (MMWR)–type sensor yields the 1D reflectivity profile, 1D or 2D polarization image, and Doppler modulation as the signatures. The useful features are the distribution and range extent of scatter number, locations of the odd and even bounce-scatterers, propulsion frequencies, and line widths (rpm, blade passing, and tread slapping).

The signature format of an RF interferometer is a spectrum- and time-dependent microwave emission. Its useful features are frequency, frequency modulation, amplitude modulation, pulse duration, and pulse intervals. The signature format of the thermal image is a 2D thermal image, and the useful features are shape (perimeter and area, aspect ratio, moments), texture, maximum and minimum emission, number and location of hot spots, and context. The MMWR is placed midway between microwave sensors and electro-optical tracking systems (EOTs). They have a smaller size, better resolution, and higher level of immunity to the effect of the atmospheric attenuation.

An electro-optical system gives intensity and color to the pixel-image, which are used for determination of the location, size, and shape of the object, and ultimately, the position and identity of the object. EOTs are superior in terms of classification and identification of the objects. They have a much higher resolution but are affected by atmospheric attenuation. Their data-processing demands are relatively high. Electronic-support measures yield amplitude, frequency, and time, which are further used for deriving SNR, polarization, and pulse shape to finally estimate the position, velocity, and identity of the target, providing both general and specific emitter IDs. An acoustic sensor's signature format is the spectrum- and time-dependence of the acoustic emission, and its useful features are the propulsion frequencies, harmonics, frequency ratios, pump and generator frequencies, and peculiar noise sources. We obtain a 2D reflectivity (contrast) image. The useful features are shape, texture, internal structure, and context.

2.5.1.4 Sensor Characteristics

It is also important to know the characteristics of various sensors. These characteristics are recounted below [4,29]:

1. Detection performance: false alarm rate, detection probabilities, and ranges for the calibrated target characteristics in a given noisy background.
2. Spatial and temporal resolution: ability to distinguish between two or more targets that are very close to each other in space or time.
3. Spatial coverage: spatial volume encompassed by the sensor. For scanning sensors, this may be described by the instantaneous field-of-view (FOV), the scan-pattern volume, and the total field-of-regard achievable by moving the scan pattern.

4. Detection and tracking modes: search and tracking modes undertaken are (a) starting or scanning, (b) single or multiple target tracking, and (c) single or multimode (track-while-scan and stare).
5. Target revisit rates: rate at which a specific target is revisited by the sensor to carry out a sample measurement.
6. Measurement accuracy: accuracy of sensor measurements in terms of statistics.
7. Measurement dimensionality: number of measurement variables (range, range rate, and spectral features) between the target categories.
8. Hard vs. soft data: sensor outputs either as hard-decision (threshold) reports or as preprocessed reports with quantitative measures of evidence for possible decision hypothesis.
9. Detection and track reporting: sensor reports each individual target detection or maintains a time-sequence representation (track) of the target's behavior.

2.5.2 Sensor Management

Several sensor NW–based systems have increased functionality and are characterized by a high degree of autonomy, reconfigurability, and redundancy [30]. The sensor can be mounted on a platform and used in various modes. When the sensors move on platforms that are either teleoperated or autonomous, the movements of the sensors should be coordinated to align them to the required positions. Hence, the goals of sensor management are to provide high-level decisions and coordinate the movement of the sensors. This might need real-time information about the system's state. Certain criteria and objectives need to be defined, so that decision making algorithms can be used to decide the future strategy of the sensor NW. The decisions could analyze the following questions: (1) where the platform carrying the sensors should move; (2) how the sensors should be utilized; and (3) what measurements would be needed.

A typical sensor-management problem could be very complex [30]: (1) collecting information from all the subsystems (sensors, platforms, and targets), which could be heterogeneous and distributed; and (2) combining this information to obtain an integrated approach for the final decision making process. One approach is to model all the subsystems and their interactions, using graphical model formalism for this purpose [30]. This approach combines probability and graph theories and can lead to a stochastic adaptive algorithm for sensor management. A typical sensor NW management scenario could be (1) static or dynamic, cooperative or noncooperative targets; (2) some environmental factors that could influence both targets and sensors; (3) sensor suite or NW; and (4) the actual sensor DF process.

The sensor-targets-environment-DF scenario is here called the STEDF assessment and consists of (1) the SA process, (2) the process of sensor fusion, (3) target classification, (4) platform deployment, and finally (5) sensor management; a possible integrated system is shown in Figure 2.10 [30]. Sensor management, based on the current SA and the available resources, provides a feedback control to the suite of sensors, and here some decision making takes place. In normal systems, this feedback is provided by a human operator. Use of intelligent decision making algorithms and automating the decisions facilitate the operator's task. An effective sensor-management algorithm or strategy applies the current information and schedules the measurement process. Rule-based expert systems and fuzzy rule–based logic schemes could be used for such purposes; such methods are called descriptive approaches. The normative or decision-theoretic methods try to establish formal decision criteria based on the gathered experimental data and subsequent analyses, involving the utility theory and probability theoretic reasoning processes. A combination of these methods is also feasible, and an approach based on the combination of Bayesian NWs (BNWs) and other graphical methods is proposed in a report by Vaghi [30]. The tactical situation is inferred in the situation-assessment block. The link between the fusion association and surveillance-system model (SSM) is fused data (Figure 2.10). If any sensor fails, this information can proceed to the SSM. The sensor types could be heterogeneous, mobile, and reconfigurable,

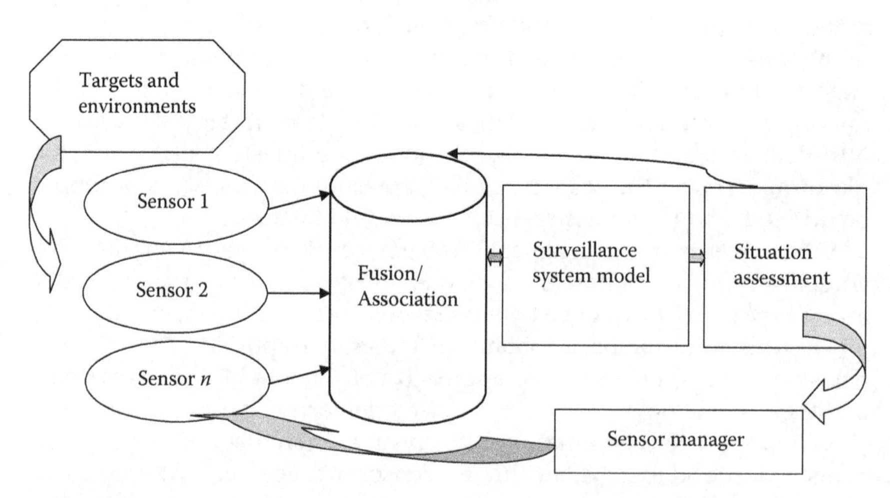

FIGURE 2.10
Integrated sensor-management system.

and the target types could be simple, maneuvering, heterogeneous, or enclosed in a clutter.

2.5.2.1 Sensor Modeling

Sensors have varying accuracies and might not provide one single accurate value for a physical variable. This is not the case with inherently very accurate and high-precision sensors, but these sensors are very costly. Even if a sensor could deliver this type of reading, the physical values being measured generally vary statistically over a period of time and will be affected by measurement noises or errors. A sensor reading that covers a finite period would thus have to return a range of values, rather than one number, to accurately represent the environment; thus, limited accuracy is a factor in all measurement-data processing. Data representations in common use have upper and lower bounds for the values that can be represented and a maximum number of digits of accuracy; if these limits on accuracy are not handled properly, important information can be lost or the results of calculations can be greatly in error. There is a possibility that a sensor might fail during the system's lifetime or may temporarily give inaccurate readings for various reasons. Regardless of how well the systems are designed, the components will eventually deteriorate or stop working perfectly. Engineers and analysts must try to minimize the impact of such component failures through proper design procedures and decisions. In addition to the normal wear and tear, there are limitations to the general capabilities of any component. This is especially true when the system is placed in hostile and/or noisy environments, such as nuclear reactors, battlefields, and outer space.

Mathematical models of the activities of these sensors can be used for characterization of the sensors and their data. Algorithms that handle sensor data efficiently and accurately can be developed, and these algorithms should be efficient in that (1) they extract the maximum information from the sensors; (2) they avoid making wrong and absurd conclusions based on faulty data; and (3) they do not waste costly system resources. The mathematical model of the sensors can be subsequently derived. A sensor can be represented as a mathematical function with two arguments, *viz.*, the environment and time, as follows:

$$S(E, t) = \{V(t), e(t)\} \tag{2.51}$$

Here, E is an abstract entity that contains several physical measurable attributes, and S is a function that maps the environment to numeric values. These attributes can be continuous-valued (temperature, velocity, range, and so on), Boolean (presence of a vehicle, temperature > given limit, and so on), integer-valued (such as number of aircraft, tanks, and number of persons in a queue), and so on. The values recorded will change over time, because the system's environment also changes over time. Assume V is a point in an n-dimensional space and is affected by some uncertainty. Knowing everything about the sensor in use and the measured signal, as much as possible *a priori*, would help in the fusion process because of the following relationship between the measurement and the object's state: measurements = sensor's dynamic math model times the state (or parameter) + noise; or,

$$z = Hx + v \tag{2.52}$$

Here, the aim is to estimate the state vector x based on y, and hence any information regarding H and the statistics of v, such as the sensor type, "range band," and "range" of the measurement signal, would help in the estimation of x. The scale factor and bias errors can be related as follows:

$$z(\text{measured data}) = ay + \text{bias} \tag{2.53}$$

Here, a is the scale factor. General aspects regarding the preprocessing of measurement considering the sensor filter are (1) filter dynamics, because a low-pass analog filter might have been used to remove the high-frequency noise, (2) sensor-mounting errors (rate gyros detect signals from other axes, if misaligned), and (3) the position of the linear accelerometer, which must be known. All of the above aspects will help the analyst understand the performance results of the DF system or to correct the data at the outset itself. The analyst should know as much as possible of the entire chain, including how the data get to him/her from the transducer or sensors, because the screening, editing of the data, and signal processing, together with system identification, are important functions to be performed to obtain very accurate results from the DF systems. The important aim of sensor modeling is to incorporate the dominant modes of the sensor dynamics, if any, into the modeling effort for DF systems, e.g., in state models used in the KF. An SSM could be represented as a nonlinear state-space model, as shown below:

$$x(t+1) = f(x(t), u(t)) \tag{2.54}$$

Here, x, the state vector, consists of the variables related to sensors, platforms, targets, and environmental conditions; u is the decision vector, representing controllable inputs (or decisions in the surveillance system, e.g., sensor allocation and platform deployment). Thus, the SSM is a stochastic model. The system output y contains the measurements, all control elements, and decisions. The platform states and the sensor measurements are observables and are part of the system output y. The state vector x is estimated from the observable z by the SA process. Basically, solving the decision making problem aims to find a policy or control law, $u(t)$, that optimizes an expected utility function within the constraints of the system dynamics [30]. Hence, the flow of the main information in such management architectures is from the sensed environment → SSM → fusion/association → ← SA → sensor management → and back to SSM [30]. The SSM could consist of devices, such as EMI sensors, GPRs, IR sensors mounted on autonomous ground vehicles and aircraft (UAVs).

The targets are represented by their anticipated features within the sensor models. The IR-detected target is characterized by the size and shape of the target, because the IR sensor measures the size and shape. The sensed information is acquired in two stages: (1) target detection, and (2) target classification. In target classification, the target is assigned a model: a static BNW or a dynamic linear-Gaussian model, and the BNW structure would be composed of feature nodes (features of the targets) based on the BNW model of the sensor that might have taken the measurement of the target [30]. For the environment, the available models or directly measurable environmental states can be used. The sensor platforms can be modeled using ordinary differential equations. Performance optimization, motion planning and inner-loop control should take into consideration the sensed information and SA. The sensor manager should also consider the kinematic and dynamic constraints of the sensor platform. Subsequently, the SSM can incorporate the platform models and specific terrain maps. The measurements processed by the fusion process are also included in the model as and when they are available. Finally, the sensor manager uses this expanded SSM to allocate the sensor resources and deploy the platforms while optimizing the desired objectives. When the model of a sensor is available, inference algorithms can be used to compute the unknown variables based on the observables. In fact, the measurements, the observations about the environment, and the parameters used by the sensors become the evidence to the sensor BNW. Then, inference algorithms can be used to estimate the target state. When multiple sensors are used, an appropriate DF process or method should be used to obtain an improved estimation of the target states. This information is then used for SA and possible sensor management. The SA algorithm would produce a picture of the tactical situation as inferred from the SSM, with the aim of updating the estimated state of the surveillance system with reference to time. This is based on inference and sensor fusion and the subsequent inference of the values of the hidden states. The SA picture thus obtained is used by the sensor manager to make decisions regarding the allocation of resources and optimization of the objectives.

2.5.2.2 Bayesian Network Model

BNWs use the concept of mapping cause-and-effect relationships among all relevant variables [24,30]. Inferences can be drawn by building a graph that represents these causal relationships. BNWs represent the extent to which variables are expected to affect each other by incorporating the probabilities, in a manner similar to that explained in Section 2.3. Thus, the fundamental formulae are similar. Learning in BNWs implies learning the probabilities and causal links from the available data, thus making BNWs adaptive in nature. Thus, a BNW is a graphical model applicable for probabilistic relationships among a set of variables. BNWs are used to encode selected aspects of the knowledge and beliefs about a domain, and after the NW is constructed, it induces a probability distribution for its variables. The learning in a BNW is related to learning the conditional probabilities within a specified structure: given a dataset, the idea is to compute the posteriori distribution and maximize it with reference to certain parameters. In a BNW, the problem of learning a structure is more complex than the learning of NW parameters. The merits of a BNW are [30] that (1) it can handle incomplete datasets; (2) the causal relationships can be learned; and (3) in conjunction with the Bayesian statistical technique, BNW can use a combination of domain knowledge and data. To handle incomplete data, a small degree of approximation is required, because the learning is about conditional probabilities. One can use the Gaussian approximation for this purpose. Let X_i represent a set of n variables called events. Then, a BNW over this set consists of two ingredients [30]: (1) a directed acyclic graph, G: each vertex of the graph is called a node and represents a variable that takes on the value x_i. The directed edge from one node to another node represents a causal relation between the two nodes; and (2) quantification of G: each variable is assigned a conditional probability given its cause (parents). Thus, $P(X_i \mid parents(X_i))$ represents the strength of the causal link. A binary BNW is shown in Figure 2.11 [30] with the associated tables (inset) that represent conditional probabilities. One can see that S (sprinkler) and R (raining) are the parents of W, the event that the grass is wet. The probability that the grass is wet when the sprinkler is off (false) but it is raining (true) is 0.9. The BNW models for the EMI, GPR, and IR sensors are described in the study by Vaghi [30]. These comprise the graphical model sensors, target, and environment. In the BNW model, it is possible to consider many variables. The interactions, if not completely known, can be learned and quantified.

2.5.2.3 Situation Assessment Process

This segment of the sensor management produces a picture of the tactical situation that is inferred from the SSM. The chain works as follows [30]: (1) The sensors are deployed on their platforms. (2) The sensors operate in a certain mode ($u(t)$) under certain environmental conditions to obtain

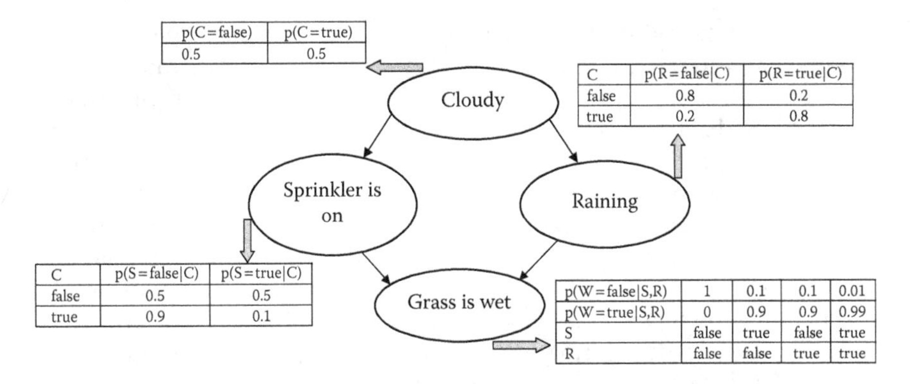

FIGURE 2.11
Binary Bayesian network—an illustrative example. (From Vaghi, A. 2003–2004. Thesis, Politecnico di Milano. http://www.fred.mems.duke.edu/posters/Alberto_GMA_thesis.pdf. Accessed December 2008. With permission.)

the measurements of a certain number of targets. (3) The information is passed onto the corresponding BNW sensor models, which obtain the estimates of the features of the detected target; if the estimates and measurements are available from multiple sensors, these are fused. (4) The fusion results and the associated probabilities for each feature are calculated using the inference and fusion methods. The information is updated in real-time. The past information is summarized and called the current surveillance-system state, and the current state is the last updated picture of the situation, also referred as the current situation. The current situation comprises the current estimates and the associated probabilities of each feature of each target. The objective of the SA is to update the estimated state over time. This SA information is used by the sensor manager to make decisions. The current SA picture also provides classification as an object or clutter for the detected target. This is carried out using a BNW classifier. The state of the surveillance system consists of the states of all its components, including sensors, platforms, and targets, and these targets are presumed to have been detected and classified. Thus, the main role of the SA is to update the target's state based on the new forthcoming information.

The SA produces a picture of the tactical situation at each time instant T. It tells the sensor manager and the human operator how many targets have been detected until time T. This is accomplished using a classifier algorithm. It provides estimates of the features of each detected target, along with the associated confidence measures. This information is used by the sensor manager to make decisions on the allocation of available sensors and to optimize the objectives. Due to the fusion of the information from multiple sensors, the number of undetected targets is reduced.

This is one of the merits of the DF process used in the sensor-management SA system. The numerical values of the associated probabilities can be used to confirm whether the estimate and classification results should be considered reliable (low values, say, $p < .5$) [30]. Low values indicate that there might be only a few sensors, and that further resource reallocation would be required. Extensive simulation studies can be carried out to assess the situation, and after reallocating the resources, the performance evaluation can be again carried out to check whether the numerical values of the computed probabilities have increased or not. In this manner, the sensor management and optimization process can be utilized to obtain more information and solve some (more) uncertain cases. Thus, a BNW helps in sensor management, optimization, and reallocation of resources for enhancing the efficiency of task-directed multisensor DF systems. The other major benefits of the sensor-management process using a BNW are that we can study the effects of the operational environment of the sensors, carry out mathematical modeling of the sensors, and learn from large datasets on the performance efficiency achievable in the MSDF system. As time elapses and increasing amounts of data are gathered and used to update the estimates of the features of the targets and their states, the SA improves in its decision making abilities. However, some saturation (due to excessive learning) might occur after some time, because the system, estimators, or classifiers would have learned the states sufficiently well and new data might not be able to provide any new information to the MSDF-based SA scheme; the system reaches stability at this point. For simulation and validation, the BNW toolbox of MATLAB® can be used [30], which supports many kinds of nodes, exact and approximate inferences, learning of parameters and structures, and static and dynamic models.

2.5.3 Information-Pooling Methods

For managed DF, there are a few important methods depending on how the information from various sensors is pooled and managed [2,31]. There are mainly three methods: (1) the linear opinion pool, (2) the independent opinion pool, and (3) the independent likelihood pool. The probabilistic elements are necessary for the theoretical development and analysis of the DF process and sensor-management methods.

2.5.3.1 Linear Opinion Pool

In this approach, the posterior probabilities from each information source are combined in a linear fashion as follows:

$$p(x \mid Z^k) = \sum_i w_i p\left(x \mid z_i^k\right) \tag{2.55}$$

Here, the weights (w) should add up to unity, and the individual weights range between 0 and 1. This means that the posteriors are evaluated for the variable to be fused, based on the corresponding measurements. The weights signify reliability, faith, or trustworthiness of the information source and are assumed known *a priori*. In the sensor-management scheme, a faulty sensor can be "weighted out" using a proper weight. The number of models is k.

2.5.3.2 Independent Opinion Pool

Here, the observation set is assumed independent. The expression for the same is written as

$$p(x \mid Z^k) = \alpha \prod_i p\left(x \mid z_i^k\right) \tag{2.56}$$

This method is suitable if the priors (probabilities) are obtained independently according to the subjective prior information at each source.

2.5.3.3 Independent Likelihood Pool

Here, each information source has common prior information. The representation of the independent likelihood pool (ILP) is shown below:

$$\begin{aligned} &p\left(z_i^k \mid x\right) -> \{ \prod_i -> p(x \mid Z^k) \\ &\ldots \\ &p\left(z_N^k \mid x\right) -> \{ \\ &p(x) -> \{ \end{aligned} \tag{2.57}$$

The ILP is consistent with the Bayesian approach involving the DF updates, and more appropriate for MSDF applications if the conditional distributions of the measurements are independent.

These "information-pooling" methods can also be extended to include the reliability aspects of the sensor or data into their formulations, thereby helping with the sensor-management problem.

3

Strategies and Algorithms for Target Tracking and Data Fusion

The problem of tracking moving objects—including targets, mobile robots, and other vehicles—using measurements from sensors is of considerable interest in many military and civil applications that use radar, sonar systems, and electro-optical tracking systems (EOTs) for tracking flight testing of aircrafts, such as missiles, unmanned aerial vehicles, micro- or mini-air vehicles, and rotorcrafts. It is also useful in nonmilitary applications such as robotics, air traffic control and management, air surveillance, and ground vehicle tracking. In practice, scenarios for target tracking could include maneuvering, crossing, and splitting (meeting and separating) targets. Various algorithms are available to achieve target tracking for such scenarios. The selection of the algorithms is generally application dependent and is also based on the merits of the algorithm, complexity of the problem (data corrupted by ground clutter, noise processes, and so on), and computational burden.

Target tracking comprises estimation of the current state of a target, usually based on noisy measurements. The problem is complex even for single target tracking because of uncertainties in the target's mathematical model, especially for maneuvering targets (which need more than one model and one transition model, and so on), and process/state and measurement noises. The complexity of the tracking problem increases for multiple-targets using measurements from multiple sensors. In this chapter, we discuss several algorithms and strategies for target tracking and associated data fusion (DF) aspects where applicable, and in Chapter 4, we present performance evaluation results for a few DF systems and algorithms.

The importance of DF for tracking stems from the following considerations. For an aircraft observed by pulsed radar and an infrared (IR) imaging sensor, the radar provides the ability to accurately determine the aircraft's range; however, it has only a limited ability to determine the angular direction, whereas the IR sensor can accurately determine the aircraft's angular direction, but not the range. If these two types of observations are correctly associated and fused, this will provide an improved determination of range and direction than could be obtained by either of the two sensors alone. Based on an observation of the attributes of an object, the identity of the object can be determined and the observations of angular direction, range, and range rate may be converted to estimate the target's position and velocity. This could be achieved using a Kalman filter (KF) for linear models, and an extended KF (EKF) [32,33] and derivative-free KF (DFKF) for nonlinear systems [34–37]. Then, the estimated states can be combined, as in state-vector fusion (Section 3.1). Observations of the target's attributes, such as the radar cross section, IR spectra, and visual images, may be used to classify the target and assign a label to its identity. The raw data from the sensors may be directly combined if the sensors are of the same type, that is, commensurate and similar, and measure the same physical phenomenon such as target range. However, in a KF the sensor data from different types of sensors (as well as the same type of sensors) can be combined appropriately via measurement models (and associated measurement covariance matrices) and one state vector can be estimated. In this mode, the KF is a data fuser (Section 2.2). In the case of dissimilar sensors (from where the data are noncommensurate and cannot be combined in a conventional way), for example, IR and acoustic data, feature and state-vector fusion may be used to combine data which are noncommensurate.

In the process of target tracking, the target track is updated by correlating measurements with the existing tracks or by initiating new tracks using measurements from different sensors. The process of gating and data association (DA) enables proper and accurate tracking in multisensor, multitarget (MSMT) scenarios [4,38,39]. Gating helps in deciding if an observation (which could include clutter, false alarms, and electronic counter measures) is a probable candidate for track maintenance or track updates. DA associates the measurements with the targets with certainty when several targets are in the same neighborhood. In practice, measurements from the sensors may not be true due to the effect of clutter, false alarms, interference from other targets, limited resolution capability (a spatial coverage limitation of the sensor), or if there are several targets in neighborhood.

Gating is used to screen out false signals such as clutter, whereas association algorithms are used for automatic track initiation, measurement-to-track correlation, and track-to-track correlation. In measurement-to-track correlation, the sensors' data are associated with an existing number of tracks to determine which sensor's data or observations belong to which target (this is the beginning of the decision level fusion; Section 7.1). Once a determination has been made that there is more than one observation for a particular target, these observations are combined at the raw level using measurement fusion, in a typical sequential estimation technique such as KF. The steps needed to implement the measurement fusion algorithm are described below. This is a technique for eliminating unlikely observation-to-track pairings. The gating process determines if an observation belongs to a previously established target or to a new target. Gates (rectangular, circular, or ellipsoidal in shape) are defined for one or more existing tracks. Figure 3.1 shows the process of gating and elliptical gates [40]. If an observation satisfies the gate, meaning if it falls in the gate, it becomes a candidate for association with that track. The region enclosed by the gate is called the *validation* or *confirmation region*.

Some situations encountered during gating are as follows: (1) more than one observation may satisfy the gate of a single track; (2) one observation may satisfy the gates of more than one existing tracks; (3) the observation might not ultimately be used to update an existing track even if it falls within the validation region; thus, it might be used to initiate a new track; and (4) the observation might not fall within the validation region of any of the existing tracks; in such a case, it is used to initiate a tentative new track. If the probability of detection is unity or there are no expected extraneous returns (unwanted stuff), the gate size should be infinite. If the target state is observed with a detection probability (DP) less than unity in the presence of clutter, it gives rise to false measurements. An observation is said to satisfy the gate of a given track if all elements of the residual error vector are less than the gate size times the standard deviation of the residuals. If $z(k)$ is the measurement at scan k (discrete time index) given by

$$z(k) = Hx(k) + v(k) \tag{3.1}$$

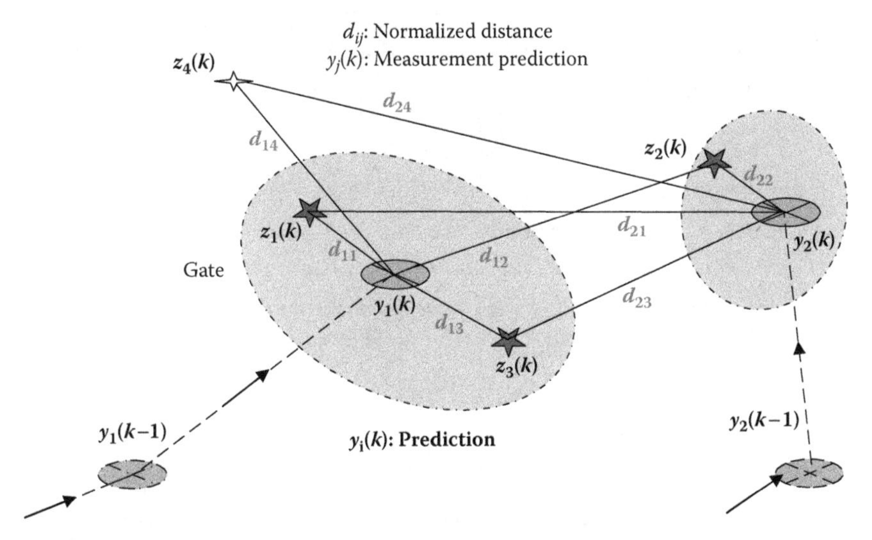

FIGURE 3.1
Depiction of principle of gating and data association.

and $y = H\hat{x}(k|k-1)$ is the predicted value with $\hat{x}(k|k-1)$ representing the predicted state at scan $(k - 1)$, then the residual vector (or innovation sequence) is given by

$$\nu(k) = z(k) - y(k) \tag{3.2}$$

The innovations covariance matrix S (Section 2.2) is given by

$$S = HPH^T + R \tag{3.3}$$

Here, R is the measurement noise covariance matrix. Assuming the measurement vector of dimension M, a distance d^2 representing the norm of the residual vector is defined as

$$d^2 = \nu^T S^{-1} \nu \tag{3.4}$$

A correlation between the observation and track is allowed if the distance d^2 is less than a certain gate threshold value G [38,39]:

$$d^2 = \nu^T S^{-1} \nu \leq G \tag{3.5}$$

The observation falling within the above-defined gate is more likely to be from the track rather than from any other extraneous source. A simple method to choose G is based on χ^2 distribution with M degrees of freedom (*dof*). The distance d^2 is the sum of the squares of M independent Gaussian variables with zero means and unit standard deviations. As such, the quadratic form of d^2 has a χ^2 distribution and a gate on d^2 can be determined using χ^2 tables.

EXAMPLE 3.1

Let $xo = 100$ and $yo = 200$ with $\sigma_o^2 = 16$. Also $xp = 85$ and $yp = 217$ with $\sigma_p^2 = 20$ [39]. Then the combined standard deviation is $\sigma_r = \sqrt{(\sigma_o^2 + \sigma_p^2)} = \sqrt{(16+20)} = 6$; hence $|xo - xp| = abs(100 - 85) = 15 = 2.5\sigma_r$ and $|yo - yp| = abs(200 - 217) = 17 = 2.83\sigma_r$, and we see that at the component level we have each value less than $3\sigma_r$. Hence, the gating test is satisfied at the component level. However, the normalized distance is given by $d^2 = \frac{(2.5\sigma_r)^2 + (2.83\sigma_r)^2}{\sigma_r^2} = 14.26$. The theoretically normalized distance is a χ^2 variable. Hence, from the χ^2 tables the probability of such a variable exceeding 13.815 is 0.001 (0.1%). However, the computed value exceeds the theoretical value. Thus, the value of the normalized distance of 14.26 is unlikely for a correct measurement-to-track association, since it exceeds the table value. This gating test is not satisfied and association need not be performed. The test rejects such an association. Thus, the individual component level test is not that powerful.

In a multitarget scenario (MTT), gating provides only a part of the solution to the problem of track maintenance and track update. Additional logic is required when an observation falls within the gates of multiple tracks or when multiple observations fall within the gate of a single track. In Figure 3.2, the process of sensing, DA, and gating or tracking [39] is shown. Systems with one or more sensors operate in multitarget tracking and surveillance, and there are possibilities of true targets and false alarms, the latter produced by noise and radar clutter. The objective of the MTT is to partition the sensor into sets of observations or tracks produced by the same source [38,39]. Once the tracks are formed and confirmed so that background and other false targets are reduced, the number of targets can be estimated and target velocity, future predicted position, and target classification characteristics can be computed for each target. The track-while-scan (TWS) is a special case of MTT. The data are received at regular intervals as the radar or other sensor regularly scans a predicted sensor volume. The most important element of the MTT system is DA (often called *correlation*). Radars, IR, and/or sonar sensors provide measurements from targets, and background noise sources—such as radar ground clutter or thermal noise—affect the true observations. The measurements in such systems are position or radar Doppler (range rate), attributes such as target type, identification (ID) numbers, length, or shape at regular intervals of time (scans or data frames), or the data could be available at irregular intervals at scan k or at every sampling interval T. The electronically scanned antennae (ESA) radars can conveniently switch back and forth between the functions of searching for new targets and illuminating existing targets, for which the time interval between scan k and scan $k + 1$ need not be same for all k. The process of DA is required because of the uncertainty of the measurement origin, which is also compounded by noise. These

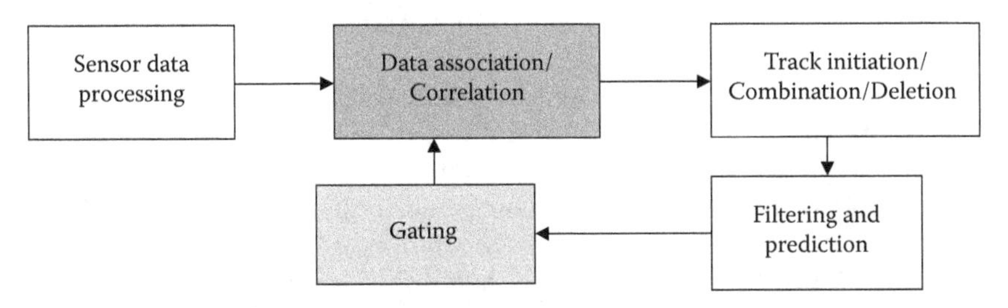

FIGURE 3.2
Sensing, data association, and gating for tracking.

uncertainties occur in an environment where there is clutter or where the false-alarm rate is high. This could be due to low observable targets or due to the presence of several targets in the same neighborhood. The persistent clutter, which would be due to spurious energy reflectors or emitters, can be identified and removed, however the random clutter needs to be handled properly.

As such, there is no perfect certainty about the origin of the observations obtained by a particular sensor. This is because one deals with remote sensors that sense energy emitted from or reflected by one object or several objects of interest. There may be other spurious sources of energy as well. The measurement to be used in the tracking algorithm may not have originated from the target of interest. This can happen when a radar, sonar, or optical sensor is operating in the presence of clutter, countermeasures, and/or false alarms. It can also happen when there are several targets in the same neighborhood, and even though one can resolve (separate) the observed detections, one cannot associate them with certainty to the targets. This difficulty is compounded when there are several targets but their number is unknown and some of the measurements may be spurious or interfering, and there could be delays in obtaining the measurements.

Hence, the procedure adopted is as follows: (1) the incoming observations are first considered for the update of the existing tracks; (2) gating is used to determine which observations-to-track pairings are reasonable; and (3) a more refined DA algorithm is used to determine final pairings. The observations not assigned to existing tracks can initiate new tentative tracks. A tentative track is confirmed when the number and quality of the observations included in the track satisfy confirmation criterion. The low-quality tracks are deleted based on a maintained or updated score. After the inclusion of the new observations, the tracks are predicted ahead to the arrival time for the next set of observations.

There are several DA-cum-filtering algorithms to handle gating and DA problems [38,39,41–43]. The most popular are the nearest neighborhood KF (NNKF) and probabilistic data association filter (PDAF) algorithms. The NNKF uses the measurement that is nearest to the predicted measurement, assuming it to be correct. The strongest NNKF considers the signal intensity if it is available. The PDAF uses all the measurements in the current time validation region (gate) weighted by the association probability computed by the PDAF. Thus, in PDAF the state estimate is updated with all the measurements weighted by their probability of having originated from the target—a combined innovation is used. In MTT, the measurement-to-measurement association is called *track formation*, measurement-to-track association is called *track maintenance* or *updating*, and the track-to-track association is called *track fusion*. In Sections 3.1 to 3.11, we describe the theory of several tracking algorithms and some practical results using some of these tracking algorithms.

3.1 State-Vector and Measurement-Level Fusion

There are generally two broad approaches for the fusion of data: measurement fusion and state-vector fusion. Theoretically, the former is superior, since the measurements are basically combined without much processing, and the optimal state vector of target position is obtained [44,45]. In practical situations, however, this may not be feasible, since the volume of data to be transmitted to the fusion center would be very large, and might create problems with the channel's transmission capacity. Hence, state-vector fusion is preferable in such practical situations. In such a system, each sensor (or sensor node) uses an estimator that obtains an estimate of the state vector and its associated covariance matrices (of the tracked target) from the data of that associated sensor. Then these state vectors are transmitted over a data link to the fusion center. This certainly would reduce the channel's overloads and overheads. At the fusion center, track-to-track correlation (better known as the DA) is carried out and the fused state vector is obtained. This is a composite vector of all the states (i.e., from all the sensors' data processing by individual KF). However, one interesting problem could arise: the process noise associated with the target is common to the filter dynamics, and hence the target tracks would be correlated. This might be true despite the measurement errors being noncorrelated.

In this section, the basic mathematical development for the fusion of data level and state-vector level is given. For similar sensor DF, the performance of measurement fusion and state-vector fusion is studied and the results of a comparison of two radars are given. The study is conducted with computer-simulated data and real data obtained from a flight test range facility. The results of the fusion of measurements obtained from the two radars from a remote sensing agency (RSA) sortie are also presented. In general, the target motion is modeled as a state-space model

$$\mathbf{X}_{(k+1)} = \Phi \mathbf{X}_{(k)} + G w_{(k)} \tag{3.6}$$

Here, X is state vector consisting of target position and velocity, Φ is state transition matrix given as $\Phi = \begin{bmatrix} 1 & T \\ 0 & 1 \end{bmatrix}$, G is gain matrix associated with process noise $G = \begin{bmatrix} T^2/2 \\ T \end{bmatrix}$, and w is process noise with $E\left[w_{(k)}\right] = 0$ and $\text{Var}\left[w_{(k)}\right] = Q$. The measurement model is given as

$$Z_{(k)} = H\mathbf{X}_{(k)} + v_{(k)} \tag{3.7}$$

Here, H is observation matrix $H = [1\ \ 0]$ for one sensor, $H = \begin{bmatrix} 1 & 0 \\ 1 & 0 \end{bmatrix}$ for two sensors, and v is measurement noise with $E[v_{(k)}] = 0$ and $\text{Var}[v_{(k)}] = R$.

3.1.1 State-Vector Fusion

The KF is given for each set of observations, meaning that the algorithm is applied independently for each sensor (data) and generates state estimates. This approach is very suitable for the data from the outputs of noncommensurate sensors, since the data cannot be directly combined as it can in direct measurement data–level fusion process [44,45].

State and covariance time propagation:

$$\tilde{X}_{(k+1)} = \Phi \hat{X}_{(k)} \tag{3.8}$$

$$\tilde{P}_{(k+1)} = \Phi \hat{P}_{(k)} \Phi^T + GQG^T \tag{3.9}$$

State and covariance measurement update:

$$K_{(k+1)} = \tilde{P}_{(k+1)} H^T \left[H\tilde{P}_{(k+1)} H^T + R \right]^{-1} \tag{3.10}$$

$$\hat{X}_{(k+1)} = \tilde{X}_{(k+1)} + K_{(k+1)} \left[Z_{(k+1)} - H\tilde{X}_{(k+1)} \right] \tag{3.11}$$

$$\hat{P}_{(k+1)} = \left[I - K_{(k+1)}\, H \right] \tilde{P}_{(k+1)} \tag{3.12}$$

First, the state estimate is generated by processing the measurement data from each sensor. Fusion is obtained by combining the state estimates using a weighted sum of the two independent state estimates. The weight factors used are the appropriate covariance matrices. Thus, these state estimates and the corresponding covariance matrices are fused as follows, that is, the fused state and covariance matrix are computed using the following expressions:

$$\hat{X}^f = \hat{\mathbf{X}}^1 + \hat{P}^1 (\hat{P}^1 + \hat{P}^2)^{-1} (\hat{\mathbf{X}}^2 - \hat{\mathbf{X}}^1) \tag{3.13}$$

$$\hat{P}^f = \hat{P}^1 - \hat{P}^1 (\hat{P}^1 + \hat{P}^2)^{-1} \hat{P}^{1^T} \tag{3.14}$$

Here, $\hat{\mathbf{X}}^1$ and $\hat{\mathbf{X}}^2$ are the estimated state vectors of filters 1 and 2 with measurements from sensor 1 and sensor 2, respectively, and $\hat{P}^1$ and $\hat{P}^2$ are the corresponding estimated state error covariances from filters 1 and 2.

3.1.2 Measurement Data–Level Fusion

In the measurement fusion approach, the algorithm fuses the sensor observations directly via a measurement model and uses one KF to estimate the fused state vector. The equations describing this process are given next (see also Section 2.2):

State and covariance time propagation:

$$\tilde{X}^f_{(k+1)} = \Phi \hat{X}^f_{(k)} \tag{3.15}$$

$$\tilde{P}^f_{(k+1)} = \Phi \hat{P}^f_{(k)} \Phi^T + GQG^T \tag{3.16}$$

State and covariance measurement data update:

$$K^f_{(k+1)} = \tilde{P}^f_{(k+1)} H^T \left[H \tilde{P}^f_{(k+1)} H^T + R \right]^{-1} \tag{3.17}$$

$$\hat{X}^f_{(k+1)} = \tilde{X}^f_{(k+1)} + K^f_{(k+1)} \left[Z_{(k+1)} - H \tilde{X}^f_{(k+1)} \right] \tag{3.18}$$

$$\hat{P}^f_{(k+1)} = \left[I - K^f_{(k+1)} H \right] \tilde{P}^f_{(k+1)} \tag{3.19}$$

The measurement vector Z is a composition of the measurement data from the two sensors. The process is also applicable to more than two sensors; this is a natural way of fusing the data. Here, the observation matrix will be 2×2, taking both sensors together as $H = \begin{bmatrix} 1 & 0 \\ 1 & 0 \end{bmatrix}$, the measurement covariance matrix $R = \begin{bmatrix} R1 & 0 \\ 0 & R2 \end{bmatrix}$, with $R1$ and $R2$ as the measurement error covariance of respective sensors.

3.1.3 Results with Simulated and Real Data Trajectories

The trajectory of a moving target is simulated using Equations 3.6 and 3.7, and measurement noises with variances 1 and 100 are added to the outputs of the data from two sensors [45]. The data are used in both of the algorithms for evaluating their performance using the percentage fit error (PFE) of the trajectory, computed as follows:

$$PFE = 100 * \frac{\text{norm}(\hat{x} - x_{gt})}{\text{norm}(x_{gt})} \tag{3.20}$$

Here, $\hat{x}$ is the filter estimated state, and x_{gt} is the ground truth, that is, the simulated state before adding noise. The results are shown in Table 3.1.

TABLE 3.1

Simulated Data Results for Two Fusion Strategies

State-Vector Fusion			Measurement Data–Level Fusion		
PFE (Position)	PFE (Velocity)	Norm (Error Covariance)	PFE (Position)	PFE (Velocity)	Norm (Error Covariance)
0.048	11.81	0.069	0.047	6.653	0.069

TABLE 3.2

Results for Real Flight Test Data

	State-Vector Fusion			Measurement Data–Level Fusion		
	PFE (Position)	PFE (Velocity)	Norm (Error Covariance)	PFE (Position)	PFE (Velocity)	Norm (Error Covariance)
x	0.079	2.80	222.84	0.071	2.182	217.73
y	0.084	2.703	251.43	0.092	1.807	239.16
z	0.555	5.431	218.26	0.574	3.855	214.57

The performance of both the filter algorithms is almost identical except in the case of the velocity fit error, where the measurement fusion shows lower value compared to the state-vector fusion. Real data from two radars K1 and K2 are used to estimate the track trajectory and perform fusion using both state-vector fusion and measurement data–level fusion. The performance with real data is shown in Table 3.2 for all the three axes. These results show that the performances of both of the fusion methods are nearly similar, or that measurement-level fusion is somewhat better than state-vector fusion, which confirms the intuitive interpretation of the data-level fusion performance.

3.1.4 Results for Data from a Remote Sensing Agency with Measurement Data–Level Fusion

Some trajectory data from a RSA in the country were available for DF analysis [45]. Since the time stamp on the K1 and K2 radars was not accurate, the data segment for which the time stamp appeared to be proper was used for measurement-level fusion. The fusion strategy used is (1) at any instant of time if the data from both K1 and K2 are available, fusion is performed using measurement-level fusion; (2) if only one of them is available, the states are estimated using the available data; and (3) when both radar data are not available, only predicted states are used. Table 3.3 gives the performance metrics of this fusion, which is compared with individual tracking data in terms of fit error computed with respect to GPS. The fit error is less than 1% in all three positions, namely x, y, and z.

TABLE 3.3

Results of RSA Sortie Data Using Measurement Data–Level Fusion

	PFE (x position)	PFE (y position)	PFE (z position)
Fused K1 and K2	0.097	0.006	0.012
K1	0.066	0.003	0.013
K2	0.141	0.003	0.010

It should be noted here that the covariance matrices estimated from the sensor characterization studies (Section 3.2) are used in the filters. One can observe from the results that in most cases measurement data–level fusion is more accurate than state-vector–level fusion. However, for dissimilar sensors or data, the state-vector fusion is more advisable. In most cases, the trajectory match was quite good and therefore these plots are not shown here.

3.2 Factorization Kalman Filters for Sensor Data Characterization and Fusion

Sensor DF using data from ground-based target tracking radars, EOTs, and inertial navigation system (INS) sensors is an involved problem, as these data are often affected by different systematic errors, time stamp errors, time delays, and random errors [46]. A practical solution is required to: (1) correct these errors; (2) estimate measurement and process noise statistics; and (3) handle time-stamp and time-delay errors. The effect of random errors such as process and measurement noises on the trajectory, in particular, is reduced to a large extent by using filtering and estimation algorithms. A method to correct systematic errors using an alignment algorithm has been suggested [47], wherein the radar measurements are mapped to the earth-centered, earth-fixed (ECEF) coordinates using a geodetic transformation, and the radar errors are estimated using least-squares technique.

3.2.1 Sensor Bias Errors

Global positioning system (GPS) data can be used as a reference to obtain an estimate of sensor bias errors and measurement noise covariance for the various sensors. The technique used is KF with an error-state space and is known as *error-state KF* (ESKF) *formulation* [32]. The ESKF estimates the bias errors in the sensor data using the difference between the actual

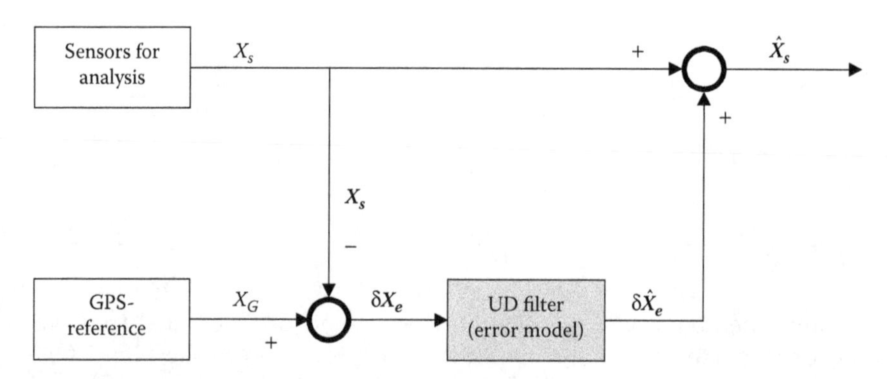

FIGURE 3.3
UD filter in error model form for sensor characterization.

measured data and the reference GPS data. The covariance of the residuals of the ESKF gives an estimate of the measurement noise covariance of a particular sensor. The estimated biases are used to correct the sensor data before these data are used for state estimation and fusion. Figure 3.3 depicts ESKF block with UD filter [46]. Sensor characterization is carried out in the ECEF coordinates system. The GPS data in the World Geodetic System 1984 (WGS-84) frame are first converted to the ECEF frame using the following transformation:

$$\begin{aligned} X_{\text{eccf}} &= (r+h)\cos\lambda\cos\mu \\ Y_{\text{eccf}} &= (r+h)\cos\lambda\sin\mu \\ Z_{\text{eccf}} &= \{(1-e^2)r+h\}\sin\lambda \end{aligned} \tag{3.21}$$

where λ, μ, and h are latitude, longitude and altitude obtained from GPS data, $r = a/(1-e^2\sin^2\lambda)^{1/2}$ is effective radius of earth, a is the semi major axis (equatorial radius) = 6378.135 km, and e is the eccentricity = 0.08181881. The radar data measured in polar frame are converted to local Cartesian coordinates in the East–North–Vertical (ENV) frame using the following transformations:

$$\begin{aligned} X_{\text{env}} &= R\sin\varphi\cos\theta \\ Y_{\text{env}} &= R\cos\varphi\cos\theta \\ Z_{\text{env}} &= R\sin\theta \end{aligned} \tag{3.22}$$

Here, R is range in meters, φ is azimuth in degrees, and θ is elevation in degrees. The sensor data are then converted from the ENV to the ECEF frame using the transformation matrix

$$\begin{bmatrix} X_{\text{eccf}} \\ Y_{\text{eccf}} \\ Z_{\text{eccf}} \end{bmatrix} = \begin{bmatrix} X_k \\ Y_k \\ Z_k \end{bmatrix} + \begin{bmatrix} -\sin\mu & -\sin\lambda\cos\mu & \cos\lambda\cos\mu \\ \cos\mu & -\sin\lambda\sin\mu & \cos\lambda\sin\mu \\ 0 & \cos\lambda & \sin\lambda \end{bmatrix} \begin{bmatrix} X_{\text{env}} \\ Y_{\text{env}} \\ Z_{\text{env}} \end{bmatrix} \quad (3.23)$$

Here, λ and μ are the latitude and longitude of the respective sensors, andX_k, Y_k, and Z_k give the location of the sensors in the ECEF frame. This is obtained from the latitude (λ), longitude (μ), and altitude (h) of the tracking station using Equation 3.21. In the case of INS, the measured down range, cross range, and altitude are converted to a local ENV frame and then to the ECEF frame. In the case of EOTs, the measured azimuth and elevation are transformed to a local ENV frame using a least-square algorithm, and then these data are transformed to the ECEF frame. It is essential to time-synchronize the GPS data and sensors data for estimating the bias errors.

3.2.2 Error State-Space Kalman Filter

The UD factorization [33] implementation form of the KF is used for estimating the sensor biases and measurement noise covariance. In sensor characterization, the KF is implemented with the "error state-space" model instead of the actual state-space model. This is known as an *indirect method* [32]. This error model is given as

$$\delta\mathbf{X}(k+1) = \Phi\delta\mathbf{X}(k) + Gw(k) \quad (3.24)$$

$$\delta\mathbf{Z}(k) = H\delta\mathbf{X}(k) + v(k) \quad (3.25)$$

Here, $\delta\mathbf{X}$ is the vector of position and velocity error states in all the three axis, $\delta\mathbf{Z}$ is the vector of measured position error (i.e., GPS—sensor) in all three axis, Φ is the transition matrix, H is the observation matrix, w is the process noise with zero mean and covariance matrix Q, and v is the measurement noise with zero mean and covariance matrix R. The complete error model for characterizing the sensors in the ECEF frame is given as

$$\begin{bmatrix} \delta x(k+1) \\ \delta v_x(k+1) \\ \delta y(k+1) \\ \delta v_y(k+1) \\ \delta z(k+1) \\ \delta v_z(k+1) \end{bmatrix} = \begin{bmatrix} 1 & T & 0 & 0 & 0 & 0 \\ 0 & 1 & 0 & 0 & 0 & 0 \\ 0 & 0 & 1 & T & 0 & 0 \\ 0 & 0 & 0 & 1 & 0 & 0 \\ 0 & 0 & 0 & 0 & 1 & T \\ 0 & 0 & 0 & 0 & 0 & 1 \end{bmatrix} \begin{bmatrix} \delta x(k) \\ \delta v_x(k) \\ \delta y(k) \\ \delta v_y(k) \\ \delta z(k) \\ \delta v_z(k) \end{bmatrix} + \begin{bmatrix} T^2/2 \\ T \\ T^2/2 \\ T \\ T^2/2 \\ T \end{bmatrix} w(k) \quad (3.26)$$

$$\begin{bmatrix} \delta x_m(k) \\ \delta y_m(k) \\ \delta z_m(k) \end{bmatrix} = \begin{bmatrix} 1 & 0 & 0 & 0 & 0 & 0 \\ 0 & 0 & 1 & 0 & 0 & 0 \\ 0 & 0 & 0 & 0 & 1 & 0 \end{bmatrix} \begin{bmatrix} \delta x(k) \\ \delta v_x(k) \\ \delta y(k) \\ \delta v_y(k) \\ \delta z(k) \\ \delta v_z(k) \end{bmatrix} + \begin{bmatrix} v_1(k) \\ v_2(k) \\ v_3(k) \end{bmatrix} \quad (3.27)$$

Here, δx is the position error in x-direction, δv_x is the velocity error in x-direction, δy is the position error in y-direction, δv_y is the velocity error in y-direction, δz is the position error in z-direction, δv_z is the velocity error in z-direction, and T is the sampling interval. The subscript m represents measured position error (GPS data – sensor data).

3.2.3 Measurement and Process Noise Covariance Estimation

Achieving adequate fusion results requires estimates of the measurement noise covariance matrix R and process noise covariance matrix Q. The sliding window method is used for adaptive estimation of R for each measurement channel. The estimate of R is obtained by determining the covariance of the residuals from ESKF over a chosen window width. Once the R is estimated, Q is adaptively estimated during the state estimation using the method [32] given below:

$$\sum_{k=i-N+1}^{i} \left[\Phi P(k-1/k-1)\Phi^T + GQ(k-1)G^T - P(k/k) - \Delta x(k)\Delta x(k)^T \right] = 0 \quad (3.28)$$

Here,

$$\Delta x(k) = \hat{x}(k/k) - \hat{x}(k/k-1) = K(k)r(k) \quad (3.29)$$

$$P(k/k) = P(k-1/k-1) - K(k)HP(k-1/k-1) \quad (3.30)$$

$$P(k-1/k-1) = K(k)\hat{A}(k)H^T \quad (3.31)$$

$$\hat{A}(k) = \frac{1}{N} \sum_{k=i-N+1}^{i} r(k)r(k)^T \quad (3.32)$$

If G is invertible for all k, then an estimate of $Q(k)$ can be defined as

$$\hat{Q}(i) = \frac{1}{N} \sum_{k=i-N+1}^{i} \left\{ G^{-1}\left[\Delta x(k)\Delta x(k)^T + P(k/k) - \Phi P(k-1/k-1)\Phi^T \right] G^{-T} \right\} \quad (3.33)$$

If G is not invertible, then pseudo-inverse is used, computed as

$$G^{\#} = [G^T G]^{-1} G^T \tag{3.34}$$

3.2.4 Time Stamp and Time Delay Errors

The fusion of data from two or more sensors requires that the measurements are available at the same instant of time and that the data are received with an accurate time stamp, corresponding to the time at which the data were sensed or acquired by the sensor or data acquisition system. This is not always the case—the data could come with erroneous time stamps either at the transmitting end or at the receiving end. There could be drifts in the time recorded on any channel. This time drift could either be of a constant value on any channel or could be different (random) at each instant of time. Some practical aspects on this issue are as follows: (1) the first few samples of data can be used to ascertain constant time drift on any channel and the subsequent time stamps can be corrected by the computed time drift; (2) to handle random time drift, at any instant of time, any data coming within one-half of the sampling time on that channel can be treated as if it has arrived at that instant for purpose of fusion; and (3) for state-vector fusion, it is expected that the output of the filter is available at every T seconds (reference sampling time). This is achieved by using the data arriving at each instant for updating the states of the filter and propagating the estimated states to the nearest T, so that it is available for fusion there. This method of synchronizing the state vectors for fusion presupposes that the sampling time requirements are appropriately chosen to suit the dynamics of the target being tracked.

3.2.5 Multisensor Data Fusion Scheme

Figure 3.4 shows a block diagram of the multisensor DF (MSDF) scheme [46]. The first step is time synchronization of various sensors' data. For this step, the data arrival on each of the tracking sensor data (including GPS) is checked and a reference time signal is initiated using the time stamp on the sensor data that arrives first. This time signal is incremented at a uniform rate of T seconds. At each step, the sensor time stamp on each of the input data channels is compared with the reference time and appropriate action is initiated using "decision logic" based on the time delay error handling procedure mentioned above. Each data set is appropriately transformed to the ECEF frame using the standard transformation equations mentioned earlier. With a GPS signal as a reference, all the sensors' data are characterized by estimating bias and measurement and process noise covariance as mentioned earlier. These data are

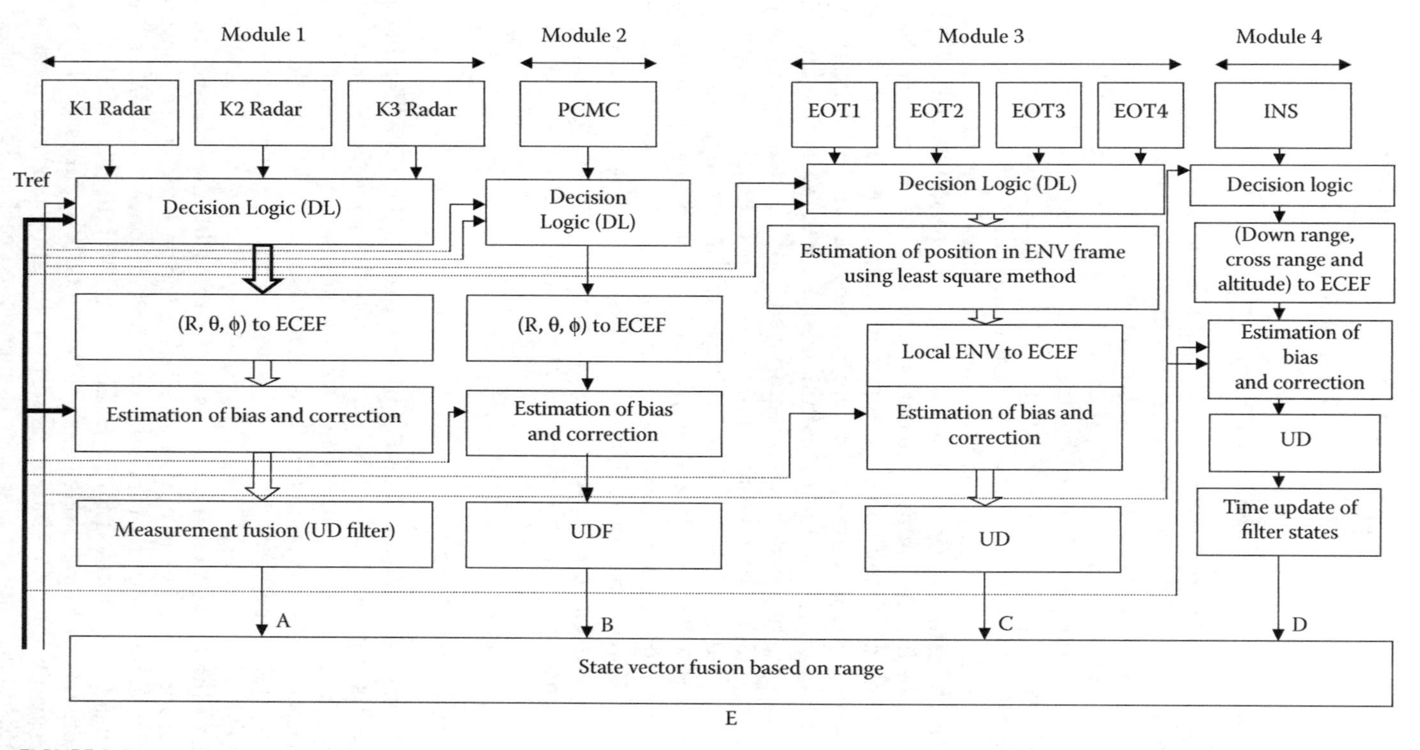

FIGURE 3.4
The data fusion scheme.

then used for filtering and fusion with appropriate models and governing equations. The target motion model is given as

$$\mathbf{X}(k+1) = \Phi\mathbf{X}(k) + Gw(k) \tag{3.35}$$

Here, X is a state vector consisting of target positions, velocities, and accelerations:

$$\mathbf{X} = [x_p \quad x_v \quad x_a \quad y_p \quad y_v \quad y_a \quad z_p \quad z_v \quad z_a]'$$

The state transition matrix is given as

$$\Phi = \begin{bmatrix} 1 & T & T^2/2 & 0 & 0 & 0 & 0 & 0 & 0 \\ 0 & 1 & T & 0 & 0 & 0 & 0 & 0 & 0 \\ 0 & 0 & 1 & 0 & 0 & 0 & 0 & 0 & 0 \\ 0 & 0 & 0 & 1 & T & T^2/2 & 0 & 0 & 0 \\ 0 & 0 & 0 & 0 & 1 & T & 0 & 0 & 0 \\ 0 & 0 & 0 & 0 & 0 & 1 & 0 & 0 & 0 \\ 0 & 0 & 0 & 0 & 0 & 0 & 1 & T & T^2/2 \\ 0 & 0 & 0 & 0 & 0 & 0 & 0 & 1 & T \\ 0 & 0 & 0 & 0 & 0 & 0 & 0 & 0 & 1 \end{bmatrix}$$

and the process noise coefficient matrix is given as

$$G = \begin{bmatrix} T^3/6 & 0 & 0 \\ T^2/2 & 0 & 0 \\ T & 0 & 0 \\ 0 & T^3/6 & 0 \\ 0 & T^2/2 & 0 \\ 0 & T & 0 \\ 0 & 0 & T^3/6 \\ 0 & 0 & T^2/2 \\ 0 & 0 & T \end{bmatrix}$$

Here, w is the process noise with $E[w(k)] = 0$ and $\text{cov}[w(k)] = Q$. The measurement model is given as

$$\mathbf{Z}(k) = H\mathbf{X}(k) + v(k) \tag{3.36}$$

Here, Z is the measurement vector given by $\mathbf{Z} = [x_p \quad y_p \quad z_p]'$, and the observation matrix is given as

$$H = \begin{bmatrix} 1 & 0 & 0 & 0 & 0 & 0 & 0 & 0 & 0 \\ 0 & 0 & 0 & 1 & 0 & 0 & 0 & 0 & 0 \\ 0 & 0 & 0 & 0 & 0 & 0 & 1 & 0 & 0 \end{bmatrix}$$

Here, v is the measurement noise with $E[v(k)] = 0$ and $\text{cov}[v(k)] = R$

3.2.5.1 UD Filters for Trajectory Estimation

KF is implemented in UD factorization form for the trajectory estimation using the target motion models described above. The UD factor time propagation algorithm is given next. State-estimate time propagation or extrapolation is given as

$$\tilde{x}(k+1) = \Phi\hat{x}(k) \tag{3.37}$$

Error covariance time propagation or extrapolation is given as

$$\tilde{P}(k+1) = \Phi\hat{P}(k)\Phi^T + GQG^T \tag{3.38}$$

Given initially $\hat{P} = \hat{U}\hat{D}\hat{U}^T$ and Q, the time update factors $\tilde{U}$ and $\tilde{D}$ are obtained using the modified Gram–Schmidt orthogonalization process:

$$W = [\Phi\hat{U} \,|\, G_A]\, \bar{D} = diag[\hat{D}, Q] \text{ with } W^T = [w1, w2, \ldots, wn]$$

Here, P is reformulated as $\tilde{P} = \tilde{W}\tilde{D}\tilde{W}^T$. Then U and D factors of $\tilde{W}\tilde{D}\tilde{W}^T$ are computed. For $j = n, n-1, \ldots, 2$, evaluate the following equations recursively [33]:

$$\begin{aligned} \tilde{D}_j &= < w_j^{(n-j)}, w_j^{(n-j)} >_D \\ \tilde{U}(i,j) &= < w_i^{(n-j)}, w_j^{(n-j)} >_D / \tilde{D}_j \quad i = 1,\ldots,(j-1) \end{aligned}$$

The above equation gives the weighted product of the two vectors of w.

$$\begin{aligned} w_i^{(n-j+1)} &= w_i^{(n-j)} - \tilde{U}(i,j) w_j^{(n-j)} \quad i = 1,\ldots,(j-1) \\ \tilde{D}_1 &= < w_1^{(n-1)}, w_1^{(n-1)} >_D \end{aligned} \tag{3.39}$$

Here, subscript D denotes the weighted inner product with reference to D. The UD factor measurement update part of the filter is given next. This update part combines *a priori* estimate $\tilde{x}$ and error covariance $\tilde{P}$ with scalar observation $z = a^T x + v$ to construct an updated (filtered state) estimate and covariance as follows [33]:

$$\begin{aligned} K &= \tilde{P}a/\alpha \\ \hat{x} &= \tilde{x} + K(z - a^T\tilde{x}) \\ \alpha &= a^T\tilde{P}a + r \\ \hat{P} &= \tilde{P} - Ka\tilde{P} \end{aligned} \tag{3.40}$$

Here, $\tilde{P} = \tilde{U}\tilde{D}\tilde{U}^T$, a is the measurement vector or matrix, r is the measurement noise covariance, and z is the string of noisy measurements. The gain K and updated covariance factors $\hat{U}$ and $\hat{D}$ are obtained using the following equations [33]:

$$\begin{aligned} &f = \tilde{U}^T a; \quad f^T = (f_1, \ldots, f_n); \; v = \tilde{D}f; \; v_i = \tilde{d}_i f_i \;\; i = 1,2,\ldots,n \\ &\hat{d}_1 = \tilde{d}_1 r/\alpha_1; \;\; \alpha_1 = r + v_1 f_1 \\ &K_2^T = (v_1 0 \ldots 0) \end{aligned} \tag{3.41}$$

Next, for $j = 2, \ldots, n$, the following equations are evaluated:

$$\begin{aligned} \alpha_j &= \alpha_{j-1} + v_j f_j \\ \hat{d}_j &= \tilde{d}_j \alpha_{j-1}/\alpha_j \\ \hat{u}_j &= \tilde{u}_j + \lambda_j k_j; \lambda_j = -f_j/\alpha_{j-1} \\ K_{j+1} &= K_j + v_j \tilde{u}_j \end{aligned} \tag{3.42}$$

Here, $\tilde{U} = [\tilde{u}_1, \ldots, \tilde{u}_n]$, $\hat{U} = [\hat{u}_1, \ldots, \hat{u}_n]$, and the Kalman gain is given by $K = K_{n+1}/\alpha_n$, wherein $\tilde{d}$ is the predicted diagonal element and $\hat{d}_j$ is the updated diagonal element of the D matrix.

3.2.5.2 Measurement Fusion

The data from similar radars, such as S-band radars (GR1, GR2, and GR3) are fused using measurement fusion. The UD filtering algorithm described fuses the sensor observations directly and estimates the fused state vector, by taking all sensor measurements together as per the observation matrix H:

$$H = \begin{bmatrix} H_1 \\ H_2 \\ . \\ . \\ H_n \end{bmatrix}$$

Here, $H_1, H_2, \ldots, H_n$ are observation matrices of n individual sensors. The measurement covariance matrix is given as

$$R = \begin{bmatrix} R_1 & 0 & 0 & 0 & 0 \\ 0 & R_2 & 0 & 0 & 0 \\ 0 & 0 & . & 0 & 0 \\ 0 & 0 & 0 & . & 0 \\ 0 & 0 & 0 & 0 & R_n \end{bmatrix}$$

Here, $R_1, R_2, \ldots, R_n$ are the measurement error covariance values of respective sensors.

3.2.5.3 State-Vector Fusion

Fusion is performed as follows:

$$\hat{X}^f = \hat{X}^1 + \hat{P}^1(\hat{P}^1 + \hat{P}^2)^{-1}(\hat{X}^2 - \hat{X}^1) \tag{3.43}$$

$$\hat{P}^f = \hat{P}^1 - \hat{P}^1(\hat{P}^1 + \hat{P}^2)^{-1}\hat{P}^{1^T} \tag{3.44}$$

Here, $\hat{\mathbf{X}}^1$ and $\hat{\mathbf{X}}^2$ are the state vectors of estimated trajectories 1 and 2, and $\hat{P}^1$ and $\hat{P}^2$ are the state error covariance matrices of estimated trajectories 1 and 2. In this method, only two trajectories are fused at a time.

3.2.5.4 Fusion Philosophy

All the tracking sensors are grouped into four major groups (Figure 3.4) [46] based on the type, sensitivity, and accuracy of the sensors. The four major groups are as follows:

1. Module 1: The tracking data from the three radars, after coordinate transformation and characterization, are fused to get single trajectory data (trajectory A) by direct measurement fusion using UD-KF.
2. Module 2: The tracking data from the precision coherent monopulse C-band (PCMC) radar, after coordinate transformation and characterization, are filtered (trajectory B) using UD-KF.
3. Module 3: The tracking data (azimuth and elevation) from EOTs are first transformed to local ENV frame using least-square method and then transformed to ECEF. These data are then filtered (trajectory C) using UD-KF.
4. Module 4: The tracking data from the onboard inertial navigation system (INS) are telemetered at a sampling interval of 72 ms.

These data (down range, cross range and altitude) are transformed to ECEF, characterized, and filtered (trajectory D). The data are then time propagated in ECEF coordinates using target dynamic model to time synchronize with other track data. These trajectories (trajectory A, trajectory B, trajectory C, and trajectory D) are then fused using state-vector fusion.

A hierarchical order, based on the accuracies of sensors, is chosen for generating the final trajectory estimates. EOT sensors are accurate up to a maximum range of 40 km. The following criteria are used to decide the priority for fusion:

1. If estimated range is less than 40 km, trajectory C is fused with trajectory A. The resultant trajectory is fused with trajectory B and finally this resultant trajectory is fused with trajectory D (INS).
2. If the estimated range is greater than 40 km, trajectory A is fused with trajectory B and the resultant trajectory is finally fused with trajectory D. That is, if the estimated range is less than 40 km, trajectory C is used for fusion but for ranges beyond 40 km trajectory C is not included for state-vector fusion.

Sensor characterization, trajectory filtering, and fusion—as described above—have been developed in "C" on the UNIX platform. The scheme is validated with simulated data of all the sensors, generated using a graphical user interface-based "simulator program" that generates noisy measurement data of a moving target launched from a given location: (1) R, θ, and φ for S- and C-band radars; (2) cross range, down range, and altitude for INS; (3) θ and φ for EOTs; and (4) WGS-84 for GPS. The results are evaluated in terms of (1) mean of residuals, (2) percent autocorrelation values out of the 2σ theoretical error bounds, (3) percent innovation values out of bounds, and (4) PFE with reference to true states. Also, several types of plots have been used (not all are shown): (1) estimated position with GPS data, (2) innovation sequence with 2σ bounds, (3) autocorrelation of residuals with bounds, (4) root sum squares of position error (RSSPE), and (5) state error covariance (trace of matrix P). Table 3.4 shows various performance metrics. Figure 3.5a, b, and c shows the plot of (a) estimated position with GPS data, (b) innovation sequence with 2σ bounds, and (c) autocorrelation of residuals with bounds (obtained at fusion level A) [46]. One can see that the performance of the tracking filters is satisfactory in terms of innovations and autocorrelations being within their theoretical bounds. Figure 3.6 shows the comparison of RSSPE [46]. It is clear that the error in the final fused trajectory is quite low. There is also a decrease in the error covariance because of an increase in information (not shown

TABLE 3.4

Results of Data Fusion and Performance Evaluation

Level[a]		A* (S-Band Radars)	B (C-Band Radars)	C (EOTs)	D (INS)	E (Fused)
Residuals mean	x	0.0038	0.01282	0.00009	−0.0122	—
	y	−0.0043	−0.0140	−0.0017	0.00624	—
	z	0.0004	−0.0009	−0.0002	0.00179	—
% autocorrelation values out of the bounds	x	3.6407	3.2967	2.9735	2.1969	—
	y	4.58855	5.2144	4.0293	2.2184	—
	z	2.9513	3.1028	3.21051	7.4521	—
% innovation values out of bounds	x	1.89575	1.6591	1.6591	2.3261	—
	y	0.3662	1.5945	0.1939	2.1753	—
	z	0.4524	0.6679	0.6464	0.3877	—
PFE with respect to true data	x	*0.0987*	0.1981	0.0124	0.1123	*0.1011*
	y	*0.0050*	0.0097	0.0007	0.0044	*0.0052*
	z	*0.0019*	0.0025	0.0007	0.0010	*0.0013*

[a] Data fusion levels from Figure 3.4

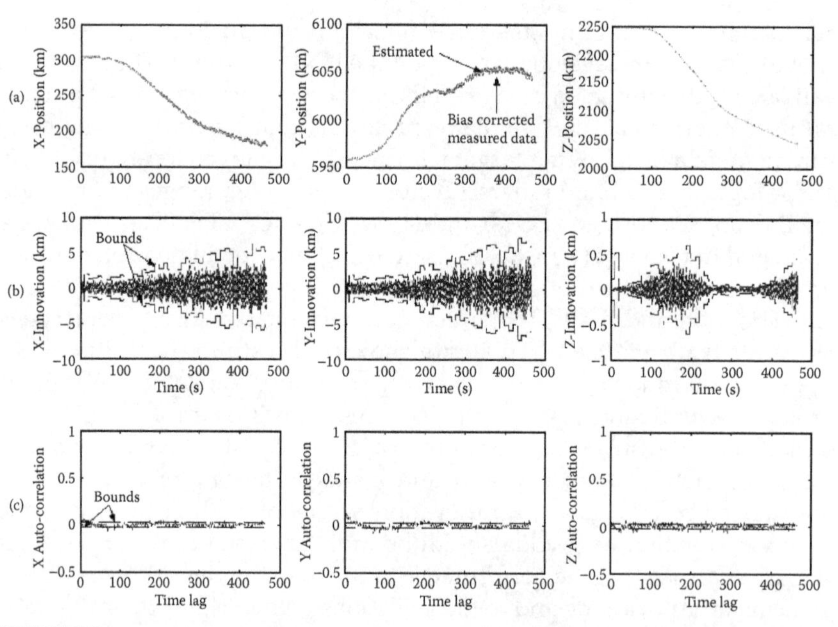

FIGURE 3.5

Filter performance results for data fusion level A (see Figure 3.4): (a) Estimated position with GPS data; (b) innovation sequence with 2σ bounds; and (c) autocorrelation of residuals with bounds (obtained at fusion level).

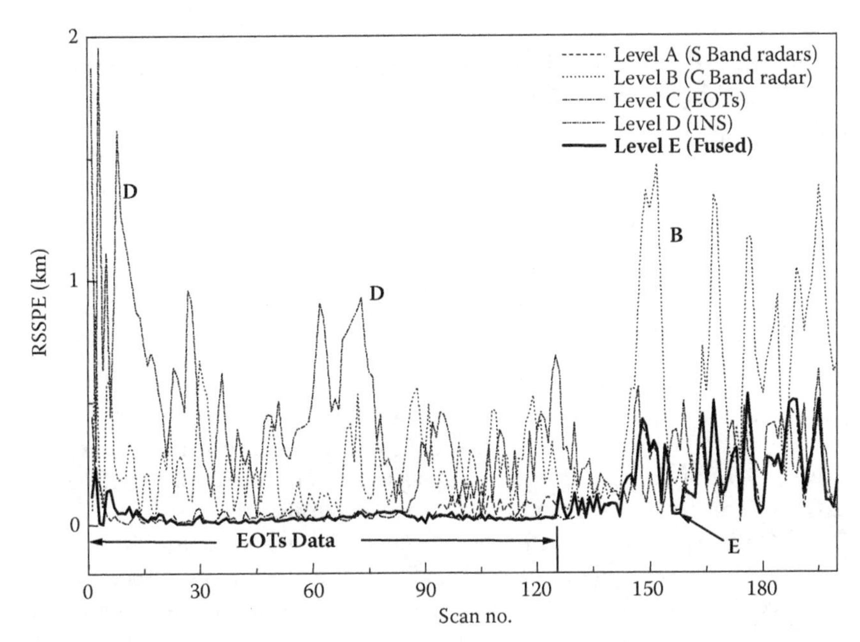

FIGURE 3.6
Root sum squares of position error for simulated data-fusion scheme.

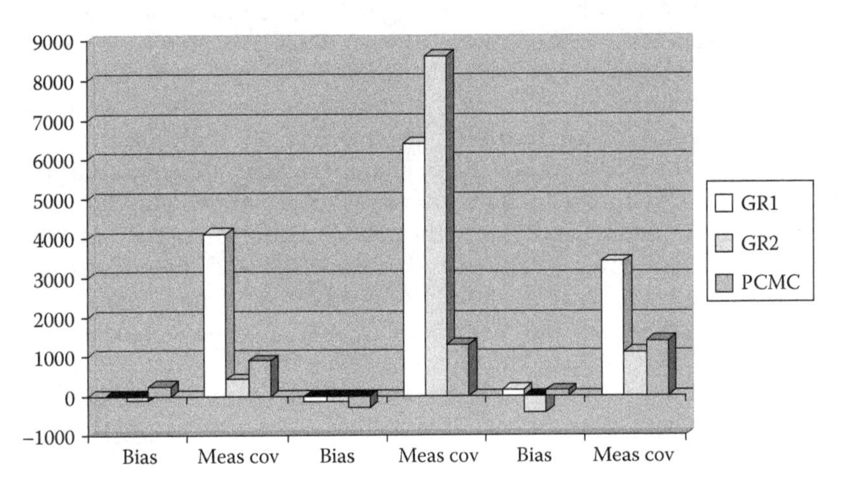

FIGURE 3.7
Bias (in x, y, and z in → x direction) and measurement covariance values for the real data (x, y, and z positions).

here) by fusing multiple trajectories. Figure 3.7 shows the comparison of bias (in meters) and measurement covariance (in square meters) for the real data for x, y, and z (the axis directions) positions, respectively, for the sensor characterization of two S bands and one PCMC radar from the real data of an aircraft tracked by these ground-based radars. The real data is generated by flying an aircraft for purposes of sensor characterization and fusion [46].

3.3 Square-Root Information Filtering and Fusion in Decentralized Architecture

The DF architecture chosen when attempting to build an MSDF system plays a significant role in deciding the algorithms for DF. The choice is made with a view toward balancing computing resources, available communication bandwidth, desired accuracy, and the capabilities of the sensors. The three most commonly used architectures are as follows: (1) centralized, (2) distributed or decentralized (there could be a subtle differences between the two), and (3) hybrid, as discussed in Section 2.1. In a decentralized DF network (NW), the sensor nodes have a processing facility for each node, and do not need any central fusion facility [31]. Fusion occurs locally at each node using the local measurements and the data communicated from the neighboring nodes. Hence, there is no single fusion center; this means that there is no central node for the NW to operate successfully. Communication in the NW is on a node-to-node basis, since the nodes only know the connections in their vicinity. There are other architectures such as distributed or hierarchical which are also described as "decentralized." The merits of decentralized DF NW architecture are as follows: (1) the system is scalable and communication bottlenecks and bandwidth issues are less important here; (2) the system is quite fault tolerant and can withstand dynamic changes; and (3) all fusion processes occur locally at each sensor site and the nodes can be created and programmed in a modular fashion [31]. Thus, such a system is scalable, survivable, and modular. One way to achieve decentralized sensor fusion network architecture is to efficiently decentralize the existing centralized DF system and algorithms with proper care, as this is quite possible and many such DF algorithms are more efficient than conventional DF algorithms.

For decentralized fusion networks, the information filter (IF) is generally advocated (instead of covariance-based KF), since the IF is a more direct and natural method of dealing with MSDF problems than the conventional KF. The IF has certain advantages in decentralized sensor networks because the IF provides a direct interpretation of node observations and contribution in terms of information. The IF, if implemented in the conventional form, could be sensitive to computer round-off errors which in turn could degrade the performance of the filter. This is a crucial aspect if the algorithm is to be used for DF and in real-time online mode (say on board computers of aircraft, missiles, and spacecrafts). Square root information filter (SRIF) algorithm offers a solution to this problem of numerical accuracy, stability of the filtering algorithm, and general overall reliability. This improved behavior of SRIF is due to the reduction of the numerical ranges of the variables compared to the nonsquare root implementation of the corresponding IF [33]. Some ramifications of SRIF to obtain SRI sensor-fusion algorithms (SRISFA) are described [48], and SRIF applied to fully decentralized sensor DF NW can be considered a new development which can offer improved numerical accuracy over conventional methods.

3.3.1 Information Filter

IF is a natural way of dealing with MSDF problems and the related numerical data processing methods than the conventional KF because it provides a direct interpretation of the node observations and contributions in terms of information [31].

3.3.1.1 Information Filter Concept

In IF, a system state is updated based on a sensor observation containing relevant information about the state and the measurements are modeled in terms of their information-content value using a linear system:

$$\mathbf{z} = H\mathbf{x} + \mathbf{v} \tag{3.45}$$

Here, z is an m-vector of observations, x is n-vector of variables to be estimated, $H(m, n)$ is measurement model, and v is an m-vector of measurement noise with zero mean and identity covariance matrix. The LS estimate of x is obtained by minimizing the mean square measurement error:

$$J(x) = (z - Hx)^T (z - Hx) \tag{3.46}$$

In addition to the system of Equation 3.45, we have an *a priori* unbiased estimate $\tilde{x}$ of x, and an *a priori* information matrix (the inverse of covariance matrix from the KF) forming an *a priori* state-information matrix pair: $(\tilde{x}, \tilde{\Lambda})$. Incorporating the *a priori* information pair in Equation 3.46, we get the modified performance functional [33]

$$J_1(x) = (x - \tilde{x})^T \tilde{\Lambda}(x - \tilde{x}) + (z - Hx)^T (z - Hx) \tag{3.47}$$

3.3.1.2 Square Root Information Filter Algorithm

By factoring the information matrix into its square roots, the following form of J is obtained [33,48]

$$J_1(x) = (x - \tilde{x})^T \tilde{R}^T \tilde{R}(x - \tilde{x}) + (z - Hx)^T (z - Hx)$$
$$J_1(x) = (\tilde{z} - \tilde{R}x)^T (\tilde{z} - \tilde{R}x) + (z - Hx)^T (z - Hx) \quad (3.48)$$

where $\tilde{z} = \tilde{R}x$. The first term of Equation 3.48 can be written as $\tilde{z} = \tilde{R}x + \tilde{v}$. It can be readily seen from Equation 3.48 that the performance functional J represents the composite system

$$\begin{bmatrix} \tilde{z} \\ z \end{bmatrix} = \begin{bmatrix} \tilde{R} \\ H \end{bmatrix} x + \begin{bmatrix} \tilde{v} \\ v \end{bmatrix} \quad (3.49)$$

Thus, it can be seen that the *a priori* information is an additional observation in the form of a data equation such as measurement Equation 3.45. This provides the basis of the SRIF algorithm. Then, the LS solution is obtained by applying the method of orthogonal transformations. This solution is likely to be less susceptible to computer round-off errors. Using an orthogonal Householder transformation matrix T, one can obtain the solution to the least squares functional as

$$T \begin{bmatrix} \tilde{R}_{j-1} & \tilde{z}_{j-1} \\ H_j & z_j \end{bmatrix} = \begin{bmatrix} \hat{R}_j & \hat{z}_j \\ 0 & e_j \end{bmatrix}; \quad j = 1, \ldots, N \quad (3.50)$$

With e_j as the sequence of residuals, it can be seen that a new information pair is generated $(\hat{z}_j, \hat{R}_j)$ and the process can be repeated with the incorporation of next measurement (z_{j+1}) to obtain the recursive SRIF, which can now form the basis for decentralized SRIF.

3.3.2 Square Root Information Filter Sensor Data Fusion Algorithm

Let a system with two sensors, H1 and H2, be specified; then, using the earlier formulation, we can fuse the sensor measurements at the data level:

$$T \begin{bmatrix} \tilde{R}_{j-1} & \tilde{z}_{j-1} \\ H_{1j} & z_{1j} \\ H_{2j} & z_{2j} \end{bmatrix} = \begin{bmatrix} \hat{R}_j & \hat{z}_j \\ 0 & e_j \end{bmatrix}; \quad j = 1, \ldots, N \quad (3.51)$$

This process results in the state estimate with the effect of two-sensor data taken into account for fusion, and it can be easily extended to more than two sensors. It is also possible to process the measurements from each sensor individually to obtain the estimate of the information-state vectors and then fuse these vectors to obtain the combined IF–state-vector fusion (as done in KFSVF). The fusion equations are given as [48]

$$\hat{z}_f = \hat{z}_1 + \hat{z}_2 \quad \text{and} \quad \hat{R}_f = \hat{R}_1 + \hat{R}_2 \tag{3.52}$$

Here, $\hat{z}$ is the information state from the point of view of the square-root information concept. If required, the fused covariance-state is obtained by

$$\hat{x}_f = \hat{R}_f^{-1}\hat{z}_f \tag{3.53}$$

Hence, it is observed that the data equation formulation of the information pair and the orthogonal transformation yield very simple and yet very useful solutions to the sensor DF problem with enhanced numerical reliability and stability.

3.3.3 Decentralized Square Root Information Filter

A scheme of decentralized square-root information is shown in Figure 3.8 [48]. It consists of a network of nodes with its own processing facility, and fusion occurs locally at each node on the basis of local observations and

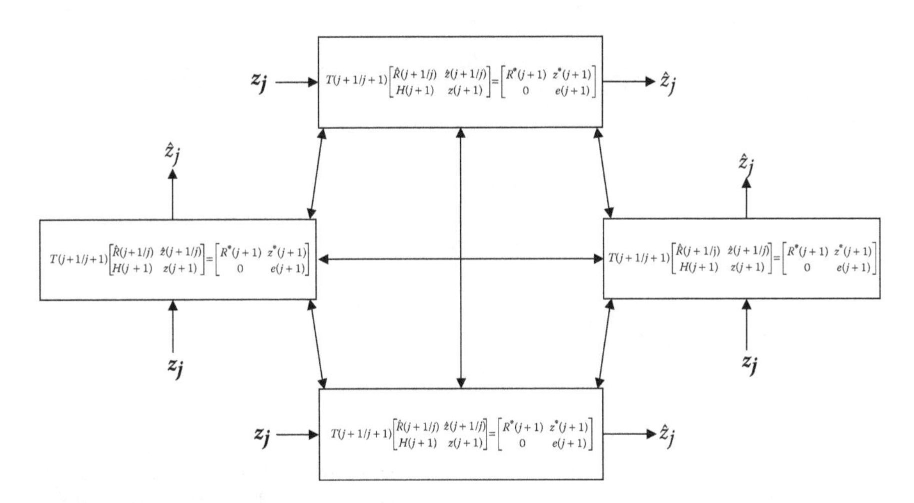

FIGURE 3.8
Decentralized architecture for square root information filter (each rectangular block contains a square root information filter).

the information communicates from neighboring nodes [31,48]. The processing node is a sensor-fusion node that takes local observations and shares information with other fusion nodes. It assimilates the communicated information and computes the estimate.

A linear system is given in state equation form as

$$x(j+1) = \Phi x(j) + Gw(j) \tag{3.54}$$

Here, w is the white Gaussian process noise with zero mean and covariance Q. It is assumed that the *a priori* information given about x_0 and w_0 can be put in data equation as [48]

$$z_w = R_w w_0 + \nu_w \tag{3.55}$$

$$\tilde{z}_0 = \tilde{R}_0 x_0 + \tilde{\nu}_0 \tag{3.56}$$

Here, the variables $\nu_0, \tilde{\nu}_0$, and ν_w are assumed to be zero mean, independent, and with unity covariance matrices. By incorporating the *a priori* information, the time propagation part of the SRIF obtains the following form. The local mapping (time propagation) is given as [48]

$$T(j+1)\begin{bmatrix} \tilde{R}_w(j) & 0 & \tilde{z}_w(j) \\ -R^d(j+1)G & R^d(j+1) & \tilde{z}(j) \end{bmatrix} = \begin{bmatrix} \hat{R}_w(j+1) & \hat{R}_{wx}(j+1) & \hat{z}_w(j+1) \\ 0 & \hat{R}(j+1) & \hat{z}(j+1) \end{bmatrix} \tag{3.57}$$

The subscript w signifies the variables related to the process noise. We have

$$R^d(j+1) = \hat{R}(j)\varphi^{-1}(j+1) \tag{3.58}$$

Local estimates are generated using measurement update of SRIF [48]

$$T(j+1/j+1)\begin{bmatrix} \hat{R}(j+1/j) & \hat{z}(j+1/j) \\ H(j+1) & z(j+1) \end{bmatrix} = \begin{bmatrix} R^*(j+1) & z^*(j+1) \\ 0 & e(j+1) \end{bmatrix} \tag{3.59}$$

Here, * denotes the local updated estimates. These estimates can be communicated between all nodes in a fully connected network, and at each node the estimates can be assimilated to produce global SRIF estimates. Assimilation equations to produce global SRI estimates at the ith node (with $k = 1, \ldots, N-1$ representing the remaining nodes) are given by [48]

$$\hat{z}_i(j+1/j+1) = z^*(j+1/j+1) + \sum_{k=1}^{N-1} z_k^*(j+1/j+1) \quad (3.60)$$

$$\hat{R}_i(j+1/j+1) = R^*(j+1/j+1) + \sum_{k=1}^{N-1} R_k^*(j+1/j+1) \quad (3.61)$$

This formulation of the information pair and the orthogonal transformation T, in the data equation format, yields in a natural way very elegant, simple, and useful solutions to the decentralized sensor DF problem. The algorithm would possess better numerical reliability and stability due to the use of the square-root filtering formulation compared to the normal IF-based schemes. The basic ramifications, such as inclusion of correlated process noise and bias parameters (from the IF/SRIF [33]), are equally applicable to the decentralized SRISFA (DSRISFA) and in SRIF structure it is easy to derive algorithm modifications and/or approximations with the data equation framework in a straightforward manner. Decentralized IF for fusion requires communication of the local "information states" and information matrices to all the neighboring nodes for computing the global estimates, whereas in the case of the SRIF for fusion the information states and information matrices are estimated together. Communication between local nodes involves the transmission of the first row of the matrix (RHS of Equation 3.59) as a whole in a fully decentralized network with a number of nodes. The smaller range of numbers in the SRIF formulation enables the results to be represented by fewer bits. This would result in communications overhead savings.

3.3.4 Numerical Simulation Results

Numerically simulated data of the position of a target moving with constant acceleration are generated. The decentralized SRISFA is validated for a two-node interconnection network. The system has position, velocity, and acceleration as its states

$$x^T = [x\ \dot{x}\ \ddot{x}] \quad (3.62)$$

With the transition matrix as follows:

$$\Phi = \begin{bmatrix} 1 & \Delta t & \Delta t^2/2 \\ 0 & 1 & \Delta t \\ 0 & 0 & 1 \end{bmatrix}; \quad G = \begin{bmatrix} \Delta t^3/6 \\ \Delta t^2/2 \\ \Delta t \end{bmatrix} \quad (3.63)$$

The Δt is sampling interval of 0.5 s, and w is white Gaussian noise with zero mean and standard deviation of 0.0001. The measurement model for each sensor is given by

$$z_m(j+1) = Hx(j+1) + \mathbf{v}_m(j+1) \quad (3.64)$$

Here, H = [1 0 0] for each sensor. The vector v is measurement noise, and white Gaussian noise with zero mean. The position data pertaining to the two nodes or measuring sensors are generated by adding random noise with $\sigma_{v_1} = 1$ and $\sigma_{v_2} = 5$. The whitened measurements with appropriate measurement equations are used for generating fused global estimates at both the nodes. This is done by utilizing the local estimates at the node and the communicated information from the neighboring nodes. The performance of the SRIFDF algorithm in terms of numerical accuracy is compared with the decentralized information filter algorithm for two nodes implemented in PC MATLAB® using the simulated data. Both the algorithms were started with identical initial conditions (for the state and information matrices) and with the same values of process and measurement noise covariance. Table 3.5 gives the PFE of the state estimation error for three states obtained using SRI filter and IF algorithms. The fit error is calculated using $PFE = 100 * \text{norm}(\hat{x} - x_t)/\text{norm}(x_t)$ where x_t is the true state. It is clear that the PFEs are lower when the SRIF algorithm is used for estimation. It is also clear that the SRIF is relatively less sensitive to variation in Q when compared with IF. Subsequently, the SRIF was applied to certain real-flight test data of a launch vehicle obtained from two ground radars (placed at different locations). The radars measure the range, azimuth, and elevation of the moving vehicle. These data are converted to Cartesian coordinate frame so that linear-state and measurement models can be used. The data are converted to a common reference point and whitened before use in state estimation or fusion exercises. Kinematic fusion is performed after processing the data using SRIF at each node. The norms of the SRI matrix, which is a direct outcome of the SRIF algorithm, are compared for individual sensors and for fused data for all the axes. The norms for the fused data were higher than those of the individual sensors. The detailed results for four nodes of SRIFDF can be found in [48].

TABLE 3.5

Performance Metrics for Local Nodes and Global Fused Estimate (Q_{ii} = 1.0e-4)

	PFE for SRIF for $[x\ \dot{x}\ \ddot{x}]$			PFE for IF $[x\ \dot{x}\ \ddot{x}]$		
Node 1	0.0212	0.3787	3.8768	0.0342	0.8637	13.8324
Node 2	0.0931	0.6771	4.6231	0.1274	2.2168	20.9926
Fused	0.0294	0.4012	3.9470	0.0338	0.8456	13.4540

3.4 Nearest Neighbor and Probabilistic Data Association Filter Algorithms

We have seen that gating and DA operations are required to enable efficient tracking in MSMT scenario. Gating helps in deciding if an observation (which might include clutter, false alarms, electronic counter measures, and so on) is a probable candidate for track maintenance or track update. DA is a step to associate the measurements to the targets with certainty when several targets are in the same neighborhood. Now, two approaches to DA and related filtering algorithms will be discussed [38,39]:

1. The nearest neighbor (NN) approach, in which a unique pairing is determined. At most one observation can be paired with an earlier established track; the goal is to minimize an overall distance function that considers all observation-to-track pairings that satisfy a preliminary gating test.
2. The PDAF algorithm, in which a track is updated based on the decision using a weighted sum of innovations from multiple validated measurements.

A program based on gating and DA using both NNKF and PDAF approaches has been developed in PC MATLAB for handling the problem of tracking in a MSMT scenario. It is primarily an adapted version of the commercially available software package [41] and is updated and modified for the present application. The salient features of the upgraded MSMT packages are [40] (1) data are transformed to common reference point; (2) both NNKF and PDAF are used; (3) similar tracks are combined; (4) direction feature is included; (5) performance metrics Singer–Kanyuck (S–K), RMSPE, and RSSPE are evaluated; and (6) the track loss feature is also included. In Figure 3.9, the steps of the MSMT program are shown [40]. The test scenarios considered are

1. Three targets launched from different sites and nine sensors located at different locations tracking these targets, and the three sensors are configured to track one target. The program generates information on the target-sensor lock status. The performance is evaluated by adding clutter to the data and simulating the data loss in one or more of the sensors for a short time period.
2. Each of the three sensors looks at six targets and then all the three sensor results are fused, with the possibility of some data loss.

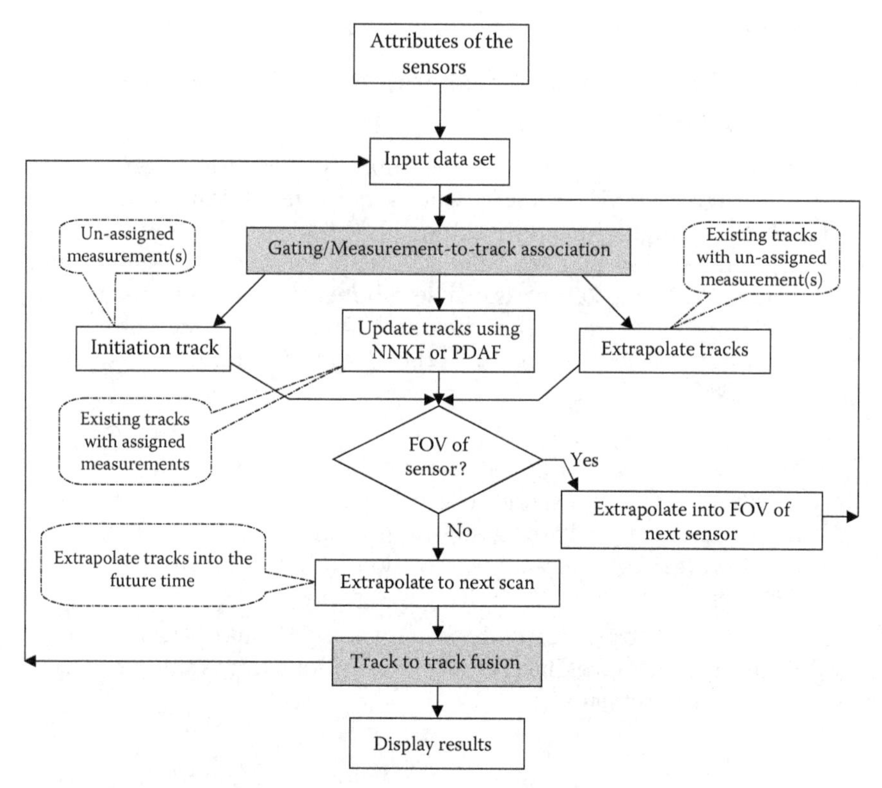

FIGURE 3.9
Multisensor, multitarget tracking program.

3.4.1 Nearest Neighborhood Kalman Filter

In NNKF, the measurement that is nearest to the track (within the gate size after gating operation is performed) is chosen for updating the track. Each measurement can only be associated with one track. No two tracks can share the same measurement. If a valid measurement exists, the track should be updated using NNKF. The time propagation follows the normal KF equations

$$\tilde{X}(k/k-1) = \Phi\hat{X}(k-1/k-1) \tag{3.65}$$

$$\tilde{P}(k/k-1) = \Phi\hat{P}(k-1/k-1)\Phi^T + GQG^T \tag{3.66}$$

The state estimate is updated using

$$\begin{aligned}\hat{X}(k/k) &= \tilde{X}(k/k-1) + K\nu(k)\\ \hat{P}(k/k) &= (I-KH)\tilde{P}(k/k-1)\end{aligned} \tag{3.67}$$

The Kalman gain is given as $K = \tilde{P}(k/k-1)H^T S^{-1}$, the residual vector as $\nu(k) = z(k) - \tilde{z}(k/k-1)$, and the residual covariance as $S = H\tilde{P}(k/k-1)H^T + R$. The measurement error covariance matrix is given by $R = \text{diag}[\sigma_x^2 \;\; \sigma_y^2 \;\; \sigma_z^2]$ where three observables x, y, z are considered. In case of no valid measurement, the track will have the propagated estimates

$$\begin{aligned}\hat{X}(k/k) &= \tilde{X}(k/k-1) \\ \hat{P}(k/k) &= \tilde{P}(k/k-1)\end{aligned} \tag{3.68}$$

Figure 3.10 shows the information flow in NNKF DA procedure [39,40].

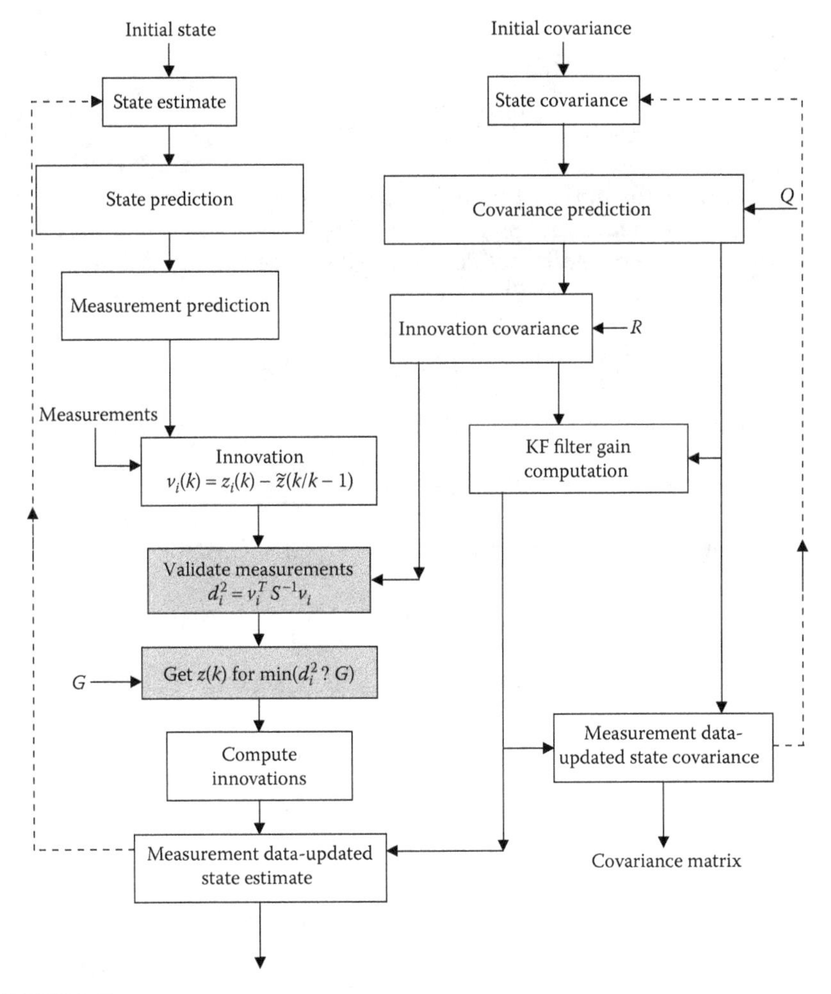

FIGURE 3.10
Diagram of the information flow in a nearest neighborhood Kalman filter.

3.4.2 Probabilistic Data Association Filter

The PDAF algorithm computes the association probabilities for each valid measurement at the current instant for the target; this information is used in the filter. If it is assumed that there are m measurements falling within a particular gate, that there is only one target of interest, and that the track has been initialized [38,40–43], then the association events as below are mutually exclusive and exhaustive for $m \geq 1$.

$$z_i = \{ y_i \text{ is the target originated measurement}\}, \quad i = 1,2,\ldots,m, \qquad (3.69)$$
$$\{\text{none of the measurements is target originated}\}, \quad i = 0$$

The conditional mean of the state is written as

$$\hat{X}(k/k) = \sum_{i=0}^{m} \hat{X}_i(k/k)p_i \qquad (3.70)$$

Here, $\hat{X}_i(k/k)$ is the updated state that is conditioned on the event that the ith validated measurement is correct and p_i is the conditional probability of the event. The estimates conditioned on the measurement i, being assumed to be correct, are given by

$$\hat{X}_i(k/k) = \tilde{X}(k/k-1) + K\nu_i(k), \quad i = 1,2,\ldots,m \qquad (3.71)$$

and the conditional innovation are given by

$$\nu_i(k) = z_i(k) - \hat{z}(k/k-1) \qquad (3.72)$$

For $i = 0$, if none of the measurements is valid ($m = 0$) then

$$\hat{X}_0(k/k) = \tilde{X}(k/k-1) \qquad (3.73)$$

The state update equation of the PDAF is given as

$$\hat{X}(k/k) = \tilde{X}(k/k-1) + K\nu(k) \qquad (3.74)$$

The merged innovation expression is given by

$$\nu(k) = \sum_{i=1}^{m} p_i(k)\nu_i(k) \qquad (3.75)$$

The covariance matrix associated with the updated state is given as

$$\hat{P}(k/k) = p_0(k)\tilde{P}(k/k-1) + \left(1 - p_0(k)\right)P^c(k/k) + P^s(k) \qquad (3.76)$$

Here, the covariance of the state updated with correct measurement is given as

$$P^c(k/k) = \tilde{P}(k/k-1) - KSK^T \tag{3.77}$$

The spread of the innovations is given as

$$P^s(k/k) = K\left(\sum_{i=1}^{m} p_i(k)\nu_i(k)\nu(k)_i^T - \nu(k)\nu(k)^T\right)K^T \tag{3.78}$$

The conditional probability is computed using the Poisson clutter model [38]

$$\begin{aligned} p_i(k) &= \frac{e^{-0.5v_i^T S^{-1} v_i}}{\lambda\sqrt{|2\Pi S|}\dfrac{(1-P_D)}{P_D} + \sum_{j=1}^{m} e^{-0.5v_j^T S^{-1} v_j}}, \quad \text{for } i = 1,2,\ldots,m \\ &= \frac{\lambda\sqrt{|2\Pi S|}\dfrac{(1-P_D)}{P_D}}{\lambda\sqrt{|2\Pi S|}\dfrac{(1-P_D)}{P_D} + \sum_{j=1}^{m} e^{-0.5v_j^T S^{-1} v_j}}, \quad \text{for } i = 0 \end{aligned} \tag{3.79}$$

Here, λ = false alarm probability, and P_D = DP. The computational steps in the PDAF algorithm are depicted in Figure 3.11 [38,40,43,49] and the features of these algorithms are given in Table 3.6 [40].

3.4.3 Tracking and Data Association Program for Multisensor, Multitarget Sensors

Two approaches for multitarget tracking are "target oriented" and "track oriented." In the target-oriented approach, the number of targets is assumed to be known. Here, all the DA hypotheses are combined into one for each target. In the track-oriented approach, each track is treated individually while it is initiated, updated, and terminated based on the associated measurement history. The track-oriented approach is pursued here and in the track-oriented algorithm, a score is assigned to each track and is updated according to the track's association history. A track is initiated based on a single measurement data and is eliminated when the computed score is below an assigned threshold. A brief description of the steps in the MSMT program is given next [40].

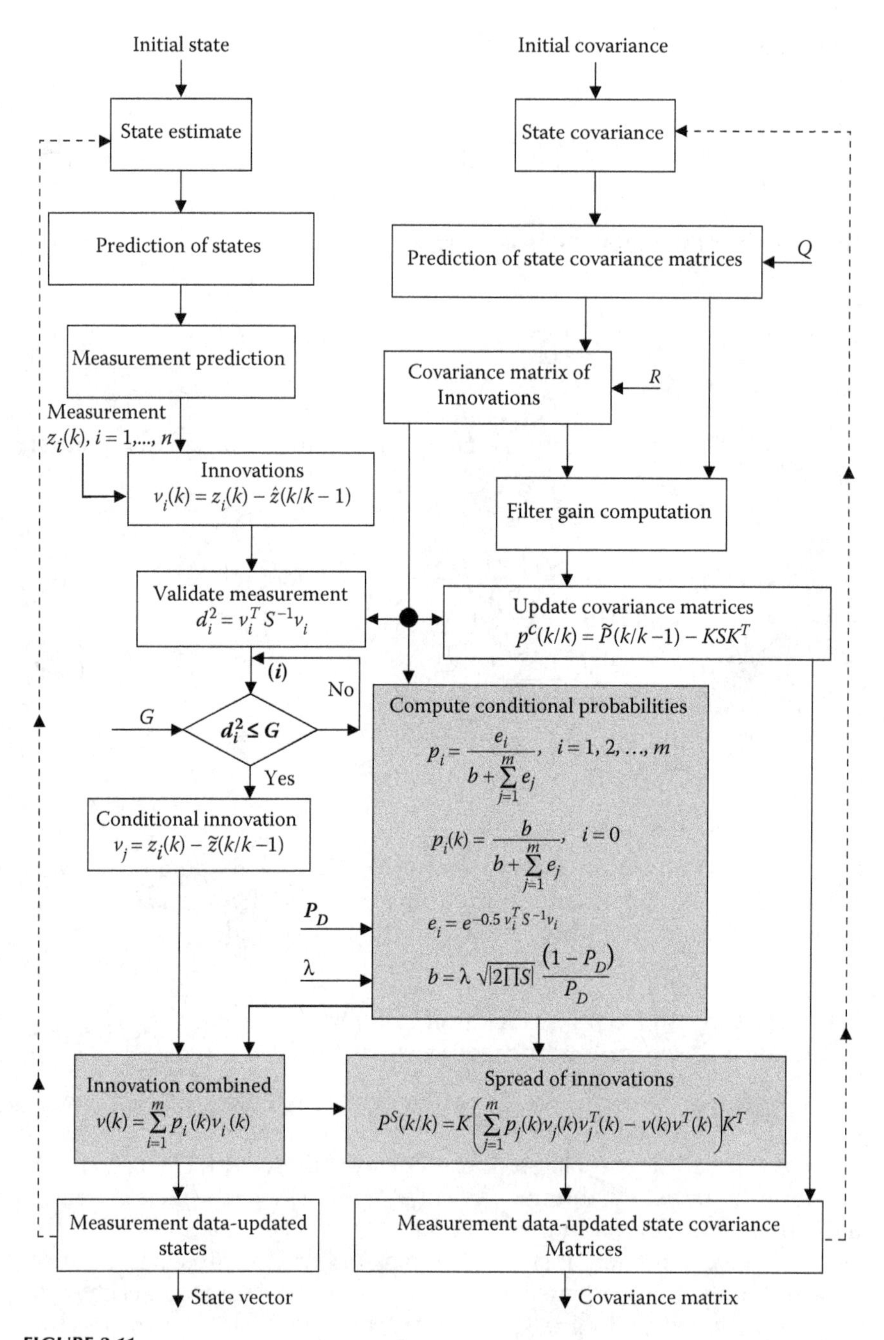

FIGURE 3.11
Probabilistic data association filter computational steps—the filled boxes show the major differences between the nearest neighborhood Kalman filter and the probabilistic data association filter.

TABLE 3.6

Important Features of NNKF and PDAF

Main Features	NNKF	PDAF
Filter type	Linear KF	Linear KF
DA	Measurement nearest to the predicted measurement in validation gate or region used	Association probabilities for each measurement lying in the validation gate or region used
Degree of possibility of track loss	High	Less
Possibility of false track	High	Less
Data loss	Degradation due to some uncertainty in estimation of earlier state	Better performance due to better estimation of earlier states
Computational cost and time	Low	High (nearly one and half that of the KF)
Capability of tracking	Less reliable in cluttered environment	Relatively more reliable in clutter environment

3.4.3.1 Sensor Attributes

In sensor attributes, the sensor location, resolution, field of view (FOV), DP, and false-alarm probability (Pfa) are considered. Using Pfa, the number of false alarms is computed by

$$Nfa = Pfa * \mu FOV \tag{3.80}$$

Here, Nfa is the expected number of false alarms and μFOV is FOV volume.

3.4.3.2 Data Set Conversion

The measurements are converted to a common reference point in a Cartesian coordinate frame using the following formulae:

$$x_{\text{ref}} = x_{\text{traj}} - x_{\text{loc}}$$
$$y_{\text{ref}} = y_{\text{traj}} - y_{\text{loc}}$$
$$z_{\text{ref}} = z_{\text{traj}} - z_{\text{loc}}$$

Here, x_{ref}, y_{ref}, and z_{ref} are x, y, and z coordinates of target with respect to common reference. The x_{loc}, y_{loc}, and z_{loc} are x, y, and z coordinates of corresponding sensor location. The x_{traj}, y_{traj}, and z_{traj} are x, y, and z coordinates of measured target trajectory.

3.4.3.3 Gating in Multisensor, Multitarget

For the measurement vector of dimension m, a distance d^2, (normalized distance) representing the norm of the residual vector is computed using

$$d^2 = \nu^T S^{-1} \nu \tag{3.81}$$

Consider two tracks ($y_i(k-1)$, $i = 1, 2$) at scan ($k - 1$). At scan k, as shown in Figure 3.1, if four measurements $z_j(k)$, $j = 1, 2, 3, 4$ are available, then the track to measurement distance d_{ij} (from ith track to jth measurement) for each of the predicted tracks ($y_i(k-1)$, $i = 1, 2$) is computed. A DA between the measurement and track is allowed if the distance $d^2 \leq G$, where G is the χ^2 threshold. The χ^2 threshold is obtained from the χ^2 tables, since the validation region is χ^2 distributed with number of *dof* equal to the dimension of the measurement [38]. For the measurements falling within the gate, the likelihood value computed using $\log(|2\pi S|) + d^2$ is entered in the correlation matrix. This matrix is called the *track-to-measurement correlation matrix* (TMCR), formed with the measurements along the rows and tracks along the columns. For the measurements falling outside the gate, a high value is entered in the TMCR matrix as shown in Table 3.7 [40].

3.4.3.4 Measurement-to-Track Association

In NNKF, a measurement nearest to the track is chosen for updating the track. After the particular measurement-to-track association pair is chosen, based on the correlation and DA matrix for updating track, both the corresponding entries can be removed from this matrix and the next track with the least association uncertainty can be processed. In Figure 3.1, measurements $z_1(k)$ and $z_3(k)$ fall within the gate region of predicted track $y_1(k)$, the measurement $z_2(k)$ falls within the gate region of the predicted track $y_2(k)$. The measurement $z_4(k)$ falls outside $y_1(k)$ and $y_2(k)$ gate regions. This is verified in Table 3.7. The measurement $z_1(k)$ is now considered for updating the track $y_1(k)$ because it is nearer than $z_3(k)$. In the PDAF, all the

TABLE 3.7

Tract-to-Measurement Correlation Values for Two Tracks ($i = 1, 2$) and Four Measurements ($j = 1, 2, 3, 4$) at Scan k

Measurements	Tracks	
	y_1	y_2
$z_1(k)$	d_{11}	1000 (or high number)
$z_2(k)$	1000 (or high number)	d_{22}
$z_3(k)$	d_{13}	1000 (or high number)
$z_4(k)$	1000 (or high number)	1000 (or high number)

measurements that fall within the gate (formed around the extrapolated track and their associated probabilities) are used for updating the track. The measurements $z_1(k)$ and $z_3(k)$ are taken for updating track $y_1(k)$. The measurement $z_2(k)$ is taken for updating the track $y_2(k)$, thus, continuing the process until all the tracks have been considered. Measurements not assigned to any track can be used to initiate a new track. A track score is obtained for each track based on the association history. This score is used in the decision to eliminate or confirm the tracks.

3.4.3.5 Initiation of Track and Extrapolation of Track

A new track can be initiated with a measurement that is not associated with any of the existing tracks, and a track score is assigned to each new track. A new track is initiated by position measurements (x, y, z) and the related velocity vector. The initial score for a new track is determined using

$$p = \frac{\beta_{\mathrm{NT}}}{\beta_{\mathrm{NT}} + \beta_{\mathrm{fa}}} \tag{3.82}$$

Here, β_{NT} = expected number of true targets and β_{fa} = expected number of false alarms (per unit surveillance volume per scan). $z_4(k)$ is used for track initiation. If a track does not have any validated measurement, then it is not updated, but existing tracks are extrapolated for processing at next scan.

3.4.3.6 Extrapolation of Tracks into Next Sensor Field of View

The surviving tracks in the present sensor FOV are taken into next sensor FOV [38]. This is so because it is assumed that in the MSMT scenario all sensors are tracking all targets. The track score is propagated into the next sensor FOV using the Markov chain transition matrix. Two models are used, one for "observable target" (true track) designated as model O and one for "unobservable target" (a target outside the sensor coverage or erroneously hypothesized target) designated as model U [41]. For both the models, the target measurements, with DP P_D, and the clutter are to be considered. $P_D = 0$ for model U. The models O and U are given by a Markov chain model with the transition probabilities [41,43]

$$\begin{aligned} P(M_O|\bar{M}_O) &= 1-\varepsilon_O, \quad P(M_U|\bar{M}_O) = \varepsilon_O \\ P(M_U|\bar{M}_U) &= 1-\varepsilon_U, \quad P(M_O|\bar{M}_U) = \varepsilon_U \end{aligned} \tag{3.83}$$

Here, M_x denotes the event that the model '*x*' is in effect during the current sampling interval and $\bar{M}_X$ is for the earlier interval. The exact values of ε_A and ε_D can be chosen based on the problem scenario under study.

3.4.3.7 Extrapolation of Tracks into Next Scan

The valid and surviving tracks are extrapolated at the next scan using a target model given as follows:

$$X(k+1) = FX(k) + Gw(k) \tag{3.84}$$

Here, the target dynamic state transition matrix is given as

$$F = \begin{bmatrix} 1 & 0 & 0 & \Delta t & 0 & 0 \\ 0 & 1 & 0 & 0 & \Delta t & 0 \\ 0 & 0 & 1 & 0 & 0 & \Delta t \\ 0 & 0 & 0 & 1 & 0 & 0 \\ 0 & 0 & 0 & 0 & 1 & 0 \\ 0 & 0 & 0 & 0 & 0 & 1 \end{bmatrix}$$

The state vector is given by $X(k) = [x(k)\ y(k)\ z(k)\ \dot{x}(k)\ \dot{y}(k)\ \dot{z}(k)]^T$. The process noise $w(k)$ is assumed to be a zero-mean white Gaussian noise with covariance $E[w(k)w(k)^T] = Q(k)$. Δt is the sampling interval. The process noise gain matrix is given as

$$G = \begin{bmatrix} \Delta t^2/2 & 0 & 0 \\ 0 & \Delta t^2/2 & 0 \\ 0 & 0 & \Delta t^2/2 \\ \Delta t & 0 & 0 \\ 0 & \Delta t & 0 \\ 0 & 0 & \Delta t \end{bmatrix}$$

The time propagation of the states and covariance matrices is performed using the KF-prediction equation.

3.4.3.8 Track Management Process

A scoring threshold is used to eliminate false tracks, and is one of the system design parameters. It can be adjusted based on the scenario and required performance. Similar tracks are fused to avoid redundant tracks. The direction of the tracks also has to be considered while merging similar tracks. An N_D-scan approach is used [39]—the tracks that have the last N_D observations in common are merged. Based on the value of N_D, the approach would automatically take the velocity as well as acceleration into account for merging similar tracks; for example, $x(2) - -x(1)$ can be regarded as velocity. A 3-scan approach is used by the MSMT program for merging the tracks. Consider two tracks with the state-vector estimates and the covariance matrices given at scan k

$$\begin{aligned} &\text{track } i: \hat{X}_i(k/k),\ \hat{P}_i(k/k) \\ &\text{track } j: \hat{X}_j(k/k),\ \hat{P}_j(k/k) \end{aligned} \tag{3.85}$$

The merged or combined state vector is the usual state-vector fusion with

$$X_c(k) = \hat{X}_i(k/k) + \hat{P}_i(k/k)\hat{P}(k)_{ij}^{-1}\left[\hat{X}_j(k/k) - \hat{X}_i(k/k)\right] \tag{3.86}$$

and the combined covariance matrix is given as

$$P_c(k) = \hat{P}_i(k/k) - \hat{P}_i(k/k)\hat{P}(k)_{ij}^{-1}\hat{P}_i(k/k) \tag{3.87}$$

$$\hat{P}(k)_{ij} = \hat{P}_i(k/k) + \hat{P}_j(k/k) \tag{3.88}$$

The program logic would finally generate information regarding the surviving tracks and sensors to target lock status. The graphical display module displays the true trajectory and measurements and also performance measures such as true and false track detections, number of good and false tracks, and good and false track probabilities. It also displays the sensor and target lock status.

3.4.4 Numerical Simulation

The performance of the NNKF and PDAF is evaluated by computing: (1) the PFE in x, y, and z positions; (2) the RMSPE; (3) the RSSPE; (4) S–K track association metric $C_{ij} = \left\|\hat{x}_i - \hat{x}_j\right\|^2_{(P_i+P_j)^{-1}}$, $C_{ij} = (\hat{x}_i - \hat{x}_j)^T(P_i + P_j)^{-1}(\hat{x}_i - \hat{x}_j)$ (the metric C_{ij} [42] can be viewed as the square of the [normalized] distance between two Gaussian distributions with mean vectors $\hat{x}_i$ and $\hat{x}_j$ and a common covariance matrix $P_i + P_j$); and (5) percentage root mean square position error (RMSPE).

$$RMSPE = \frac{RMSPE}{\sqrt{\dfrac{1}{N}\displaystyle\sum_{i=1}^{N}\frac{x_i^2 + y_i^2 + z_i^2}{3}}} * 100 \tag{3.89}$$

Some of the performance metrics are defined in Chapter 4. The program steps for DA and tracking and estimation discussed above are used to identify which of the sensors in the given scenario are tracking the same targets using the scenario of nine sensors located at different points in space and their related measurements. The results are shown in Figure 3.12 [40], Figure 3.12a shows the trajectories as seen from these nine sensors. The MSMT program displays the target ID and the sensors which are tracking that particular target. Initially nine tracks survived before similar tracks were combined using a predetermined distance threshold. After the merger, only three tracks survived (they have been assigned three target ID numbers T1, T2, and T3). The sensors, which track a particular target, are shown in Table 3.8 [40]. It is clear that three sensors track one target. It is also clear from Table 3.9 [40] that the performance of these two DA algorithms in the

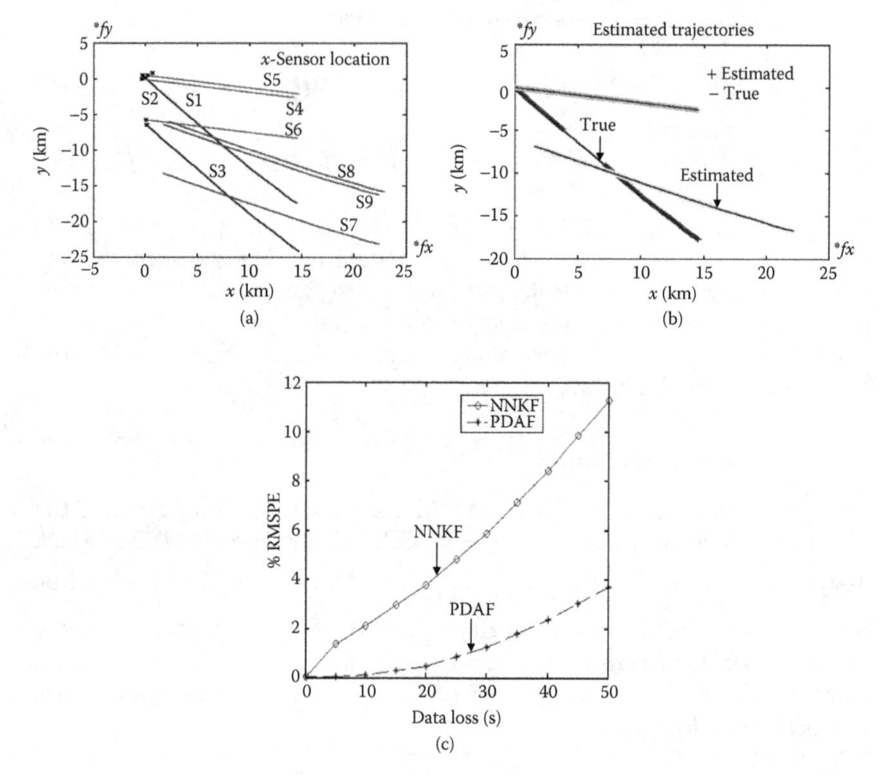

FIGURE 3.12
View of trajectories seen by respective sensors: (a) seemingly nine trajectories, (b) actually three trajectories identified by the multisensor, multitarget program, (c) effect of data loss on the performance of nearest neighborhood Kalman filter and probabilistic data association filter. (From Naidu, V. P. S., G. Girija, and J. R. Raol. 2005. *J Inst Eng I* 86:17–28. With permission.)

presence of clutter for the given scenario is nearly the same. Comparison of true tracks and estimated tracks using NNKF is shown in Figure 3.12b. The track score, the innovations with bounds, and the χ distance measure on the x-axis data for target and track 1 (where there is data loss) and for target and track 2 (where there is no data loss) are also computed. The track score was zero during the measurement data loss, innovations were within the theoretical bounds, and the χ^2 distance values at each scan were below the threshold values obtained from the χ^2 tables. The S–K association metrics for ith track and jth track from the same target were almost zero, which means that the association was found to be feasible. Track loss is simulated in data from sensors 1–3 during 100–150 s. The PFE and RMSPE for data loss in track 1 are shown in Table 3.10 [40]. It is observed that the PFE and RMSPE increase as the duration of data loss increases. It is also seen from Figure 3.12c that the performance of PDAF is better than that of

TABLE 3.8

Sensor ID Numbers

Target Number (Sensor ID Number—ID)
T1 (S1, S2, S3)
T2 (S4, S5, S6)
T3 (S7, S8, S9)

TABLE 3.9

PFE Metrics for Track Positions

Track No.	NNKF DA and Tracking		PDAF DA and Tracking	
	PFE in x	PFE in y	PFE in x	PFE in y
Track 1	0.060	0.056	0.08	0.075
Track 2	1.039	1.049	1.039	1.049
Track 3	0.052	0.028	0.053	0.029

TABLE 3.10

PFE and RMSPE (Data Loss in Track 1 and Distance in Meters)

Data Loss	NNKF			PDAF		
	PFE		% RMSPE	PFE		% RMSPE
	x	y		x	y	
0 s	0.06	0.05	0.067	0.081	0.075	0.056
5 s	1.32	1.4	1.37	0.083	0.078	0.059
20 s	3.62	3.87	3.77	0.48	0.448	0.463

NNKF in the presence of data loss. The data loss for longer time might be acceptable if PDAF is used. However, this depends on how much accuracy is required in overall tracking. The results of DF of 3-sensors and 6-targets (another test scenario) and associated performance aspects such as track probability, good tracks [38,43], and so on were obtained using the same MSMT program. Correct association and fusion of trajectories from different sensors were also obtained, and more details are given in [40].

3.5 Interacting Multiple Model Algorithm for Maneuvering Target Tracking

Target tracking is more difficult if the target in motion is also maneuvering. KF has some ability to adapt to maneuvers by tuning the process noise covariance (and perhaps to some extent measurement noise-covariance) matrices, if the target is mildly maneuvering. The filter tuning could be either manual or automatic. Manual tuning means that the KF has been properly evaluated offline with postflight data already available. However, for any new scenario this might not be the case. In such situations, trial and error method or automatic tuning has to be used. Thus, in a multitarget tracking system, a KF can be used for relatively benign maneuvers and an adequate noise reduction can be used for the periods when target is not maneuvering. In most situations, an interacting multiple-model KF (IMMKF) has been found to perform better than a KF [50,51]. The IMMKF uses a few target motion models (constant velocity, constant acceleration, coordinate turn model, and so on). For example, the IMMKF may use one model for a straight and level flight and different models for maneuvers or turns. The IMMKF always maintains the repository of all the models and blends their outputs with certain weighting coefficients that are computed probabilistically as part of the algorithm. In addition to the state estimates for each motion model, the IMMKF maintains an estimate of the probability with which the target is moving in accordance with each model.

3.5.1 Interacting Multiple Model Kalman Filter Algorithm

Thus, the IMMKF uses (1) several possible models for the target's motion, and (2) a probabilistic switching mechanism between these models. It is implemented with multiple parallel (not in the real sense of parallelism or on parallel computers, however this is feasible and should be implemented in this way for saving computational times) filters, where each of the filters would correspond to one of the multiple models. Because of the switching between different models, there is an exchange of some information between the filters. During each sampling period, it is likely that all the filters of the IMMKF are in operation. The overall state estimate is a combination of the state estimates from the individual filters. Let us consider $M_1,\ldots,M_r$ as the r models of IMMKF and let $M_j(k)$ mean that the model M_j is in effect during the sampling period ending at frame k. During the event $M_j(k+1)$, the state of the target evolves according to the following equation

$$X(k+1) = F_j X(k) + w_j(k) \tag{3.90}$$

The measurement equation is given by

$$z(k+1) = H_j X(k+1) + v_j(k+1) \tag{3.91}$$

The variables have the usual meanings. Figure 3.13 depicts the information flow of one iteration-cycle of IMMKF [52]. For simplicity, a two-model IMMKF is considered. The IMMKF algorithm has four major steps [52,53].

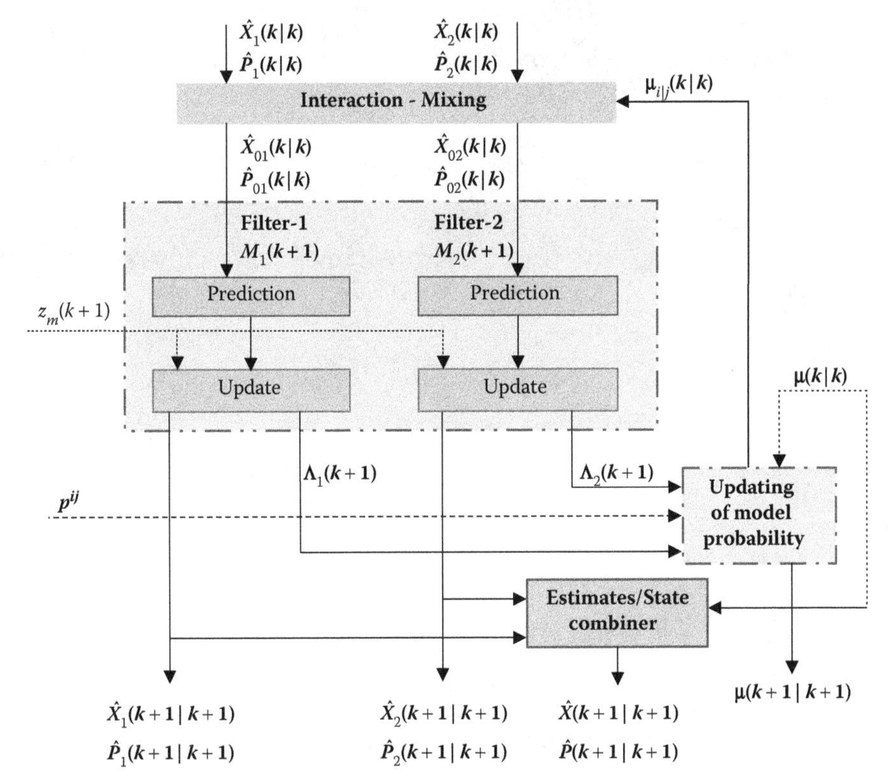

FIGURE 3.13
Interactive multiple models Kalman filter—one cycle of operation.

3.5.1.1 Interaction and Mixing

For the event $M_j(k+1)$, the mixed estimate $X_{0j}(k|k)$ and the covariance matrix $P_{0j}(k|k)$ are computed as

$$\hat{X}_{0j}(k|k) = \sum_{i=1}^{r} \mu_{i|j}(k|k)\hat{X}_i(k|k)$$
$$\hat{P}_{0j}(k|k) = \sum_{i=1}^{r} \mu_{i|j}(k|k)\Big\{\hat{P}_i(k|k) + \big[\hat{X}_i(k|k) - \hat{X}_{0j}(k|k)\big] \times \big[\hat{X}_i(k|k) - \hat{X}_{0j}(k|k)\big]^t\Big\} \tag{3.92}$$

The mixing probabilities $\mu_{i|j}(k|k)$ are given by

$$\mu_{i|j}(k|k) = \frac{1}{\mu_j(k+1|k)} p_{ij}\mu_i(k|k) \tag{3.93}$$

Here, the predicted mode probability $\mu_j(k+1|k)$ is computed by

$$\mu_j(k+1|k) = \sum_{i=1}^{r} p_{ij}\mu_i(k|k) \tag{3.94}$$

The mode switching process is done by Markov process and is specified by the following mode transition probabilities

$$p_{ij} = \Pr\{M_j(k+1)|M_i(k)\} \tag{3.95}$$

Here, Pr{.} denotes the probability of an event, which means p_{ij} is the probability at which M_i model at kth instant is switching over to M_j model at $(k + 1)$th instant. This is used to calculate the model probabilities for the final output.

3.5.1.2 Kalman Filtering

The usual KF equations are used with appropriate target motion models to update the mixed state estimates with current measurement as shown in Figure 3.13. The innovation covariance is given by

$$S_j = H_j\tilde{P}_j(k+1|k)H_j^T + R_j \tag{3.96}$$

The innovations sequence is given by

$$\upsilon_j = z(k+1) - \tilde{z}_j(k+1|k) \tag{3.97}$$

The likelihood function for matched filter j is a Gaussian pdf of innovation $\upsilon_j(k+1)$ with zero mean and covariance S_j. It is used for updating the probabilities of the various models. It is computed as

$$\Lambda_j = \frac{1}{(2\Pi)^{0.5n}\sqrt{|S_j|}}\exp\left\{-0.5\upsilon_j^T S_j^{-1}\upsilon_j\right\} \tag{3.98}$$

Here, n denotes the dimension of the innovation vector $\boldsymbol{\upsilon}$.

3.5.1.3 Mode Probability Update

Once each model has been updated with measurement $z(k+1)$, the mode probability $\mu_j(k+1|k+1)$ is updated using mode likelihood Λ_j and the predicted mode probabilities $\mu_j(k+1|k)$ for $M_j(k+1)$

$$\mu_j(k+1|k+1) = \frac{1}{c}\mu_j(k+1|k)\Lambda_j \tag{3.99}$$

Here, the normalization factor is

$$c = \sum_{i=1}^{r}\mu_i(k+1|k)\Lambda_i \tag{3.100}$$

3.5.1.4 State Estimate and Covariance Combiner

The estimated states $\hat{X}_j(k+1|k+1)$ and covariance $\hat{P}_j(k+1|k+1)$ from each filter are combined using the updated mode probability $\mu_j(k+1|k+1)$ to produce overall state estimate $\hat{X}(k+1|k+1)$ and the associated covariance $\hat{P}(k+1|k+1)$ as given below:

$$\hat{X}(k+1|k+1) = \sum_{j=1}^{r}\mu_j(k+1|k+1)\hat{X}_j(k+1|k+1) \tag{3.101}$$

$$\hat{P}(k+1|k+1) = \sum_{j=1}^{r}\mu_j(k+1|k+1)\{\hat{P}_j(k+1|k+1) + [\hat{X}_j(k+1|k+1) - \hat{X}(k+1|k+1)][\hat{X}_j(k+1) - \hat{X}(k+1|k+1)]^T\} \tag{3.102}$$

3.5.2 Target Motion Models

The most common forms of target motion models are (1) 2-degree of freedom (DOF) kinematic models (constant velocity model), and (2) 3-DOF kinematic model (constant acceleration model) [43,52,53].

3.5.2.1 Constant Velocity Model

The 2-DOF model with position and velocity components in each of the three Cartesian coordinates x, y, and z has the following transition and process noise gain matrices

$$F_{CV} = \begin{bmatrix} \Phi_{CV} & 0 & 0 \\ 0 & \Phi_{CV} & 0 \\ 0 & 0 & \Phi_{CV} \end{bmatrix} \quad G_{CV} = \begin{bmatrix} \varsigma_{CV} & 0 & 0 \\ 0 & \varsigma_{CV} & 0 \\ 0 & 0 & \varsigma_{CV} \end{bmatrix} \tag{3.103}$$

Here, $\Phi_{CV} = \begin{bmatrix} 1 & T & 0 \\ 0 & 1 & 0 \\ 0 & 0 & 0 \end{bmatrix} \quad \varsigma_{CV} = \begin{bmatrix} T^2/2 \\ T \\ 0 \end{bmatrix}$

The variations in velocity are modeled as zero-mean white noise accelerations. Low noise variance Q_{cv} is used with the model to represent the constant course and speed of the target in a nonmaneuvering mode. The process noise intensity in each coordinate is generally assumed to be small and equal $\left(\sigma_x^2 = \sigma_y^2 = \sigma_z^2\right)$, which accounts for air turbulence, slow turns, and small linear acceleration. Although the 2-DOF model is primarily used to track the nonmaneuvering mode of a target, use of higher level of process noise variance will allow the model to track maneuvering targets as well, of course to a limited extent. The model can be easily extended to more DOF.

3.5.2.2 Constant Acceleration Model

The 3-DOF model with position, velocity, and acceleration components in each of the three Cartesian coordinates x, y, and z has the following transition and process noise gain matrices

$$F_{CA} = \begin{bmatrix} \Phi_{CA} & 0 & 0 \\ 0 & \Phi_{CA} & 0 \\ 0 & 0 & \Phi_{CA} \end{bmatrix} \quad G_{CA} = \begin{bmatrix} \varsigma_{CA} & 0 & 0 \\ 0 & \varsigma_{CA} & 0 \\ 0 & 0 & \varsigma_{CA} \end{bmatrix} \tag{3.104}$$

Here, $\Phi_{CA} = \begin{bmatrix} 1 & T & T^2/2 \\ 0 & 1 & T \\ 0 & 0 & 1 \end{bmatrix} \quad \varsigma_{CA} = \begin{bmatrix} T^3/6 \\ T^2/2 \\ T \end{bmatrix}$

The acceleration increments over a sampling period are a discrete time zero-mean white noise. A low value of process noise variance Q_2 (but relatively higher than Q_1) will yield nearly a constant acceleration motion. The noise variances in each coordinate are assumed to be equal $\left(\sigma_x^2 = \sigma_y^2 = \sigma_z^2\right)$. Studies have shown that the use of higher process noise levels combined with 3-DOF model can help track the onset and termination of a maneuver to a certain extent.

3.5.3 Interacting Multiple Model Kalman Filter Implementation

The algorithm implemented in MATLAB consists of both IMMKF and conventional KF. Two-model IMMKF (one for nonmaneuver mode and another for maneuver mode, i.e., $r = 2$) is considered to describe the principles and steps of IMMKF algorithm. The observation matrix used is given as

$$H = \begin{bmatrix} 1 & 0 & 0 & 0 & 0 & 0 & 0 & 0 & 0 \\ 0 & 0 & 0 & 1 & 0 & 0 & 0 & 0 & 0 \\ 0 & 0 & 0 & 0 & 0 & 0 & 1 & 0 & 0 \end{bmatrix}$$

The initial mode probabilities $\mu = [\mu_1 \ \mu_2]^T$ corresponding to nonmaneuver and maneuver mode are taken as 0.9 and 0.1 respectively. p_{ij} is the Markov chain transition matrix, which takes care of the switching from mode i to mode j. This is a design parameter and can be chosen by the user. The switching probabilities are generally known to depend upon sojourn time. For example, consider the following Markov chain transition matrix between the two modes of the IMMKF [50]

$$p_{ij} = \begin{bmatrix} 0.9 & 0.1 \\ 0.33 & 0.67 \end{bmatrix}$$

The basis for selecting $p_{12} = 0.1$ is that initially the target is likely to be in nonmaneuvering mode and the probability to switch over to maneuvering mode will be relatively low, whereas p_{22} is selected based on the number of sampling periods during which the target is expected to maneuver (target's sojourn time). If the target maneuver lasts for three sampling periods ($\tau = 3$), the probability p_{22} is

$$p_{22} = 1 - \frac{1}{\tau} = 0.67 \tag{3.105}$$

The performance of the IMMKF with respect to conventional KFcv (KF with only constant velocity model) and KFca (KF with only constant acceleration model) is evaluated by computing certain performance metrics, in addition to the root sum variance in position

$$\text{RSvarP} = \sqrt{P_x + P_y + P_z} \tag{3.106}$$

Here, P_x, P_y, and P_z are the diagonal elements in P corresponding to positions x, y, and z, similarly it can be used for other elements. Similar computations are carried out for root sum variance in velocity (RSvarV) and acceleration (RSvarA).

3.5.3.1 Validation with Simulated Data

The data for maneuvering target are simulated using 2-DOF and 3-DOF models [53] with a sampling interval of 1 second for 500 seconds. The simulation proceeds with the following parameters: (1) initial state $(x, \dot{x}, \ddot{x}, y, \dot{y}, \ddot{y}, z, \dot{z}, \ddot{z})$ of the target = (11097.6, –6.2, 0, 3425, –299.9, 0, 40, 0, 0); (2) a low-level process noise variance of $Q_1 = 0.09$ is considered for model 1; (3) a higher-level process noise variance of $Q_2 = 36$ is considered for model 2; (4) noise variances for both models in each coordinate are assumed to be equal, that is, $Q_{xx} = Q_{yy} = Q_{zz} = Q$; and (5) a measurement noise variance of $R = 100$ is considered, also assumed to be equal $R_{xx} = R_{yy} = R_{zz} = \text{R}$.

At scan 100, additional acceleration $\ddot{x} = 27.4\,\text{m/s}^2$ is given in x direction to simulate first maneuver and at scan 350, $\ddot{y} = -99.4\,\text{m/s}^2$ is given in y direction to simulate the second maneuver. The acceleration profiles in x- and y-coordinates are shown in Figure 3.14 [53]. The target spends most of the time in nonmaneuvering state. The first maneuver starts at scan 100 and ends at scan 150 and the second maneuver starts at scan 350 and ends at scan 400. KFcv uses a single constant velocity model with $Q_1 = 0.09$ and $R = 100$, whereas KFca uses a single constant acceleration model with $Q_2 = 36$ and $R = 100$, and IMMKF with two-mode (constant velocity model with $Q_1 = 0.09$ and $R = 100$ for nonmaneuvering mode and constant acceleration model with $Q_2 = 36$ and $R = 100$ for maneuvering mode) filters to track the target through both the modes. The Markov chain transition matrix is

$$p_{ij} = \begin{bmatrix} 0.9 & 0.1 \\ 0.05 & 0.95 \end{bmatrix}$$

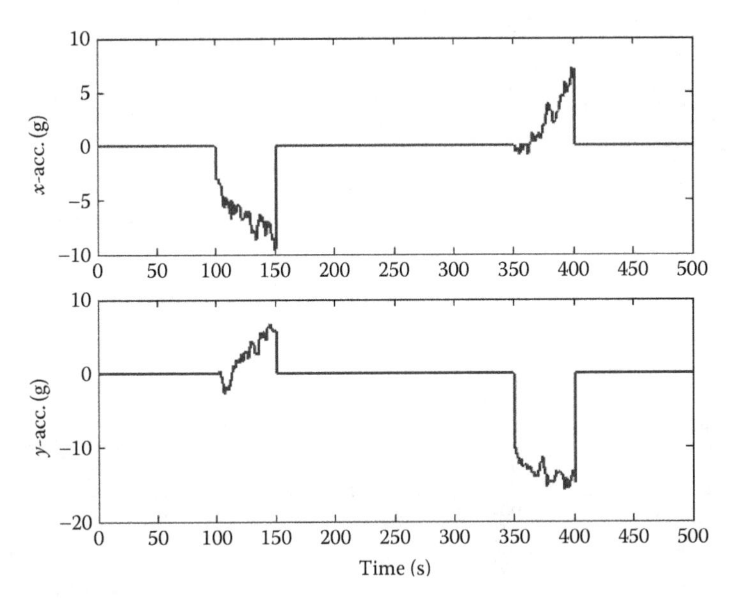

FIGURE 3.14
Acceleration profiles of the x- and y-axes.

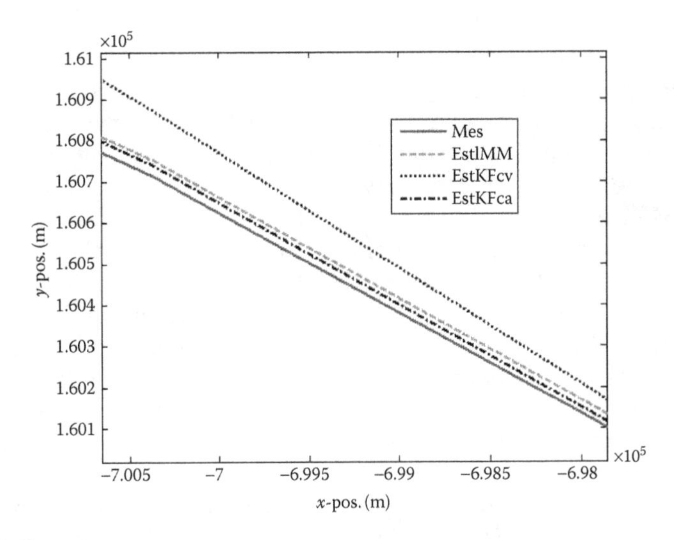

FIGURE 3.15
An expanded view of the trajectories.

This choice is made, keeping the sojourn time ($\tau = 50$) of the target in view. Equation 3.105 is used to arrive at the p_{22} value in the above transition matrices. The enlarged view of a portion of the trajectories using KFcv, KFca, and IMMKF is shown in Figure 3.15 [53]. It is observed

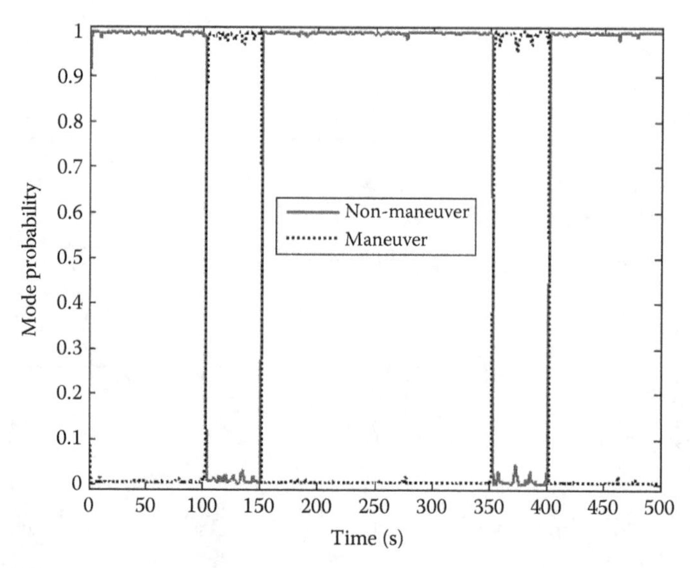

FIGURE 3.16
Maneuver mode probabilities.

that the track tracked by KFcv is away from the true track during the maneuver. The tracks that are tracked by both KFca and IMMKF are very close to the true track. The mode probabilities are shown in Figure 3.16. There are two maneuvers in the trajectory with start and end time of the maneuvers, that is, which model or the weighted combination of models is active at a given time. The RSVar from KFcv, KFca, and IMMKF are shown in Figure 3.17. The RSvar in KFcv and KFca are constant throughout the trajectory. Because of the high process noise covariance, the RSVar is high in KFca than in KFcv, whereas it is high during maneuver and low during nonmaneuver in case of IMMKF. This information could be useful in track fusion and is one of the features of IMMKF. This shows that maneuvers can detect using the IMMKF. The RSSPE from KFcv, KFca, and IMMKF are shown in Figure 3.18. It is very high during the maneuver in case of KFcv, and low throughout the trajectory in case of KFca. The mean in x-, y- and z-position errors and PFE in x, y and z positions are shown in Table 3.11. These results are obtained from 50 Monte Carlo simulations. From these results, it is observed that the tracking performance is worst in the case of KFcv during maneuver. The tracking performance of both KFca and IMMKF are almost similar during maneuver. Overall the tracking performance of IMMKF is acceptable.

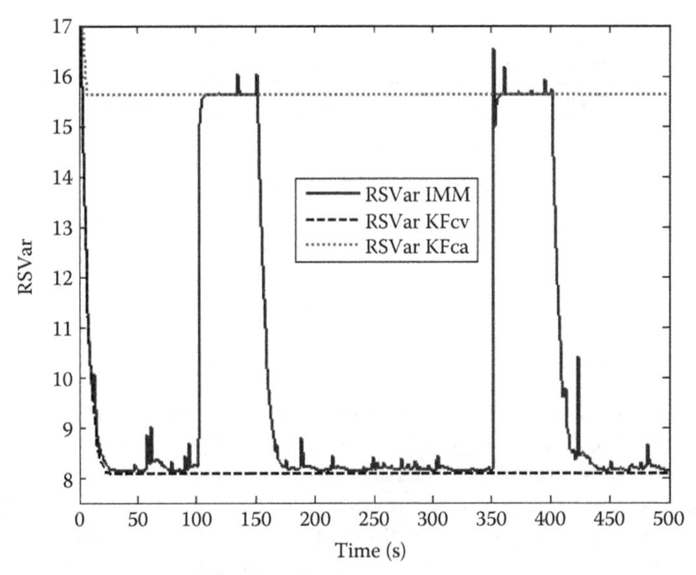

FIGURE 3.17
The root sum variance in velocity plots for KFcv, KFca, and interacting multiple model filters.

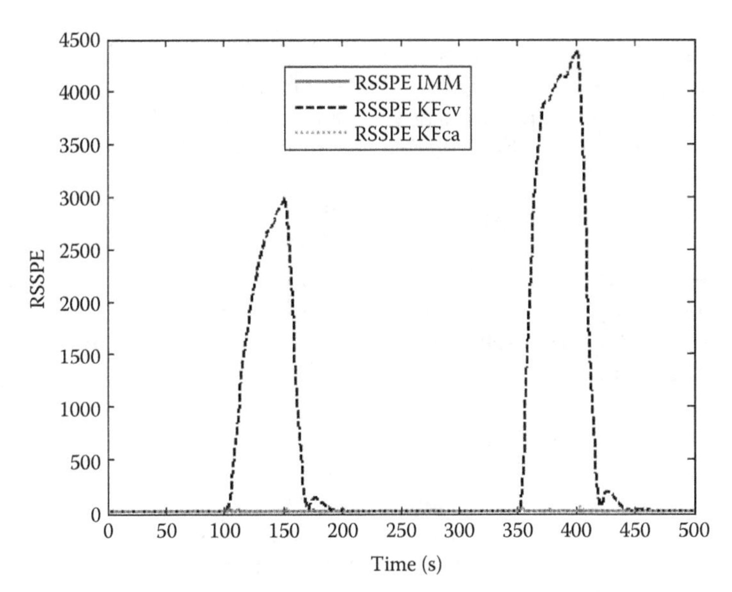

FIGURE 3.18
The root sum squares of position error for KFcv, KFca, and interacting multiple model filters.

TABLE 3.11

Performance Metrics for KFcv, KFca, and IMMKF

Filter	PFEx	PFEy	PFEz	RMSPE	RMSVE	RMSAE
KFcv	0.11	0.84	0.11	824.4	220.78	29.49
KFca	~0	~0	~0	2.75	9.15	7.72
IMMKF	~0	~0	~0	2.78	6.88	5.83

3.6 Joint Probabilistic Data Association Filter

In target tracking applications and mobile robotics, the main requirement is to track the targets and the positions of the humans, respectively [54]. For the latter application in the robotics field, knowledge of the positions of the humans and the vehicles or obstacles can greatly improve the behavior of the robot or robotic system. This information will allow the robot to adapt its motion variables in tune with the speed of the people in its vicinity. An accurate determination of the position and velocity of the moving objects would help a robot (1) improve its performance in map-building algorithms, and (2) improve its collision avoidance. The problem of estimating the position of multiple moving objects, targets, and humans has certain ramifications: (1) the problem is more difficult than that of a single object, (2) the number of objects being tracked must be estimated, (3) the ambiguities of the measurements must be resolved, and (4) there might be more features than objects, and some features may not be distinguishable. Hence, the problem in DA of assigning the observed features to the tracked objects needs to be also handled (Section 3.4).

In this section, we will discuss the problem of tracking multiple objects using the joint probabilistic DA filter (JPDAF) in the context of a mobile robot [54]. The development is equally applicable to any moving objects. The JPDAF computes a Bayesian estimate of the correspondence between features (detected in the sensor data) and the various objects being tracked. The basic filtering mechanism used in many JPDAF algorithms is that of the KF. However, particle filters have recently found applications for estimation of non-Gaussian and nonlinear dynamic systems. The main idea in a particle filter is that the state is represented by sets of samples which are called particles and the multimodal state (probability) densities can then be represented in a more favorable manner so that the robustness of the estimation process improves. When the particle filter is applied to track multiple objects, then the combined state (space) is estimated. The S-sample–based JPDAF (SJPDAF) utilizes particle filter and existing approach to multiple-target tracking. Basically, the JPDAF is used to assign the measurements to the individual moving objects and the particle filter is used for estimating the states of these objects, that is, to track the objects. The probability distribution over the number of objects being tracked is maintained.

3.6.1 General Version of a Joint Probabilistic Data Association Filter

To keep track of multiple moving targets, objects, and humans, we need to estimate the joint probability distribution of the states of all the objects. We need to know which object has generated what measurement. The JPDA filtering algorithm is used for this purpose. Let there be N objects that are being tracked. Let $X(k) = \{x_1(k), \ldots, x_N(k)\}$ be the state vector of these moving objects at time k. Let $Y(k) = \{y_1(k), \ldots, y_{mk}(k)\}$ denote a measurement (at k). Here, $y_j(k)$ is one feature of such a measurement. Then Y^k is the sequence of all the measurements up to time instant k. The main aspect for tracking multiple objects is to assign the measured features to the individual objects. A joint association event θ is defined, which is a set of pairs $(j, i) \in \{0, \ldots, m_k\} \times \{1, \ldots, N\}$. In this case, each θ determines which feature is assigned to which object. Let Θ_{ji} be the set of all joint association events with feature j assigned to the object i. The JPDAF computes the posterior probability in which feature j has occurred because of the object i

$$\beta_{ji} = \sum_{\theta \in \Theta_{ji}} P(\theta \mid Y^k) \tag{3.107}$$

The probability P is given as [54]

$$\begin{aligned} P(\theta \mid Y^k) &= P(\theta \mid Y(k), Y^{k-1}) \\ &= \int P\left(\theta \mid Y(k), Y^{k-1}, X(k)\right) p\left(X(k) \mid Y(k), Y^{k-1}\right) \mathrm{d}X(k) \\ &= \int P\left(\theta \mid Y(k), X(k)\right) p\left(X(k) \mid Y(k), Y^{k-1}\right) \mathrm{d}X(k) \end{aligned} \tag{3.108}$$

From Equation 3.108, it is clear that we need to know the state of the moving objects for us to determine θ. In this implicit problem of the assignments and the state, we also need to know θ to determine the states of the objects (e.g., positions). To resolve this issue we can use the incremental method, whereby p is approximated by the belief $p(X(k) \mid Y^{k-1})$. This means that the predicted state of the objects is computed using all the measurements before time instant k. With this assumption, we obtain the following equation for P:

$$\begin{aligned} P(\theta \mid Y^k) &\approx \int P\left(\theta \mid Y(k), X(k)\right) p\left(X(k) \mid Y(k), Y^{k-1}\right) \mathrm{d}X(k) \\ &= \alpha \int P\left(Y(k) \mid \theta, X(k)\right) P\left(\theta \mid X(k)\right) p\left(X(k) \mid Y^{k-1}\right) \mathrm{d}X(k) \end{aligned} \tag{3.109}$$

The second term in the integral sign corresponds to the probability of the θ given the current states of the objects, whereas the first term is the probability of making an observation given the states and a specific assignment between the observed features and the tracked objects. That a feature may not be caused by any object must also be considered. Let γ be the probability of an observed feature being a false alarm. Then the number of false alarms in the association θ is $(m_k - |\theta|)$. Thus, the $\gamma^{(m_k - |\theta|)}$ is the probability assigned to all false alarms in $Y(k)$ given θ. With the features being detected independently of each other, we obtain

$$P\left(Y(k)|\theta, X(k)\right) = \gamma^{(m_k - |\theta|)} \prod_{(j,i)\in\theta} \int p\left(z_j(k)|x_i(k)\right) p\left(x_i(k)|Y^{k-1}\right) \mathrm{d}x_i(k) \tag{3.110}$$

By using Equation 3.110 in Equation 3.109, we obtain the following equation:

$$\beta_{ji} = \sum_{\theta\in\Theta_{ji}} \left[\alpha\gamma^{(m_k - |\theta|)}\right] \prod_{(j,i)\in\theta} \int p\left(z_j(k)|x_i(k)\right) p\left(x_i(k)|Y^{k-1}\right) \mathrm{d}x_i(k) \tag{3.111}$$

Because of the large number of possible assignments, the problem will be complex and hence only an event with substantial assignments should be considered. This aspect is handled by gating and DA, as is done in the case of NNKF and PDAF.

Next, the belief about the states of the individual objects must be updated. In conventional JPDAF and in many such algorithms, the probability densities are assumed to be described by their first and second order moments, and then KF is used for state prediction and filtering (essentially updating the underlying densities via the updating of the moments). In case of the Bayesian filtering, this updating is given as

$$p\left(x_i(k)|Y^{k-1}\right) = \int p\left(x_i(k)|x_i(k-1), t\right) p\left(x_i(k-1)|Y^{k-1}\right) \mathrm{d}x_i(k-1) \tag{3.112}$$

With the new measurements the state is updated as follows:

$$p\left(x_i(k)|Y^k\right) = \alpha\, p\left(Z(k)|x_i(k)\right) p\left(x_i(k)|Y^{k-1}\right) \tag{3.113}$$

Next, the single features are integrated with assignment probabilities β_{ji}

$$p\left(x_i(k)|Y^k\right) = \alpha \sum_{j=0}^{m_k} \beta_{ji} p\left(z_j(k)|x_i(k)\right) p\left(x_i(k)|Y^{k-1}\right) \tag{3.114}$$

From the foregoing development it is clear that we need the models $p(x_i(k)|X_i(k-1), t)$ and $p(z_i(k)|x_i(k))$. We can see that these models depend on the moving objects being detected and the sensors used for tracking. The update cycle of the JPDAF is as follows: (1) the states are $x_i(k)$ predicted based on the earlier obtained estimates and the available motion model of the objects; (2) the association probabilities are computed; and (3) the latter are used to incorporate the individual state estimates.

3.6.2 Particle Filter Sample–Based Joint Probabilistic Data Association Filter

The main idea in particle filters is to represent the density $p(x_i(k)\,|\,Y^k)$ by a set S_i^k of M weighted random samples (often called particles) $s_{i,n}^k$, ($n = 1, \ldots, M$). The sample set constitutes a discrete approximation of a probability distribution, and each sample is a $(x_{i,n}(k), \omega_{i,n}(k))$ consisting of the states and an importance factor. The prediction step is realized by pulling samples from the set computed from the preceding iteration and by evolving (updating) their state based on the prediction model $p(x_i(k)\,|\,x_i(k-1),t)$. In the correction (measurement update) stage, a measurement $Z(k)$ is integrated into the samples obtained in the prediction stage. At this stage, we have to consider assignment probabilities β_{ji}, which are obtained by integrating over the states, weighted by the probabilities $p(x_i(k)\,|\,Y^{k-1})$. In the particle sample–based approach, this integration is replaced by the summation over all samples generated after the prediction step [54]:

$$\beta_{ji} = \sum_{\theta \in \Theta_{ji}} \left[\alpha \gamma^{(m_k - |\theta|)} \prod_{(j,i) \in \theta} \frac{1}{M} \sum_{n=1}^{M} p(z_j(k)\,|\,x_{i,n}(k)) \right] \tag{3.115}$$

Once the assignment probabilities are determined, one can compute the weights of the particles as

$$\omega_{i,n}(k) = \alpha \sum_{j=0}^{m_k} \beta_{ji}\, p\big(z_j(k)\,|\,x_{i,n}(k)\big) \tag{3.116}$$

In the foregoing developments α is a normalizing factor. Then M new samples are obtained by bootstrapping the resampling process from the current samples and selecting every sample with the probability $\omega_{i,n}(k)$. The JPDAF algorithm presumes that the number of objects to be tracked is known. This is not always true in practice. The solution to this problem for JPDAF is given in [54]. The SJPDAF can include a recursive Bayesian filter to deal with handling of varying number of moving objects. Because of the use of the particle filter in the SJPDAF the arbitrary densities over the state spaces of the individual objects can be represented. This would give more accurate estimates in the prediction stage. This can also handle nonlinear systems. This will result in more robust performance and perhaps fewer DA errors. The SJPDAF can also handle situations with a large amount of clutter.

3.7 Out-of-Sequence Measurement Processing for Tracking

In many tracking situations, at times we get *out-of-sequence measurements* (OOSMs), that is, the measurements arrive at the processing center later than the time they were supposed to arrive. In a networked sensors architecture interconnected via complex communication network modules, the measurements are received in out-of-time-order at the fusion node or center. In a multisensor tracking system, the observations are sent to a fusion center over communication networks, which can introduce random time delays [55]. The conventional KF can be easily extended to multisensor systems wherein the data arrive at known times and in correct time sequence. The problem of time delay between the sensor and tracking computer is defined as follows: When a measurement corresponding to time, τ, arrives at time t_k after the states and covariance matrices in the KF have been computed, then the problem of updating these estimates with the delayed measurement arises.

3.7.1 Bayesian Approach to the Out-of-Sequence Measurement Problem

Let $x(t_k)$ be the target state at time t_k, $Z(\tau)$ be the set of delayed measurements at time τ, and Z^k be the set of sensor measurement sequence received up to time t_k. Assume that all the measurements, Z^k, have been processed. Then the information about the target state $x(t_k)$ is given by the probability density function (pdf) $p(x(t_k)|Z^k)$. In the OOSM problem, the delayed data are received at time t_k, but these data correspond to time $\tau < t_k$. The solution to this problem is to update $p(x(t_k)|Z^k)$ with $Z(\tau)$ to obtain $p\left(x(t_k)|Z^k, Z(\tau)\right)$. Invoking Bayes rule, we get

$$p\left(x(t_k)|Z^k, Z(\tau)\right) = \frac{p\left(Z(\tau)|x(t_k), Z^k\right)p\left(x(t_k)|Z^k\right)}{p\left(Z(\tau)|Z^k\right)} \tag{3.117}$$

Introducing the target state at time τ, $x(\tau)$ finally [55] we have

$$p\left(x(t_k)|Z^k, Z(\tau)\right) = \int p\left(x(\tau), x(t_k)|Z^k, Z(\tau)\right)dx(\tau) \tag{3.118}$$

We see that the solution to OOSM problem needs the joint density of the current target state and the target state related to the delayed measurement.

3.7.2 Out-of-Sequence Measurement with Single Delay and No Clutter

An exact solution for the OOSM problem with single delay is the so-called Y algorithm [55]. The Y algorithm considers the process noise between the time of the delayed measurement and the current time and the correlation of the current state with the delayed measurement $z(\tau)$.

3.7.2.1 Y Algorithm

We assume that the measurement delay is less than one sampling period, $t_{k-1} \le \tau < t_k$. A joint Gaussian random variable $y(t_k)$ is defined as

$$y(t_k) = \begin{bmatrix} x(t_k) \\ z(\tau) \end{bmatrix} \text{ with } P_y = \begin{bmatrix} P_{xx} & P_{xz} \\ P_{xz} & P_{zz} \end{bmatrix} \tag{3.119}$$

Here,

$$P_{xx} = E\left\{ \left(x(t_k) - \hat{x}(t_{k/k}) \right) \left(x(t_k) - \hat{x}(t_{k/k})^T \right) | Z^k \right\} = P_{k/k} \tag{3.120}$$

$$P_{zz} = E\left\{ \left(z(\tau) - \hat{z}(\tau) \right) \left(z(\tau) - \hat{z}(\tau) \right)^T | Z^k \right\} = S_{\tau/k} \tag{3.121}$$

$$P_{xz} = E\left\{ \left(x(t_k) - \hat{x}(t_{k/k}) \right) \left(z(\tau) - \hat{z}(\tau) \right)^T | Z^k \right\} = P_{zx}^T \tag{3.122}$$

The solution to this problem requires the conditional density $p\left(x(t_k) | z(\tau), Z^k\right)$, which is known to be Gaussian with mean [55]

$$\hat{x}(t_{k|\tau,k}) = \hat{x}(t_{k|k}) + P_{xz} P_{zz}^{-1} \left(z(\tau) - \hat{z}(\tau) \right) \tag{3.123}$$

and the associated covariance matrix

$$P(t_{k|\tau,k}) = P_{xx} - P_{xz} P_{zz}^{-1} P_{zx} \tag{3.124}$$

Here, the backward predicted measurement is given as

$$\hat{y}(\tau) = H_\tau F_{\tau|k} \left\{ \hat{x}(t_{k|k}) - Q_k(\tau) H_\tau^T S_{\tau|k}^{-1} \left(z(t_k) - \hat{z}(t_{k|k-1}) \right) \right\} \tag{3.125}$$

where, H_τ is the measurement model matrix at time τ and $F_{\tau|k}$ is the system backward transition matrix from t_k to τ. The last term accounts for the effect of process noise with covariance Q_k on the estimate $\hat{x}(t_{k|k})$. The cross covariance P_{xz} is given by

$$P_{xz} = \{P_{k|k} - P_{x\tilde{z}}\}F_{\tau|k}{}^{T}H_{\tau}^{T} \tag{3.126}$$

Here,

$$P_{x\tilde{z}} = \text{cov}\{x(t_k), w_k(\tau) \mid Z^k\} = Q_k(\tau) - P(t_{k|k-1})H_{\tau}^{T}S^{-1}(t_k)H_{\tau}P(t_{k|k-1}) \tag{3.127}$$

This algorithm requires the storage of the last innovations and hence it is regarded as a kind of nonstandard smoothing filter.

3.7.2.2 Augmented State Kalman Filters

With a single delay, the OOSM problem assumes that this delayed measurement $z(\tau)$ is received at time t_k and also the current measurement $z(t_k)$. The vector $[x(t_k), x(\tau)]^T$ is the augmented state. Then, the following equations are considered [55]:

$$\begin{bmatrix} x(t_k) \\ x(\tau) \end{bmatrix} = \begin{bmatrix} F_{t_{k|\tau}} & 0 \\ I & 0 \end{bmatrix}\begin{bmatrix} x(t_{k-1}) \\ x(\tau) \end{bmatrix} + \begin{bmatrix} w(t_{k|\tau}) \\ 0 \end{bmatrix} \tag{3.128}$$

$$\begin{bmatrix} z(t_k) \\ z(\tau) \end{bmatrix} = \begin{bmatrix} H_k & 0 \\ 0 & H_{\tau} \end{bmatrix}\begin{bmatrix} x(t_k) \\ x(\tau) \end{bmatrix} + \begin{bmatrix} v(t_k) \\ v(\tau) \end{bmatrix} \tag{3.129}$$

where, $F_{t_k|\tau}$ is from the system dynamic equation and $t_{k-1} = \tau$. KF recursion is utilized to obtain the estimate of the augmented state that is further updated using both delayed measurement $z(\tau)$ and current measurement $z(t_k)$. Without the delayed measurement, the measurement equation will be

$$z(t_k) = H_k x(t_k) + v(t_k)$$

The correlation of the state of the target and process noise is implicitly handled in this algorithm, whereas it is explicitly handled in the Y algorithm.

EXAMPLE 3.2

For this numerical example from [55], the discrete time system equation is

$$x(k) = \begin{bmatrix} 1 & T \\ 0 & 1 \end{bmatrix} x(k-1) + v(k) \tag{3.130}$$

Here, $T = 1$ is the sampling interval and $v(k)$ is a zero-mean white Gaussian noise with covariance matrix

$$\text{cov}\{v(k)\} = Q(k) = \begin{bmatrix} T^3/3 & T^2/2 \\ T^2/2 & T \end{bmatrix} q \tag{3.131}$$

The measurement model is given by

$$z(k) = [1\ 0]\ x(k) + w(k) \tag{3.132}$$

Here, $w(k)$ is a zero-mean white Gaussian noise with covariance $\text{cov}\{w(k)\} = R(k) = 1$. On the basis of these models, a 2D target state model was developed in [50] and used for simulation. The maneuvering index [55] is given as $\lambda = \sqrt{\frac{qT^3}{R}}$. In [55] two cases (with process noise $q = 0.1$, and 1 corresponding to $\lambda = 0.3$, and 1, respectively) were examined: The target performs a straight line motion or is highly maneuvering, and the data were generated randomly for each run with $x(t = 0) = [200\text{ km}\ 0.5\text{ km/s}\ 100\text{ km}\ -0.08\text{ km/s}]$. The filters were initialized in [55] with

$$P(0/0) = \begin{bmatrix} P_0 & 0 \\ 0 & P_0 \end{bmatrix} \text{ with } P_0 = \begin{bmatrix} R & R/T \\ R/T & 2R/T^2 \end{bmatrix} \tag{3.133}$$

It was assumed that the OOSM can only have a maximum of one lag delay. The data delay was assumed to be uniformly distributed (for the entire simulation period) with a probability P_r that the current measurement is delayed. The comparison of computational delay is given in Table 3.12 [55]. It was found that the augmented state algorithms outperformed Y algorithm, however the computational load for these filters was greater than that of the Y algorithm as can be seen from Table 3.12 [55].

TABLE 3.12

Comparison of Computational Load (In Terms of Number of Floating Point Operations with respect to KF)

Algorithm|P_r	0.0	0.25	0.50	0.75
Y algorithm	1	2.26	2.30	4.41
Augmented state algorithm and augmented state smoother algorithm	5.57	5.57	5.57	5.57

Source: Challa, S., R. J. Evans, and X. Wang. 2003. *Inf Fusion* 4(3):185–199. With permission.

3.8 Data Sharing and Gain Fusion Algorithm for Fusion

In certain works on estimation fusion, the results could be the global estimate of the state in a decentralized system; however, this would require extensive calculation of the local and global inverse covariance matrices. This can be avoided in the data sharing and gain fusion scheme [56]. In a decentralized architecture, all the information is processed locally and central fusion may not be needed. There is information exchange between several nodes.

3.8.1 Kalman Filter–Based Fusion Algorithm

The estimates of the state vectors are obtained by each sensor using the optimal linear KFs:

$$\tilde{x}^m(k+1) = F\hat{x}^m(k) \tag{3.134}$$

State and covariance time propagation is given as

$$\tilde{x}^m(k+1) = F\hat{x}^m(k)$$

$$\tilde{P}^m = F\hat{P}^m F^t + GQG^t \tag{3.135}$$

State and covariance update equations are given as

$$r(k+1) = z^m(k+1) - H\tilde{x}^m(k+1) \tag{3.136}$$

$$K^m = \tilde{P}^m H^t [H\tilde{P}^m H^t + R_v^m]^{-1} \tag{3.137}$$

$$\hat{x}^m(k+1) = \tilde{x}^m(k+1) + K^m\left[z^m(k+1) - H\tilde{x}^m(k+1)\right] \tag{3.138}$$

$$\hat{P}^m = [I - K^m H]\tilde{P}^m \tag{3.139}$$

$$\hat{P}^f = \hat{P}^1 - \hat{P}^1(\hat{P}^1 + \hat{P}^2)^{-1}\hat{P}^{1^t} \tag{3.140}$$

The filters for both the sensors use the same state dynamics model. The measurement models and the measurement noise statistics could be different. The fusion algorithm is then given as

$$\hat{x}^f = \hat{x}^1 + \hat{P}^1(\hat{P}^1 + \hat{P}^2)^{-1}(\hat{x}^2 - \hat{x}^1) \tag{3.141}$$

We know from the above equations that the fused state vector and the covariance of the fused state utilize the individual estimate state vectors (of each sensor) and covariance matrices.

3.8.2 Gain Fusion–Based Algorithm

We know from the equations in Section 3.8.1 that the KF-based fusion algorithm (KFBFA) requires calculations of inverse covariance to obtain the global results, Equations 3.140 and 3.141. The recent fusion algorithm [56] does not require calculations of covariance inverses and has parallel processing capability. The dynamic system equations are the same as also used in KF equations. The gain fusion algorithm for multisensor integration involves information feedback from the global filter to the local filters. The filtering algorithm is given next [56].

Time propagation of global estimates is

$$\tilde{x}^f(k+1) = F\hat{x}^f(k) \tag{3.142}$$

$$\tilde{P}^f(k+1) = F\hat{P}^f(k)F^t + GQG^t \tag{3.143}$$

The local filters are reset as

$$\tilde{x}^m(k+1) = \tilde{x}^f(k+1) \tag{3.144}$$

$$\tilde{P}^m(k+1) = \tilde{P}^f(k+1) \tag{3.145}$$

The measurement update of local gains and states is obtained by

$$K^m = (1/\gamma^m)\tilde{P}^f(k+1)H'\left[H\tilde{P}^f(k+1)H' + (1/\gamma^m)R^m\right]^{-1} \tag{3.146}$$

$$\hat{x}^m(k+1) = \tilde{x}^f(k+1) + k^m\left[z^m(k+1) - H\tilde{x}^f(k+1)\right] \tag{3.147}$$

The global fusion of m local estimates is given as

$$\hat{x}^f(k+1) = \sum^m \hat{x}^m(k+1) - (m-1)\tilde{x}^f(k+1) \tag{3.148}$$

$$\hat{P}^f(k+1) = \left[I - \sum^m K^m H\right]\tilde{P}^f(k+1)\left[I - \sum^m K^m H\right]^t + \sum^m K^m R^m K^{m^t} \tag{3.149}$$

We see that there is information feedback from the global filter to the local filters. The GFBA does not need the measurement update of the local covariances to obtain the global estimates. Because the global *a priori* estimates are fed back to the local filter, there is implicit measurement data sharing between the local filters [56]. This feature is evaluated when there is data loss in either of the two sensors.

3.8.3 Performance Evaluation

The individual local filters and fusion algorithms are implemented in PC MATLAB [57]. The flight data of a moving target are used wherein the target is tracked by two ground-based S-band radars (which measure the range, azimuth, and elevation of the target). The radar data are converted to Cartesian coordinates frame so that the linear state and measurement models can be used, and the coupling between three axes is assumed to be nil. The data are sampled at 0.1 seconds and the simulated data loss is for 50 seconds. The performance of the fusion filters can also be evaluated in terms of the ratio

$$\frac{\sum_{k=0}^{N}\left(\hat{x}^f(k)-x(k)\right)^t\left(\hat{x}^f(k)-x(k)\right)}{\left(\hat{x}_0^f-x_0^f\right)^t P_0^f\left(\hat{x}_0^f-x_0^f\right)+\sum_{k=0}^{N} w^t(k)w(k)+\sum^{m}\sum_{k=0}^{N} v^{m^t}(k)v^m(k)} \tag{3.150}$$

Basically this ratio (the H-infinity [H-I] norm) should be less than square of gamma (a scalar parameter) that can be considered an upper bound on the maximum energy gain from the input to the output. It can be observed from the above norm that the input to the filter consists of energies due to the error in the initial condition, state disturbance (process noise), and measurement noise for both the sensors. The output energy of the filter is due to the error in fused state. For the GFBA, the value of gamma for each local filter is equal to 2 ($m = 2$). The data sets are from two tracking radars. The performance metrics are given in Table 3.13.

TABLE 3.13

Residual (%) Fit Errors and H-Infinity Norms

PFE	No Data Loss (Normal)			Data Loss in Sensor 1			Data Loss in Sensor 2		
	x	y	z	x	y	z	x	y	z
KFBFA Trajectory 1	0.308	0.140	0.553	1.144	1.543	2.802	0.308	0.139	0.553
KFBFA Trajectory 2	0.129	0.124	0.180	0.129	0.124	0.180	0.762	1.157	6.199
GFBA Trajectory 1	0.376	0.622	1.306	0.379	0.627	1.328	0.392	0.610	1.604
GFBA Trajectory 2	0.131	0.142	0.226	0.131	0.142	0.219	0.220	0.205	1.142
H_∞ Norm for fusion filters									
KFBFA	0.604	2.759	1.037	0.704	3.051	0.941	5.076	19.55	69.85
GFBA	0.546	2.325	0.821	0.562	2.38	0.913	0.625	2.203	2.952

The PFE is evaluated with respect to the ground truth. We see that when GFBA is used the H-infinity norm is below the theoretical limit $\gamma^m (2^2 = 4)$. Figures 3.19 and 3.20 show the state estimates and residuals obtained by KFBFA and GFBA, respectively, when there is data loss in sensor 1 [57]. It can be observed from these results that the performance of KFBFA suffers when there is loss of data, whereas that of the GFBA remains largely unaffected.

3.9 Global Fusion and H-Infinity Filter–Based Data Fusion

The H-I concept is related to the theory of optimal control synthesis in frequency domain [58]. The H-I optimal control is a frequency domain optimization and synthesis theory. The theory explicitly addresses question of modeling errors and the basic philosophy is to treat the worst case scenario, that is, plan for the worst and optimize (i.e., minimize the maximum error). The framework must have the following properties: (1) it must be capable of dealing with plant modeling errors and unknown disturbances; (2) it must represent a natural extension to existing theory; (3) it must be amenable to meaningful optimization; and (4) it must be applicable to the multivariable problem. The H-I concept involves RMS value of a signal, that is, a measure of a signal that reflects eventual average size of root-mean-square (rms) value, a classical notion of the size of a signal used in many areas of engineering. Equation 3.150, the H-I norm, is used in deriving a robust filtering algorithm. The basic H-I filters used for fusion algorithms are based on this H-I norm. The results of sensor data fusion using the H-I filters are relatively new. The results of simulation validation are presented for the situation in which two local individual filters track a moving object.

3.9.1 Sensor Data Fusion using H-Infinity Filters

An object is tracked with a KF (or similar local filter) associated with the sensor. The kinematic model of a tracked object is described by

$$x(k+1) = Fx(k) + Gw(k) \tag{3.151}$$

with

$$F = \begin{bmatrix} 1 & T \\ 0 & 1 \end{bmatrix}; \quad G = \begin{bmatrix} T^2/2 \\ T \end{bmatrix}$$

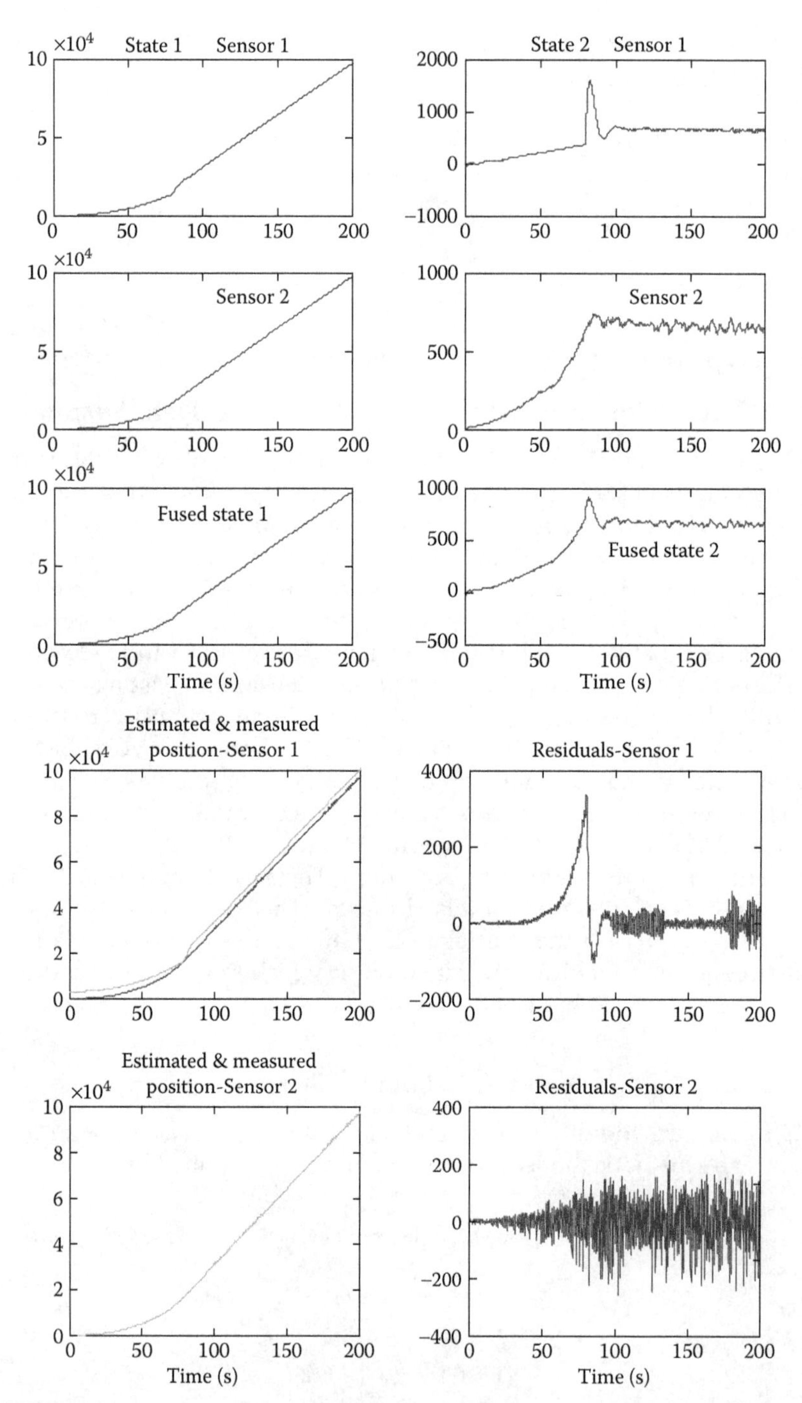

FIGURE 3.19
State fusion with Kalman filter–based fusion algorithm—loss of data in sensor 1.

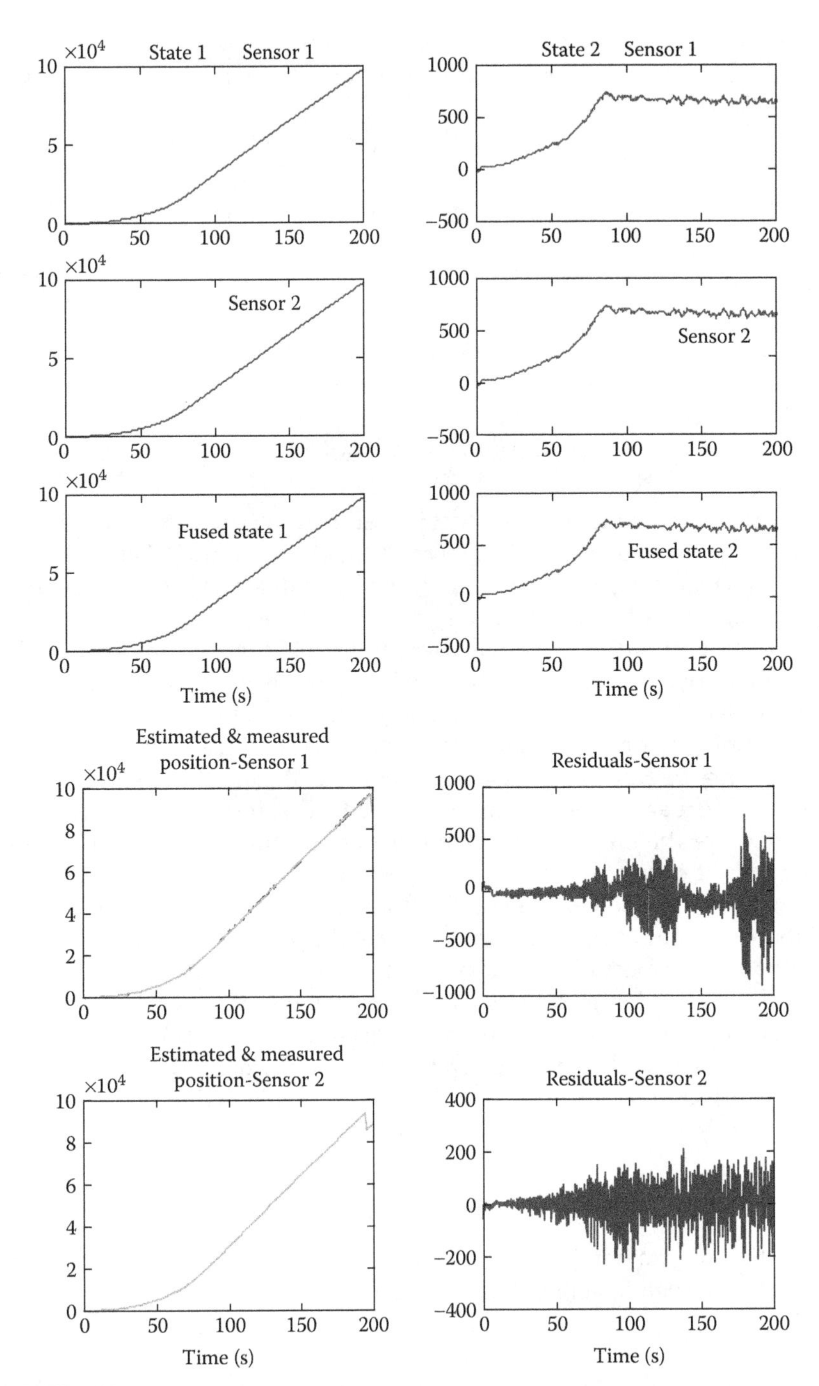

FIGURE 3.20
State fusion with a gain fusion–based algorithm—loss of data in sensor 1.

where the object's state vector has two components: position and velocity. Also, we have w as the zero-mean white Gaussian noise:

$$E\{w(k)\} = 0; \quad \text{Var}\{w(k)\} = Q$$

The measurement at each sensor is given by

$$z^m(k) = Hx(k) + v^m(k) \tag{3.152}$$

with $m = 1, 2$ (number of sensors). The measurement noise is assumed to be zero-mean white Gaussian with the statistics

$$E\{v^m(k)\} = 0; \quad \text{Var}\{v^m(k)\} = R_v^m$$

In KF, the signal generating system is assumed to be a state space driven by a white noise process with known statistical properties. The measurement signals are assumed to be corrupted by white noise processes with known statistical properties. The aim of the filter is to minimize the variance of the terminal state estimation error. The H-I filtering problem differs from KF in two respects [58]: (1) the white noise is replaced by unknown deterministic disturbance of finite energy; and (2) a prespecified positive real number (gamma, a scalar parameter) is defined. The aim of the filter is to ensure that the energy gain (i.e., H-I norm) from the disturbance to the estimation error is less than this number. This number can be called a threshold for the magnitude of the transfer function between estimation error and the input disturbance energies. One important aspect is that the KF evolves from the H-I filter as the threshold tends to infinity. From the robustness point of view, we see that the H-I concept, at least in theory, would yield a robust filtering algorithm.

Two H-I filter–based fusion algorithms are considered when there is specific data loss in either of the two sensors used for tracking a moving object. The sensor locations or stations use individual H-I filter to create two sets of track files. The performance is evaluated in terms of state errors and H-I-based norm using simulated data.

3.9.2 H-Infinity a Posteriori Filter–Based Fusion Algorithm

The estimates are obtained for each sensor ($i = 1, 2$) using H-I *a posteriori* filter as described next [59]. The covariance time propagation is given as

$$P_i(k+1) = FP_i(k)F' + GQG' - FP_i(k)\begin{bmatrix} H_i^t & L_i^t \end{bmatrix} R_i^{-1} \begin{bmatrix} H_i \\ L_i \end{bmatrix} P_i(k)F' \tag{3.153}$$

$$R_i = \begin{bmatrix} I & 0 \\ 0 & -\gamma^2 I \end{bmatrix} + \begin{bmatrix} H_i \\ L_i \end{bmatrix} P_i(k) \begin{bmatrix} H_i^t & L_i^t \end{bmatrix} \quad (3.154)$$

The H-I filter gain is given as

$$K_i = P_i(k+1)H_i^t \left(I + H_i P_i(k+1)H_i^t\right)^{-1} \quad (3.155)$$

The measurement update of states is obtained by

$$\hat{x}_i(k+1) = F\hat{x}_i(k) + K_i\left(y_i(k+1) - H_i F\hat{x}_i(k)\right) \quad (3.156)$$

The conditions for the existence of the H-I filters are given in [59]. The fusion of the estimates from the two sensors then can be obtained by

$$\hat{x}_f(k+1) = \hat{x}_1(k+1) + \hat{P}_1(k+1)\left(\hat{P}_1(k+1) + \hat{P}_2(k+1)\right)^{-1} \times \left(\hat{x}_2(k+1) - \hat{x}_1(k+1)\right) \quad (3.157)$$

$$\hat{P}_f(k+1) = \hat{P}_1(k+1) - \hat{P}_1(k+1)\left(\hat{P}_1(k+1) + \hat{P}_2(k+1)\right)^{-1} \hat{P}_1^t(k+1) \quad (3.158)$$

The fused state vector and the covariance of the fused state utilize the individual estimate state vectors (of each sensor) and covariance matrices.

3.9.3 H-Infinity Global Fusion Algorithm

This filtering algorithm is based on [60,61]. The local filters are given for each sensor ($i = 1, 2, \ldots, m$).

State and ovariance time propagation is given by

$$\tilde{x}_i(k+1) = F\hat{x}_i(k) \quad (3.159)$$

$$\tilde{P}_i(k+1) = F\hat{P}_i(k)F' + GQG' \quad (3.160)$$

The covariance update is given by

$$\hat{P}_i^{-1}(k+1) = \tilde{P}_i^{-1}(k+1) + \begin{bmatrix} H_i^t & L_i^t \end{bmatrix} \begin{bmatrix} I & 0 \\ 0 & -\gamma^2 I \end{bmatrix}^{-1} \begin{bmatrix} H_i \\ L_i \end{bmatrix} \quad (3.161)$$

The local filter gains are given as

$$A_i = I + 1/\gamma^2 \hat{P}_i(k+1)L_i^t L_i; \quad K_i = A_i^{-1}\hat{P}_i(k+1)H_i^t \quad (3.162)$$

The measurement update of local states is given as

$$\hat{x}_i(k+1) = \tilde{x}_i(k+1) + K_i\left(y_i(k+1) - H_i\tilde{x}_i(k+1)\right) \tag{3.163}$$

The time propagation of fusion state and covariance is given as

$$\tilde{x}_f(k+1) = F\hat{x}_f(k) \tag{3.164}$$

$$\tilde{P}_f(k+1) = F\hat{P}_f(k)F' + GQG' \tag{3.165}$$

The measurement update of the fusion states and covariance is

$$\hat{P}_f^{-1}(k+1) = \tilde{P}_f^{-1}(k+1) + \sum_{i=1}^{m}\left(\hat{P}_i^{-1}(k+1) - \tilde{P}_i^{-1}(k+1)\right) + \frac{m-1}{\gamma^2}L^tL \tag{3.166}$$

The global gain is

$$A_f = I + 1/\gamma^2\hat{P}_f(k+1)L^tL \tag{3.167}$$

The global (measurement update) fused state is

$$\hat{x}_f(k+1) = \left[I - A_f^{-1}\hat{P}_f(k+1)H_f^tH_f\right]\tilde{x}_f(k+1) + A_f^{-1}\hat{P}_f(k+1)$$
$$\sum_{i=1}^{m}\left\{\hat{P}_i^{-1}(k+1)A_i\hat{x}_i(k+1) - \left(\hat{P}_i^{-1}(k+1)A_i + H_i^tH_i\right)F\hat{x}_i(k)\right\} \tag{3.168}$$

3.9.4 Numerical Simulation Results

The simulated data are generated in PC MATLAB for a tracking problem [62]. The normalized random noise is added to the state vector and the measurements of each sensor are corrupted with random noise. The sensors could have dissimilar measurement noise variances (e.g., sensor 2 having higher variance than sensor 1). The initial condition for the state vector is $x(0)$ = [200 0.5]. The performance of the fusion filters is also evaluated in terms of H-I norm. This ratio (the H-I norm) should be less than square of gamma, which can be considered an upper bound on the maximum energy gain from the input to the output. The output

TABLE 3.14

Percentage Fit Errors

	Normal	Data Loss in Sensor 1	Data Loss in Sensor 2
HIPOFA-F1	0.443	0.442	0.443
HIPOFA-F2	0.435	0.435	0.427
HIGFA-F1	0.443	0.442	0.443
HIGFA-F2	0.436	0.436	0.427

TABLE 3.15

Percentage State Errors

	Normal		Data Loss in Sensor 1		Data Loss in Sensor 2	
	Position	Velocity	Position	Velocity	Position	Velocity
HIPOFA-F1	0.210	5.56	0.202	5.55	0.210	5.55
HIPOFA-F2	0.210	5.99	0.207	5.99	0.188	5.98
HIPOFA	0.151	5.54	0.146	5.54	0.142	5.53
HIGFA-L1	0.211	5.55	0.203	5.55	0.211	5.55
HIGFA-L2	0.210	5.94	0.207	5.94	0.188	5.92
HIGFA	0.065	6.24	0.066	6.24	0.064	6.24

TABLE 3.16

H-Infinity Norm (Fusion Filter)

	Normal	Data Loss in sensor 1	Data Loss in Sensor 2
HIPOFA	0.0523	0.0523	0.0523
HIGFA	0.0151	0.0155	0.0145

energy of the filter is due to the error in fused state. Tables 3.14 to 3.16 give performance indices for these two filters (with Rv1 = Rv2): under normal condition and with data loss in either sensor for about a few seconds. The theoretical covariance norm of the fused state was found to be lower than that of the individual filters. Figures 3.21 and 3.22 depict the time histories of state errors with the bounds for H-I posteriori fusion algorithm (HIPOFA) and H-I global fusion algorithm (HIGFA) schemes when there is data loss in sensor 1 [62]. One can observe that the two fusion algorithms are fairly robust to the loss of data and that the errors are lower when the fusion filters are used. The important aspect for the

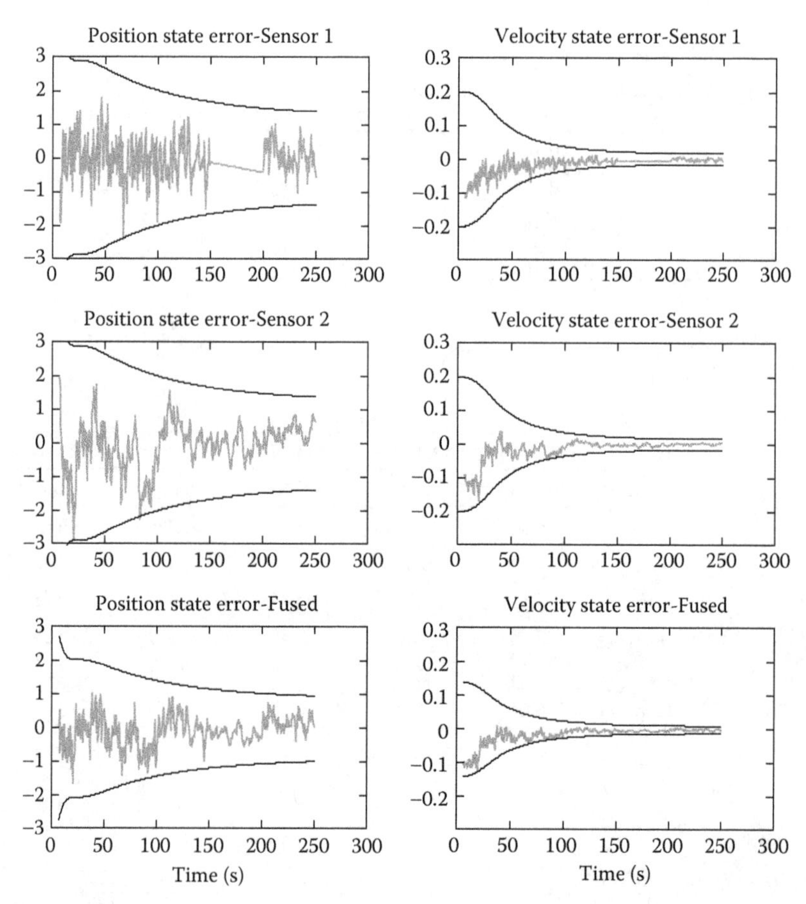

FIGURE 3.21
State errors with bounds for HIPOFA with var($v2$) = 9*var($v1$); data loss in sensor 1.

H-I filters is the adjustment of the scalar parameter gamma to obtain the desirable results. From the numerical results presented in the tables, one can see that very satisfactory accuracy of position and velocity estimation has been obtained using the H-I filter. Also the H-I norms have acceptable values as required by the theory.

3.10 Derivative-Free Kalman Filters for Fusion

In general, extended KF provides a suboptimal solution to a given nonlinear estimation problem. EKF has two major limitations [34–37]: (1) the

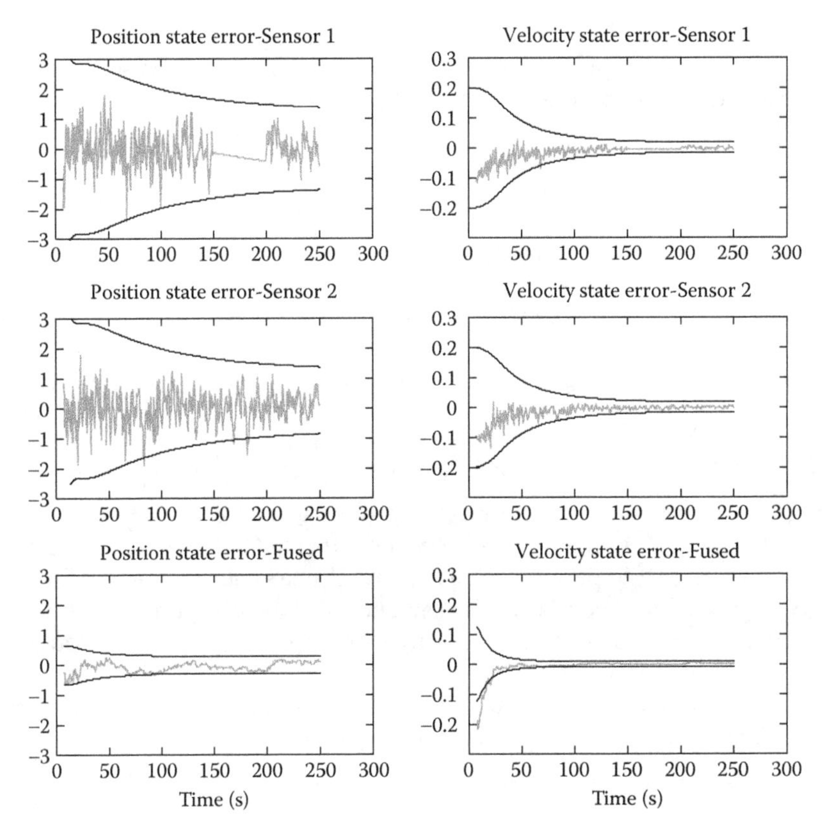

FIGURE 3.22
State errors with bounds for HIGFA with var($v2$) = var($v1$); data loss in sensor 1.

derivations of the Jacobian matrices (in the case of linearization of the nonlinear sensor model) are often nontrivial, leading to some implementation problems; and (2) linearization can lead to highly unstable filters and divergence of the solution trajectory for highly nonlinear systems. In many tracking applications, sensors often provide nonlinear measurements in a polar frame, that is, range, bearing, or azimuth and elevation. State estimation is performed in Cartesian frame. To alleviate these problems, a technique called DFKF has been developed [34,35]. DFKF yields a similar performance when compared to EKF when the assumption of local linearity is not violated. It does not require any linearization of the nonlinear systems or functions, and it uses a deterministic sampling approach to compute the mean and covariance estimates with a set of sample points. These points are called sigma points. Thus, the emphasis is shifted from linearization of nonlinear systems, such as in EKF and many higher-order EKF-type filters, to a sampling approach of pdf. In this section, the concept of DFKF [35] is extended to DF for similar sensors. Let a nonlinear system model in discrete domain be given as

$$x(k+1) = f\left[x(k), u(k), w(k), k\right] \tag{3.169}$$

The sensor-measurement model is given by

$$z_m(k) = h\left[x(k), u(k), k\right] + v(k) \tag{3.170}$$

Here, the variables have the usual meanings.

3.10.1 Derivative-Free Kalman Filters

In EKF, the nonlinear system models are linearized to parameterize the pdf in terms of its mean and covariance. In DFKF, linearization is not required and pdf is parameterized via nonlinear transformation chosen sigma points. These points are chosen deterministically. Consider the time propagation of a random variable x (of dimension L, say $L = 2$) through a nonlinear function $y = f(x)$. First, assume that the mean and covariance of (primarily) sigma points for random variable are given as $\bar{x}$ and P_x, respectively. The sigma points are computed as [34–37]

$$\begin{aligned} \chi_0 &= \bar{x} \\ \chi_i &= \bar{x} + \left(\sqrt{(L+\lambda)P_x}\right)_i \quad i = 1,\ldots,L \\ \chi_i &= \bar{x} - \left(\sqrt{(L+\lambda)P_x}\right)_{i-L} \quad i = L+1,\ldots,2L \end{aligned} \tag{3.171}$$

The associated weights are computed as [34–37]

$$\begin{aligned} W_0^{(m)} &= \frac{\lambda}{L+\lambda} \\ W_0^{(c)} &= \frac{\lambda}{L+\lambda} + (1 - \alpha^2 + \beta) \\ W_i^{(m)} = W_i^{(c)} &= \frac{1}{2(L+\lambda)} \quad i = 1,\ldots,2L \end{aligned} \tag{3.172}$$

To provide unbiased transformation, the weights must satisfy the condition $\sum_{i=1}^{2L} W_i^{(m \text{ or } c)} = 1$. The scaling parameters of DFKF are (1) α determines the spread of sigma points at approximately $\bar{x}$; (2) β incorporates any prior knowledge about distribution of $\bar{x}$; (3) $\lambda = \alpha^2(L+\kappa) - L$; and (4) κ is a secondary tuning parameter. Secondly, these sigma points are propagated through the nonlinear function of the system ($y_i = f(\chi_i)$, where, $i = 0,\ldots,2L$) resulting in transformed sigma points. Third, the mean and covariance of transformed points are formulated as follows [35]:

$$\bar{y} = \sum_{i=0}^{2L} W_i^{(m)} y_i \tag{3.173}$$

$$P_y = \sum_{i=0}^{2L} W_i^{(c)} \{y_i - \bar{y}\}\{y_i - \bar{y}\}^T \tag{3.174}$$

DFKF is a straightforward and simple extension of derivative-free transformation (the three-step process described above) for the recursive estimation problem. The full state of the filter can be constructed by the augmented state vector consisting of the actual system states, process noise states, and measurement noise states. The dimension of augmented state vector would be $n_a = n + n + m = 2n + m$.

3.10.2 Numerical Simulation

First, the performance of the DFKF was evaluated for the 3D trajectory simulation of a vehicle moving with constant acceleration and tracked by a sensor capable of giving vehicle data: range (meter), azimuth (radian), and elevation (radius). In the simulation, the information used [37] are (1) true initial state of the vehicle: $x_t(0/0) = [x \;\; \dot{x} \;\; \ddot{x} \;\; y \;\; \dot{y} \;\; \ddot{y} \;\; z \;\; \dot{z} \;\; \ddot{z}] = [10 \;\; 10 \;\; 0.1 \;\; 10 \;\; 5 \;\; 0.1 \;\; 1000 \;\; 0 \;\; 0]$; (2) sampling interval $T = 0.1$ s; (3) total flight time, $T_F = 100$ s; (4) process noise variance $Q = 0.01$; (5) system model (F):

$$F1 = \begin{bmatrix} 1 & T & T^2/2 \\ 0 & 1 & T \\ 0 & 0 & 1 \end{bmatrix}; \quad F = \begin{bmatrix} F1 & 0 & 0 \\ 0 & F1 & 0 \\ 0 & 0 & F1 \end{bmatrix};$$

and (6) process noise matrix (G):

$$G1 = \begin{bmatrix} T^3/6 & 0 & 0 \\ 0 & T^2/2 & 0 \\ 0 & 0 & T \end{bmatrix}; \quad G = \begin{bmatrix} G1 & 0 & 0 \\ 0 & G1 & 0 \\ 0 & 0 & G1 \end{bmatrix}$$

The polar measurements are obtained using the model

$$\left.\begin{aligned} z_m(k) &= \left[\underbrace{r(k)}\ \underbrace{\theta(k)}\ \underbrace{\varphi(k)}\right] \\ r(k) &= \sqrt{x(k)^2 + y(k)^2 + z(k)^2} + n_r(k) \\ \theta(k) &= \tan^{-1}\left(y(k)/x(k)\right) + n_\theta(k) \\ \varphi(k) &= \tan^{-1}\left(z(k)/\sqrt{x(k)^2 + y(k)^2}\right) + n_\varphi(k) \end{aligned}\right\} \tag{3.175}$$

The variables n_r, n_θ, and n_φ represent random noise sequences. The standard deviations of measurement noise for range, azimuth, and elevation are computed based on prespecified SNR (=10). The measurement noise

covariance matrix is specified as $R = \begin{bmatrix} \sigma_r^2 & 0 & 0 \\ 0 & \sigma_\theta^2 & 0 \\ 0 & 0 & \sigma_\varphi^2 \end{bmatrix}$.

To check the basic functioning of the DFKF, the state estimation was carried out using UDEKF and DFKF algorithms and the results were found to be very satisfactory [37]. This was done to establish the working of the DFKF (program code) for the same example as that was used for UDEKF and hence the results are not shown here.

After state estimation, DFKF was used for DF exercises. Consider a vehicle reentry problem [35]. We assume that while the vehicle is entering the atmosphere at high altitude and speed, it is tracked by two ground-based sensors placed nearby that have different inherent accuracies. The measurements are in terms of range and bearing. In the initial phase, the vehicle has an almost ballistic trajectory, but as the density of the atmosphere increases, the drag is more effective and the vehicle rapidly decelerates until its motion is almost vertical [35]. The state-space model of the vehicle dynamics is given as [35,37]

$$\left.\begin{aligned} \dot{x}_1(k) &= x_3(k) \\ \dot{x}_2(k) &= x_4(k) \\ \dot{x}_3(k) &= D(k)x_3(k) + G(k)x_1(k) + w_1(k) \\ \dot{x}_4(k) &= D(k)x_4(k) + G(k)x_2(k) + w_2(k) \\ \dot{x}_5(k) &= w_3(k) \end{aligned}\right\} \tag{3.176}$$

Here, (1) x_1 and x_2 are target positions; (2) x_3 and x_4 are velocities; (3) x_5 is a parameter related to some aerodynamic properties; (4) D is the drag-related term; (5) G is the gravity-related term; and (6) w_1, w_2, and w_3 are uncorrelated white Gaussian process noises with zero mean and standard deviations of $\sigma_{w_1} = 0.0049$, $\sigma_{w_1} = 0.0049$ and $\sigma_{w_1} = 4.9e-8$, respectively. The drag and gravitational terms are computed using the following equations [35,37]:

$$\left.\begin{aligned} D(k) &= -\beta(k)\exp\left\{\frac{r_0 - r(k)}{H_0}\right\}V(k) \\ G(k) &= -\frac{Gm_0}{r^3(k)} \\ \beta(k) &= -\beta_0 \exp\left(x_5(k)\right) \\ r(k) &= \sqrt{x_1^2(k) + x_2^2(k)} \\ V(k) &= \sqrt{x_3^2(k) + x_4^2(k)} \end{aligned}\right\} \tag{3.177}$$

Here, $\beta_0 = -0.59783$, $H_0 = 13.406$, $Gm_0 = 3.9860 \times 10^5$, and $r_0 = 6374$ are the parameters that reflect certain environmental and the vehicle characteristics [35]. The initial state of vehicle is $[6500.4, 349.14, -1.8093, -6.7967, 0.6932]$, and the data are generated for $N = 1450$ scans. The vehicle is tracked by two sensors in proximity (at $x_r = 6375$ km, $y_r = 0$ km), and the data rate is five samples. The sensor model equations are

$$\begin{aligned} r_i(k) &= \sqrt{(x_1(k) - x_r)^2 + \left(x_2(k) - y_r\right)^2} + v_{ir}(k) \\ \theta_i(k) &= \tan^{-}\left(\frac{x_2(k) - y_r}{x_1(k) - x_r}\right) + v_{i\theta}(k) \end{aligned} \tag{3.178}$$

Here, r_i and θ_i are the range and bearing of ith sensor, and v_{ir} and $v_{i\theta}$ are the associated white Gaussian measurement noise processes. Here, it is assumed that sensor 1 gives good angle and bearing information but has noisy range measurements, and vice versa for the second sensor (although this may not be true in practice, this is assumed here for the sake of performance evaluation of the DFKF or fusion algorithm) with the standard deviations of range and bearing noises as for sensor 1: $\sigma_{1r} = 1$ km, $\sigma_{1\theta} = 0.05°$ and for sensor 2: $\sigma_{2r} = 0.22$ km, $\sigma_{2\theta} = 1°$. To develop a fusion scheme, the assumptions or the changes made for the DFKF algorithm are (1) the sensors are of similar type and have the same data type or format; and (2) the measurements are synchronized with time.

3.10.2.1 *Initialization of the Data Fusion-Derivative Free Kalman Filter Algorithm*

The augmented state and its error covariance are given as [36,37]

$$\left.\begin{aligned}&\hat{x}(0/0) = E[x(0/0)]\\&\hat{P}(0/0) = E\left[\left(x(0/0)-\hat{x}(0/0)\right)\left(x(0/0)-\hat{x}(0/0)\right)^T\right]\end{aligned}\right\} \tag{3.179}$$

$$\left.\begin{aligned}&\hat{x}^a(0/0) = E\left[x^a(0/0)\right] = \begin{bmatrix}\hat{x}^T(0/0) & \underbrace{0,\ldots,0}_{n-\text{dim } w} & \underbrace{0,\ldots,0}_{\substack{m-\text{dim } v_1\\(\text{sensors } 1)}} & \underbrace{0,\ldots,0}_{\substack{m-\text{dim } v_2\\(\text{sensors } 2)}} & ,\ldots, & \underbrace{0,\ldots,0}_{\substack{m-\text{dim } v_{NS}\\(\text{sensors NS})}}\end{bmatrix}^T\\&\hat{P}^a(0/0) = E\left[\left(x^a(0/0)-\hat{x}^a(0/0)\right)\left(x^a(0/0)-\hat{x}^a(0/0)\right)^T\right]\\&= \begin{bmatrix}\hat{P}(0/0) & 0 & 0 & 0 & 0 & 0\\0 & Q & 0 & 0 & 0 & 0\\0 & 0 & R_1 & 0 & 0 & 0\\0 & 0 & 0 & R_2 & 0 & 0\\0 & 0 & 0 & 0 & ,\ldots, & 0\\0 & 0 & 0 & 0 & 0 & R_{NS}\end{bmatrix}_{2n+NS*m \text{ by } 2n+NS*m}\end{aligned}\right\} \tag{3.180}$$

Here, NS is the total number of sensors and the dimension of the augmented state vector is now $n_a = n + n + NS*m = 2n + NS*m$.

3.10.2.2 *Computation of the Sigma Points*

The required sigma points are computed as follows [35,37]:

$$\left.\begin{aligned}&\chi_0^a(k/k) = \hat{x}^a(k/k)\\&\chi_i^a(k/k) = \hat{x}^a(k/k) + \left(\sqrt{(n_a+\lambda)\hat{P}^a(k/k)}\right)_i \qquad i = 1,\ldots,n_a\\&\chi_i^a(k/k) = \hat{x}^a(k/k) - \left(\sqrt{(n_a+\lambda)\hat{P}^a(k/k)}\right)_{i-n_a} \qquad i = n_a+1,\ldots,2n_a\end{aligned}\right\} \tag{3.181}$$

Here, $\chi^a = \left[\underbrace{\chi}_{\text{state}} \quad \underbrace{\chi^w} \quad \underbrace{\chi^{v_1}}\underbrace{\chi^{v_2}},\ldots,\underbrace{\chi^{v_{NS}}}\right]$. The augmented states are those of the measurement data noise processes.

3.10.2.3 State and Covariance Propagation

The state and covariance equations are given as [35,37]

$$\left.\begin{aligned}
&\chi(k+1/k) = f\left(\chi(k/k), u(k), \chi^w(k/k), k\right)\\
&\tilde{x}(k+1/k) = \sum_{i=0}^{2n_a} W_i^{(m)}\chi_i(k+1/k)\\
&\tilde{P}(k+1/k) = \sum_{i=0}^{2n_a} W_i^{(c)}\left[\chi_i(k+1/k) - \tilde{x}(k+1/k)\right]\left[\chi_i(k+1/k) - \tilde{x}(k+1/k)\right]^T
\end{aligned}\right\} \tag{3.182}$$

$$\left.\begin{aligned}
W_0^{(m)} &= \frac{\lambda}{n_a + \lambda}\\
W_0^{(c)} &= \frac{\lambda}{n_a + \lambda} + (1 - \alpha^2 + \beta)\\
W_i^{(m)} = W_i^{(c)} &= \frac{1}{2(n_a + \lambda)} \qquad i = 1,\ldots,2n_a
\end{aligned}\right\} \tag{3.183}$$

3.10.2.4 State and Covariance Update

The measurement update equations for the state and covariance are given as [35,37]

$$\left.\begin{aligned}
&y^j(k+1/k) = h\left(\chi(k/k), u(k), k\right) + \chi^{v_j}(k/k)\\
&\tilde{z}^j(k+1/k) = \sum_{i=0}^{2n_a} W_i^{(m)} y_i^j(k+1/k)
\end{aligned}\right\} \tag{3.184}$$

Here, $j = 1,\ldots,\text{NS}$ and

$$\begin{aligned}
y(k+1/k) &= \left[y^1(k+1/k) \quad y^2(k+1/k) \quad ,\ldots, \quad y^{\text{NS}}(k+1/k)\right]^T\\
\tilde{z}(k+1/k) &= \left[\tilde{z}^1(k+1/k) \quad \tilde{z}^2(k+1/k) \quad ,\ldots, \quad \tilde{z}^{\text{NS}}(k+1/k)\right]^T
\end{aligned} \tag{3.185}$$

$$\left.\begin{aligned} S &= \sum_{i=0}^{2n_a} W_i^{(c)} \left[y_i(k+1/k) - \tilde{z}(k+1/k) \right] \left[y_i(k+1/k) - \tilde{z}(k+1/k) \right]^T \\ P_{xy} &= \sum_{i=0}^{2n_a} W_i^{(c)} \left[\chi_i(k+1/k) - \tilde{x}(k+1/k) \right] \left[y_i(k+1/k) - \tilde{z}(k+1/k) \right]^T \\ K &= P_{xy} S^{-1} \text{ (filter gain)} \end{aligned}\right\} \tag{3.186}$$

$$\left.\begin{aligned} \hat{x}(k+1/k+1) &= \tilde{x}(k+1/k) + K\left(z_m(k+1) - \tilde{z}(k+1/k)\right) \\ \hat{P}(k+1/k+1) &= \tilde{P}(k+1/k) - KSK^T \end{aligned}\right\} \tag{3.187}$$

The variable Z_m is restructured as

$$z_m(k+1) = \left[z_m^1(k+1) \quad z_m^2(k+1) \quad ,...., \quad z_m^{NS}(k+1) \right]^T \tag{3.188}$$

Here, z_m^1, z_m^2, …, are the measurements from sensor 1, sensor 2, …, and so on at $k+1$ scan. The results for 25 Monte Carlo simulations are generated and performances of DF-DFKF [36] and two individual-DFKFs (sensor 1 and sensor 2) are evaluated as shown in Figure 3.23. It is seen that the

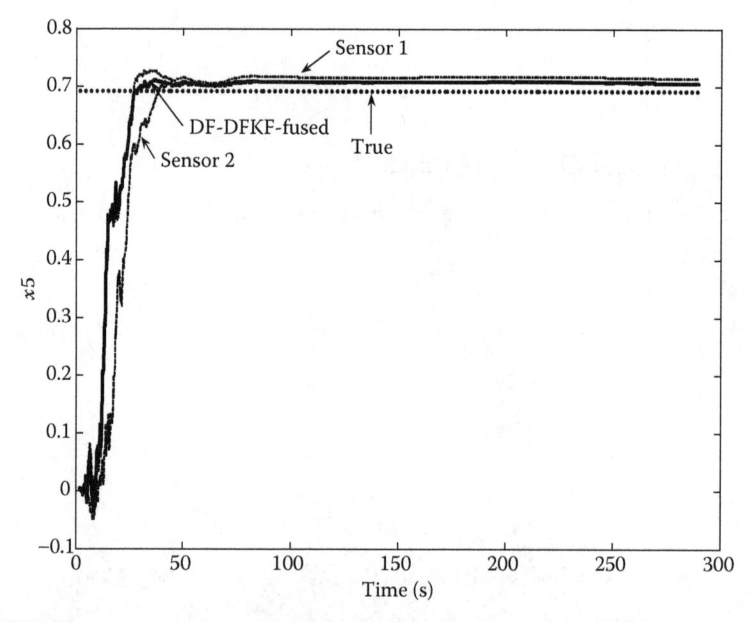

FIGURE 3.23
Individual and fused trajectories with the derivative-free Kalman filter data-fusion method.

fused state as compared to the estimated state from other two filters is closer to the true state [37].

3.11 Missile Seeker Estimator

In interceptors active radar seeker is used to measure: (1) relative range, (2) relative range rate, (3) line-of-sight (LOS) angles, and (4) rates between interceptor and evader. The measurements are contaminated by a high degree of noise due to glint, radar cross section (RCS) fluctuation, and thermal noise. The true LOS rates have to be estimated for proportional navigation (PN) guidance of the interceptor [63,64]. Advanced proportional navigation (APN) guidance control systems would also use target acceleration. An estimator or seeker filter is required, which processes the seeker measurements recursively to obtain the signals required for guidance of the interceptor towards the evader or target. The design of the estimator is complex because the cumulative effects of the noise are non-Gaussian and time correlated. There would also be a periodic loss of seeker measurements due to target eclipsing effects. Since the seeker measurements are available in the inner Gimbal frame, one would need an EKF for the seeker estimator. To handle non-Gaussian noise, an augmented EKF (AEKF) is used. In interceptor-evading target engagement scenarios or games, the evader would execute maneuvers—perhaps unpredictably—to avoid the interceptor. To track such a maneuvering target, an IMM model is generally used. The IMM is an adaptive estimator that is based on the assumption of the use of a finite number of models [65]. In this section, an IMM based on the soft switching between a set of predefined (evader target) models such as constant velocity, constant acceleration, and constant jerk [66] is described. At each sampling time, the mode probability is computed for each model using innovation-vector and covariance matrix. Each of the mode matched filter is based on the AEKF. The new filter is expected to give better performance in an engagement scenario [67,68]. The performance evaluation of an IMM-AEKF in a closed-loop, using MATLAB simulated data of a typical interceptor–evader engagement scenario, is discussed.

3.11.1 Interacting Multiple Model–Augmented Extended Kalman Filter Algorithm

The state vector of respective mode-matched filter is augmented with additional states for handling glint noise and RCS fluctuations.

3.11.1.1 State Model

The state vector consists of 18 states (with six states related to glint noise and RCS fluctuations):

$$\begin{bmatrix} \Delta x & \Delta V_x & a_{tx} & j_{tx} & \Delta y & \Delta V_y & a_{ty} & j_{ty} \\ \Delta z & \Delta V_z & a_{tz} & j_{tz} & \text{RCS and Glint states} \end{bmatrix}$$

The mathematical models for glint and RCS are given next [63,67,68]. State model 1, constant velocity (CV) model:

$$\begin{aligned} &\Delta\dot{x} = \Delta V_x; \quad \Delta\dot{V}_x = 0; \quad \dot{a}_{tx} = 0; \quad \dot{j}_{tx} = 0 \\ &\Delta\dot{y} = \Delta V_y; \quad \Delta\dot{V}_y = 0; \quad \dot{a}_{ty} = 0; \quad \dot{j}_{ty} = 0 \\ &\Delta\dot{z} = \Delta V_z; \quad \Delta\dot{V}_z = 0; \quad \dot{a}_{tz} = 0; \quad \dot{j}_{jz} = 0 \end{aligned} \tag{3.189}$$

State model 2, constant acceleration (CA) model:

$$\begin{aligned} &\Delta\dot{x} = \Delta V_x; \quad \Delta\dot{V}_x = a_{tx} - a_{mx}; \quad \dot{a}_{tx} = -\left(\frac{a_{tx}}{\tau_x}\right); \quad \dot{j}_{tx} = 0; \quad \Delta\dot{y} = \Delta V_y; \\ &\Delta\dot{V}_y = a_{ty} - a_{my}; \quad \dot{a}_{ty} = -\left(\frac{a_{ty}}{\tau_y}\right); \quad \dot{j}_{ty} = 0; \\ &\Delta\dot{z} = \Delta V_z; \quad \Delta\dot{V}_z = a_{tz} - a_{mz}; \quad \dot{a}_{tz} = -\left(\frac{a_{tz}}{\tau_z}\right); \quad \dot{j}_{jz} = 0 \end{aligned} \tag{3.190}$$

State model 3, constant jerk (CJ) model:

$$\begin{aligned} &\Delta\dot{x} = \Delta V_x; \quad \Delta\dot{V}_x = a_{tx} - a_{mx}; \quad \dot{a}_{tx} = j_{tx}; \quad \dot{j}_{tx} = -\left(\frac{j_{tx}}{\tau_x}\right) \\ &\Delta\dot{y} = \Delta V_y; \quad \Delta\dot{V}_y = a_{ty} - a_{my}; \quad \dot{a}_{ty} = j_{ty}; \quad \dot{j}_{ty} = -\left(\frac{j_{ty}}{\tau_y}\right) \\ &\Delta\dot{z} = \Delta V_z; \quad \Delta\dot{V}_z = a_{tz} - a_{mz}; \quad \dot{a}_{tz} = j_{tz}; \quad \dot{j}_{jz} = -\left(\frac{j_{tz}}{\tau_z}\right) \end{aligned} \tag{3.191}$$

In Equations 3.189 to 3.191, Δx, Δy, and Δz are the relative positions and ΔV_x, ΔV_y, and ΔV_z are the relative velocities of target with respect to missile, a_{tx}, a_{ty}, and a_{tz} are the target accelerations, J_{tx}, J_{ty}, and J_{tz} are the target jerks, a_{mx}, a_{my}, and a_{mz} are the missile accelerations, and τ_x, τ_y, and τ_z are the correlation time constants.

3.11.1.2 Measurement Model

The measurement vector is given as $[\rho \quad \dot{\rho} \quad \varphi_y \quad \varphi_z \quad \dot{\varphi}_y \quad \dot{\varphi}_z]$ during the noneclipsing period and $[\rho \quad \dot{\rho} \quad \varphi_y \quad \varphi_z]$ during the eclipsing period. Here, ρ is range-to-go, $\dot{\rho}$ is range rate, φ_y and φ_z are Gimbal angles in yaw and pitch planes, respectively, and $\dot{\varphi}_y$ and $\dot{\varphi}_z$ are the respective LOS rates in inner Gimbal frame. The relative position and velocity states of the target, with respect to the missile in the inertial frame, are transformed to the LOS frame using the following formulae [67,68]:

$$\begin{aligned}
&\rho = \sqrt{\Delta x^2 + \Delta y^2 + \Delta z^2}; \qquad && \dot{\rho} = \frac{\Delta x \Delta \dot{x} + \Delta y \Delta \dot{y} + \Delta z \Delta \dot{z}}{\rho} \\
&\lambda_e = \tan^{-1}\left(\frac{\Delta z}{\sqrt{\Delta x^2 + \Delta y^2}}\right); \qquad && \dot{\lambda}_e = \frac{\Delta \dot{z}(\Delta x^2 + \Delta y^2) - \Delta z(\Delta x \Delta \dot{x} + \Delta y \Delta \dot{y})}{\rho^2 \sqrt{\Delta x^2 + \Delta y^2}} \\
&\lambda_a = \tan^{-1}\left(\frac{\Delta y}{\Delta x}\right); \qquad && \dot{\lambda}_a = \frac{(\Delta x \Delta \dot{y} - \Delta y \Delta \dot{x})}{\sqrt{\Delta x^2 + \Delta y^2}}
\end{aligned} \tag{3.192}$$

The measurement model during the noneclipsing period is given as

$$\begin{aligned}
\begin{bmatrix} \rho \\ \dot{\rho} \end{bmatrix}_m &= \begin{bmatrix} \rho \\ \dot{\rho} \end{bmatrix} \\
\begin{bmatrix} \varphi_y \\ \varphi_z \end{bmatrix}_m &= \begin{bmatrix} \varphi_y \\ \varphi_z \end{bmatrix} \\
\begin{bmatrix} 0 \\ \dot{\varphi}_y \\ \dot{\varphi}_z \end{bmatrix}_m &= C_f^g C_b^f C_i^b C_l^i \begin{bmatrix} -\dot{\lambda}_a \sin\lambda_e \\ \dot{\lambda}_e \\ \dot{\lambda}_a \cos\lambda_e \end{bmatrix}
\end{aligned} \tag{3.193}$$

The measurement model during the eclipsing period is given as

$$\begin{aligned}
\begin{bmatrix} \rho \\ \dot{\rho} \end{bmatrix}_m &= \begin{bmatrix} \rho \\ \dot{\rho} \end{bmatrix} \\
\begin{bmatrix} \varphi_y \\ \varphi_z \end{bmatrix}_m &= \begin{bmatrix} \varphi_y \\ \varphi_z \end{bmatrix}
\end{aligned} \tag{3.194}$$

Here,

$$\begin{aligned}\varphi_y &= \tan^{-1}\left(\frac{m}{l}\right)\\ \varphi_z &= \tan^{-1}\left(\frac{n}{\sqrt{l^2+m^2}}\right)\end{aligned} \quad \text{and} \quad \begin{bmatrix} l \\ m \\ n \end{bmatrix} = C_b^f C_i^b C_l^i \begin{bmatrix} 1 \\ 0 \\ 0 \end{bmatrix} \tag{3.195}$$

The direction cosine matrix (DCM) for LOS-to-inertial frame transformation is given as

$$C_l^i = \begin{bmatrix} \cos\lambda_e \cos\lambda_a & -\sin\lambda_a & -\sin\lambda_e \cos\lambda_a \\ \cos\lambda_e \sin\lambda_a & \cos\lambda_a & -\sin\lambda_e \sin\lambda_a \\ \sin\lambda_e & 0 & \cos\lambda_e \end{bmatrix} \tag{3.196}$$

The DCM for inertial-to-body frame transformation is given as

$$C_i^b = \begin{bmatrix} q_4^2+q_1^2-q_2^2-q_3^2 & 2(q_1q_2+q_3q_4) & 2(q_1q_3-q_2q_4) \\ 2(q_1q_2-q_3q_4) & q_4^2-q_1^2+q_2^2-q_3^2 & 2(q_2q_3+q_1q_4) \\ 2(q_1q_3+q_2q_4) & 2(q_2q_3-q_1q_4) & q_4^2-q_1^2-q_2^2+q_3^2 \end{bmatrix} \tag{3.197}$$

where, q_1, q_2, q_3 and q_4 are the attitude quaternion of the missile coordinate system.

The DCM for the body-to-fin frame transformation is given as

$$C_b^f = \begin{bmatrix} 1 & 0 & 0 \\ 0 & \frac{1}{\sqrt{2}} & \frac{1}{\sqrt{2}} \\ 0 & -\frac{1}{\sqrt{2}} & \frac{1}{\sqrt{2}} \end{bmatrix} \tag{3.198}$$

The DCM for the fin-to-inner Gimbal frame transformation is given as

$$C_f^g = \begin{bmatrix} \cos\varphi_z \cos\varphi_y & \cos\varphi_z \sin\varphi_y & \sin\varphi_z \\ -\sin\varphi_y & \cos\varphi_z & 0 \\ -\sin\varphi_z \cos\varphi_y & -\sin\varphi_z \sin\varphi_y & \cos\varphi_z \end{bmatrix} \tag{3.199}$$

3.11.2 Interceptor–Evader Engagement Simulation

A closed-loop simulation is carried out for a typical interceptor–evader terminal phase scenario [68]. A typical 6-DOF missile dynamic is simulated and relative measurements containing target information are generated in inner-Gimbal frame in terms of range, range rate, two Gimbal angles, and two LOS rates in pitch- and yaw-planes. For the terminal guidance, measurement noise is added to true $(r, \dot{r}, \varphi_g, \gamma_g, \dot{\varphi}_g, \dot{\gamma}_g)$ to obtain the noisy measurements. Gaussian noise is added to the range and range rate measurements, random colored noise with bore sight error is added to Gimbal angle measurements and thermal Gaussian, correlated glint and RCS fluctuation noises are added to the LOS rate measurements. Data loss in LOS rates due to pulse repetition frequency and closing velocity is also simulated. The data is sampled at 0.01 seconds. The LOS rates are also affected by body rates. The measured data are then processed by the IMM-AEKF algorithm. The guidance commands are generated and fed back to the missile autopilot, which in turn provides sufficient fin deflections to steer the missile towards the evading target.

3.11.2.1 Evader Data Simulation

To simulate realistic missile-target engagement, the aspects considered are (1) in the presence of adversary, the target generally executes a turn and accelerates at short range to go (≈2 km); (2) target velocity to be maintained at 300 to 400 km/h; (3) target turns (velocity vector) at the rate of 20–25°/s; and (4) target under goes maximum roll rate (≈270°/s) to generate glint effect. The evader performing evasive maneuver is simulated within the permissible *g* limit of a typical fighter aircraft at different altitudes. The target speeds are chosen to achieve the desired turn rates in the range of 20°–25° per second. Different data sets have been generated by allowing the target to maneuver continuously, however only a particular result is presented hear. More results can be found in [68].

3.11.3 Performance Evaluation of Interacting Multiple Model–Augmented Extended Kalman Filter

The performance of the algorithm is evaluated in a closed-loop simulation of missile-target engagement. The filter initialization parameters are given as [68]

1. Initial state vector:

$$\Delta\hat{x} = \Delta x + 100; \quad \Delta\hat{\dot{x}} = \Delta\dot{x} + 20; \quad \hat{a}_{t_x} = 0; \quad \hat{j}_{t_x} = 0$$
$$\Delta\hat{y} = \Delta y - 50; \quad \Delta\hat{\dot{y}} = \Delta\dot{y} - 5; \quad \hat{a}_{t_y} = 0; \quad \hat{j}_{t_y} = 0$$
$$\Delta\hat{z} = \Delta z + 50; \quad \Delta\hat{\dot{z}} = \Delta\dot{z} - 5; \quad \hat{a}_{t_z} = 0; \quad \hat{j}_{t_z} = 0$$

2. Initial state error covariance: $\hat{P}(0/0) = 10000 * \text{eye}\,(ns, ns)$
3. Initial process noise covariance for CV model:

   Q.v = diag
   [0.0 0.005555 0.05 0.005555 0.0 0.005555
   0.05 0.005555 0.0 0.005555 0.05 0.005555
   0.001 0.001 0.001 0.001 0.001 0.001]
   for CA model: Q.a = 100*Q.v
   for CJ model: Q.j = 10*Q.a

4. Measurement noise covariance:

   $$R = \text{diag}[2.5e3 \quad 1e2 \quad 7.6e-5 \quad 7.6e-5 \quad 2.467e-2 \quad 2.467e-2]$$

5. Initial mode probability: $\mu = [0.3 \;\; 0.3 \;\; 0.4]$
6. Mode transition probability:

$$p = \begin{bmatrix} 0.800 & 0.100 & 0.100 \\ 0.0009 & 0.999 & 0.0001 \\ 0.0009 & 0.0001 & 0.999 \end{bmatrix}$$

A 3-DOF model–based AEKF is integrated with 6-DOF missile-simulation model in a closed-loop simulation [68]. Innovation smoothing with a low-pass filter with a cutoff frequency of 3 Hz is used in the filter. For a particular case where the maneuver is initiated at $R_{\text{to-go}} = 10$ km at an altitude of 0.5 km (other parameters being: target speed 135 m/s maneuver g's = 6 in pitch and –6 in yaw, turn rate 24.6 degrees per second) we have the following closed-loop performance result: the miss distance is 5.44 meters and the time to intercept is 15.41 seconds for the IMM-AEKF (with glint states). Figure 3.24 shows the comparison of estimated, true, and noisy measurement signals along with the estimation error and Figure 3.25 shows the comparison of typical estimated state with true values [68]. Figure 3.26 shows the mode probabilities for the three models for engagement at an altitude of 0.5 kilometers and with target maneuvers initiated at $R_{\text{to-go}} = 10$ kilometers. Comparison of noise attenuation factors (NAF) in the measurements is shown in Figure 3.27. The NAF computed using a sliding window of length 10 is, by and large, well within the stipulated limit of 0.1 [68]. Some points out of these limits, especially in the rate measurements, are perhaps due to the eclipsing effects in the measurements.

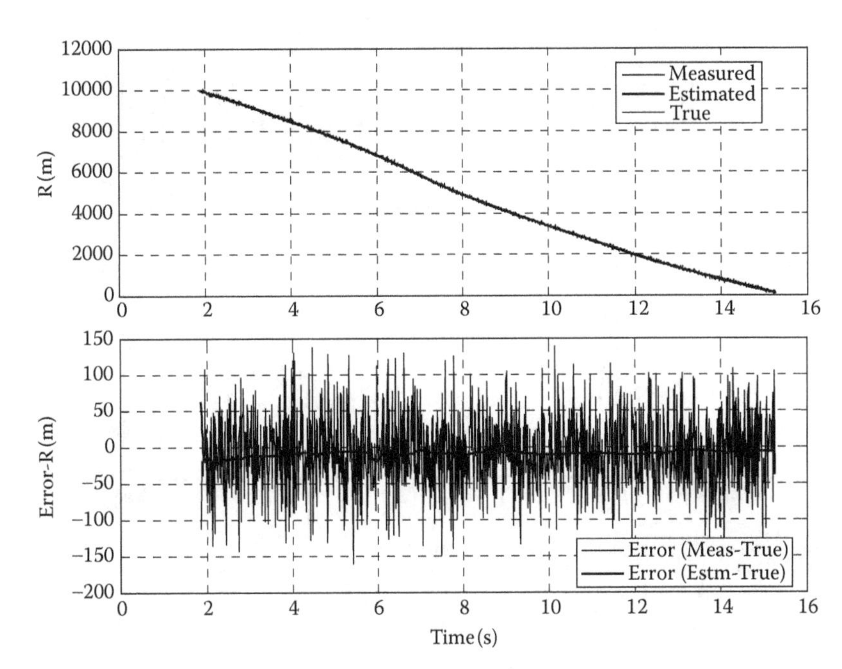

FIGURE 3.24
Measurements $(r_m, \hat{r}, r)$ and estimate error for a particular case.

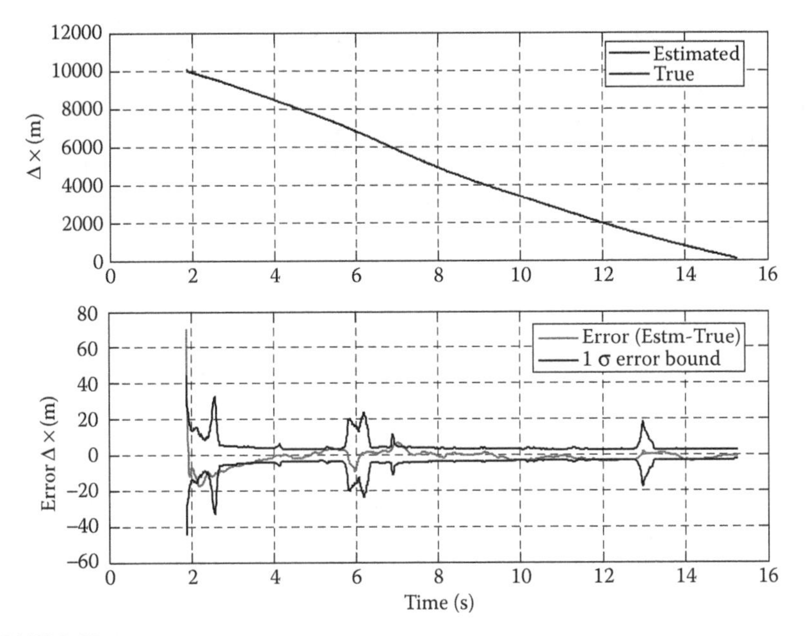

FIGURE 3.25
Estimated states and state error with bounds.

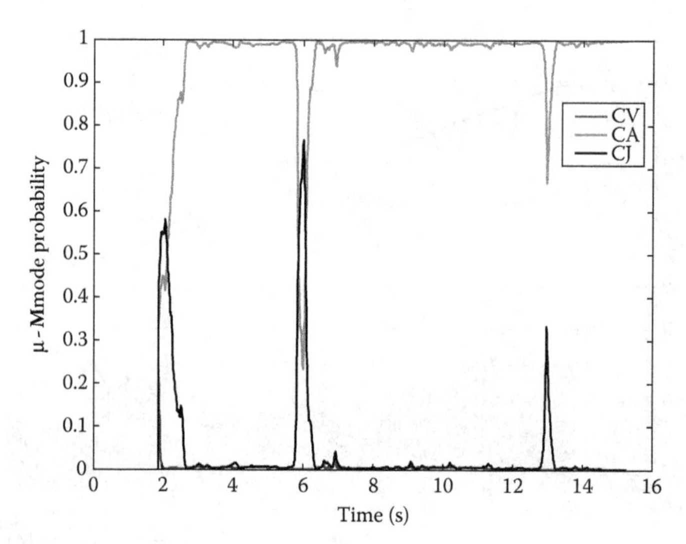

FIGURE 3.26
Mode probability for three models.

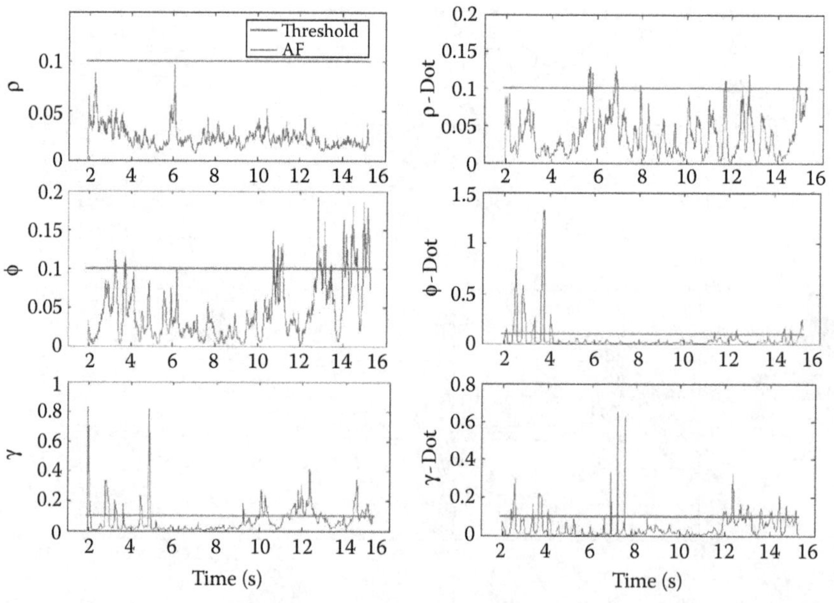

FIGURE 3.27
Noise attenuation factors for a particular case.

3.12 Illustrative Examples

EXAMPLE 3.3

Generate simulated (noisy) data using the state and measurement models of Equations 3.6 and 3.7 for two sensors in MATLAB. Use the measured data in a KF and perform state estimation using Equations 3.8 to 3.12 and state-vector fusion using Equations 3.13 and 3.14. Plot the trajectory match, norm of relevant covariance matrices for filter 1 (sensor 1), filter 2 (sensor 2), and the fused covariance, and position and velocity (state) errors within their theoretical bounds.

SOLUTION 3.3

The data using the prescribed models are generated and Gaussian process and measurement noises with zero mean and unity standard deviation (initially) are added to these data. The data are processed in KF and state-vector fusion is performed. The MATLAB .m (dot.m) file for generating these data and the fusion results with KF are given as *Ch3ExKFGF*. Use ftype = 1 (for KFA) and mdata = 0 (on prompt while running the program) to generate the results for KF with no measurement-data loss. The plots of estimated and measured position and residuals for sensors are given in Figure 3.28. The two sensors are differentiated by adding more measurement noise in sensor 2 (see the .m file) as can be seen from Figure 3.28. The norms of the covariance matrices are plotted in

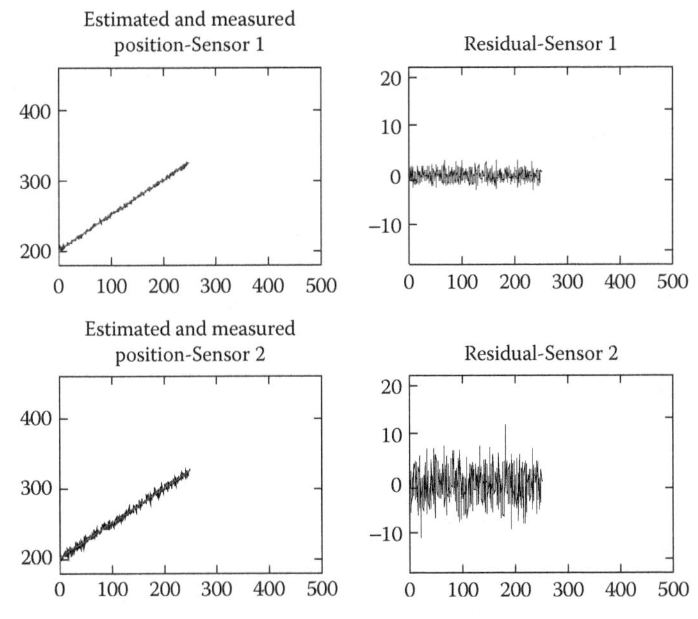

FIGURE 3.28
State estimation results using Kalman filter (see Example 3.3).

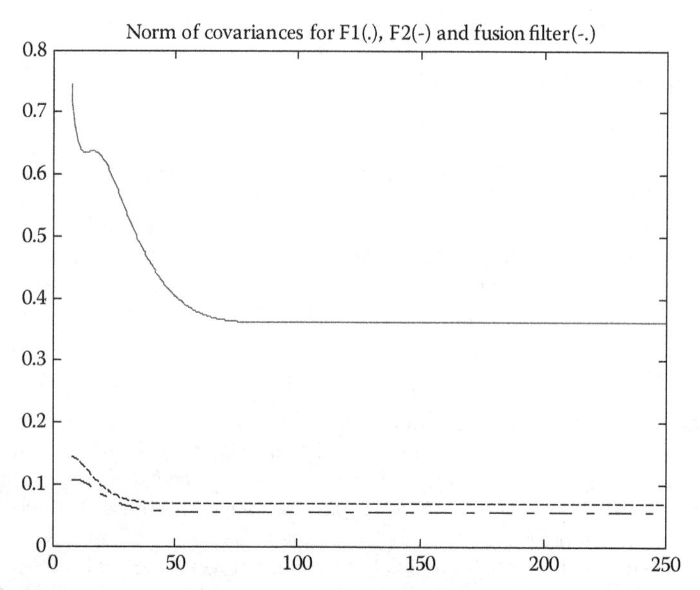

FIGURE 3.29
Norm of covariance matrices (see Example 3.3).

Figure 3.29. One can clearly see that the norm of the fused covariance is lower than both the norms. Since sensor 2 is noisier (by factor 3; see the .m file) than sensor 1, its covariance norm is higher than that of sensor 1. This example demonstrates that the fusion gives better predictive accuracy (less uncertainty). Figure 3.30 depicts the state errors within their theoretical bounds for both the sensors and the fused states. Since the state errors are within the bounds, the KF has performed well and the results correspond to those in Figure 3.29. This example illustrates the fusion concept based on KF and the state-vector fusion for two sensors with noisy measurement data.

EXAMPLE 3.4

Generate simulated (noisy) data using the state and measurement models of Equations 3.6 and 3.7 for two sensors in MATLAB. Use the measured data in a data-sharing or gain-fusion (GF) algorithm and perform state estimation-cum-DF using Equations 3.142 to 3.149. Plot the norms of relevant covariance matrices and position and velocity (state) errors within their theoretical bounds.

Solution 3.4

The data using the prescribed models are generated and Gaussian process and measurement noises with zero mean and unity standard deviation (initially) are added to these data as in Example 3.3. The data are then processed using the GF algorithm. The MATLAB .m (dot.m) file for generating these data and the fusion results is given as *Ch3ExKFGF*. Use ftype = 2 (for GFA) and mdata = 0 to generate the results for the GF algorithm with no measurement-data loss. The two sensors are differentiated by adding more measurement noise in sensor 2

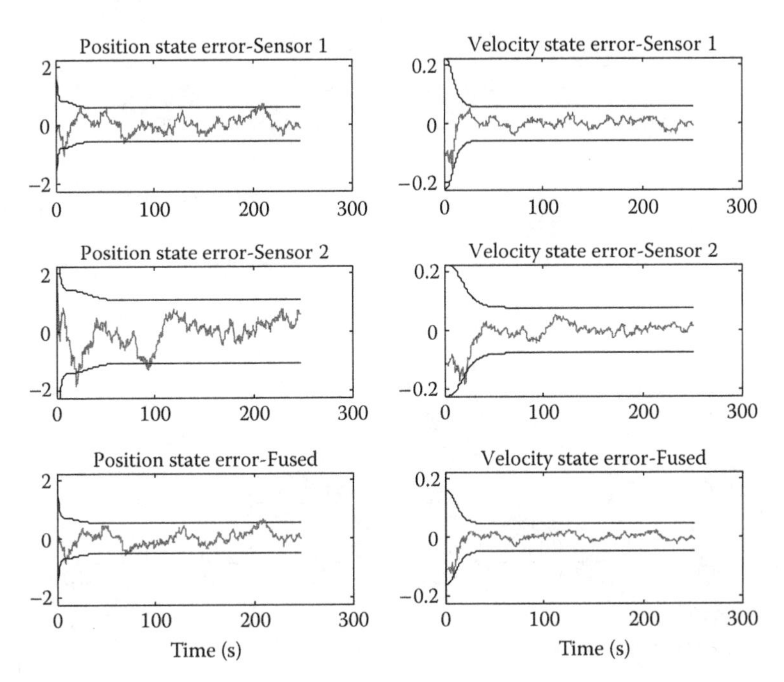

FIGURE 3.30
State errors using Kalman filter (see Example 3.3).

(by a factor 3; see the .m file). The norms of the covariance matrices are plotted in Figure 3.31. This example demonstrates that the fusion gives better predictive accuracy and less uncertainty. Figure 3.32 depicts the state errors within their theoretical bounds for both the sensors and the fused states. This example illustrates the fusion concept based on data-sharing or GF algorithm for two sensors with noisy measurement data.

EXAMPLE 3.5

Generate simulated (noisy) data using the state and measurement models of Equations 3.6 and 3.7 for two sensors in MATLAB. Process the measured data using the H-I *a posteriori* filtering algorithm, Equations 3.153 to 3.158, and perform state estimation. For fusion, use the state-vector fusion equations, 3.13 and 3.14. Plot norms of relevant covariance matrices and position and velocity (state) errors within their theoretical bounds.

Solution 3.5

The data using the prescribed models are generated and Gaussian process and measurement noises with zero mean and unity standard deviation (initially) are added to these data as in Example 3.3. The data are then processed using the H-I *a posteriori* state estimation algorithm. The MATLAB .m (dot.m) file for generating these data and the fusion results is given as *Ch3ExHIPOSTF*. The two sensors are differentiated by adding more measurement noise in sensor 2 (by factor 1.3;

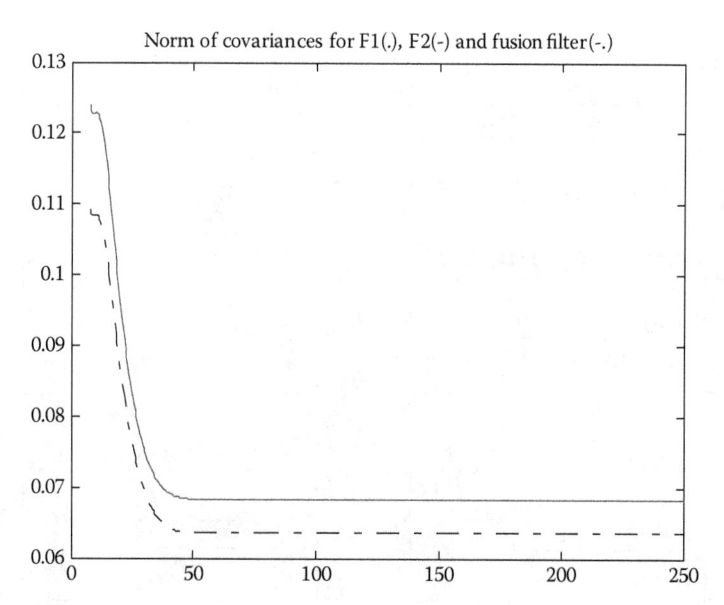

FIGURE 3.31
Norms of covariance matrices (see Example 3.4).

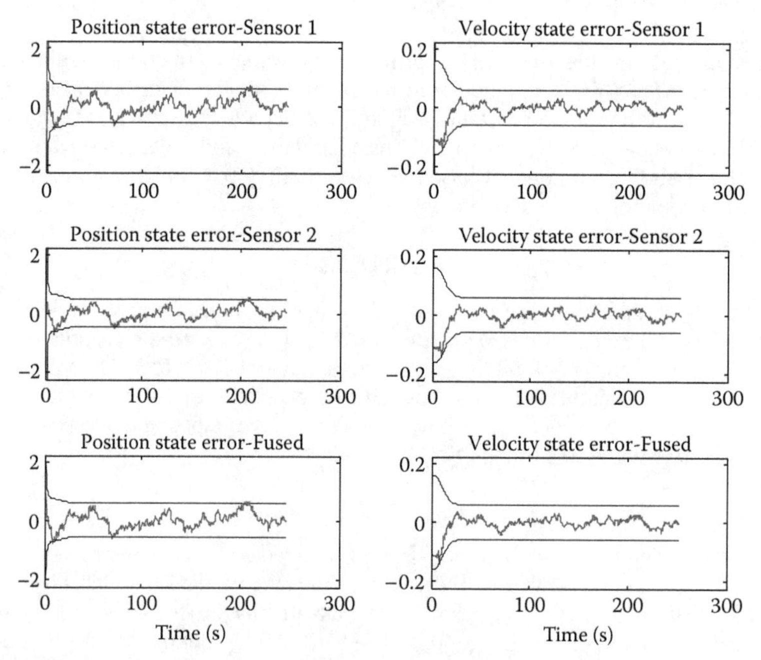

FIGURE 3.32
State errors using gain-fusion (see Example 3.4).

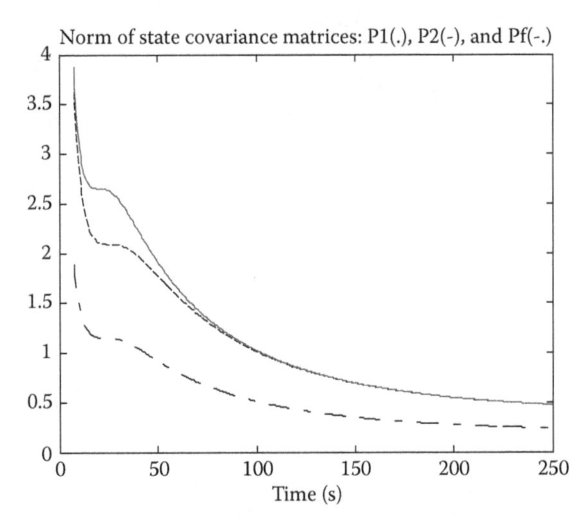

FIGURE 3.33
Norms of covariance matrices (see Example 3.5).

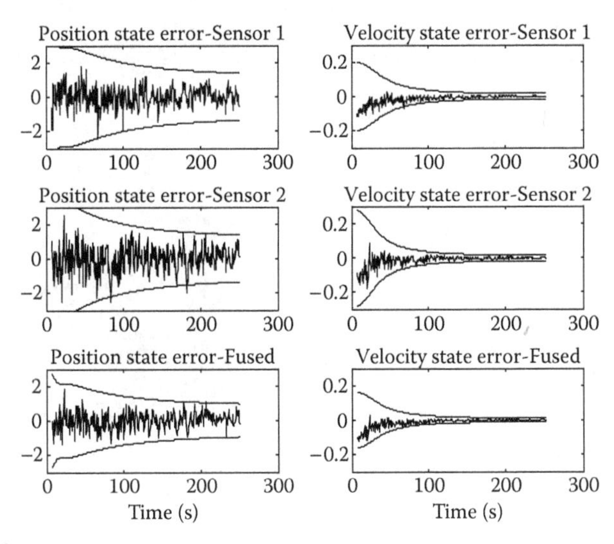

FIGURE 3.34
State errors using H-I posteriori fusion (HIPOSF; see Example 3.5).

see the .m file). The initial values of P1 = [400 0;0 0.01] and P2 = [400 0;0 0.02]. The norms of the covariance matrices are plotted in Figure 3.33 (using the program *Ch3ExHIPOSTFplot*). This example demonstrates that the fusion gives better predictive accuracy with less uncertainty. Figure 3.34 depicts the state errors within their theoretical bounds for both of the sensors and fused states. This example illustrates the fusion concept based on H-I *a posteriori* state estimation and state-vector fusion for two sensors with noisy measurement data.

4

Performance Evaluation of Data Fusion Systems, Software, and Tracking

As we discussed in the Joint Directors' Laboratories (JDL) data model, the evaluation of the performance of a data fusion (DF) system, algorithms, and software is very important for establishing the effectiveness of the DF system and its results. Filter tracking performance is evaluated by computing several metrics [14], some of which have already been described in Chapter 3:

1. Percentage fit error (PFE) in position: This is computed as the ratio of the norm of the difference between the true and estimated positions to the norm of the true positions. This will be zero when both true and estimated positions are exactly alike (at the discrete sampling points), and it will increase when the estimated positions deviate from the true positions. When comparing the performance of different tracking algorithms, the algorithm that gives the least PFE is preferable.

$$\text{PFE} = 100 \times \frac{\text{norm}(x_t - \hat{x})}{\text{norm}(x_t)} \tag{4.1}$$

   Here, x_t is the true x-position and $\hat{x}$ is the estimated x-position, and norm is the operator to find the Euclidean length of the vector (it is a MATLAB® function). A similar expression would be applicable to compute the PFE for y- and z-positions; x-, y-, and z-velocities; x-, y-, and z-accelerations; or any other variable of interest. If the true positions or states are not known, as in the real data case, one can compute this norm of PFE between the measured trajectories and the predicted or estimated trajectories.

2. Root mean square error in position: This is computed as the root mean square error of the true and estimated x-, y-, and z-positions. It produces a single number, and it will be zero when the true and estimated positions are the same. This value will increase when the estimated positions deviate from the true positions. When comparing the tracking performance of different algorithms, the algorithm that gives the minimum value is highly preferable.

$$\text{RMSPE} = \sqrt{\frac{1}{N}\sum_{i=1}^{N}\frac{(x_t(i)-\hat{x}(i))^2+(y_t(i)-\hat{y}(i))^2+(z_t(i)-\hat{z}(i))^2}{3}} \quad (4.2)$$

Here, N is the number of samples in the trajectory. Similar expressions can be used to compute root mean square error in velocity (RMSVE), root mean square error in acceleration (RMSAE), and any other variable of interest.

3. Root sum square error (RSSE) in position: This is computed as the root sum square error of the true and estimated x-, y-, and z-positions, and it is a discrete (index or time) dependent variable and can be plotted to infer the results. Thus, it produces a sequence of numbers, and these values will be zero when the corresponding true and estimated positions are exactly alike. In reality, these values will increase when the corresponding estimated positions deviate from the true positions. When comparing the tracking performance of different algorithms, the algorithm that produces the lowest time history of the RSSE is highly preferable.

$$\text{RSSPE}(i) = \sqrt{\left(x_t(i)-\hat{x}(i)\right)^2+\left(y_t(i)-\hat{y}(i)\right)^2+\left(z_t(i)-\hat{z}(i)\right)^2}$$
$$i = 1, 2, \ldots, N \quad (4.3)$$

Similar expressions can be used to compute root sum square error in velocity (RSSVE) and root sum square error in acceleration (RSSAE).

4. State error $(X-\hat{X})$ with theoretical bounds $\pm 2\sqrt{\hat{P}(i,i)}$: The state errors in x-, y-, and z-positions (and velocities, accelerations, jerk, and even surge) are plotted with their theoretical bounds. The theoretical bounds are computed as two times of square root of their covariance (matrix's diagonal elements). If 95% of the total number of computed errors are within these bounds, this means that the tracker performs well.

   The state error in x-position is

$$\text{SE}x(i) = x(i) - \hat{x}(i) \quad (4.4)$$

and the corresponding theoretical bound is

$$\pm 2\sqrt{\hat{P}_x(i)} \quad (4.5)$$

5. Innovation (or measurement residuals) sequence is given as

$$(z_m(k) - \tilde{z}(k \mid k-1)) \quad (4.6)$$

with theoretical bounds of

$$\pm 2\sqrt{S} \tag{4.7}$$

The innovation sequence is plotted with their theoretical bounds, where S is the innovations covariance matrix output from the Kalman filter (KF). The theoretical bounds are computed as two times of the square root of their covariance (matrix diagonal elements). If the errors are within these bounds, this means that the tracker performs well and it has extracted much or all information present in the measurements.

6. Normalized estimation error square

$$\left(X - \hat{X}\right) P^{-1} \left(X - \hat{X}\right)^{T} \tag{4.8}$$

with theoretical bounds [38]: This is computed as the state error square and is normalized with the corresponding covariance matrix. In general, the mean of this should be equal to the size of the state vector. If it is within the theoretical bounds, this means that the filter performs well. The theoretical bounds are computed as follows:

Lower bound:

$$\frac{\chi^{-1}(0.025, p)}{N_{\text{MCS}}} \tag{4.9}$$

Upper bound:

$$\frac{\chi^{-1}(0.975, p)}{N_{\text{MCS}}} \tag{4.10}$$

Here, degree of freedom (DOF) p is $N_X N_{\text{MCS}}$; N_X is the number of elements in the state vector, N_{MCS} is the number of Monte Carlo simulations, and χ is the chi-square operator.

7. Normalized innovation square $\vartheta S^{-1} \vartheta^T$ with theoretical bounds [39]: This is computed as the innovation square and is normalized with the corresponding covariance matrix. In general, the mean of this should be equal to the size of the measurement vector. If it is within the theoretical bounds, this means that the filter performs well. The theoretical bounds are computed using the Equations 4.9 and 4.10, with the DOF p as $N_z N_{MCS}$, where N_z is the number of elements in the measurement vector.

4.1 Real-Time Flight Safety Expert System Strategy

A part of any multisensor tracking system, the flight test range requires a comprehensive planning of flight tests and preparation for test evaluation, while keeping the objectives and test-vehicle dynamics in mind. The evaluation process stems from the proposition of the range user and the intention of the test after the submission of the following flight vehicle information [69]: (1) flight objectives, (2) details of flight vehicle and flight dynamics, (3) instrumentation support needed, (4) measurement of parameters required, and (5) information on various failure mode studies carried out by the design and simulation groups.

Such a test launch of a flight vehicle is essentially monitored in real time with the help of various tracking sensors. The sensor data are supplied to the range safety officer (RSO) at the flight test range facility through the central computer after processing. Many test ranges use the sensor data available in real time to process and present the tracking trajectory individually to the RSO. Thus, there will be a number of trajectories, one from each sensor, which monitor the flight of a single vehicle (after multiplexed or concurrent processing in the central computer). A comparison of the trajectories is made using the safety background charts that are generated on the basis of extensive failure mode analysis of the flight vehicle, keeping in mind the topographical limitations of the range in relation to its surroundings. It is very difficult for the RSO to assess the real-time flight vehicle position from the multiple trajectories presented by the sensors. She or he has to arrive at an appropriate decision almost instantaneously, which is very critical because of (1) the severity of the risk due to the errant flight vehicle, including uncertainty about its impact point, and (2) the very limited time (2–3 seconds) available for the range safety office to react.

Any human error or mistake in such crucial decision making could lead to the loss of lives and property. Humans' subjective judgment could be erroneous due to various conditions and their state of mind: mental fatigue, biased decisions, slow reactions, memory loss, mistakes in judgment, and so on. Total dependence on a human safety officer for such a crucial judgment on termination (or otherwise) of flight vehicle, which could create unwarranted loss due wrong or delayed decisions, should always be questioned. The capability of parallel processing or high-speed computer processors, the accessibility of huge amounts of memory, and the evolution of DF philosophy have opened up new avenues to meet the difficulties faced by a human safety office in a flight test range. The computation of trajectories of the vehicle from a number of sensor data streams in real time, and then fusing these trajectories, aids the RSO to a great extent and reduces his or her workload. Thus, the ranges in the world are trying best to utilize such a philosophy and arrive at better decision making in order to overcome the weak points of human psychology. The test range-tracking sensors are generally deployed and distributed over a large area along a long coastline or a range on land and in sea, and in some cases from a satellite or aircraft to cover various phases of the flight path trajectory. The range-tracking sensors used for target tracking are electro-optical transducers (EOTs), infrared (IR) trackers, laser rangers, charge-coupled devices (CCD) cameras, Doppler radar, beacon radar, and telemetry trackers. Characterization of each measurement channel (Section 3.2) is a very important task, involving data collection from postflight records or by conducting helicopter or aircraft trial sorties over the range area.

4.1.1 Autodecision Criteria

The data processing system should have certain processing capabilities to ensure accuracy, reliability, and a fast response [69]: (1) validation criteria to fix up data acceptability; (2) a dynamic tracking filter to estimate the target trajectory from the noisy measurements; (3) a multisensor DF (MSDF) to obtain fused trajectory for decision making; (4) an expert rule processor to determine the logic–based on the knowledge bank; (5) an inference engine to select the best among the alternative decisions; and (6) a display scheme to present the final decision. After processing the data in KF, the system should combine the data at two or three levels to achieve one representative trajectory of the vehicle for flight safety decision making. In the next level of fusion, expert system–based decision architecture is used to give a final decision for flight safety.

4.1.2 Objective of a Flight Test Range

The main role of a flight test range is to undertake a flight test and evaluate several weapons systems under development such as guided weapon systems, unguided rocket systems, unmanned aerial vehicles, warhead test vehicles, and special aerial delivery systems. Some of the typical aspects need to be defined, such as suitable selection of sensor site, accurate survey and deployment of tracking sensors to ensure full coverage of the trajectory from launch to impact, calibration of tracking sensors, range tracking network validation, real-time range safety monitoring and decision making for safe launch, reliable tracking and processing system, data accuracy and acceptance criteria, data security, storage and maintenance, and video data processing and analysis.

4.1.3 Scenario of the Test Range

Trajectory tracking sensors are deployed over a vast area along the coast and downrange and are connected to the central computer through a reliable communication link.

4.1.3.1 Tracking Instruments

The main tracking sensors and related information are given in Table 4.1 [69]. The given numbers are only approximate/representative values, and may not represent the actual values. The deployment of these tracking systems requires good planning and should fulfill several conditions: (1) assuring clear line of sight for optical tracking; (2) assuring correct base distance of angle measuring systems like EOTS and telemetry (TM); (3) assuring less electromagnetic interference (EMI) or electromagnetic coupling (EMC) interference; and (4) good flight trajectory coverage aspects. Fifteen tracking sources are connected to a central computer—data acquisition and processing in order to meet the test range objectives is a challenging task.

TABLE 4.1

Typical Range Tracking Instruments Scenario

Type of the Sensor	Typical Numbers (Not Optimal)	Primary Measurements Provided by the System	Derived/ Desired Measurements	Transmission (Tx) Medium to the Central Computer	Data Rate (Hz) (Samples per Second)	Capability in the Range (km)
EOT	6	Picture element (pixel) pixel intensity, pixel color	Azimuth, elevation	Cable/ microwave	10–50	35
IR	3	Pixel intensity color	-do-	Cable	-do-	30
Laser	1	Optical intensity detection	Range	-do-	-do-	20
S-band radar	2	RCS frequency time	Range, azimuth, elevation	Cable	4	2000
C-band radar	1	-do-	Range, rate, azimuth, elevation	Cable	10–100	4000
TM	3	Frequency	Azimuth, elevation	Cable/ microwave	10	1500
DGPS	1	-do-	Position	Cable	5	200

The ranges may vary depending upon the specific device/brand.

Source: Banerjee, P., and R. Appavu Raj. 2000. IEEE International Conference on Industrial Technology (ICIT), Goa, India.

4.1.3.2 Data Acquisition

Missile coordinate data from the tracking instruments is transmitted to the central computer through RS-232 C digital interface, connected through a modem to an audio frequency-dedicated line. The computer also acquires central and countdown timing data through parallel input interfaces. The raw data from tracking instruments are validated using certain criteria [69]: (1) communication error check (longitudinal redundancy check), (2) tracking station check, and (3) validation by 3s criteria. Other criteria are also applied: (1) systematic error correction, (2) refraction error correction, (3) correction due to earth curvature, and (4) correction due to transmission delay. Other processing tasks are (1) trajectory computation, (2) filtering, (3) coordinate conversion, (4) computer designate mode data Tx, (5) range safety computation, and (6) display task activation or graphics.

4.1.3.3 Decision Display System

This system consists of two sets of workstations, which receive current state vector, and impact footprints of flight vehicle from multiple track sources superimposed on destruct contours. The decision to terminate the flight vehicle is based on the following: (1) when the real-time trajectory of the flight vehicle parallels one of the destruct-contours; and (2) when the impact footprint intersects with the shore line. The decision is further reinforced using missile health parameters displayed on the PC system. The expert-based decision support system wherein the machine processes the source data and displays the flight safety action required during real time is established. The two levels of fusion used are (1) first-level fusion based on priority and fusion of similar sensors using the MSDF method; and (2) decision-level fusion based on voting, expert decision support system (DSS).

4.1.4 Multisensor Data Fusion System

Sensors are normally subjected to external noise, internal malfunctioning, and mechanical failures. To ensure that a system receives a realistic view of its environment, it is necessary to combine the readings of two or more sensors. This process is beneficial to human decision making, as the decision is made under time constraints and is based on a large amount of sensor information.

4.1.4.1 Sensor Fusion for Range Safety Computer

The fusion scheme evolved based on the following criteria [69]:

1. Priority-based source selection is based on
   a. Most accurate track data: I - EOTS
   b. Reliability: II - Skin Radar
   c. Trajectory coverage aspects: III- Telemetry- Skin Radar} IV - C-band- Skin Radar
2. Algorithm/state-vector fusion for dissimilar type of sensors (inertial navigation system [INS]–global positioning system [GPS], TM–S-band)
3. Algorithm/measurement-level (or state-vector) fusion of similar type of sensors (S-band and S-band)
4. Pure trajectory prediction for 4 seconds in case of track loss (when the measurement data are not available)
5. At any time, the MSDF will not exceed three types of trajectory based on (1), (2), and (3) alone.

4.1.4.2 Algorithms for Fusion

Fusion filter or scheme uses the following equations:

The state vector for a fused track is given by

$$\hat{X}^f = \hat{X}^1 + \hat{P}^1\left(\hat{P}^1 + \hat{P}^2\right)^{-1}\left(\hat{X}^2 - \hat{X}^1\right) \tag{4.11}$$

$$\hat{P}^f = \hat{P}^1 - \hat{P}^1\left(\hat{P}^1 + \hat{P}^2\right)^{-1}\hat{P}^{1^T} \tag{4.12}$$

where, $\hat{X}^1$ and $\hat{X}^2$ are the estimated state vectors of filter 1 and filter 2, respectively, with measurements from sensor 1 and sensor 2, and $\hat{P}^1$ and $\hat{P}^2$ are the estimated state error covariance matrices from filters 1 and 2, respectively. The cross-covariance matrix between any two sensors is not considered here for the sake of simplicity. In measurement data level fusion, measurements from each sensor are processed by only one KF algorithm in a sequence. The data from the first sensor are processed by one KF, then the data from the second sensor are processed by the same KF and so on. The KF fuses the sensor observations directly via the measurement model and uses only one KF to estimate the (fused) state vector. The equations for this data level fusion are the same as Equations 3.15 through 3.19, with the observation matrix being 2×2, taking both the sensor together as $H = \begin{bmatrix} 1 & 0 \\ 1 & 0 \end{bmatrix}$, and the measurement covariance matrix

TABLE 4.2

Module Number 1 for Fusion (Range Limit ≤ RL [Trajectory 3])

EOT	S-Band	TM	Action 1	Action 2	Action 3
1	1	1	EOT estimation sent to DSS	S-band estimation	TM estimation
1	1	0	EOT estimation sent to DSS	S-band estimation	TM prediction
1	0	1	EOT estimation sent to DSS	S-band prediction	TM estimation
1	0	0	EOT estimation sent to DSS	S-band prediction	TM prediction
0	1	1	EOT prediction	Fusion of TM and S-band/send to DSS	
0	1	0	EOT prediction	Fusion of estimated S-band and predicted TM and send to DSS for $t \leq$ ms,[a] $t >$ ms	
0	1	0	EOT prediction	S-band estimation send to DSS	TM prediction
0	0	1	EOT prediction	Fusion of predicted S-band and estimated TM and send to DSS for $t \leq$ ms, $t >$ ms	
0	0	1	EOT prediction	S-band prediction	TM estimation/ send to DSS
0	0	0	If (ivalpcmc ! = 0:0) then PCMC estim. and send to DSS and EOT, S-band, and TM prediction else if (ivalpcmc == 0:0) then track loss, EOT, S-band, and TM prediction and EOT send to DSS		

For range limit > RL

PCMC	ACTION
1	PCMC estimation and send to DSS
0	PCMC prediction and send to DSS

[a] $m = n\Delta t$, n is the number of sampling intervals.

Source: Girija, G., J. R. Raol, R. Appavu Raj, and S. Kashyap. 2000. Tracking filter and multi-sensor data fusion, *Sadhana* 25(2):159–67. With permission.

$R = \begin{bmatrix} R1 & 0 \\ 0 & R2 \end{bmatrix}$, with $R1$ and $R2$ as the measurement error covariances of respective sensors.

4.1.4.3 Decision Fusion

The criterion for fusion is based on voting. Expert knowledge is represented in the form of production rules with a typical decision matrix (table 2 of [69]). The fused trajectory data of MSDF is processed through this expert rule base, and the final decision is displayed for action by the

RSO. The software is run in Microsoft Windows under the Visual C++ platform. Further related results are discussed in Section 4.2.

4.2 Multisensor Single-Target Tracking

The multisensor single-target (MSST) strategy is validated using postflight data. The state and measurement models used are

$$x_{j+1} = \phi_{j,j+1} x_j + G w_j \tag{4.13}$$

$$z_j = H x_j + v_j \tag{4.14}$$

with the associated state transition matrix as

$$\phi = \begin{bmatrix} 1 & \Delta t & \Delta t^2/2 \\ 0 & 1 & \Delta t \\ 0 & 0 & 1 \end{bmatrix} \tag{4.15}$$

Here, $H = [1\ 0\ 0]$ is the measurement model, and state vector x has position, velocity, and acceleration as its components. The KF is implemented in the factorized form for this application. The U-D filter (Section 3.2) is developed in C language and implemented on Unix-based Alpha-DEC computer. It is validated using simulated trajectory data and also real-flight data.

4.2.1 Hierarchical Multisensor Data Fusion Architecture and Fusion Scheme

A hierarchical MSDF architecture adopted for a typical flight test range equipped with several types of sensors for tracking a flight vehicle is given in Figure 4.1 [70]. This obtains three sets of fused positions (x, y, z) and corresponding velocity state information to a DSS (as the user's facility or in the control room of the range facility). The data from two ground-based radars (azimuth, elevation, and range) are combined via polar-to-Cartesian (PTC) coordinate transformation, used in the individual U-D filters, and then fused to obtain Trajectory 1 (Traj1/t, x, y, z, velocities in the three axes). Here, U is an upper triangular unitary matrix and D is a diagonal matrix. Trajectory 2 (Traj2) is generated by the fusion of INS (t, x, y, z, and velocities) and GPS (t, x, y, z) data (received at the ground station) by using appropriate U-D filters and the fusion filter. The angular data from the TM channels that measure only the azimuth (ϕ) and elevation (θ) are combined using least-squares (LS) method to generate position information (t, x, y, z). The azimuth and elevation data from EOTs are

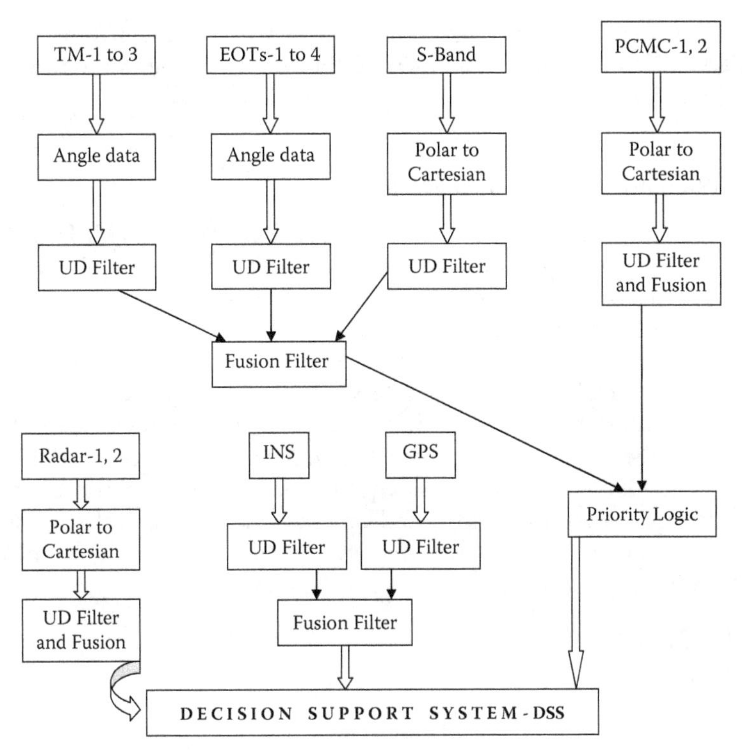

FIGURE 4.1
A typical range safety expert system's hierarchical multisensor data fusion scheme.

also combined in a similar manner to generate the position information (t, x, y, z). These trajectories and the one from the S-band radar (via PTC/from angles to t, x, y, z) are processed by U-D filters and then fused, and t, x, y, z, and velocities are sent to the priority logic. The priority logic also receives the trajectory from the precision coherent monopulse C-band (PCMC) data via PTC, U-D filters, and fusion filter (t, x, y, z, velocities), and thus, this trajectory is Traj3. These three trajectories are sent to the decision support system, which helps the RSO to take decision about the status and position of the target being tracked. We may be able to think of some other alternative fusion scheme for such a test range facility. Although the U-D filter block is shown explicitly in Figure 4.1 for various channels, only one U-D filter is used and the data are processed sequentially by the same filter. The estimated states (x, y, z, velocities, and accelerations) and the error covariance matrices are input to the fusion process to achieve a joint or fused state-vector estimate based on multiple sensors. In the state-vector fusion equations, the cross-covariance between any two state vectors estimated using data from two sensors is neglected for simplicity. The estimates of the state vectors and the covariance matrices are obtained from the U-D filtering process of each channel data and used in the fusion filter as $P = UDU^T$. This is because the fusion filter does not need any updating of the U-D factors.

4.2.2 Philosophy of Sensor Fusion

For sensor channels such as EOT, PCMC, S-Band, two TM, RADAR 1, RADAR 2, INS, and GPS for fusion, it is necessary to develop fusion logic to use the information from these sensors. This logic is mainly based on priority logic and range limit of individual sensors. Since the real data from PCMC, GPS, and TM sensors were not available for simulation and validation, the PCMC and GPS data were replaced by INS data (for the sake of the validation process) and were used to generate the synthetic trajectory data from two TM sensors using a triangulation method. The priority logic is decided based on the sensor accuracy (within the range of the sensor capability), and based on this priority logic, the following sequence could be given to the sensors (of the first module) within the range limit (of say, a certain range limit [RL] km) [70]:

- EOT PCMC
- M and S-band fusion
- S-band
- TM
- Track loss

For range (>RL km), the PCMC radar tracks the target.

For the second module (with the RADARs), the sequence followed is

- RADAR 1 and RADAR 2 fusion
- RADAR 1
- RADAR 2
- Track loss

For the third module (with INS and GPS), the sequence followed is

- GPS/INS fusion (GPS data replaced by INS)
- INS
- GPS (GPS data replaced by INS) Track loss

The rule base for the above modules that depends on the health of the sensors is then given, and Tables 4.2 through 4.4 [70] describe the possible conditions and corresponding actions that would occur in the fusion philosophy.

4.2.3 Data Fusion Software Structure

The MSDF software consists of modules and submodules for different fusion channels and fusion levels; modules are subroutines of the main software, and submodules are subroutines of modules [70]. Fusion is performed using two similar sensor channels. The MSDF software reads the sensor data from the files and performs filtering and fusion processes for each DF module, generating three trajectories: Traj1, Traj2, and Traj3; these trajectories are sent to the DSS for further necessary action.

4.2.3.1 Fusion Module 1

Submodule 1 performs angular TM DF (triangulation or LS method), filtering, and earth curvature correction if needed.

Submodule 2 performs S-band coordinate conversion to TM station, filtering, fusion, and earth curvature correction if needed.

4.2.3.2 Fusion Modules 2 and 3

Fusion module 2 (FM2) is the code for the RADAR 1 and RADAR 2 sensor fusion, and fusion module 3 (FM3) is the code for the GPS and INS sensor fusion.

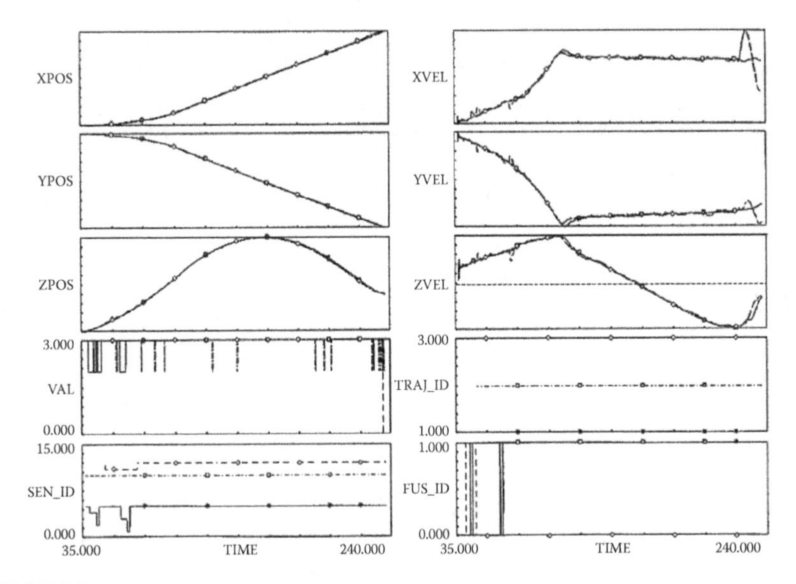

FIGURE 4.2
Trajectories from the multisensor data fusion scheme. Legend: Traj1 -.-; Traj2 -.-□-.-; Traj3 --◊--. (From Girija, G., J. R. Raol, R. Appavu Raj, and S. Kashyap. 2000. *Sadhana* 25(2):159–67. With permission.)

4.2.4 Validation

The MSDF fusion strategy in Figure 4.1 is validated with one set of post-flight data. Where data were not available, some proxy data were used to carry out and complete the validation process. Figure 4.2 shows the comparison plots of the three trajectories: Traj1, Traj2, and Traj3 [70]. The various possible combinations given in Tables 4.2 through 4.4 are exercised when generating these MSDF trajectories, and we observe that a satisfactory MSDF process has been exercised for the range safety decisions. "VAL" signifies the validity bit or health of the sensor, Traj-ID stands for the trajectory identity (to which module the trajectory belongs), "SEN-ID" gives the sensor identity during the trajectory of the target, and "FUS-ID" tells whether the trajectory is "filtered only" or is fused after filtering. The

TABLE 4.3

Module Number 2 for Fusion (Trajectory 1)

RADAR 1	RADAR 2	Action 1	Action 2
1	1	Fusion and send to DSS, $t \leq$ ms	
1	0	Fusion of estimated radar 1 and predicted radar 2, $t >$ ms	
1	0	RADAR 1 estimation and send to DSS	Predicted RADAR 2, $t \leq$ ms
0	1	Fusion of estimated radar 2 and predicted radar 1, $t >$ ms	
0	1	RADAR 1 prediction	RADAR 2 estimation/send to DSS
0	0	RADAR 1 prediction	RADAR 2 prediction/send to DSS

Source: Girija, G., J. R. Raol, R. Appavu Raj, and S. Kashyap. 2000. *Sadhana* 25(2):159–67. With permission.

TABLE 4.4

Module Number 3 for Fusion (Trajectory 2)

INS	GPS	Action 1	Action 2
1	1	Fusion and send to DSS, $t <$ ms	
1	0	Fusion of estimated INS and predicted GPS, $t >$ ms	
1	0	INS estimation and GPS prediction/send to DSS, $t <$ ms	
0	1	Fusion of estimated GPS and predicted INS, $t >$ ms	
0	1	INS prediction, GPS estimation/send to DSS	
0	0	INS and GPS prediction/send to DSS	

Source: Girija, G., J. R. Raol, R. Appavu Raj, and S. Kashyap. 2000. *Sadhana* 25(2):159–67. With permission.

strategy and algorithms of Sections 4.1 and 4.2 have also been further evaluated at the actual test range.

4.3 Tracking of a Maneuvering Target—Multiple-Target Tracking Using Interacting Multiple Model Probability Data Association Filter and Fusion

The performance of interacting multiple model probability data association filters (IMMPDAF) for the estimation of multiple maneuvering target trajectories in clutter is investigated using simulated data, the targets being tracked by a single sensor. The scheme can be extended to more than one sensor, and evaluates the sensitivity of the filter to the choice of the models and process noise covariance matrix values. The use of the IMMPDAF algorithm for the tracking and fusion of simulated data for four targets whose positions are measured by two sensors located at different spatial positions is also discussed in this section. Thus, the algorithm is evaluated for a multisensor multitarget scenario using the fusion method. The IMMPDAF gives a realistic confidence in estimates during maneuvers and a low RSSPE during the nonmaneuvering phase of the targets.

4.3.1 Interacting Multiple Model Algorithm

The interactive multiple model (IMM) algorithm and related concepts [38,39,43,50,51,53] are versatile methods for adaptive state estimation in systems whose behavior changes with time, and which obey one of a finite number of models. Two or more models are run in concurrence (of course this is mainly true in the case of parallel computer-based implementations) to achieve better performance of maneuvering targets. The features of IMM can be combined with probability data association filters (PDAF) to obtain the IMMPDAF algorithm. Figure 4.3 [71] depicts one cycle of a two-model IMMPDAF algorithm [50,51], which is applicable to single-sensor multitarget scenario, and can also be extended to multiple sensors. Figure 4.4 depicts the block schematic of the multisensor–multitarget (MSMT) fusion algorithm [71].

4.3.1.1 Automatic Track Formation

The objective of track initiation is to compute initial state estimates of all possible tracks along with the computations of the associated state covariance matrices. There is a possibility of false tracks being generating due to the presence of spurious measurements and multiple targets. The following technique to initiate a track based on a logic that requires M detections out of N scans in the gate can be used: A tentative track is formed on all two-measurement pairs with the expected motions of the targets.

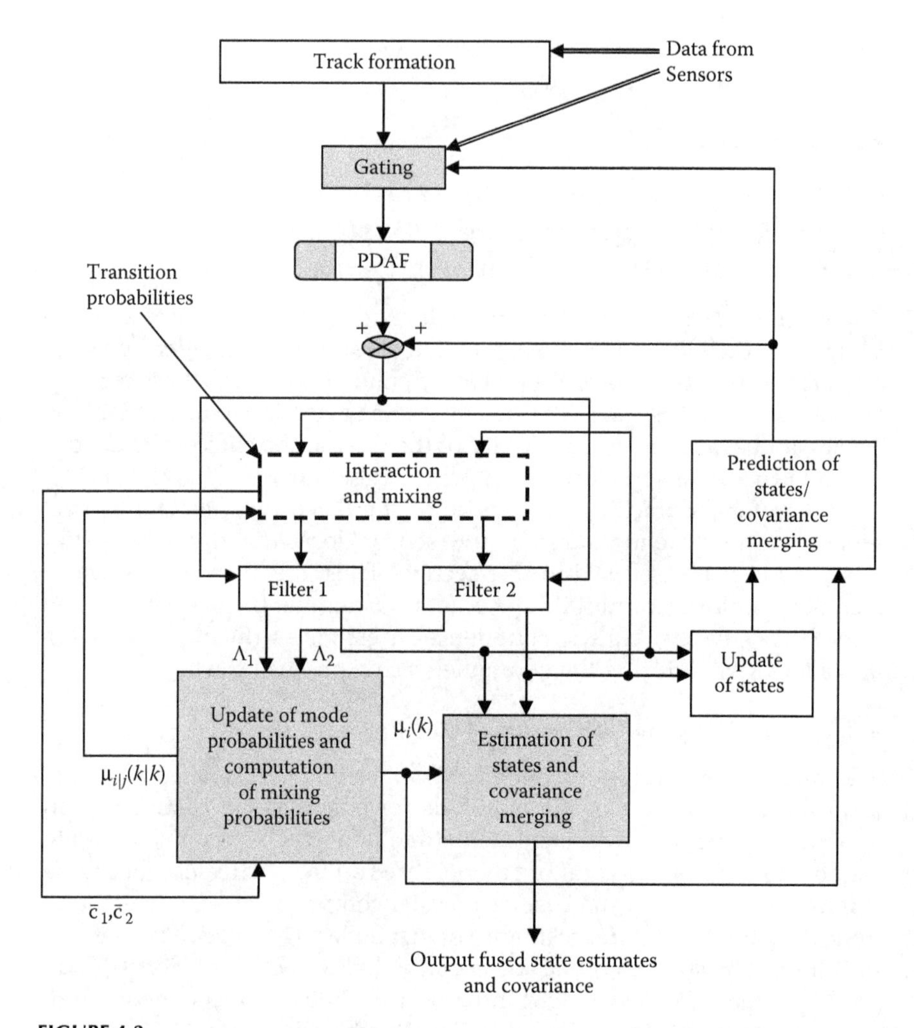

FIGURE 4.3
Interacting multiple models probability data association filter block diagram for two models.

4.3.1.2 Gating and Data Association

After the formation of the track, the measurement screening is carried out using the procedure outlined in Chapter 3 (Equations 3.1 through 3.5) to determine potential candidates for corresponding tracks. The combined innovation sequence v is computed using the steps, as in the PDAF (see Section 3.4), which has two additional blocks (compared to nearest-neighbor KF [NNKF]/KF) for the computation of association probabilities $\beta_i(k)$ and combined innovations sequences $v(k)$. If $v_i(k)$ corresponds to innovation on measurement i, then the merged innovation is given by

$$v(k) = \sum_{i=1}^{m(k)} \beta_i(k) v_i(k) \tag{4.16}$$

Here, $m(k)$ denotes the number of detections in k-th scan, and the probability $\beta_i(k)$ that the i-th validated measurement is correct one is given by

$$\beta_i(k) = \begin{cases} \dfrac{e_i}{b + \sum_{j=1}^{m(k)} e_j} & i = 1, 2, \ldots, m(k) \\ \dfrac{b}{b + \sum_{j=1}^{m(k)} e_j} & i = 0 \end{cases} \tag{4.17}$$

Here, $\beta_0(k)$ is the probability that none of the measurements is correct, and in Equation 4.17,

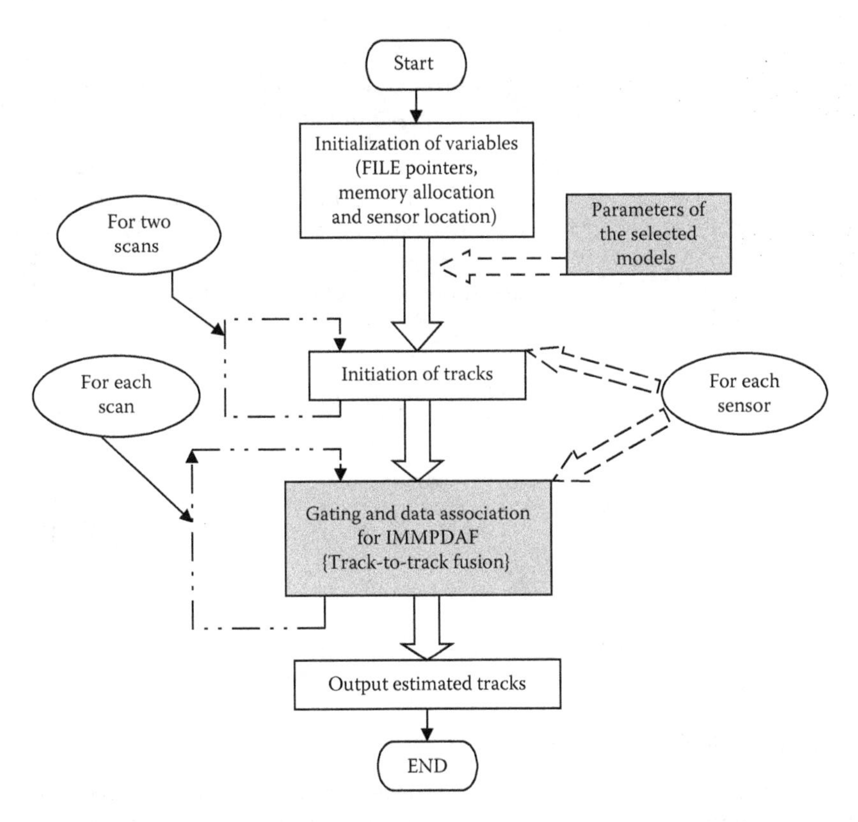

FIGURE 4.4
Program structure of multisensor–multitarget fusion.

$$e_i \cong e^{-\frac{1}{2}v_i(k)'S(k)^{-1}v_i(k)} \tag{4.18}$$

$$b \cong \left(\frac{2\pi}{\gamma}\right)^{\frac{n_z}{2}} m(k)c_{n_z}^{-1}\frac{(1-P_DP_G)}{P_D} \tag{4.19}$$

Here, n_z is the dimension of the measurement, c_{n_z} is the volume of the unit hypersphere of dimensions ($c_1 = 2$, $c_2 = \pi$, $c_3 = 4\pi/3$, and so on), and P_G and P_D represent the gate probability and probability of target detection. The computed $\beta_i(k)$ and $v(k)$ are used to estimate the states and covariance matrices of the updating cycle. The resulting measurement used next in mode-conditioned filtering is computed with the following equation:

$$Z(k) = H\tilde{X}\left(\frac{k}{k-1}\right) + v(k) \tag{4.20}$$

Here, $\tilde{X}(k/k-1)$ is the predicted target state at k-th scan and is computed from the state and covariance prediction blocks (Figure 4.3).

4.3.1.3 Interaction and Mixing in Interactive Multiple Model Probabilistic Data Association Filter

This step uses the mixing probabilities $\mu_{i|j}$ $(k-1|k-1)$ as weighting coefficients, the estimates of $\hat{X}_{\text{and}_i}(k-1|k-1)$ and $\hat{P}_i(k-1|k-1)$ from the previous cycle are used to obtain the initial conditions $\hat{X}_{0j}(k-1|k-1)$ and $\hat{P}_{0j}(k-1|k-1)$ for the mode-matched filters M_1 and M_2 of the present cycle. For all $i, j \in M$, the initial conditions for the filters are computed from the following equations:

$$\hat{X}_{0j}(k-1|k-1) = \sum_{i=1}^{r}\hat{X}_i(k-1|k-1)\mu_{i|j}(k-1|k-1) \tag{4.21}$$

$$P_{0j}(k-1|k-1) = \sum_{i=1}^{r}\begin{bmatrix} P_i(k-1|k-1) \\ +[\{\hat{X}_i(k-1|k-1) - \hat{X}_{0j}(k-1|k-1)\} \\ \times\{\hat{X}_i(k-1|k-1) - \hat{X}_{0j}(k-1|k-1)\}^T] \end{bmatrix}\mu_{i|j}(k-1|k-1) \tag{4.22}$$

Here, mode-matched filters run as $j = 1, \dots r$; the models run as $i = 1, \dots r$; and $r = 2$ for the two-model IMM method.

4.3.1.4 Mode-Conditioned Filtering

With r parallel (though used in a sequential mode only!) mode-matched KFs, the states and covariance matrices are estimated using the normal prediction and update steps.

$$\begin{aligned}
\hat{X}_j(k|k-1) &= F_j(k-1)\hat{X}_{0j}(k-1|k-1) + G_j(k-1)w_j(k-1) \\
P_j(k|k-1) &= F_j(k-1)P_{0j}(k-1|k-1)F_j(k-1)^T + G_j(k-1)Q_j(k-1)G_j(k-1)^T \\
\hat{X}_j(k|k) &= \hat{X}_j(k|k-1) + K_j(k)v_j(k) \\
P_j(k|k) &= P_j(k|k-1) - K_j(k)S_j(k)K_j(k)^T
\end{aligned} \tag{4.23}$$

The measurement prediction is computed by

$$\hat{Z}_j(k|k-1) = H_j(k)\hat{X}_j(k|k-1) \tag{4.24}$$

The residual $v_j(k)$, residual covariance $S_j(k)$ and the filter gain are computed by

$$\begin{aligned}
v_j(k) &= Z(k) - \hat{Z}_j(k|k-1) \\
S_j(k) &= H_j(k)P_j(k|k-1)H_j(k)^T + R_j(k) \\
K_j(k) &= P_j(k|k-1)H_j(k)^T S_j(k)^{-1}
\end{aligned} \tag{4.25}$$

The process and measurement noise covariance matrices are Q and R, and these might differ from mode to mode. The likelihood function for mode-matched filter j is a Gaussian probability density function (PDF) of residual v with zero mean and covariance matrix S and is computed as

$$\Lambda_j(k) = \frac{1}{\sqrt{|S_j(k)|}\,(2\pi)^{n/2}} e^{-0.5[v_j(k)^T S_j(k)^{-1} v_j(k)]} \tag{4.26}$$

Here, n is the dimension of the measurement vector Z.

4.3.1.5 Probability Computations

The mixing probabilities are computed as follows:

$$\mu_{i|j}(k-1|k-1) = \frac{1}{\bar{c}_j} p_{ij}\mu_i(k-1) \tag{4.27}$$

$$\bar{c}_j = \sum_{i=1}^{r} p_{ij}\mu_i(k-1) \tag{4.28}$$

Here, $\mu_i(k)$ is the mode probability at time k, $\bar{c}_j$ is a normalization factor, and p_{ij} is the Markov transition probability to take care of switching from one mode to another. This is a design parameter chosen by the user. The switching probabilities depend on *sojourn* time, or the time the target is expected to be in a maneuver. Consider the Markov chain transition matrix between the two modes of the IMM: $p_{ij} = \begin{bmatrix} 0.9 & 0.1 \\ 0.33 & 0.67 \end{bmatrix}$, then the basis for selecting $p_{12} = 0.1$ is that the target is likely to be in nonmaneuvering mode in the initial stages, and the probability to switch over to maneuvering mode will be relatively low. The p_{22} is selected based on the number of sampling periods for sojourn time. Thus, if the target maneuver lasts for three sample periods ($\tau = 3$), the probability p_{22} is given by

$$p_{22} = 1 - \frac{1}{\tau} = 0.67 \tag{4.29}$$

The initial mode probabilities $\mu_i(k)$, corresponding to nonmaneuver and maneuver mode, are 0.9 and 0.1, respectively, to compute $\mu_{i|j}(k|k)$ and $\bar{c}_j$ in the first cycle of the estimation algorithm. This is based on the assumption that the target is more likely to be in the nonmaneuver mode during the initial stages of its motion. Subsequently, the mode probabilities are updated using the following equation:

$$\mu_j(k) = \frac{1}{c} \Lambda_j(k) \bar{c}_j \quad j = 1, \ldots, r \tag{4.30}$$

Here, $\Lambda_j(k)$ represents the likelihood function corresponding to filter j, and the normalizing factor c is given as

$$c = \sum_{j=1}^{r} \Lambda_j(k) \bar{c}_j \tag{4.31}$$

4.3.1.6 Combined State and Covariance Prediction and Estimation

In the prediction stage, the average mode probabilities obtained in Equation 4.30 are used as weighting factors to combine the predicted the state and covariance matrices for all filters ($j = 1, \ldots, r$), in order to obtain the overall state estimate and covariance prediction:

$$\begin{aligned} \tilde{X}_j(k+1|k) &= F_j(k)\hat{X}_j(k|k) \\ \tilde{P}_j(k+1|k) &= F_j(k)\hat{P}_j(k|k)F_j(k)^T + G_j(k)Q_j(k)G_j(k)^T; \quad j = 1, \ldots, r \end{aligned} \tag{4.32}$$

$$\begin{aligned} \tilde{X}(k+1|k) &= \sum_{j=1}^{r} \tilde{X}_j(k+1|k)\mu_j(k) \\ \tilde{P}(k+1|k) &= \sum_{j=1}^{r} \Big[\tilde{P}_j(k+1|k) + \{\tilde{X}_j(k+1|k) - \tilde{X}(k+1|k)\} \\ &\qquad \times \{\tilde{X}_j(k+1|k) - \tilde{X}(k+1|k)\}^T \Big] \mu_j(k) \end{aligned} \tag{4.33}$$

In the estimation stage, the average mode probabilities are also used as weighting factors to combine the updated state and covariance matrices for all filters ($j = 1,\ldots, r$), in order to obtain the overall state estimate and covariance as follows:

$$\begin{aligned} \hat{X}(k\mid k) &= \sum_{j=1}^{r} \hat{X}_j(k\mid k)\mu_j(k) \\ P(k\mid k) &= \sum_{j=1}^{r}\left[P_j(k\mid k) + \{\hat{X}_j(k\mid k) - \hat{X}(k\mid k)\}\{\hat{X}_j(k\mid k) - \hat{X}(k\mid k)\}^T \right]\mu_j(k) \end{aligned} \tag{4.34}$$

4.3.2 Simulation Validation

Simulation is carried out in PC MATLAB [71]. The target motion is simulated using a 2-DOF constant velocity model (CVM) during the nonmaneuvering phase and a 3-DOF constant acceleration model (CAM) during the maneuvering phase of the target. The models are described in Cartesian coordinate system by

$$\mathbf{X}(k+1) = F\mathbf{X}(k) + Gw(k) \tag{4.35}$$

$$\mathbf{Z}(k) = H\mathbf{X}(k) + v(k) \tag{4.36}$$

Here, the Cartesian state vector $\mathbf{X}$ consists of the position and velocity of the target: $\mathbf{X} = [x \quad \dot{x} \quad y \quad \dot{y}]$ when the target is in nonmaneuvering phase, and $\mathbf{X} = [x \quad \dot{x} \quad \ddot{x} \quad y \quad \dot{y} \quad \ddot{y}]$ when it is maneuvering. The process noise w and measurement noise v are white and are zero-mean Gaussian noises with covariances Q and R, respectively.

4.3.2.1 Constant Velocity Model

The 2-DOF model with position and velocity components in each of the two Cartesian coordinates x and y is given as

$$F = \begin{bmatrix} 1 & T & 0 & 0 & 0 & 0 \\ 0 & 1 & 0 & 0 & 0 & 0 \\ 0 & 0 & 0 & 0 & 0 & 0 \\ 0 & 0 & 0 & 1 & T & 0 \\ 0 & 0 & 0 & 0 & 1 & 0 \\ 0 & 0 & 0 & 0 & 0 & 0 \end{bmatrix} \quad G = \begin{bmatrix} T^2/2 & 0 \\ T & 0 \\ 0 & 0 \\ 0 & T^2/2 \\ 0 & T \\ 0 & 0 \end{bmatrix} \tag{4.37}$$

The acceleration component in the above model (though equal to zero) is retained for compatibility with the 3-DOF model, the variations in velocity are modeled as zero-mean white noise accelerations, and low noise variance Q_1 is used with the model to represent the constant speed of the target in nonmaneuvering mode. The process noise intensity is assumed to be equal:

$$Q_1 = \sigma_x^2 = \sigma_y^2 \tag{4.38}$$

4.3.2.2 Constant Acceleration Model

The 3-DOF model with position, velocity, and acceleration components in each of the two Cartesian coordinates x and y is given as follows:

$$F = \begin{bmatrix} 1 & T & T^2/2 & 0 & 0 & 0 \\ 0 & 1 & T & 0 & 0 & 0 \\ 0 & 0 & 1 & 0 & 0 & 0 \\ 0 & 0 & 0 & 1 & T & T^2/2 \\ 0 & 0 & 0 & 0 & 1 & T \\ 0 & 0 & 0 & 0 & 0 & 1 \end{bmatrix} \quad G = \begin{bmatrix} T^2/2 & 0 \\ T & 0 \\ 1 & 0 \\ 0 & T^2/2 \\ 0 & T \\ 0 & 1 \end{bmatrix} \tag{4.39}$$

The acceleration increments are taken as a discrete time zero-mean white noise, and a low value of process noise variance Q_2 ($> Q_1$) will yield nearly a constant acceleration motion. The noise variances in each coordinate are

$$Q_2 = \sigma_x^2 = \sigma_y^2 \tag{4.40}$$

The following are the components of the target data simulation scenario [71].

For the data simulation for a single sensor, we have (1) number of targets = 2; (2) target initial states:

$$\text{Target 1:} [x \quad \dot{x} \quad \ddot{x} \quad y \quad \dot{y} \quad \ddot{y}] = [0 \quad 5 \quad 0 \quad 100 \quad 5 \quad 0]$$
$$\text{Target 2:} [x \quad \dot{x} \quad \ddot{x} \quad y \quad \dot{y} \quad \ddot{y}] = [1200] \quad 5 \quad 0 \quad -1200 \quad 5 \quad 0];$$

(3) sampling time T = 1 second; (4) false alarm density = 1.0e-005; (5) measurement noise covariance R = 25; (6) process noise covariance (for model 1) Q_1 = 0.1; (7) process noise covariance (for model 2) Q_2 = 0.1; (8) number of data points N = 150; and (9) maneuvering injection time and magnitude of injection for two targets (Table 4.5).

For data simulation for multiple sensors, we have (1) number of sensors = 2; (2) sensor locations—sensor 1: [0,0,0], sensor 2: [1000,1000,0];

TABLE 4.5

Maneuvering Times and Magnitudes

Target Number	1	2	3
Maneuvering time (s) (start, end)	(30, 50)	(70, 100)	(50, 80)
Maneuvering magnitude (m/s²) (x, y, z)	(1 g, −1 g, 0)	(1 g, −1 g, 0)	(−1 g, 1 g, 0)

TABLE 4.6

Q Range for Sensitivity Study (IMMPDAF)

Combinations	Case 1	Case 2
Q_1	Low	Low
Q_2	High	Low

(3) number of targets = 3 for sensor 1, and 3 for sensor 2; (4) target initial states

$$\begin{aligned}
&\text{Target 1: } [x \;\; \dot{x} \;\; \ddot{x} \;\; y \;\; \dot{y} \;\; \ddot{y}] = [0 \;\; 5 \;\; 0 \;\; 100 \;\; 5 \;\; 0] \\
&\text{Target 2: } [x \;\; \dot{x} \;\; \ddot{x} \;\; y \;\; \dot{y} \;\; \ddot{y}] = [1200 \;\; 5 \;\; 0 \; -1200 \;\; 5 \;\; 0] \\
&\text{Target 3: } [x \;\; \dot{x} \;\; \ddot{x} \;\; y \;\; \dot{y} \;\; \ddot{y}] = [1200 \;\; 5 \;\; 0 \; -1200 \;\; 5 \;\; 0] \\
&\text{Target 4: } [x \;\; \dot{x} \;\; \ddot{x} \;\; y \;\; \dot{y} \;\; \ddot{y}] = [1200 \;\; 5 \;\; 0 \; -1200 \;\; 5 \;\; 0]
\end{aligned}$$

(5) sampling time $T = 1$ second; (6) false alarm density = 1.0e-007; (7) measurement noise covariance $R = 5$; (8) process noise covariance (for model 1) $Q_1 = 0.01$; (9) process noise covariance (for model 2) $Q_2 = 0.01$; (10) number of data points $N = 100$; and (11) target 1 maneuvers with 1 g acceleration from 40 to 50 second, while there is no acceleration in the other three targets.

4.3.2.3 Performance Evaluation and Discussions

The performance of IMMPDAF for single and multiple sensors or targets is evaluated. The KF with a CVM can be used for tracking maneuvering targets if a relatively large value of Q is used during maneuver phase. During the nonmaneuver phase, higher Q might result in a degraded performance. However, if a lower value of Q is used for tracking during maneuvers, there could be filter divergence and track loss due to the filter being unable to jack its performance during the maneuver. The sensitivity of the association algorithms to Q values and models is evaluated by considering the two combinations of models and Q values as per Table 4.6. The constant parameters used are (1) probability of detection $P_D = 0.99$; (2) gate probability P_G = 0.99998; (3) gating threshold G = 25; (4) sojourn time τ = 15 seconds; and (5) onset mode probability $P_{12} = 0.12$.

4.3.2.3.1 *Evaluation of Interacting Multiple Models Probability Data Association Filter*

Figure 4.5 shows the x-position data of the two maneuvering targets in clutter [71]. The performance of the algorithm is evaluated for two conditions of Q in terms of estimated and true x-position in clutter, estimated standard deviations ($\sigma_{x\text{-pos}}$), and RSSPE. For case 1, process noise covariances for CVM and CAM are 0.1 and 30, respectively, whereas for case 2, process noise covariance values of 0.1 and 2 are used. Figures 4.6 and 4.7 show the comparison of the performances of case 1 (IMMPDAF 1) and case 2 (IMMPDAF 2), respectively, in terms of estimated and measured tracks with clutter, mode probability of tracks, $\sigma_{x\text{-pos}}$, and RSSPE [71]. The tracking performance for both combinations of Q is good, and the mode probability indicates the switch from the nonmaneuver to the maneuver mode. The $\sigma_{x\text{-pos}}$ has a higher value during the maneuver, which reflects the correct situation that the filter is adaptively tracking the target while maneuvering. In case 2, there is a delay in maneuver detection as compared with case 1.

4.3.2.3.2 *Multiple Sensors—Fusion of Data*

We consider the simulated data of four targets seen by two sensors for fusion using the IMMPDAF (case 1 algorithm), where we use the data from each of the sensors to initiate tracks using the first two scans of measurements, and the tracks are updated with the valid measurements

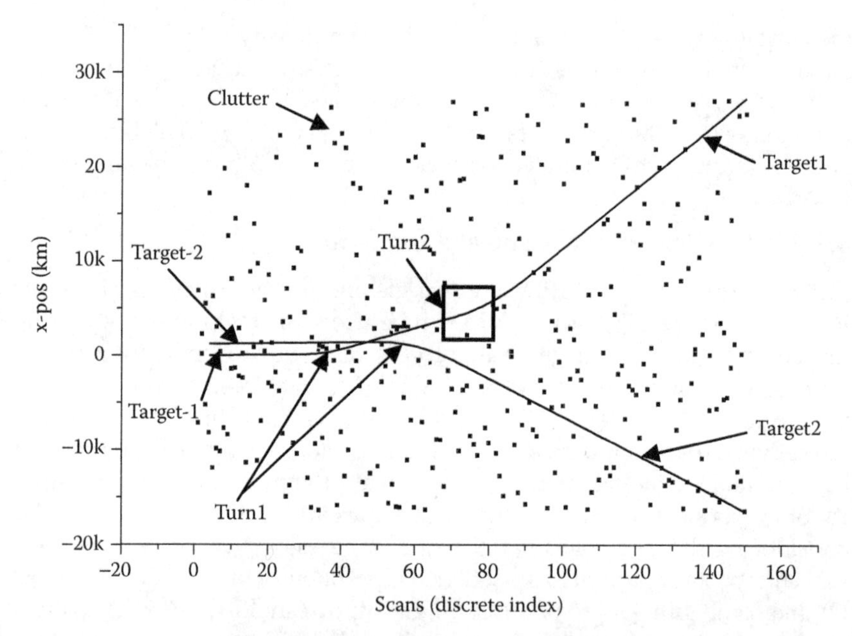

FIGURE 4.5
Simulated/true trajectories in clutter (target 1 turns twice and target 2 turns once).

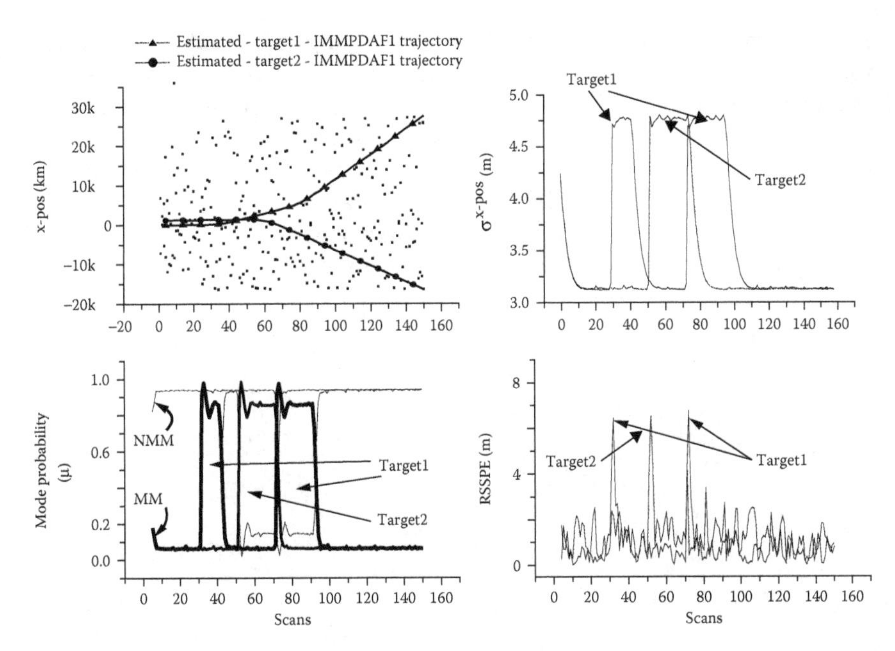

FIGURE 4.6
Performance results of interacting multiple models probability data association filter ($Q_1 = 0.1$, and $Q_2 = 30$; NMM, nonmaneuver mode; MM, maneuver mode).

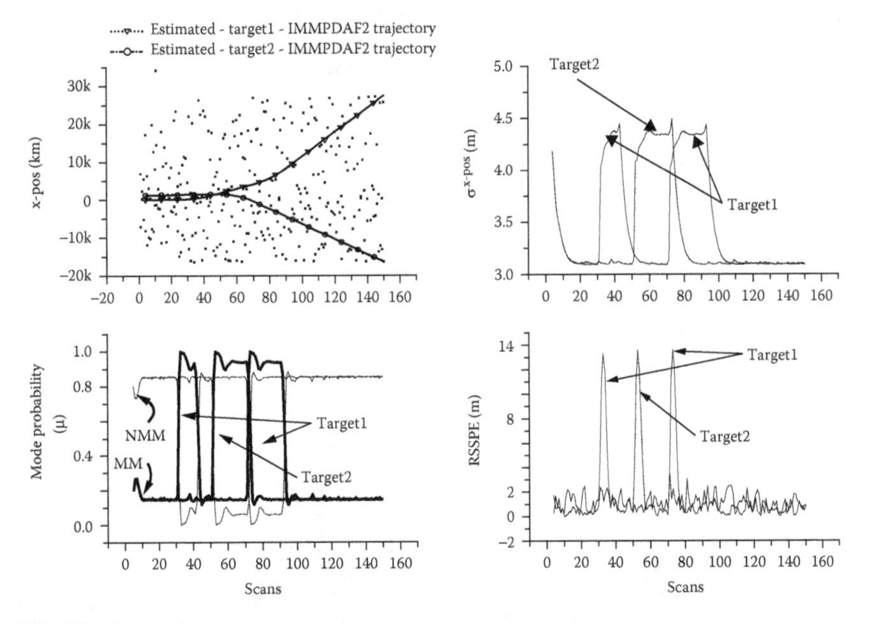

FIGURE 4.7
Performance results of interacting multiple models probability data association filter ($Q_1 = 0.1$, and $Q_2 = 20$).

after the measurement-to-track association using IMMPDAF. It is necessary to transform the data into a common reference before fusion is carried out. Track-to-track association is used for combining similar tracks. In this scenario (amongst the four targets), the three targets are seen by both sensors. Figure 4.8 shows measurements for sensor 1 and sensor 2, the estimated tracks, and mode probability. The four targets are seen from the estimated positions after combining similar tracks using track-to-track fusion, and we can conclude that target maneuvering occurs in track 1 only. We

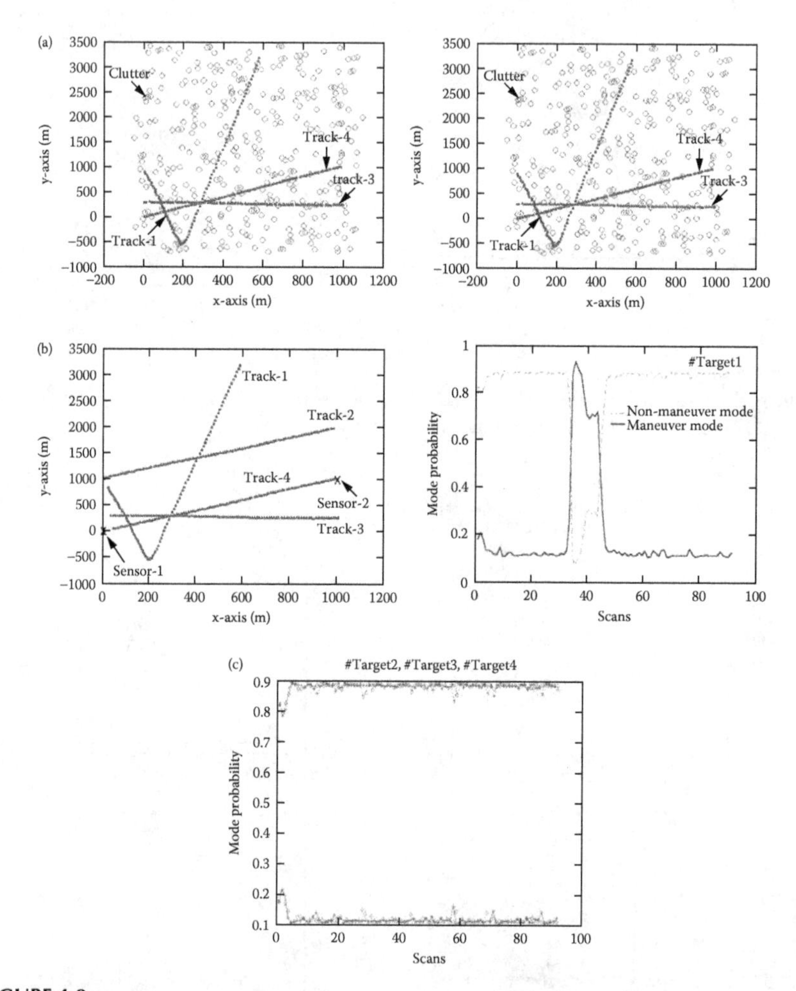

FIGURE 4.8
Performance evaluation results of interacting multiple models probability data association filter for a multisensor data fusion task. (a) Measurements from sensor 1 and sensor 2; (b) the estimated tracks for each sensor; and (c) mode probability. (From Kashyap, S. K., V. P. S. Naidu, J. Singh, G. Girija, and J. R. Raol. 2006. *Aerosp Sci Technol Aeronaut Soc India* 58(1): 65–74. With permission.)

find that the IMMPDAF gives realistic estimates during maneuvers and a lower RSSPE during the nonmaneuvering phase of the targets.

4.4 Evaluation of Converted Measurement and Modified Extended Kalman Filters

Many algorithms are available for obtaining improved accuracy with radar measurements in target tracking applications using KF [72,73]; however, it would be desirable to use a scheme with lower computational complexity when the data rate is high and the available computing power is limited. Target dynamics are better described in Cartesian coordinates, since the dynamics are uncoupled and linear, whereas radars measure the target range, elevation, and azimuth, yielding nonlinear relations between the states and measurements. The inaccuracies of the measurements will have a direct effect on the performance of the tracking algorithms. For target tracking using the radar measurements, two approaches are commonly used:

1. Linear KF or converted measurements KF (CMKF): In this method, the measurements used for updating the states are generated by converting the raw measurements in polar frame to the Cartesian frame, rendering the measurements as the linear functions of the states. The converted measurement errors would become correlated. The cross-range measurement errors would be large, the mean of the errors would be high, and a debiasing process would be required. Thus, the measurement noise covariance matrix should include cross-covariance terms to account for correlated measurement errors. Analytical expressions for the debiased CMKF (CMKF-D) are available in [72]. This would necessitate evaluation of complex equations.
2. EKF approach: In this method, the measurements used for updating the states are the range, azimuth, and elevation in polar frame. The measurements are nonlinear functions of the states yielding a mixed coordinates filter. In the EKF, the initial covariance depends on the initial converted measurements and the gains depend on the accuracy of the subsequent linearization. A simple way to handle the nonlinearities [74] is to process the radar measurements sequentially in the order of elevation, azimuth, and range, while linearizing the nonlinear equations with respect to the estimated states. This results in considerable computational savings. When the nonlinearities are strong, modified expressions for the mean and covariance errors can be used and a modified EKF (MEKF; measurements are still sequentially processed) was proposed [73].

To achieve better accuracy, both methods require certain modifications in order to handle the bias and measurement error covariance in the conventional linear KF and EKF. In this section, an alternative method of achieving debiasing and obtaining an estimate of the measurement error covariance using converted radar measurements is proposed [75,76]. The method based on error model converted measurement KF (ECMKF) presupposes the availability of very accurate reference data from an independent measurement source. The GPS or differential GPS (DGPS) method gives very accurate measurements of position, and hence, these data can be used to get accurate estimates of the bias and measurement noise covariance of the converted measurements in the Cartesian frame using KF with error state-space formulation [77]. Here, the estimated bias is used for correction in the converted measurements and the estimated covariance values used in the ECMKF. The performance of the ECMKF is compared with the error model MEKF (EMEKF) algorithm, and the latter handles the measurements of range, azimuth, and elevation directly. The simulated data of a target with different measurement accuracies are used. The algorithms are implemented in PC MATLAB and evaluated in terms of RSSPE and PFE with respect to true data. These algorithms are also used for tracking a moving target from ground-based radar measurements when GPS measurements of position of the aircraft are available. For the sake of comparison, the results of CMKF-D [72] are also presented.

4.4.1 Error Model Converted Measurement Kalman Filter and Error Model Modified Extended Kalman Filter Algorithms

The target-tracking model is described in Cartesian coordinate system with additive noise:

$$\mathbf{X}_{k+1} = F\mathbf{X}_k + Gw_k \tag{4.41}$$

Here, the vector $\mathbf{X}$ consists of the position and velocity of the target ($\mathbf{X} = [x \; y \; z \; \dot{x} \; \dot{y} \; \dot{z}]$), the process noise w is assumed to be white, and the zero mean with covariance matrix Q. The ground-based radar provides measurements of range (r_m), azimuth (θ_m) and elevation (ϕ_m). The measurement model is given as:

$$\begin{bmatrix} r_m \\ \theta_m \\ \phi_m \end{bmatrix} = h(\mathrm{X}_k) + v_k = \begin{bmatrix} \sqrt{x^2+y^2+z^2} \\ \tan^{-1}\left(\dfrac{y}{x}\right) \\ \tan^{-1}\dfrac{z}{\sqrt{x^2+y^2}} \end{bmatrix} + \begin{bmatrix} v_r \\ v_\theta \\ v_\phi \end{bmatrix} \tag{4.42}$$

Here, v_r, v_θ, v_ϕ are mutually uncorrelated and zero-mean white Gaussian noises with variances $\sigma_r^2, \sigma_\theta^2, \sigma_\phi^2$, respectively.

4.4.1.1 Error Model Converted Measurement Kalman Filter Algorithm

The range, azimuth, and elevation data are converted to positions in Cartesian frame using the following equations:

$$\left.\begin{aligned} x &= r_m \cos(\phi_m)\cos(\theta_m) \\ y &= r_m \cos(\phi_m)\sin(\theta_m) \\ z &= r_m \sin(\phi_m) \end{aligned}\right\} \tag{4.43}$$

4.4.1.1.1 Error Model Kalman Filter

Here, in KF, we use the error state-space formulation in place of the actual state-space formulation (Figure 4.9) [75,76]. The error state-space KF estimates the errors in the converted sensor data using the difference between the measured position data and the reference data, which are supposed to be accurate and from an independent source. The error model Kalman filter (EMKF) gives the estimates of the errors ($\delta\hat{X}$) in the sensors, the difference

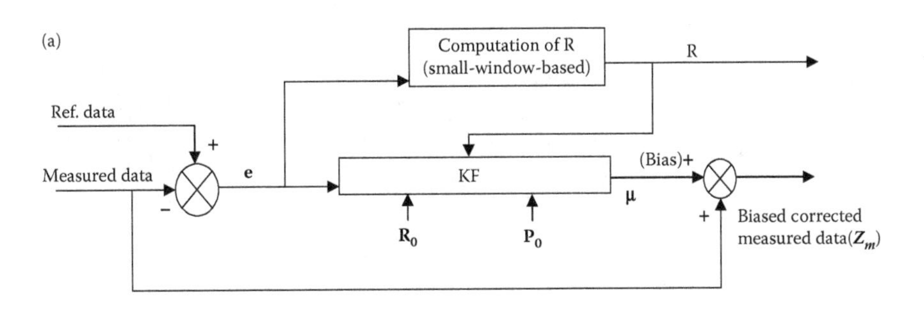

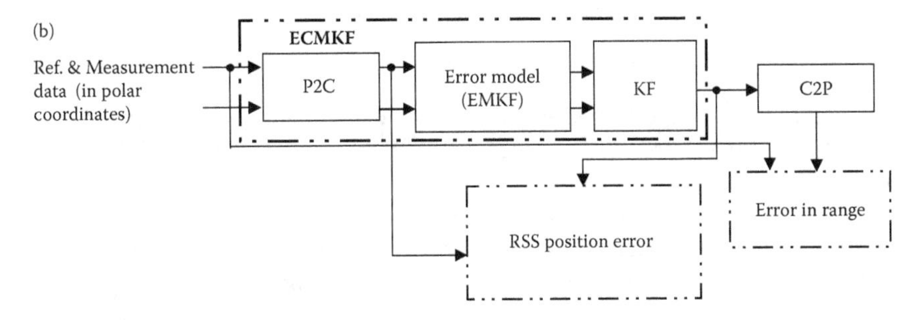

FIGURE 4.9
Error model–based converted measurements Kalman Filter. (a) EMKF-error model KF; (b) ECMKF Schematic.

between the converted radar data and the GPS/DGPS reference data, and the noise statistics. These error estimates can be used for the correction of actual measurements. The error state model is given as

$$\delta \mathbf{X}_{k+1} = F\,\delta \mathbf{X}_k + G w_k \tag{4.44}$$

$$\delta \mathbf{Z}_k = H \delta \mathbf{X}_k + v_k \tag{4.45}$$

Here, $\delta \mathbf{X}$ is the vector of position and velocity errors in all the three axes and $\delta \mathbf{Z}$ is a vector of computed position error in all the three axes (GPS–radar measurement). This estimated position error is used to correct the converted data that are used for measurement update in the KF. The ECMKF algorithm follows the conventional linear KF equations:

Initialization:

$$\hat{X}_{0|0} = \bar{X}_0, \quad \hat{P}_{0|0} = P_{X_0} \tag{4.46}$$

Time propagation:

$$\begin{aligned} \tilde{X}_{k|k-1} &= F_{k-1}\hat{X}_{k-1|k-1} \\ \tilde{P}_{k|k-1} &= F_{k-1}\hat{P}_{k-1|k-1}F_{k-1} + G_{k-1}Q_{k-1}G_{k-1}^T \end{aligned} \tag{4.47}$$

State update by measurements:

$$\begin{aligned} S &= H\tilde{P}_{k|k-1}H^T + R \\ K &= \tilde{P}_{k|k-1}H^T(S)^{-1} \\ \hat{X}_{k|k} &= \tilde{X}_{k|k-1} + K(Z_m - H\tilde{X}_{k|k-1}) \\ \hat{P}_{k|k} &= \tilde{P}_{k|k-1} - KSK^T \end{aligned} \tag{4.48}$$

The scheme for ECMKF is given in Figure 4.9 [76].

4.4.1.2 Error Model Modified Extended Kalman Filter Algorithm

In this scheme, the radar measurements are processed one component at a time in a preferred order: elevation, azimuth, and range. The initialization and time propagation are done using Equations 4.46 and 4.47. The error model states are r, θ, and ϕ. The measurement update with respect to the range measurement includes extra terms in the measurement covariance part to account for nonlinear cross-coupling among the range, azimuth, and elevation measurements. Figure 4.10 shows the scheme of EMEKF [75,76]. It is assumed that the measurements are processed starting with ϕ and predicted states and then θ and range [73,76].

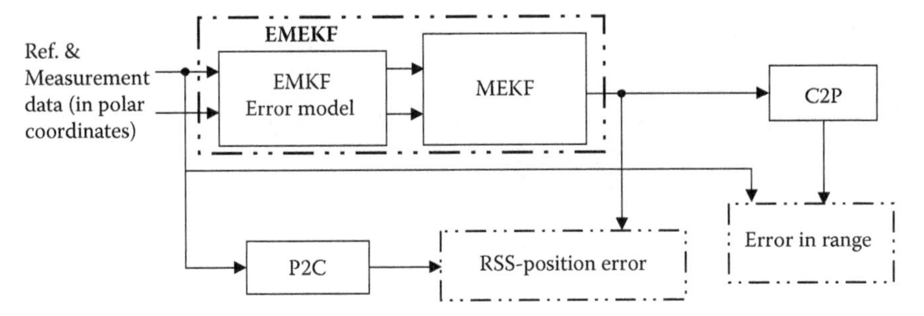

FIGURE 4.10
The error model modified extended Kalman filter scheme.

The update by elevation measurement is

$$\tilde{X}_{k|k-1} = \left[\tilde{x}_{k|k-1} \tilde{y}_{k|k-1} \tilde{z}_{k|k-1} \tilde{\dot{x}}_{k|k-1} \quad \tilde{\dot{y}}_{k|k-1} \quad \tilde{\dot{z}}_{k|k-1} \right]^T$$

$$\begin{aligned}
\overline{r}_k &= \sqrt{\left(\tilde{x}^2_{k|k-1} + \tilde{y}^2_{k|k-1} + \tilde{z}^2_{k|k-1}\right)} \\
\overline{\theta}_k &= \tan^{-1}\left(\frac{\tilde{y}_{k|k-1}}{\tilde{x}_{k|k-1}}\right) \\
\overline{\phi}_k &= \tan^{-1}\left(\frac{\tilde{z}_{k|k-1}}{\sqrt{\tilde{x}^2_{k|k-1} + \tilde{y}^2_{k|k-1}}}\right)
\end{aligned} \tag{4.49}$$

$$H_{k,1} = \left[\frac{-\cos\overline{\theta}_k \sin\overline{\phi}_k}{\overline{r}_k} \quad \frac{-\sin\overline{\theta}_k \sin\overline{\phi}_k}{\overline{r}_k} \quad \frac{\cos\overline{\phi}_k}{\overline{r}_k} \quad 0 \; 0 \; 0 \right]$$

$$S_{k,1} = H_{k,1}\tilde{P}_{k|k-1}H^T_{k,1} + \sigma^2_\phi$$

$$K_{k,1} = \frac{\tilde{P}_{k|k-1}H^T_{k,1}}{S_{k,1}}$$

$$\hat{X}_{k,1} = \tilde{X}_{k|k-1} + K_{k,1}(\phi^m_k - \overline{\phi}_k)$$

$$\hat{P}_{k,1} = \tilde{P}_{k|k-1} - K_{k,1}S_{k,1}K^T_{k,1}$$

The update by azimuth measurement is

$$\begin{aligned}
\hat{X}_{k,1} &= \left[\hat{x}_{k,1}\,\hat{y}_{k,1}\,\hat{z}_{k,1}\,\hat{\dot{x}}_{k,1}\quad \hat{\dot{y}}_{k,1}\quad \hat{\dot{z}}_{k,1}\right]^T \\
\bar{r}_k &= \sqrt{\left(\hat{x}_{k,1}^2 + \hat{y}_{k,1}^2 + \hat{z}_{k,1}^2\right)} \\
\bar{\theta}_k &= \tan^{-1}\left(\frac{\hat{y}_{k,1}}{\hat{x}_{k,1}}\right) \\
\bar{\phi}_k &= \tan^{-1}\left\{\frac{\hat{z}_{k,1}}{\sqrt{\hat{x}_{k,1}^2 + \hat{y}_{k,1}^2}}\right\} \\
H_{k,2} &= \left[\frac{-\sin\bar{\theta}_k}{\bar{r}_k\cos\bar{\phi}_k}\quad \frac{\cos\bar{\theta}_k}{\bar{r}_k\cos\bar{\phi}_k}\quad 0\quad 0\quad 0\quad 0\right] \\
S_{k,2} &= H_{k,2}\hat{P}_{k,1}H_{k,2}^T + \sigma_\theta^2 \\
K_{k,2} &= \frac{\hat{P}_{k,1}H_{k,2}^T}{S_{k,2}} \\
\hat{X}_{k,2} &= \hat{X}_{k,1} + K_{k,2}\left(\theta_k^m - \bar{\theta}_k\right) \\
\hat{P}_{k,2} &= \hat{P}_{k,1} - K_{k,2}S_{k,2}K_{k,2}^T
\end{aligned} \tag{4.50}$$

The update by range measurement is

$$\begin{aligned}
\hat{X}_{k,2} &= \left[\hat{x}_{k,2}\,\hat{y}_{k,2}\,\hat{z}_{k,2}\,\hat{\dot{x}}_{k,2}\quad \hat{\dot{y}}_{k,2}\quad \hat{\dot{z}}_{k,2}\right]^T \\
\bar{r}_k &= \sqrt{\left(\hat{x}_{k,2}^2 + \hat{y}_{k,2}^2 + \hat{z}_{k,2}^2\right)} \\
\bar{\theta}_k &= \tan^{-1}\left(\frac{\hat{y}_{k,2}}{\hat{x}_{k,2}}\right) \\
\bar{\phi}_k &= \tan^{-1}\left\{\frac{\hat{z}_{k,2}}{\sqrt{\hat{x}_{k,2}^2 + \hat{y}_{k,2}^2}}\right\} \\
H_{k,3} &= \left[\cos\bar{\theta}_k\cos\bar{\phi}_k\quad \sin\bar{\theta}_k\cos\bar{\phi}_k\quad \sin\bar{\phi}_k\quad 0\quad 0\quad 0\right] \\
S_{k,3} &= H_{k,3}\hat{P}_{k,2}H_{k,3}^T + \sigma_r^2 + \left(\frac{\bar{r}_k^2(\sigma_\phi^4 + \sigma_\theta^4)}{2}\right) \\
K_{k,3} &= \frac{\hat{P}_{k,2}H_{k,3}^T}{S_{k,3}}
\end{aligned} \tag{4.51}$$

$$\mu_k^r = \frac{\bar{r}_k\left[\left(\theta_k^m+\bar{\theta}_k\right)^2+\left(\phi_k^m+\bar{\phi}_k\right)^2+\sigma_\theta^2+\sigma_\phi^2\right]}{2}$$
$$\hat{X}_{k|k} = \hat{X}_{k,2}+K_{k,3}\left(r_k^m+\bar{r}_k+\mu_k^r\right)$$
$$\hat{P}_{k|k} = \hat{P}_{k,2}+K_{k,3}S_{k,3}K_{k,3}^T$$

Here, r_k^m, θ_k^m, and ϕ_k^m are the radar measurements at k-th scan. For clarity and completeness, the details of CMKF-D [72] are also given in Section 4.4.2.1.1.

4.4.2 Discussion of Results

Two sets of simulated data (set 1 and set 2) with different measurement accuracies are generated for validation [75,76]. The appropriate models for simulation are given as follows.

The state transition matrix F is defined as

$$F = \begin{bmatrix} 1 & 0 & 0 & T & 0 & 0 \\ 0 & 1 & 0 & 0 & T & 0 \\ 0 & 0 & 1 & 0 & 0 & T \\ 0 & 0 & 0 & 1 & 0 & 0 \\ 0 & 0 & 0 & 0 & 1 & 0 \\ 0 & 0 & 0 & 0 & 0 & 1 \end{bmatrix}$$

Here, T is sampling time interval in seconds. The process noise gain matrix G is defined as

$$G = \begin{bmatrix} T^2/2 & 0 & 0 & 0 & 0 & 0 \\ 0 & T^2/2 & 0 & 0 & 0 & 0 \\ 0 & 0 & T^2/2 & 0 & 0 & 0 \\ 0 & 0 & 0 & T & 0 & 0 \\ 0 & 0 & 0 & 0 & T & 0 \\ 0 & 0 & 0 & 0 & 0 & T \end{bmatrix}$$ and the observation matrix H is defined as $H = \begin{bmatrix} 1 & 0 & 0 & 0 & 0 & 0 \\ 0 & 1 & 0 & 0 & 0 & 0 \\ 0 & 0 & 1 & 0 & 0 & 0 \end{bmatrix}$. The data are generated with the following initial conditions: (1) [100 –100 100] (m) for position; (2) [5 –5 5] (m/s) for velocity; (3) Q = 0.25; and (4) 500 data points with a sampling

interval of 1.0 seconds. Noise is added to the true data with the following standard deviations for set 1 data: $\sigma_r = 30$ m; $\sigma_\theta = 0.015°$; $\sigma_\phi = 0.015°$; and for set 2 data: $\sigma_r = 30$ m; $\sigma_\theta = 1.5°$; $\sigma_\varphi = 1.5°$.

For set 1 and set 2 data, the range, azimuth, and elevation errors for ECMKF and EMEKF were studied. The range errors were found well within the theoretical bounds. Here, the bounds vary because the computation is based on the windowing method. Initially, the azimuth and elevation errors for ECMKF were outside the theoretical bounds. The performance of the two algorithms showed comparable RSSPE for data set 1, and EMEKF indicated lower RSSPE than the ECMKF for data set 2. Tables 4.6, 4.7, and 4.8 show performance metrics for data sets 1, 2, and 3, respectively. When the angular accuracies of the measurements from the radar were low, the EMEKF performed better than the ECMKF. The performance of the two algorithms is nearly similar when the angular radar measurements are accurate as seen from Table 4.7 and 4.8. For a Monte-Carlo simulation of 25 runs for set 2 data, the seed number for process noise was kept constant. We see from Table 4.7 that, on average, EMEKF performs better than ECMKF algorithm. The performance of the EMEKF and ECMKF algorithms for data set 3, independently generated by another agency, for a moving aircraft tracked by a ground-based radar for which accurate GPS position measurements were available, was also

TABLE 4.7

Metrics for Set 1 Data for Single Run

	PFE in Polar Frame (w.r.t. Reference)			PFE in Cartesian Frame (w.r.t. Reference)		
Methods	**Range**	**Azimuth**	**Elevation**	**X-Position**	**Y-Position**	**Z-Position**
ECMKF	0.204	0.369	0.446	0.237	0.247	0.233
EMEKF	0.221	0.008	0.009	0.214	0.219	0.226

Source: Kashyap, S. K., G. Girija, and J. R. Raol. 2006. *Def Sci J* 56(5):679–92. With permission.

TABLE 4.8

Metrics for Set 2 Data (Single/Multiple Runs)

	PFE in Polar Frame (w.r.t. Reference)			PFE in Cartesian Frame (w.r.t. Reference)		
Methods	**Range**	**Azimuth**	**Elevation**	**X-Position**	**Y-Position**	**Z-Position**
ECMKF	0.388(0.299)	0.974(0.649)	0.777(0.544)	1.343(0.894)	0.7998(0.575)	0.551(0.307)
EMEKF	0.239(0.182)	0.5235(0.581)	0.481(0.237)	0.728(0.749)	0.563(0.465)	0.369(0.212)

Values in parentheses are computed based on Monte-Carlo simulation of 25 runs.

Source: Kashyap, S. K., G. Girija, and J. R. Raol. 2006. *Def Sci J* 56(5):679–92. With permission.

TABLE 4.9

Metrics for Set 3 Data (Data Supplied by Some Agency)

	PFE in Polar Frame (w.r.t. Reference)			PFE in Cartesian Frame (w.r.t. Reference)		
Methods	**Range**	**Azimuth**	**Elevation**	***X*-Position**	***Y*-Position**	***Z*-Position**
ECMKF	0.0010	0.0013	0.0409	0.0001	0.0215	0.0606
EMEKF	0.0004	0.0017	0.0047	0.0011	0.0283	0.0062

Source: Kashyap, S. K., G. Girija, and J. R. Raol. 2006. *Def Sci J* 56(5):679–92. With permission.

studied (Table 4.9). However, the EMEKF shows somewhat better performance in terms of RSSPE. When the angular accuracies of the measuring radar are low, the EMEKF performs better than the ECMKF.

4.4.2.1 Sensitivity Study on Error Model Modified Extended Kalman Filter

The radar measurements were processed one component at a time in the following order: elevation, azimuth, and then range, assuming that the radar data will give more accurate angular data than the range. The effect on the performance of changing the order of data processing was studied for three cases: case 1—elevation, azimuth, and range; case 2—azimuth, range, and elevation; and case 3—range, elevation, and azimuth. The measurement sequencing had little effect on the performance.

4.4.2.2 Comparison of Debiased Converted Measurements Kalman Filter, Error Model Converted Measurement Kalman Filter, and Error Model Modified Extended Kalman Filter Algorithms

The accuracy of the converted measurements would not only depend on the geometry (range and bearing) but also on the original measurements. In case of a large cross-range error, i.e., range multiplied by bearing error, the converted measurements could have inherent bias that can be corrected by a debiasing technique known as CMKF-D [72]. The technique is based on the following aspects: (1) the measurement noise inaccuracies σ_r, σ_θ should be known; (2) the reference data in the polar frame should be made available, so that the computation of bias and measurement noise covariance R can be performed using a technique called CMKF-D-T (T = true data); and (3) if the reference data is not available, measurement data in the polar frame should be used with a technique called CMKF-D-M (M = measured data). The equations used in CMKF-D-T and those used in CMKF-D-M are from [72] and [76]. Figure 4.11 gives the features and differences of the various algorithms studied in Section 4.4 [75,76].

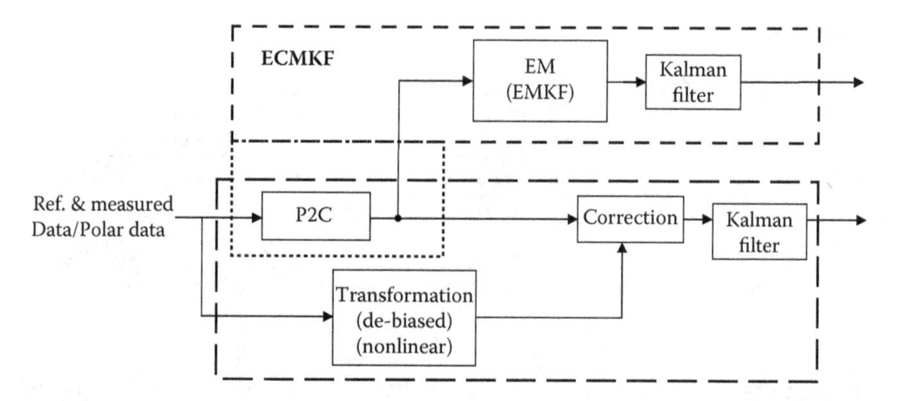

FIGURE 4.11
Error model converted measurement Kalman filter and debiased converted measurements Kalman filter error model schemes. (The lower block is for CMKF-D*; CMKF-D-T; and CMKF-D-M; T = true [reference data]; and M = measured data.)

The salient features of CMKF-D are as follows: (1) it uses linear KF; (2) it converts the measurements from polar to Cartesian using nonlinear transformation (P2C); (3) it uses the debiased technique (through nonlinear transformation) for bias estimation and measurement noise covariance estimation in Cartesian frame; (4) it uses nonlinear transformation (debiased) at every instant of measurement; and (5) it is computationally complex. The salient features of ECMKF are as follows: (1) it uses also linear KF; (2) it converts the measurements from polar to Cartesian using nonlinear transformation (P2C); (3) it uses the error model (EMKF) technique for bias and measurement noise covariance estimation in Cartesian frame; (4) it does not use nonlinear transformation (debiased); and (5) it is computationally less complex. The salient features of EMEKF are as follows: (1) it uses EKF; (2) it does not convert the measurements; (3) it uses the error model (EMKF) technique for bias and measurement noise covariance estimation in the polar frame; (4) it does not use nonlinear transformation (de-biased); and (5) it is computationally less complex.

The performance evaluation of these algorithms is carried out for data set 4. For set 4 data, we have $\sigma_r = 30\,\text{m}$; $\sigma_\theta = 1.5°$, $X(0) = [100\ -100\ 5\ -5]$; $Q = 0.25$; $N = 500$; $T = 1.0$. The results for data set 4 (Table 4.10) show that (1) ECMKF shows better performance as compared to CMKF-D-*T*/CMKF-D-M in terms of PFE; (2) the estimated measurement noise covariance R (for ECMKF) as a time history (not shown here) was, on the average, comparable with that of CMKF-D-*T* and CMKF-D-M methods for the window length of 10 used for R (ECMKF); and (3) EMEKF shows overall better performance in terms of PFE. The use of EMEKF can be further pursued for DF applications.

TABLE 4.10

Metrics for CMKF-D, ECMKF, and EMEKF (Set 4 Data for Single Run)

	PFE in Polar Frame (w.r.t. Reference)		PFE in Cartesian Frame (w.r.t. Reference)	
Methods	**Range**	**Azimuth**	***X*-Position**	***Y*-Position**
CMKF-D-*T*	0.5205	1.4889	0.8432	1.2734
CMKF-D-M	0.5267	1.4960	0.8404	1.2764
ECMKF	0.4867	1.2271	0.7741	1.1584
EMEKF	0.4781	0.6715	0.4062	

Source: Kashyap, S. K., G. Girija, and J. R. Raol. 2006. *Def Sci J* 56(5):679–92. With permission.

4.5 Estimation of Attitude Using Low-Cost Inertial Platforms and Kalman Filter Fusion

Inertial measurement units (IMUs), when combined with GPS, can provide critical information needed for navigation [78]. An IMU system consists of a set of spinning-mass vertical gyros that provide analog outputs and is usually heavy and expensive to maintain. Because of its analog nature, the IMU requires data conversion when interfaced to a digital navigation system. This increases the overall cost and complexity. High-performance inertial-grade IMUs of avionics systems show superior performance specifications. However, because of the high costs, their use is not preferred in many applications. A suitable combination of low-cost microelectrical mechanical sensors (MEMS) and digital signal processing techniques can provide an inexpensive and adaptable alternative to existing IMUs. The closely coupled integration of the sensors, data acquisition systems, and KF-based fusion algorithms allow the MEMS-based IMU to provide an accurate estimation of the attitude of a flight vehicle with a performance comparable to some of the older IMUs.

Two small and extensively instrumented vehicles, ALEX-I and ALEX-II, were developed by and flight-tested at the Institute of Flight Systems and Research, at the German Aerospace Center in Germany, to identify the dynamic behavior of a parafoil-load system (Figure 4.12) [78–81]. The system is also suited for the investigation of guidance, navigation, and control (GNC) concepts for the autonomous landing. A small sensor platform, called a miniaturized inertial platform (MIP), was utilized, consisting of small, low-cost accelerometers, rate gyros, magnetometers, and a temperature transducer. Deriving the position data as the data from the GPS receiver that were available from this small sensor box was very critical. The attitudes must be computed from the output of the three angular

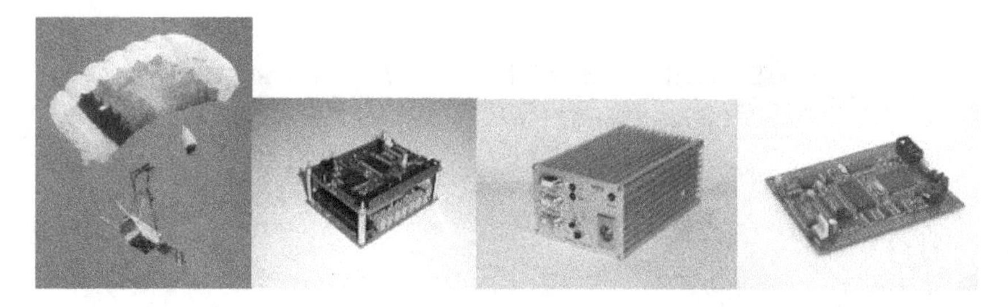

FIGURE 4.12
ALEX system: ALEX, MIPS, and HC12 compact board. (Adapted from Shanthakumar, N., and T. Jann. 2004. In *Sadhana, special issue on multi-sensor data fusion*, ed. J. R. Raol, IAS, Bangalore.)

rate sensors using the strap down algorithm. Measurement errors would accumulate due to numerical integration if no corrections were applied. The KF provides this correction with the fusion of data from other sensors.

The accelerometers provide a roll and pitch attitude reference using gravity (the effect of the Earth). The magnetometer V2**X** (vector 2X) is a two-axis magnetometer that measures the magnetic field in a single plane created by its two sensors (which are perpendicular to each other on the board). It measures the Earth's magnetic field along the body *X*-*Y* plane of the vehicle. If the Earth's magnetic field is known at a particular place (in the geodetic plane), it can be related to the measured magnetic field along the body axis through Euler angles. The accelerometer measurements and the magnetometer measurements provide sufficient information to estimate all three attitudes by the fusion process in EKF: the accelerometers and magnetometers provide an attitude reference and the filter provides corrections to the attitude trajectory computed from the integration of the rate gyros output. The low-cost sensors are modeled according to the calibration tables and used in the filter to compensate for the inaccuracies. The EKF-based fusion algorithm for the estimation of attitudes from low-cost MIP was first realized and studied in a MATLAB/Simulink® environment. Then the algorithm was implemented on the hardware by a programming microcontroller (the Motorola HC12 compact) enclosed inside the MIP box. It was tested by subjecting the MIP to pure angular motion.

The attitudes are derived from the small, low-cost inertial magnetometer sensors. An EKF-based estimation scheme is used to fuse the information coming from three orthogonally mounted accelerometers and a two-axis magnetometer. The accelerometers, together with the magnetometer, provide attitude reference, and the filter provides corrections to the attitude trajectory generated from the integration of the rate gyros output. The accuracy of the attitude estimation is improved by including the sensor models in the estimation algorithm.

Finally, the KF-based fusion algorithm is implemented in the MIP by programming the microcontroller using an ICC12 compiler. Because of the computational speed limitations of the microcontroller, no augmented states (sensor bias states) are included in the model. The decision to implement this minimal set of states (only four quaternion states) model is arrived at based on the assumption that the sensor model used in the filter model is almost accurate and the output is bias free. This four-state model is thoroughly validated in the MATLAB/Simulink setup before implementing on microcontroller. The algorithm is tested and evaluated by subjecting the MIP to pure angular motion and holding it at different orientations.

4.5.1 Hardware System

The MIP is a miniaturized multisensor inertial platform (shown in Figure 4.12 [78]) with two motherboards stacked one above the other; the sensor board consists of all the sensors, and the other board consists of a microcontroller HC12 compact, which is used to acquire and process the sensors' output data. The sensor board consists of (1) three orthogonally mounted rate gyroscopes, (2) three orthogonally mounted linear accelerometers, (3) a temperature transducer, and (4) a two-axis magnetometer with two sensors along the body *X*-*Y* axis. The temperature transducer records the temperature variation effect on the other sensors. These sensors are of a low cost and small size (MEMS class from aerospace consumer bulk market). A GPS receiver connected to the MIP via one of the serial communication interfaces (SCI) can provide the position information. The data processing unit of MIP is the HC12 compact with a 16 Mhz clock speed, and is an universal microcontroller module on the basis of a Motorola MC68HC812A4 microcontroller unit (MCU) [82]. The HC12 compact has the following peripheral units: (1) 512 kB flash memory and 256 kB random access memory (RAM), (2) 12 bit, 11 channels analog-to-digital conversion (ADC), (3) 12 bit, 2 channels digital-to-analog conversion (DAC), (4) controller area network (CAN), (5) RS232 interface driver, (6) beeper, and (7) light-emitting diode (LED) indicator. The outputs of rate gyros, accelerometers, and temperature transducer are analog and are connected to the ADC of the microcontroller board. The output of magnetometer is digital and is connected directly to the microcontroller (figure 2d in [78]).

4.5.2 Sensor Modeling

Low-cost sensors suffer from inaccuracy and are greatly influenced by temperature variation. Such inaccuracies and subjectivity to temperature variations are modeled and compensated for in order to get quality measurements. Based on the calibration data, the sensors' misalignment, temperature drift, and center of gravity (CG) offsets are modeled (see Part 5.2). These models represent the true sensor model and are included as Simulink blocks for generating the sensor output; the inverse model is used in the filter for estimating the attitudes [78].

4.5.2.1 Misalignment Error Model

This model is given as follows:

$$\begin{bmatrix} A_x \\ A_y \\ A_z \end{bmatrix}_m = \begin{bmatrix} 1.0021 & 0.0188 & -0.0146 \\ -0.0093 & 1.0008 & 0.0160 \\ 0.0401 & -0.0137 & 0.9998 \end{bmatrix} \begin{bmatrix} A_x \\ A_y \\ A_z \end{bmatrix}_t$$

$$\begin{bmatrix} p \\ q \\ r \end{bmatrix}_m = \begin{bmatrix} 1.0294 & -0.0024 & 0.0157 \\ 0.0139 & 1.0137 & 0.0100 \\ -0.0091 & -0.0254 & 0.9838 \end{bmatrix} \begin{bmatrix} p \\ q \\ r \end{bmatrix}_t \tag{4.52}$$

4.5.2.2 Temperature Drift Model

This model is given as follows:

$$\begin{bmatrix} A_x \\ A_y \\ A_z \end{bmatrix}_m = \begin{bmatrix} A_x \\ A_y \\ A_z \end{bmatrix}_t + \begin{bmatrix} -0.0049T + 0.2955 \\ -0.0179T + 0.4348 \\ -0.0005T^2 + 0.0425T - 1.0268 \end{bmatrix}$$

$$\begin{bmatrix} p \\ q \\ r \end{bmatrix}_m = \begin{bmatrix} p \\ q \\ r \end{bmatrix}_t + \begin{bmatrix} 4.1420 \times 10^{-5}T^3 - 0.0039T^2 + 0.0876T - 0.3803 \\ 0.0018T^2 - 0.0236T - 0.5370 \\ 1.7881 \times 10^{-4}T^3 - 0.0150T^2 + 0.3564T - 2.2526 \end{bmatrix} \tag{4.53}$$

Here, T is the temperature in °C.

4.5.2.3 CG Offset Model

The offset model is given by:

$$\begin{bmatrix} A_x \\ A_y \\ A_z \end{bmatrix}_m \cong \begin{bmatrix} A_x \\ A_y \\ A_z \end{bmatrix}_t + \begin{bmatrix} -(q^2 + r^2) X_{A_x} \\ -(p^2 + r^2) Y_{A_y} \\ -(p^2 + q^2) Z_{A_z} \end{bmatrix} \tag{4.54}$$

4.5.3 MATLAB®/Simulink Implementation

The attitudes must be computed from the output of the three rate sensors using the strap-down algorithm. However, because of the numerical

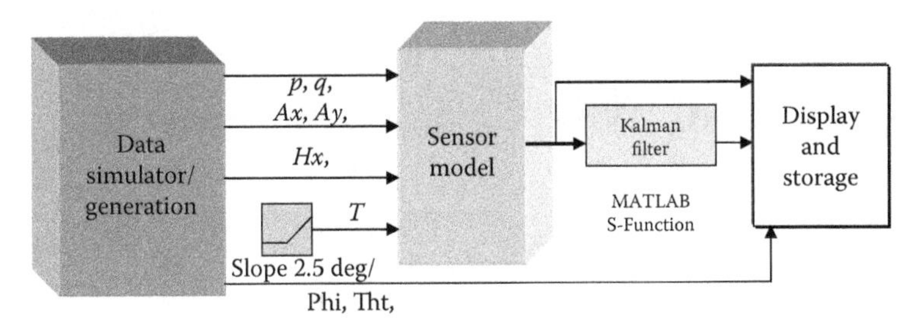

FIGURE 4.13
Blocks with Kalman Filter as MATLAB S-function. (Adapted from Shanthakumar, N., and T. Jann. 2004. In *Sadhana, special issue on multi-sensor data fusion*, ed. J. R. Raol, IAS, Bangalore.)

integration process, the measurement errors will be accumulated if no corrections are made. EKF provides these corrections and is used as a tool for fusing output data from multiple sensors to arrive at an accurate estimation of attitudes and states. Joint estimation of the states and unknown parameters requires augmenting the state vector with the unknown parameters in the EKF. The EKF estimates attitudes and sensor bias by using the measurements from the accelerometers and magnetometers and the known input U_m (rates). The Simulink block is shown in Figure 4.13 (and figures 3a,b,c of [78]), and the EKF algorithm is integrated into the Simulink block as S-function [78]. To avoid singularities while dealing with the cosine rotation matrix, a quaternion formulation is used in the filter model for attitude propagation. A 12-state mathematical model (four quaternion states and eight augmented states) with six observations is initially proposed for attitude estimation as given below [78].

4.6.3.1 State Model

The state model for KF is given as follows:

$$\dot{q}_0 = \frac{1}{2}\left[-q_1\ p_b - q_2\ q - q_3\ r_b\right]$$

$$\dot{q}_1 = \frac{1}{2}\left[q_0\ p_b + q_2\ r_b - q_3\ q_b\right]$$

$$\dot{q}_2 = \frac{1}{2}\left[q_0\ q_b - q_1\ r_b + q_3\ p_b\right]$$

$$\dot{q}_3 = \frac{1}{2}\left[q_0\ r_b + q_1\ q_b - q_2\ p_b\right]$$

$$\dot{B}_p = 0.0$$

$$\dot{B}_q = 0.0$$

$$\dot{B}_r = 0.0$$

$$\dot{B}_{Ax} = 0.0$$

$$\dot{B}_{Ay} = 0.0$$

$$\dot{B}_{Az} = 0.0$$

$$\dot{B}_{Hx} = 0.0$$

$$\dot{B}_{Hy} = 0.0$$

Here, to account for the misalignment (the inverse of Equation 4.52), we have:

$$\begin{bmatrix} p_b \\ q_b \\ r_b \end{bmatrix} = \begin{bmatrix} 0.9713 & 0.0019 & -0.0155 \\ -0.0134 & 0.9862 & -0.0098 \\ 0.0086 & 0.0255 & 1.0161 \end{bmatrix} \begin{bmatrix} p_2 \\ q_2 \\ r_2 \end{bmatrix}$$

To account for the temperature variation (Equation 4.53) we have:

$$\begin{bmatrix} p_2 \\ q_2 \\ r_2 \end{bmatrix} = \begin{bmatrix} p_1 \\ q_1 \\ r_1 \end{bmatrix} - \begin{bmatrix} 4.1420 \times 10^{-5} T_m^3 - 0.0039 T_m^2 + 0.0876 T_m - 0.3803 \\ 0.0018 T_m^2 - 0.0236 T_m - 0.5370 \\ 1.7881 \times 10^{-4} T_m^3 - 0.0150 T_m^2 + 0.3564 T_m - 2.2526 \end{bmatrix}$$

To account for estimated sensor bias, we have:

$$\begin{bmatrix} p_1 \\ q_1 \\ r_1 \end{bmatrix} = \begin{bmatrix} p_m \\ q_m \\ r_m \end{bmatrix} - \begin{bmatrix} B_p \\ B_q \\ B_r \end{bmatrix}$$

4.6.3.2 Measurement Model

The measurement models are given as follows:

$$1.0 = \left[q_0^2 + q_1^2 + q_2^2 + q_3^2 \right]$$

$$A_{x_m} = A_{x_3} + B_{Ax}$$

$$A_{y_m} = A_{y_3} + B_{Ay}$$

$$A_{z_m} = A_{z_3} + B_{Az}$$

$$H_{x_m} = H_{x_b} + B_{Hx}$$

$$H_{y_m} = H_{y_b} + B_{Hy}$$

Here, we have to account for temperature variation (Equation 4.53):

$$\begin{bmatrix} A_{x_3} \\ A_{y_3} \\ A_{z_3} \end{bmatrix} = \begin{bmatrix} A_{x_2} \\ A_{y_2} \\ A_{z_2} \end{bmatrix} + \begin{bmatrix} -0.0049T_m + 0.2955 \\ -0.0179T_m + 0.4348 \\ -0.0005T_m^2 + 0.0425T_m - 1.0268 \end{bmatrix}$$

To account for misalignment (Equation 4.52), we have:

$$\begin{bmatrix} A_{x_2} \\ A_{y_2} \\ A_{z_2} \end{bmatrix} = \begin{bmatrix} 1.0021 & 0.0188 & -0.0146 \\ -0.0093 & 1.0008 & 0.0160 \\ 0.0401 & -0.0137 & 0.9998 \end{bmatrix} \begin{bmatrix} A_{x_1} \\ A_{y_1} \\ A_{z_1} \end{bmatrix}$$

To account for CG offset (Equation 4.54), we have:

$$\begin{bmatrix} A_{x_1} \\ A_{y_1} \\ A_{z_1} \end{bmatrix} = \begin{bmatrix} A_{x_b} \\ A_{y_b} \\ A_{z_b} \end{bmatrix} - \begin{bmatrix} (q^2 + r^2) X_{A_x} \\ (p^2 + r^2) Y_{A_y} \\ (p^2 + q^2) Z_{A_z} \end{bmatrix}$$

$$\begin{bmatrix} A_{x_b} \\ A_{y_b} \\ A_{z_b} \end{bmatrix} = \begin{bmatrix} q_0^2 + q_1^2 - q_2^2 - q_3^2 & 2(q_0q_3 + q_1q_2) & 2(q_1q_3 - q_0q_2) \\ 2(q_1q_2 - q_0q_3) & q_0^2 - q_1^2 + q_2^2 - q_3^2 & 2(q_0q_1 + q_2q_3) \\ 2(q_0q_2 + q_1q_3) & 2(q_2q_3 - q_0q_1) & q_0^2 - q_1^2 - q_2^2 + q_3^2 \end{bmatrix} \begin{bmatrix} A_{x_g} \\ A_{y_g} \\ A_{z_g} \end{bmatrix}$$

$$\begin{bmatrix} H_{x_b} \\ H_{y_b} \end{bmatrix} = \begin{bmatrix} q_0^2 + q_1^2 - q_2^2 - q_3^2 & 2(q_0q_3 + q_1q_2) & 2(q_1q_3 - q_0q_2) \\ 2(q_1q_2 - q_0q_3) & q_0^2 - q_1^2 + q_2^2 - q_3^2 & 2(q_0q_1 + q_2q_3) \end{bmatrix} \begin{bmatrix} H_{x_g} \\ H_{y_g} \\ H_{z_g} \end{bmatrix}$$

Although the sensor models are derived accurately from the calibration data, additional states are provided in the KF model to account for errors in the sensor model, such as sensor bias. The first observable element (also known as measurement variable) in the measurement model is the quaternion constraint equation [83]. It contains the quaternion states in the range of –1 to +1 and eases the scaling problems in the computation. It is assigned the highest priority by assigning a low measurement covariance.

Filter performance, with the attitudes changing at a rate of 0.1 rad/second, and the filter update time of 0.1 second were obtained; although it gave a reasonably good performance when it was implemented on the microcontroller, the performance deteriorated, mainly because of the slower computational speed of the microcontroller (16 MHz clock speed). With the 12-state model, each update cycle on the microcontroller took nearly 2 seconds. Also, because the system was nonlinear, the filter performance deteriorated when the filter update rate was slow. To have a faster filter update, the filter model size was reduced from 12-state to 4-state, retaining only the quaternion states with an assumption that the sensor model is accurate enough. The estimates from the 12-state model and the 4-state model were reasonably close. This showed the validity of the 4-state model because it estimates attitudes reasonably well. Complete results are presented in [78]. Hence, only the 4-state model was implemented on the microcontroller, and it models the sensors as accurately as possible, reducing the unaccounted bias in their outputs.

4.5.4 Microcontroller Implementation

The outputs of rate gyros, accelerometers, and temperature sensors are analog and are acquired through a 12-bit ADC; an assembly code was developed to acquire the ADC output data. The magnetometer V2X, which is operational in raw mode, gives the digital output and hence was connected directly to the HC12 compact. The magnetometer gives out the data after certain events take place at certain intervals of time. The EKF algorithm was implemented in U-D factorization form, which was coded in embedded C programming. The EKF code, before implementation on the microcontroller, was compiled using PC compiler Borland C++, executed, and tested to reproduce the results obtained from MATLAB/Simulink to ensure that there were no errors in filter code conversion from MATLAB to C. The rate outputs from the gyros, accelerometers, and temperature transducer were acquired via ADC by including the relevant assembly code. All the files were compiled to a machine code using an Image Crafts C (ICC12) compiler [84]. This compiled M/C language code was downloaded onto the microcontroller using TwinPEEKs monitor program [85] residing in the internal EEPROM of the HC12 compact. The ADC outputs acquired were in 12-bit resolution and were converted into engineering units. Similarly, the magnetometer readings were converted into microTesla, the rate gyros and temperature transducer outputs were passed onto the KF algorithm as inputs, and the accelerometers and magnetometer outputs as observations. The estimated quaternion states were then converted to Euler angles. While the algorithm was running, the MIP box was rotated and held at a different orientation, with the estimated attitudes displayed on the terminal. From this experiment, we observed that (1) the estimates of roll and pitch attitude match the orientation of the MIP at an accuracy of approximately ±3°; (2) the heading angle estimated mainly from the magnetometer output, though it seemed to be working, showed large variations due to the effect of the local magnetic field (hard iron effect); when the box was held at a distance from the ferromagnetic elements (local magnetic field), the estimates of heading angle were comparatively better; (3) both roll attitude and headings were tested for a full 360° rotation; though the roll attitude estimation was reasonably accurate at all orientations, the heading estimation at 0° and 180° showed a jump of approximately ±10°; and (4) the pitch attitude was found to be working well at ±90°. The estimates of attitudes from both the 12-state and 4-state models were found comparable (particularly roll and pitch attitudes) as shown in Figure 4.14 [78]. The heading angle was not compensated in the 4-state model, hence the difference.

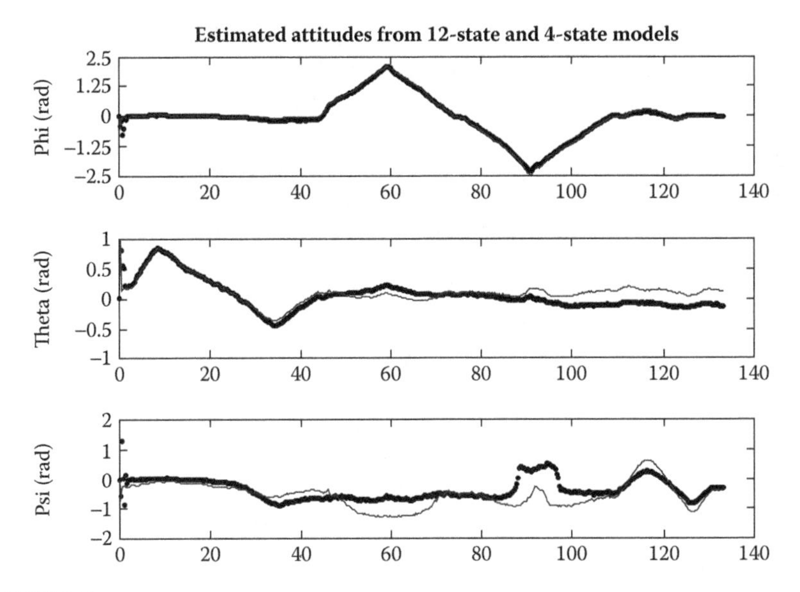

FIGURE 4.14
Attitude estimates from the 12-state model (thick line) and 4-state model (thin line), from real data. (Adapted from Shanthakumar, N., and T. Jann. 2004. In *Sadhana, special issue on multi-sensor data fusion*, ed. J. R. Raol, IAS, Bangalore.)

Epilogue

Reference 86 gives the fusion equations for the state-vector fusion process when common process noise is present in the system. It gives analytical evaluations of the possible effects on data fusion due to such noise, which are ignored in many studies. [87] is one of the earliest satisfactory books on mathematics in data fusion. Work on the evaluation of the H-I filters for systems of the state-vector dimension higher than two (i.e., to include the acceleration state of the moving object) and more than two sensors would be very interesting. Future work could be pursued in the validation of these fusion algorithms with real data from practical applications such as tracking problems, decentralized estimations and control of flexible structures, and mechatronics. Some discussion on data association and fusion algorithms for tracking in the presence of measurement loss can be found in reference 88. Research on the use of multisensor data fusion for sensor failure detection and health monitoring systems is gradually catching up (reference 89 is just an indication). A direct approach to data fusion that estimates the joint distribution of just the variables of interests is presented in reference 90. This approach yields solutions that can be implemented along with conventional statistical models. Interestingly, for marketing applications, it considers variables such as common psychographic and demographic data and the variables to be fused such as media viewing and product purchases. However, the common variables are not used for the estimation of the joint distribution of the fusion variables. "Pooling information" methods can be used for data fusion and decision fusion, as well as for incorporating "reliability" into the fusion process (see Appendix). Some details and development related to data association and interactive multiple model applications can be found in references 91 and 92.

Exercises

I.1. What is the motivation for sensor fusion and what are the general merits and demerits of a multiple sensor suit for data fusion?

I.2. Give specific names for the advantages of sensor data fusion system. (Hint: e.g., robust performance.)

I.3. What are the specific merits and demerits of various sensor data fusion models (Section 2.1)?

I.4. What are the merits and demerits of sensor-network configurations and architectures?

I.5. What is the fundamental difference between the weighted LS and KF *per se* and how are they different from the sensor data fusion point of view?

I.6. In state-vector fusion, is it necessary to invoke the cross-covariance between KFs for fusion? Why or why not?

I.7. What are the merits of using the Bayesian approach for fusion as well as for parameter/state estimation?

I.8. Which fusion method, the data level or the state-vector level, will generally be more accurate?

I.9. Which part of the KF can be regarded as very similar to the recursive least-squares (RLS) algorithm?

I.10. What are the merits of a factorization approach for filtering compared to the basic KF algorithm?

I.11. Establish by simple algebraic steps the relationship between the fusion equations from covariance-state and information-state points of view.

I.12. What are the possible failure states of the sensor suit and actuator suit?

I.13. How can these failures be detected and identified?

I.14. In a sensor data fusion system, one cannot actually improve the accuracy of any single or all sensors. Why not?

I.15. Regarding Exercise I.14, what is really meant by saying that the sensor data fusion improves the accuracy of the overall system? (When fusion is performed, people say they have obtained more accurate results.)

I.16. What is the distinction between sensor fusion and data fusion?

I.17. What are data fusion functions?

I.18. List some military and civilian applications of MSDF.

I.19. Sequence through the data acquisition chain from energy to the acquired knowledge.

I.20. How is the utility of the data fusion system measured?

I.21. What is the traditional view of the KF?

I.22. What is a Bayesian network? What is the use of a dynamic Bayesian network (DBN)?

I.23. What are the constraints of a decentralized data fusion system (DDFS)?

I.24. What are the specific characteristics of a DDFS?

I.25. Is fusion only an additive process? Or could it be otherwise?

I.26. Derive from first principle the state-level fusion expression (refer Section 3.1.1) using the estimated states from two channels and associated covariance matrices as the weighting factors. (Hint: A good estimate would have a smaller covariance and should be given a greater weighting—that could be supplied by the not-so-good estimate!)

I.27. What are the kinematic and dynamic constraints of a mobile sensor platform?

I.28. What are the merits of square-root information filter (SRIF) when compared to the information filter (IF)?

I.29. How is a model transition matrix for the IMM-based filter for tracking a maneuvering target decided?

I. 30. When measurements are converted, why and how are their statistics affected?

I.31. How could some errors creep into the fusion process?

I.32. What is the use of so-called sensor integration functions? (Hint: abstract sensors.)

I.33. What is the advantage gained by the fusion of data from a millimeter-wave radar (MMWR) and IR image?

I.34. Generate simulated (noisy) data using the state and measurement models of Equations 3.6 and 3.7 for two sensors in MATLAB using ftype = 1 (for KFA) and mdata = 1 (with measurement loss in sensor 1), in the MATLAB SW provided with the MSDF book. Use the measured data in a KF and perform state estimation using Equations 3.8 through 3.12 and state-vector fusion using Equations 3.13 and 3.14. Plot position and velocity (state) errors within their theoretical bounds.

I.35. Generate simulated (noisy) data using the state and measurement models of Equation 3.6 and 3.7 for two sensors in MATLAB using ftype = 2 (for GFA) and mdata = 1 (with measurement loss in sensor 1). Use the measured data in GFA and perform both state estimation and fusion using Equations 3.142 through 3.149. Plot position and velocity (state) errors within their theoretical bounds. Observe the difference between the results of Exercise I.34 and this one.

I.36. Generate simulated (noisy) data using the state and measurement models of Equations 3.6 and 3.7 for two sensors in MATLAB (with measurement loss in sensor 2). Use the measured data in H-infinity *a posteriori* estimation (filtering) algorithm and perform state estimation using Equations 3.153 through 3.156 and state-vector fusion using Equations 3.157 and 3.158. Plot position and velocity (state) errors within their theoretical bounds.

I.37. Derive the standard BLUE estimation rule from the equations of the BLUE FR with *a priori* knowledge. What assumption is required?

I.38. If the standard deviation of the angular error of the radar is σ_{ar} and the current estimated range is ρ_{rr}, then what is the variance of the radar range? What is the significance of this relationship?

I.39. What are the background noise sources?

I.40. What is a false alarm?

I.41. What are the probable sources of a false alarm?

I.42. Why is faster data and scan sampling required in target tracking?

I.43. When does an observation satisfy the gate of a given track?

I.44. What is the relation between the Fisher information matrix and the KF covariance matrix?

I.45. What is the merit of the IF compared to the KF?

I.46. What are the features of clutter?

I.47. How might clutter be related to the false alarm?

I.48. What are the major differences between the NNKF, PDAF, and the strongest neighbor KF (SNKF) data association algorithms?

I.49. What is the major problem with NNKF?

I.50. How does PDAF try to circumvent this problem with NNKF (see Exercise I.49)?

I.51. How would you obtain the formulae for optimal WLS FR from the generalized weighted least-squares fusion rule?

References

1. Wilfried, E. 2002. *An introduction to sensor fusion. Research report 47/2001*. Austria: Institute fur Technische Informatik, Vienna University of Technology.
2. Abidi, M. A., and R. C. Gonzalez, eds. 1992. *Data fusion in robotics and machine intelligence*. USA: Academic Press.
3. Dasarathy, B. V. 2001. Information fusion—what, where, why, when, and how? (Editorial) *Inf Fusion* 2:75–6.
4. Hall, D. L. 1992. *Mathematical techniques in multi-sensor data fusion*. Norwood, MA: Artech House.

5. Grossmann, P. 1998. Multisensor data fusion. *GEC J Technol* 15:27–37.
6. Nahin, P. J., and J. L. Pokoski. 1980. NCTR plus sensor fusion equals IFFN or can two plus two equal five? *IEEE Trans Aerosp Electron Syst* 16:320–37.
7. Dasarathy, B. V. 2000. More the merrier ... or is it? Sensor suite augmentation benefits assessment. In: *Proceedings of the 3rd International Conference on Information Fusion*, 2: 20–25. Paris, France.
8. Mutambra, A. G. O. 1998. *Decentralized estimation and control for multisensor systems*. Florida, USA: CRC Press.
9. Dasarathy, B. V. 2000. Industrial applications of multi-sensor information fusion. *Proceedings of IEEE International Conference on Industrial Technology*, Goa, India (Invited Plenary Session Talk).
10. Markin, M., C. Harris, M. Bernhardt, J. Austin, M. Bedworth, P. Greenway, R. Johnston, et al. 1997. *Technology foresight on data fusion and data processing*. Publication of The Royal Aeronautical Society.
11. Shulsky, A. N. 1991. *Silent warfare: Understanding the world of intelligence*. New York: Brassey's.
12. Boyd, J. R. 1987. *A discourse on winning and losing*. Maxwell AFB, Alabama: Unpublished set of briefing slides available at Air University Library.
13. Bedworth, M. D., and J. O'Brien. 1999. The omnibus model: A new architecture for data fusion? In *Proceedings of the 2nd International Conference on Information Fusion* (FUSION'99).
14. Raol, J. R., G. Girija, and J. Singh. 2004. *Modelling and parameter estimation of dynamic systems (IEE control engg,)* Vol. 65. London: IEE.
15. Rong Li, X., Y. M. Zhu, and C. Z. Han. Unified optimal linear estimation fusion—Part I: Unified models and fusion rules. In *Proceedings of 2000 International Conference on Information Fusion*, MoC2.10–MoC2.17, Paris, France.
16. Rong Li, X., and J. Wang. 2000. Unified optimal linear estimation fusion—Part II: Discussions and examples. In *Proceedings of 2000 International conference on information fusion*, MoC2.10–MoC2.17, Paris, France.
17. Rong Li, X., and P. Zhang. Optimal linear estimation fusion—Part III: Cross-correlation of local estimation errors. In *Proceedings of 2000 International Conference on Information Fusion*, WeB1.11–WeB1.18, Montreal, QC, Canada.
18. Rong Li, X., and K. Zhang. Optimal linear estimation fusion—Part IV: Optimality and efficiency of distributed fusion. In *Proceedings of 2000 International Conference on Information Fusion*, WeB1.19–WeB1.26, Montreal, QC, Canada.
19. Rong Li, X., K. Zhang, J. Zhao, and Y. M. Zhu. 2002. Optimal linear estimation fusion—Part V: Relationships. *Proceedings of 5th International Conf. Information Fusion*, 497–504, Annapolis, MD, USA.
20. Dempster, A. P. 1967. Upper and lower probabilities induced by a multivalued mapping. *Annu Math Stat* 38:325–39.
21. Shafer, G. 1976. *A mathematical theory of evidence*. Princeton, NJ: Princeton University Press.
22. Provan, G. M. 1992. The validity of Dempster–Shafer belief functions. *Int J Approx Reason* 6:389–99.
23. Crowley, J. L., and Y. Demazeu. 1993. Principles and techniques of sensor data fusion. *Signal Processing*, 32(2):5–27.

24. Challa, S., and D. Koks. 2004. Bayesian and Dempster–Shafer fusion. *Sadhana* 29(2):145–76.
25. Kadambe, S., and C. Daniell. Sensor/data fusion based on value of information. In *Proceedings of the Sixth International Conference on Information Fusion,* 1:25–32.
26. Leung, L. W., B. King, and V. Vohora. 2001. Comparison of image data fusion techniques using entropy and INI. In *Proceedings of the 22nd Asian conference on remote sensing,* Singapore, November 5–9. http://www.pubs.drdc.gc.ca/inbasket/Laidman.071121_1403.toronto_CR_2007_066.pdf. Accessed December 2008.
27. Iyengar, S. S., and R. R. Brooks, eds. 2005. *Distributed sensor networks.* Boca Raton, FL: Chapman & Hall/CRC Press.
28. Aggarwal, J. K., ed. 1993. *Multisensor fusion for computer vision, NATO ASI Series.* Berlin: Springer-Verlag.
29. Hu, H., and J. Q. Gan. 2005. Sensors and data fusion algorithms in mobile robotics. TR CSM-422, Dept. of Comp. Science, University of Essex, UK.
30. Vaghi, A. 2003–2004. Sensor management by a graphical model approach. Thesis, Politecnico di Milano. http://www.fred.mems.duke.edu/posters/Alberto_GMA_thesis.pdf. Accessed December 2008.
31. Manyika, J., and H. Durrant-Whyte. 1994. *Data fusion and sensor management—decentralized information theoretic approach.* New York: Ellis Horwood Series.
32. Maybeck, P. S. 1982. *Stochastic models, estimation and control,* Vol. 2. New York: Academic Press.
33. Bierman, G. J. 1977. *Factorization methods for discrete sequential estimation.* New York: Academic Press.
34. Julier, S. J., and J. K. Uhlmann. 1997. A new extension of the Kalman filter to non-linear systems. In *Proceedings of AeroSense, 11th International Symposium Aerospace/Defense Sensing, Simulation and Controls.* Bellingham, WA: SPIE.
35. Julier, S. J., and J. K. Uhlmann. 2004. Unscented filtering and non-linear estimation. *Proc IEEE* 92(3):401–22.
36. Kashyap, S. K., and J. R. Raol. 2006. Evaluation of derivative free Kalman filter and fusion in non-linear estimation. Ottawa, ON, Canada: IEEE Canadian Conference on Electrical and Computer Engineering.
37. Kashyap, S. K., and J. R. Raol. 2008. Evaluation of derivative free Kalman filter for non-linear state-parameter estimation and fusion. *J Aeronaut Soc India* 60(2):101–14.
38. Bar Shalom, Y., and X. R. Li. 1995. *Multitarget–multisensor tracking: Principles and techniques.* Storrs, CT: Academic Press.
39. Blackman, S. S. 1986. *Multiple-target tracking with radar applications.* Norwood, MA: Artech House.
40. Naidu, V. P. S., G. Girija, and J. R. Raol. 2005. Data association and fusion algorithms for tracking in presence of measurement loss. *J Inst Eng I AS* 86:17–28.
41. Chang, K. C., and Y. Bar-Shalom. 1994. FUSEDAT: A software package for fusion and data association and tracking with multiple sensors. In *Proceedings of the SPIE conference on signal and data processing of small targets,* Orlando, FL.
42. Mori, S., W. H. Barker, C.-Y. Chong, and K.-C. Chang. 2002. Track association and track fusion with nondeterministic target dynamics. *IEEE Trans Aerosp Electron Syst* 38(2):659–668.

43. Bar-Shalom, Y., and X.-R. Li. 1993. *Estimation and tracking: Principles, techniques, and software.* Boston, MA: Artech House.
44. Haimovich, A. M., J. Yosko, R. J. Greenberg, and M. A. Parisi. Fusion of sensors with dissimilar measurements/tracking accuracies. IEEE Log No. T-AES/29/1/02258.
45. Shanthakumar, N., and G. Girija. 2007. Measurement level and state-vector data fusion implementations. Personal communications and personal notes. Flight Mechanics and Control Division, National Aerospace Laboratories, Bangalore.
46. Kashyap, S. K., N. Shanthakumar, G. Girija, and J. R. Raol. 2003. Sensor data characterization and fusion for target tracking applications. International Radar Symposium India (IRSI), 3–5 December. Bangalore.
47. Zhou, Y., H. Leung, and M. Blanchette. 1999. Sensor alignment with earth-centered earth-fixed co-ordinate system. *IEE Trans AES* 35(2):410–8.
48. Raol, J. R., and G. Girija. 2002. Sensor data fusion algorithms using square-root information filtering. *IEE Proc Radar Sonar Navig* 149(2):89–96.
49. Bar-Shalom, Y. and X.-R. Li. 1993. *Estimation and tracking: Principles, techniques and software*. Boston: Artech House.
50. Blom, H. A. P., and Y. Bar-Shalom. 1988. The interacting multiple model algorithm for systems with markovian switching coefficients. *IEEE Trans Autom Control* 33:780–3.
51. Mazor, E., A. Averbuch, Y. Bar-Shalom, and J. Dayan. 1998. Interacting multiple model methods in target tracking: A survey. *IEEE Trans Aerosp Electron Syst* 34:103–23.
52. Naidu, V. P. S., G. Girija, and N. Santhakumar. 2007. Three model IMM-EKF for tracking targets executing evasive maneuvers. AIAA-66928, 45th AIAA conference on Aerospace Sciences, Reno, USA.
53. Simeonova, L., and T. Semerdjiev. 2002. Specific features of IMM tracking filter design. *Inf Secur Int J* 9:154–65.
54. Schulz, D., W. Burgard, D. Fox, and A. B. Cremers. People tracking with mobile robots using sample-based joint probabilistic data association filters. http://www.informatik.unibonn.de/~schulz/articles/people-tracking-ijrr-03.pdf.
55. Challa, S., R. J. Evans, and X. Wang. 2003. A Bayesian solution and its approximations to out-of-sequence measurement problems. *Inf Fusion* 4(3):185–99.
56. Paik, B. S., and J. H. Oh. 2000. Gain fusion algorithm for decentralized parallel Kalman filters. *IEE Proc Control Theor App* 147(1):97–103.
57. Shanthakumar, N., G. Girija, and J. R. Raol. 2001. Performance of Kalman and gain fusion algorithms for sensor data fusion with measurement loss. International Radar Symposium India (IRSI), Bangalore.
58. Green, M., and D. J. N. Limebeer. 1995. *Linear robust control.* Englewood Cliffs, NJ: Prentice Hall.
59. Hassibi, B., A. H. Sayad, and T. Kailath. 1996. Linear estimation in Krein spaces—part II: Applications. *IEEE Trans Autom Control* 41(1):34–49.
60. Jin, S. H., J. B. Park, K. K. Kim, and T. S. Yoon. 2001. Krein space approach to decentralized H_∞ state estimation. *IEE Proc Control Theor App* 148(6):502–8.
61. Lee, T. H., W. S. Ra, T. S. Yoon, and J. B. Park. 2004. Robust Kalman filtering via Krein space estimation. *IEE Proc Control Theor App* 151(1):59–63.

62. Raol, J. R., and F. Ionescu. 2002. Performance of H-Infinity filter-based data fusion algorithm with outliers. 40th AIAA Aerospace Sciences Meeting & Exhibit, 14–17 January 2002, Reno, NV, USA, A02-13598.
63. Ananthasayanam, M. R., A. K. Sarkar, A. Bhattacharya, P. Tiwari, and P. Vorha. 2005. Nonlinear observer state estimation from seeker measurements and seeker-radar measurements fusion, Paper No AIAA-2005-6066-CP.
64. Zarchan, P. 1997. *Tactical and strategic missile guidance*, Vol. 176, 3rd ed. Progress in Aeronautics and Astronautics.
65. Bar-Shalom, Y., and K. C. Chang. 1989. Tracking a maneuvering target using input estimation versus the interacting multiple model algorithm. *IEEE Trans Aerosp Electron Syst* 26(2):296–300.
66. Mehrotra, K., and P. Mahapatra. 1997. A jerk model for tracking highly maneuvering targets. *IEEE Trans Aerosp Electron Syst* 33(4):1094–105.
67. Vora, P., A. Bhattacharyya, M. Jyothi, P. K. Tiwari, and R. N. Bhattacharjee. 2004. *RF Seeker modeling and Seeker filter design*. Hyderabad, India: National Workshop on Tactical Missile Guidance.
68. Kashyap, S. K., N. Shanthakumar, V. P. S. Naidu, G. Girija, and J. R. Raol. 2007. *State estimation for Pursuer guidance using interacting multiple model based augmented extended Kalman filter*. International Radar Symposium, India, (IRSI) Bangalore.
69. Banerjee, P., and R. Appavu Raj. 2000. Multi-sensor data fusion strategies for real-time application in test and evaluation of rockets/missiles system. IEEE International Conference on Industrial Technology (ICIT), Goa, India.
70. Girija, G., J. R. Raol, R. Appavu Raj, and S. Kashyap. 2000. Tracking filter and multi-sensor data fusion. In *Special issue: Advances in Modeling, System Identification, and Parameter Estimation*, eds. J. R. Raol and N. K. Sinha, *Sadhana* 25(2):159–67.
71. Kashyap, S. K., V. P. S. Naidu, J. Singh, G. Girija, and J. R. Raol. 2006. Tracking of multiple targets using interactive multiple model and data association filter. *J Aerosp Sci Technol Aeronaut Soc India* 58(1):65–74.
72. Lerro, D., and Y. Bar-Shalom. 1993. Tracking with de-biased consistent converted measurements versus EKF. *IEEE Trans Aerosp Electron Syst* 29(3):1015–1022.
73. Park, S.-T., and J. Gyu Lee. 2001. Improved Kalman filter design for three-dimension radar tracking. *IEEE Trans Aerosp Electron Syst* 37(2):727–739.
74. Blackman, S. 1986. *Multiple target tracking with radar applications*. Dedham, MA: Artech House.
75. Kashyap, S. K., G. Girija, and J. R. Raol. 2003. Evaluation of converted measurement- and modified extended-Kalman filters for target tracking. AIAA Conference and Exhibit, Guidance, Navigation and Control, Texas, US, 11–17th August.
76. Kashyap, S. K., G. Girija, and J. R. Raol. 2006. Converted measurement- and modified extended-Kalman filter for target tracking. *Def Sci J* 56(5):679–92.
77. Peter, S. 1979. *Maybeck, stochastic models, estimation and control*, Vol. 1. New York: Academic Press.
78. Shanthakumar, N., and T. Jann. 2004. Estimation of attitudes from a low cost miniaturized inertial platform using Kalman filter based sensor fusion algorithm. In *Sadhana, special issue on multi-sensor data fusion*, ed. J. R. Raol, IAS, Bangalore.

79. Doherr, K. F., and T. Jann. 1997. Test vehicle ALEX for low cost autonomous parafoil landing experiments. 14th AIAA Aerodynamic Decelerator Systems Technology Conference and Seminar, USA, AIAA-97-1543.
80. Jann, T., K. F. Doherr, and W. Gockel. Parafoil test vehicle ALEX- further development and flight test results, AIAA-99-1751.
81. Jann, T. 2001. Aerodynamic model identification and GNC design for the Parafoil-Load system ALEX. *Proccedings of AIAA Aerodynamic Systems Decelerator Conference*, USA, AIAA-2001-2015.
82. Valvano, J. W. 2000. *Embedded microcomputer systems*. USA: Brooks/Cole.
83. Collinson, R. P. G. 1998. *Introduction to avionics*. UK: Chapman and Hall.
84. Anonymous. 2002. ImageCraft C compiler and development environment for motorola HC12 manual, version 6.
85. Anonymous. 1999. HC12Compact hardware manual, Version 1.0.
86. Saha, R. K. 1996. Effect of common process noise on two-sensor track fusion. *J Guid Control Dyn* 19(4):829–835.
87. Goodman, I. R., R. P. S. Mahler, and T. Hung. 1997. *Mathematics of data fusion*. Norwell, MA: Kluwer.
88. Naidu, V. P. S., G. Girija, and J. R. Raol. 2003. Evaluation of data association and fusion algorithms for tracking in the presence of measurement loss. AIAA Paper No. 5733, Austin (USA).
89. Girija, G., C. Zorn, and A. Koch. 2005. Multi sensor data fusion for sensor failure detection and health monitoring. AIAA Guidance, Navigation and Control Conference and Exhibit, 15–18 Aug, AIAA-2005-5843, San Francisco, CA.
90. Gilula, Z., R. E. Mcculloch, and P. E. Rossi. 2006. A direct approach to data fusion. *J Mark Res* 43(1):73–83.
91. Girija, G., and J. R. Raol. 2001. Comparison of methods for association and fusion of multi-sensor data for tracking applications. AIAA Guidance, Navigation, and Control Conference and Exhibit. Montreal, Quebec, Canada. Paper No. AIAA-2001–4107, 6–9 August 2001.
92. Kashyap, S. K., V. P. S. N. Naidu, J. Singh, G. Girija, and J. R. Raol. 2004. Multiple target tracking using IMMPDAF filter. National Workshop on Tactical Guidance, DRDL, Hyderabad.

Part II

Fuzzy Logic and Decision Fusion

J. R. Raol and S. K. Kashyap

5

Introduction

When we have multiple situations with several different analysis results, we often need to choose a situation or result that will lead us to our desired goal with reasonable accuracy. We need to use available knowledge and information as well as logic or a statistical method to arrive at an appropriate decision. Decision making is often (and perhaps always) coupled with knowledge representation, which is derived through an inference process or engine, operating with the measured data or collected raw information. In Part II, we emphasize the use of fuzzy logic (Type I FL/S; see Epilogue) for decision making and fusion.

Fuzzy logic (FL) has been understood for the last four decades [1]. Even before that, a third possibility beyond the "true" or "false" options given by binary logic might have been indicated. The third value can be translated as "possible," with a numeric value between true and false, and there can be infinite-valued logic possible to represent certain kind of uncertainty. Zadeh's FL facilitates the modeling or representation of conditions that are inherently imprecisely defined. Fuzzy techniques (based on FL—the techniques in themselves may not be fuzzy), as a part of approximate reasoning, provide decision support and expert systems with reasoning capabilities. FL techniques have been used in various applications: (1) image-analysis, e.g. detection of edges, feature extraction, classification, and clustering; (2) parameter estimation of unknown dynamic systems, e.g. aircraft; (3) home appliances, e.g. washing machines, air conditioning systems; (4) design of control systems; (5) sensor failure detection, identification, and isolation; (6) reconfiguration control; and (7) decision fusion, e.g. situation and threat assessment [2–4]. FL has an inherent ability to approximately mimic the human mind so that it can be used for reasoning that is more approximate rather than exact.

In Part II, we will summarize FL concepts and their potential applications in both engineering and nonengineering fields. Part II introduces FL concepts, fuzzy sets, membership functions for fuzzification processes, fuzzy operators, fuzzy implication methods (FIM), aggregation methods, defuzzification methods, and the steps of fuzzy inference engine, or process (FIE). It highlights: (1) the relationship between fuzzy operators such as T-norms and S-norms, (2) the interconnectivities of fuzzy operators and FIM, and (3) the contradistinctions among various operations using numerical simulation examples [3,5–7]. The use of FL in recursive state estimation and sensor data fusion is also considered. The algorithms are developed by considering the combination of FL and Kalman filter (KF). Two schemes, KF and fuzzy KF (FKF), are used for target-tracking applications and the evaluation of performance. The concept of FL is extended to state-level data fusion for similar sensors. The performances of FL-based fusion methods are compared with the conventional fusion method, that is, state-vector fusion (SVF), to track a maneuvering target.

We will need to select an appropriate implication method from the existing methods. However, if any new implication method is found, it should satisfy some of the intuitive criteria of forward chain logic/generalized *modus ponens* (GMP) and backward chain logic/generalized *modus tollens* (GMT) so that it can be fitted into the process of system development using FL. We will need to develop a procedure to find out if any of the existing implication methods satisfies the given set of intuitive criteria for GMP and GMT. To realize this scheme, MATLAB® and graphics have been used to develop a user-interactive package to evaluate implication methods with respect to these criteria [5,7]. The criteria of GMP/GMT [3,7] and the derivations and an explanation of various FIM are given. The steps required to interpret the implication methods with respect to these criteria are also given. Some new FIM are proposed and derived using material implications, propositional calculus, and fuzzy operators. These FIM are evaluated with GMP/GMT using MATLAB or a graphics-based tool, as referred to above. We will consider one situation in which decision fusion can be used: in the battlefield, where it can be used to determine a final decision based on the entire surveillance volume at any time, using outputs from different levels, e.g., level 1—object refinement, and level 2—situation refinement of the multisensor data fusion (MSDF) system. This will include some examples of decision support systems realized using a MATLAB/Simulink® environment or MATLAB-based FL toolbox. These examples are only a part of the many situations that can be encountered in combat scenarios, such as air-to-air, air-to-ground, ground-to-air, and so on. A similar process is also applicable to other nonmilitary systems, but the details of the system would be different.

6

Theory of Fuzzy Logic

Crisp logic is traditionally used in decision making and related analyses. The theory of probability is based on this logic, through the medium of classical set theory. In crisp logic, there are only two discrete states: "yes" or "no," 0 or 1, –1 or +1, and "off" or "on." There is no third possibility here—either a person is in the room or not, an event will occur on not, a light bulb is on or off, and so on. Real-life experiences prove that many more conditions than these two options are possible: the light could be dim, the day could be a certain degree of brightness or darkness, the weather could be warm, hot, hotter, or cold, very cold, and so on. This complexity necessitates variations in the degree of uncertainty, and hence, truth and falsity (1 or 0, respectively) are the two extremes of a continuous spectrum of uncertainty. This leads to multivalued logic and to fuzzy logic (FL)—a theory of sets wherein the characteristic function is generalized to assume an infinite number of values between 0 and 1 (e.g., a triangular form). In fact, the theory of possibility [4] is based on FL, in a manner similar to the theory of probability, which is based on crisp logic (via set theory).

FL-based methods have found excellent applications in industrial control systems, home appliances, robotics, and aerospace engineering. The conventional control system can be regarded as a (minimally) intelligent-control system, if it uses FL (or any other logic) and/or any learning mechanism. If, for example, $y >$ some number, then use one controller or use another one—this is conditional logic. Essentially, artificial intelligence (AI)-based systems should have strong logical capability and very good learning and adaptive mechanisms. An FL-based controller is suited (1) to keep output variables between the limits; and (2) to maintain control actuation (i.e., control input or related variable) between the limits [2]. Thus, FL is a form of logic that can be used in the design and operation of an intelligent control. FL deals with vagueness, rather than uncertainty. For example, if a patient has a "severe" headache, then there is a "good chance" that she or he has a migraine. To avoid contradictions in the rules, a certain degree of human intervention is sometimes needed at certain stages to tune various adjustable parameters in the FL system (FLS) and rule base. FL-based control is suitable when (1) the control-system dynamics are slow and/or nonlinear; (2) the models of the plant are not available; and (3) competent human operators who can provide expert rules are available [2].

The development of an FLS requires that the following tasks be accomplished: (1) select fuzzy sets and their membership functions for the fuzzification process; (2) create a rule-base with the help of a domain expert, for input-output mapping; (3) select fuzzy operators in the fuzzy implication and aggregation process; (4) select the fuzzy implication and aggregation methods; and (5) select a defuzzification method. The concepts of FL, fuzzy sets, their properties, FL operators, fuzzy proposition- and rule-based systems, fuzzy maps and inference engines, defuzzification methods, and the evaluation of fuzzy implication functions are studied in this chapter.

6.1 Interpretation and Unification of Fuzzy Logic Operations

In this section, the relationships, interconnectivities, and contradistinctions among various operations and operators used in FLS [2–8] are explained using numerical simulations and examples.

6.1.1 Fuzzy Sets and Membership Functions

In principle, a fuzzy set is an internal extension or expansion of a crisp set, within the limits of the crisp set; hence, FL is richer than a crisp set (logic). A crisp set allows only full membership (1, an event occurred) or no membership at all (0, event did not occur), whereas a fuzzy set allows partial membership for a member of a set (belonging to the set). A crisp logic–membership function is defined as follows:

$$\mu_A(u) = \begin{cases} 1 & \text{if } u \in A \\ 0 & \text{if } u \notin A \end{cases} \tag{6.1}$$

Here, $\mu_A(u)$ is a membership function that characterizes the elements u of set A, in Figure 6.1.

A fuzzy set A based on a universe of discourse (UOD) U with elements u is expressed as

$$A = \int \{\mu_A(u)/u\} \forall\, u \in U \tag{6.2}$$

or

$$A = \sum \{\mu_A(u)/u\} \forall\, u \in U \tag{6.3}$$

Here, $\mu_A(u)$ is a membership function (a fuzzy membership function or FMF) of u in the set A and provides a mapping of UOD U in the closed

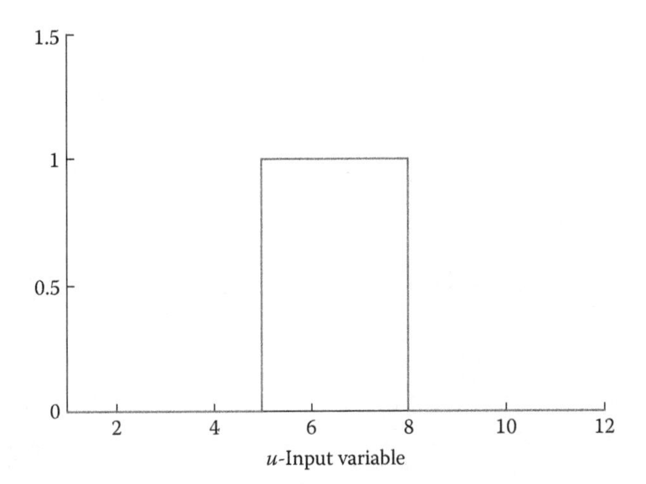

FIGURE 6.1
Membership functions of a typical crisp set.

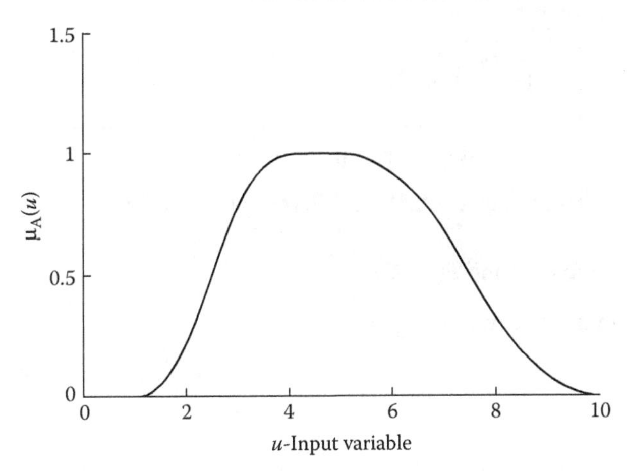

FIGURE 6.2
Membership functions of a typical fuzzy set.

interval [0,1]. The term $\mu_A(u)$ is simply a measure of the degree to or by which u belongs to the set A, i.e., $\mu_A(u)$: $U \rightarrow [0,1]$ (see Figure 6.2). It should be noted here that the notations $\int$ and Σ represent only a fuzzy set and are not related to the usual integration and summation interpretations. A fuzzy variable is a name of the element u, whose values can be considered labels of fuzzy sets; for example, *temperature* as a fuzzy variable (or linguistic variable), with a numeric value (u), can be within the range of $U = [0,100]$. The fuzzy variable can assume different labels defined by

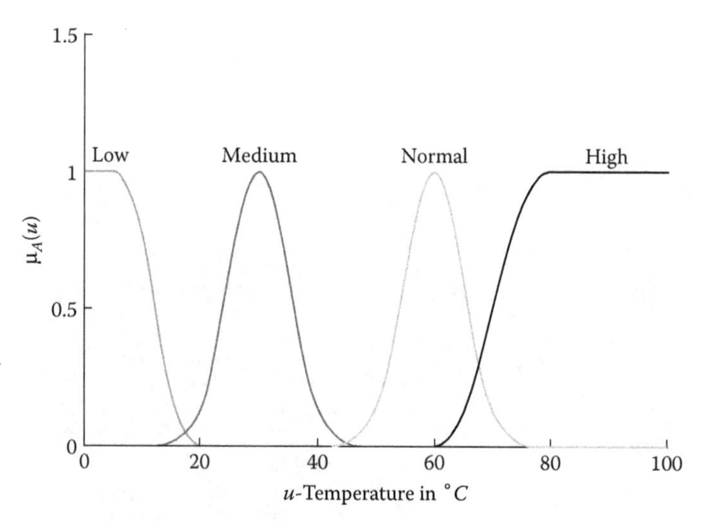

FIGURE 6.3
Membership functions of the fuzzy variable *temperature* (at various levels).

linguistic values, such as *low*, *medium*, *normal*, and *high*, with each represented by different membership functions as shown in Figure 6.3 [2].

6.1.2 Types of Fuzzy Membership Functions

There are several types of FMFs [2–7], described below.

6.1.2.1 Sigmoid-Shaped Function

This membership function is given as

$$\mu_A(u) = \frac{1}{1 + e^{-a(u-b)}} \tag{6.4}$$

Here, u is the fuzzy variable in the UOD U, and a, b are constants that shape the membership function, which is a sigmoid (or sigmoidal). The sign of parameter a decides whether the sigmoid membership function will open to the right or to the left. Figure 6.4 shows the membership function for $a = 2$, $b = 4$, and $U \in [0, 10]$. Interestingly, the sigmoid nonlinearity is also used in artificial neural networks (ANNs) as a nonlinear activation function.

6.1.2.1 Gaussian-Shaped Function

The corresponding membership function is given as

$$\mu_A(u) = e^{\frac{-(u-a)^2}{2b^2}} \tag{6.5}$$

Here, a and b signify the mean and standard deviation of the membership function, respectively; the function is distributed around parameter a, and parameter b decides the width of the function. Figure 6.5 shows the membership function for $a = 5$ and $b = 2$. In FL, the G (Gaussian)-curve is used without attaching any specific probabilistic meaning; however, the shape of the function might appear similar to that of the Gaussian probability density function.

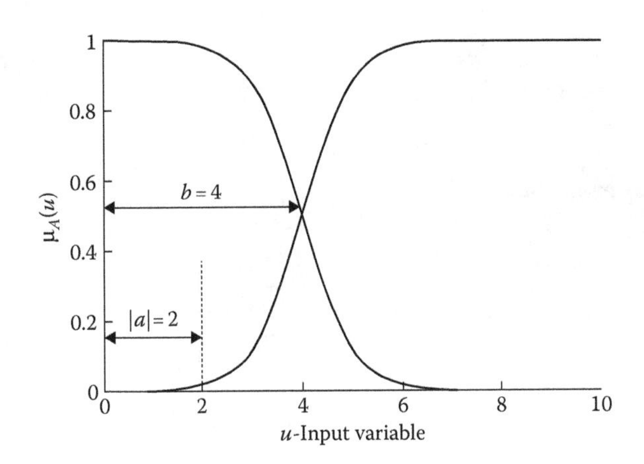

FIGURE 6.4
Sigmoid membership function (right open when $a = 2$, and left open when a = −2).

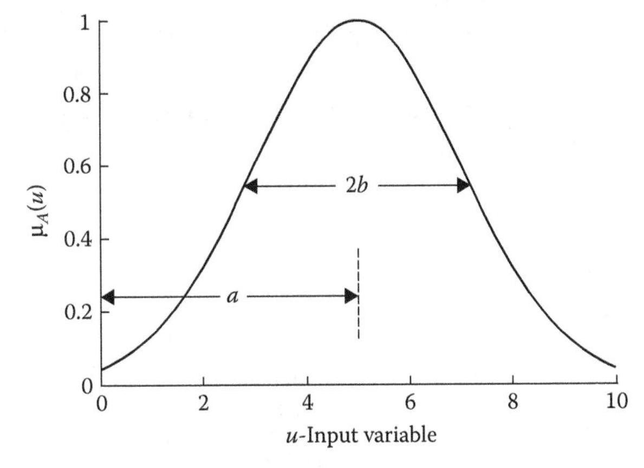

FIGURE 6.5
Gaussian membership function.

6.1.2.3 Triangle-Shaped Function

This membership function is given as

$$\mu_A(u) = \begin{cases} 0 & \text{for } u \le a \\ \dfrac{u-a}{b-a} & \text{for } a \le u \le b \\ \dfrac{c-u}{c-b} & \text{for } b \le u \le c \\ 0 & \text{for } u \ge c \end{cases} \tag{6.6}$$

Here, parameters a and c are the "bases," and b signifies the "peak" of the membership function. Figure 6.6 shows a triangular FMF for $a = 3$, $b = 6$, and $c = 8$.

6.1.2.4 Trapezoid-Shaped Function

This function is given as

$$\mu_A(u) = \begin{cases} 0 & \text{for } u \le d \\ \dfrac{u-d}{e-d} & \text{for } d \le u \le e \\ 1 & \text{for } e \le u \le f \\ \dfrac{g-u}{g-f} & \text{for } f \le u \le g \\ 0 & \text{for } u \ge g \end{cases} \tag{6.7}$$

Here, the parameters d and g define the "bases" of the trapezoid, and the parameters e and f define the "shoulders." Figure 6.7 is a trapezoidal FMF for $d = 1$, $e = 5$, $f = 7$, and $g = 8$. It should be noted here that the triangular- and the trapezoid-shaped membership functions are not generally differentiable, in the strictest sense.

6.1.2.5 S-Shaped Function

This function is given as

$$\mu_A(u) = S(u;a,b,c) = \begin{cases} 0 & \text{for } u \le a \\ \dfrac{2(u-a)^2}{(c-a)^2} & \text{for } a \le u \le b \\ 1 - \dfrac{2(u-c)^2}{(c-a)^2} & \text{for } b \le u \le c \\ 1 & \text{for } u \ge c \end{cases} \tag{6.8}$$

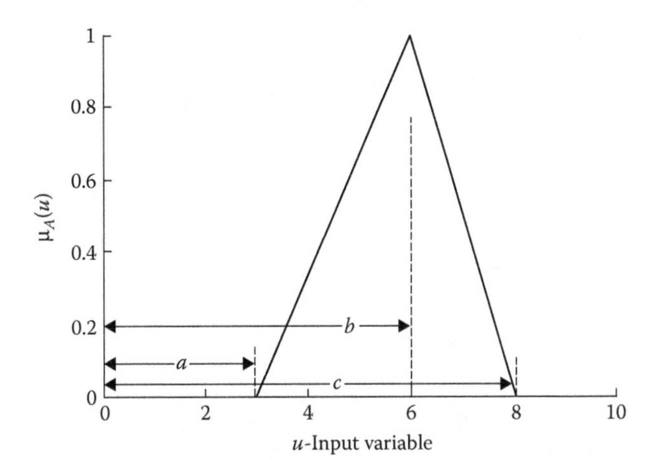

FIGURE 6.6
Triangle-shaped membership function.

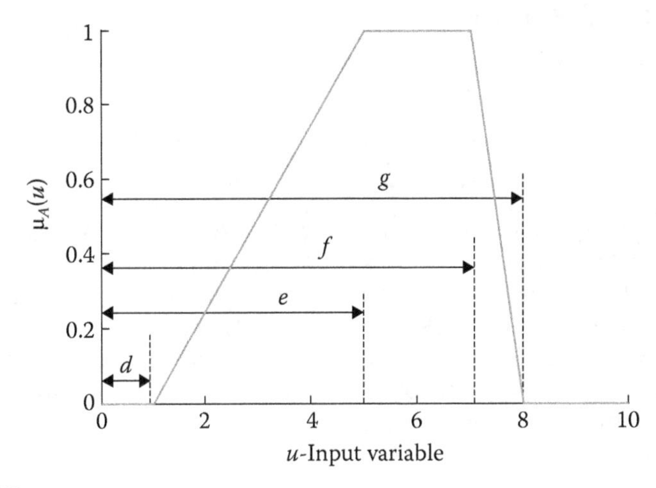

FIGURE 6.7
Trapezoid-shaped membership function.

Here, the parameters a and c define the extremes of the sloped portion of the function, and b signifies the point at which $\mu_A(u) = 0.5$. Figure 6.8 shows an S-shaped FMF for $a = 1$, $b = 3$, and $c = 5$. This also looks like an s-curve as defined in literature using a combination of linear and cosine functions. In such a case, the z-curve is defined as a reflection of this s-curve. Interestingly, both the S-curve and the s-curve [5] look like sigmoid functions.

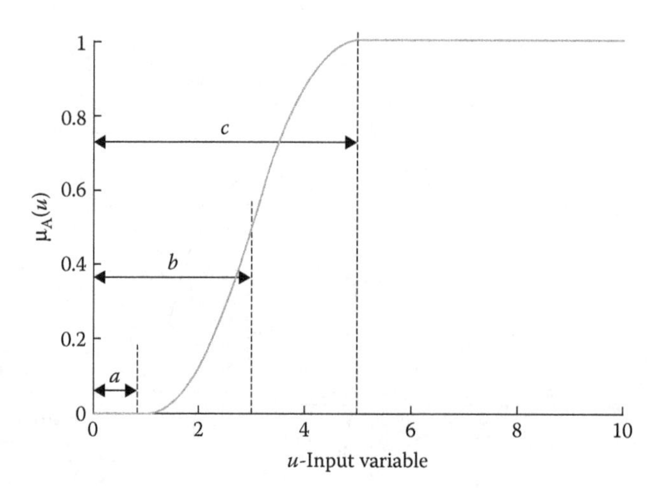

FIGURE 6.8
S-shaped membership function.

6.1.2.6 Π-Shaped Function

This membership function is given as

$$\mu_A(u) = \begin{cases} S(u; c-b, c-b/2, c) & \text{for } u \le c \\ 1 - S(u; c, c+b/2, c+b) & \text{for } u > c \end{cases} \tag{6.9}$$

Here, the parameter c locates the "peak," and the parameters $c-b$ and $c+b$ locate the extremes of the slopes (left and right) of the curve. At $u = c - b/2$ and $u = c + b/2$, the membership grade of the function is equal to 0.5 (Figure 6.9). This function can also be implemented as a combination of a z-curve and an s-curve. Although the shape looks like the bell-shaped Gaussian function, normally the top part of the curve is slightly flattened out, unlike that shown in Figure 6.9.

6.1.2.7 Z-Shaped Function

This FMF is given as

$$\mu_A(u) = \begin{cases} 1 & \text{for } u \le a \\ 1 - 2\dfrac{(u-a)^2}{(a-b)^2} & \text{for } u > a \,\&\, u \le \dfrac{a+b}{2} \\ 2\dfrac{(b-u)^2}{(a-b)^2} & \text{for } u > \dfrac{a+b}{2} \,\&\, u \le b \\ 0 & \text{for } u > b \end{cases} \tag{6.10}$$

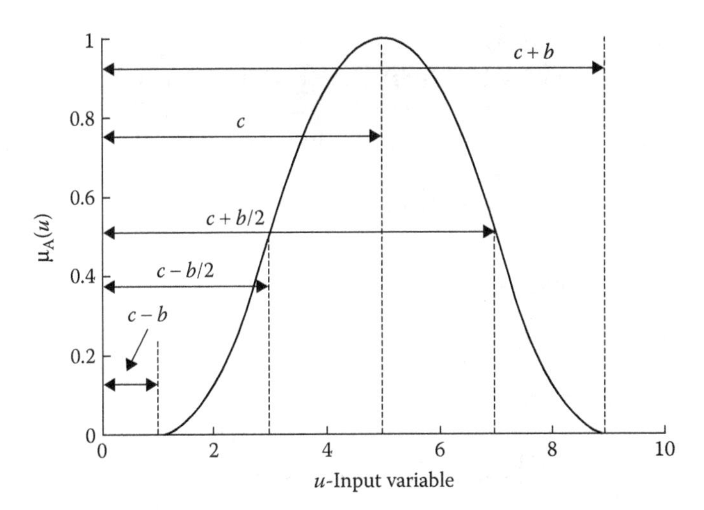

FIGURE 6.9
Π-shaped membership function.

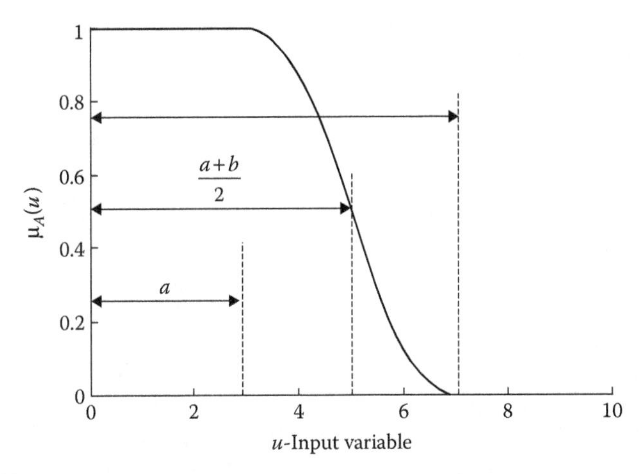

FIGURE 6.10
Z-shaped membership function.

Here, the parameters a and b define the extremes of the sloped portion of the function. Figure 6.10 is an FMF for $a = 3$ and $b = 7$.

6.1.3 Fuzzy Set Operations

The most elementary and well-known crisp-set operations and operators, such as the intersection, union, and complement, are represented by *and*, *or*, and *not*, respectively (in computer logic, Boolean logic, and algebra). Let A and B

be two subsets of U. The intersection of these subsets, denoted by $A \cap B$ (this is a new resultant set in itself), contains all the elements that are common between A and B, i.e., $\mu_{A \cap B}(u) = 1$ if $u \in A$ *and* $u \in B$. The union of A and B, denoted $A \cup B$, contains all elements either in A or in B; i.e., $\mu_{A \cup B}(u) = 1$ if $u \in A$ *or* $u \in B$. The complement of A, denoted by $\bar{A}$, contains all the elements that are *not* in A, i.e., $\mu_{\bar{A}}(u) = 1$ if $u \notin A$; and $\mu_{\bar{A}}(u) = 0$ if $u \in A$.

6.1.3.1 Fuzzy Logic Operators

Like in the conventional crisp logic described above, FL operations are also defined. For FL, the the operators *min, max,* and *complement* correspond to *and, or,* and *not* [2], and are defined as:

$$\mu_{A \cap B}(u) = \min\left[\mu_A(u), \mu_B(u)\right] \text{ (intersection)} \tag{6.11}$$

$$\mu_{A \cup B}(u) = \max\left[\mu_A(u), \mu_B(u)\right] \text{ (union)} \tag{6.12}$$

$$\mu_{\bar{A}}(u) = 1 - \mu_A(u) \quad \text{(complement)} \tag{6.13}$$

Another method of defining the operators *and* and *or* in FL [1] is as follows:

$$\mu_{A \cap B}(u) = \mu_A(u)\mu_B(u) \tag{6.14}$$

$$\mu_{A \cup B}(u) = \mu_A(u) + \mu_B(u) - \mu_A(u)\mu_B(u) \tag{6.15}$$

We can define the fuzzy sets A, using Equation 6.5 (Gaussian membership function or GMF), and B, using Equation 6.7 (trapezoidal membership function or TMF). Then the element-wise fuzzy intersection is computed by substituting the values of $\mu_A(u)$ and $\mu_B(u)$ (obtained using Equations 6.5 and 6.7, respectively) into Equations 6.11 and 6.14. Similarly, the element-wise fuzzy union is computed using Equations 6.12 and 6.15. Figures 6.11 and 6.12 [9], respectively, show the results of the fuzzy intersection and union operations for both definitions using the GMF and TMF. The resulting membership function, obtained using Equation 6.14 as shown in Figure 6.11, shrinks in comparison to that obtained using Equation 6.11; and we see from Figure 6.12 that the resulting membership function obtained using Equation 6.15 expands when compared to that obtained using Equation 6.12. This example illustrates the effect of using a particular definition for intersection and union, and it also indicates that any consistent definition is possible for defining basic FL operations.

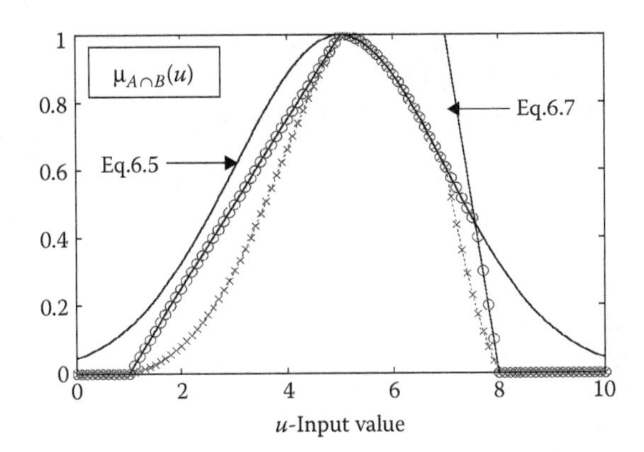

FIGURE 6.11
Fuzzy intersection-min operations.

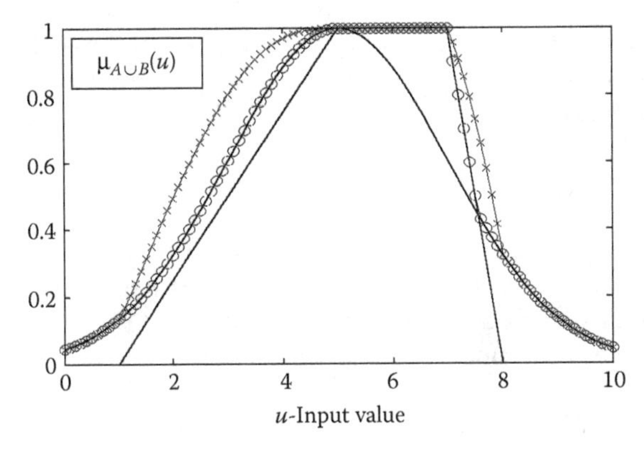

FIGURE 6.12
Fuzzy union-max operations.

6.1.4 Fuzzy Inference System

We can define a fuzzy rule as follows: "If u is A, then v is B." The *if* part of the rule, "if u is A," is called the *antecedent* or *premise*. The *then* part of the rule, "then v is B," is called the *consequence* or *conclusion*. The core is a fuzzy inference engine (FIE) that defines the mapping from input fuzzy sets into output fuzzy sets via a fuzzy implication operation. The FIE determines the degree to which the *antecedent* is satisfied for each rule. If the *antecedent* of a given rule has more than one clause (e.g., *if* u_1 is A_1 *and* u_2 is A_2, *then* v is B), fuzzy operators (T-norm and S-norm) are applied to obtain one number that represents the result of the *antecedent* for that rule. An FIE can take different forms depending on how the inference rule is defined. In addition, one or more rules may fire at the same time, in which case, the outputs of all rules are *aggregated*, i.e., the fuzzy sets that represent the output of each rule are combined into a single fuzzy set. An important aspect of a fuzzy inference system (FIS) is that fuzzy rules fire concurrently, and the order in which firing occurs does not affect the output. Figure 6.13 shows a schematic of an FIS for a multi-input and single-output (MISO) system [2,9]. The fuzzifier (or the fuzzification process) then maps the input values into corresponding memberships that are essential to activate rules in terms of linguistic variables, i.e., the fuzzifier takes the input values and, via membership functions, determines the degree to which these numbers belong to each of the fuzzy sets. The rule base contains linguistic rules provided by domain and human experts. Subsequently, defuzzification converts the output fuzzy set values (Type-I FL) into crisp numbers (Type–0 FL). A comprehensive overview of the FIS–fuzzy implication process (FIP; studied in Section 6.2) is presented in Figure 6.14 [9]. If any rule has more than one clause in the antecedent, then these clauses are combined using any one definition from either the T-norm or the S-norm, and the FIP gives the fuzzified output for each fired rule. These outputs are combined using an aggregation process (by using one of the methods shown in the S-norm block).

6.1.4.1 Triangular Norm or T-norm

In FL, the corresponding operator for the Boolean parameter *and* is *min*; another possibility is given by Equation 6.14. The intersection of two fuzzy sets A and B is specified by the binary mapping of T (the T-norm) in the unit interval, i.e., as a function of the following form:

$$\begin{gathered} T\colon [0,1]X[0,1] \rightarrow [0,1] \text{ or more specifically} \\ \mu_{A\cap B}(u) = T(\mu_A(u), \mu_B(u)) \end{gathered} \tag{6.16}$$

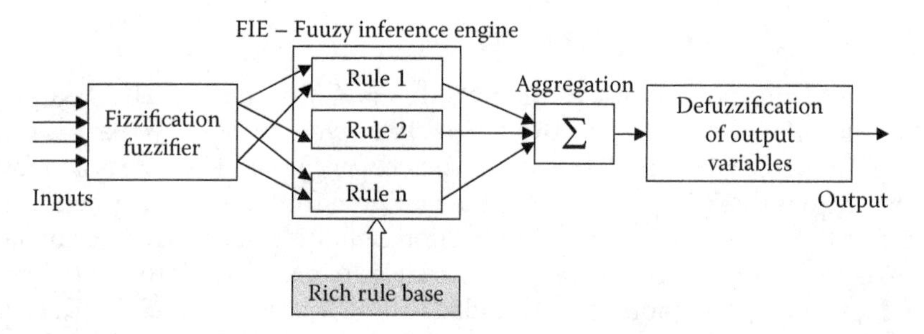

FIGURE 6.13
Overall FIS process. (Adapted from Raol, J. R., and J. Singh. 2009. *Flight Mechanics Modeling and Analysis*. Boca Raton, FL: CRC Press.)

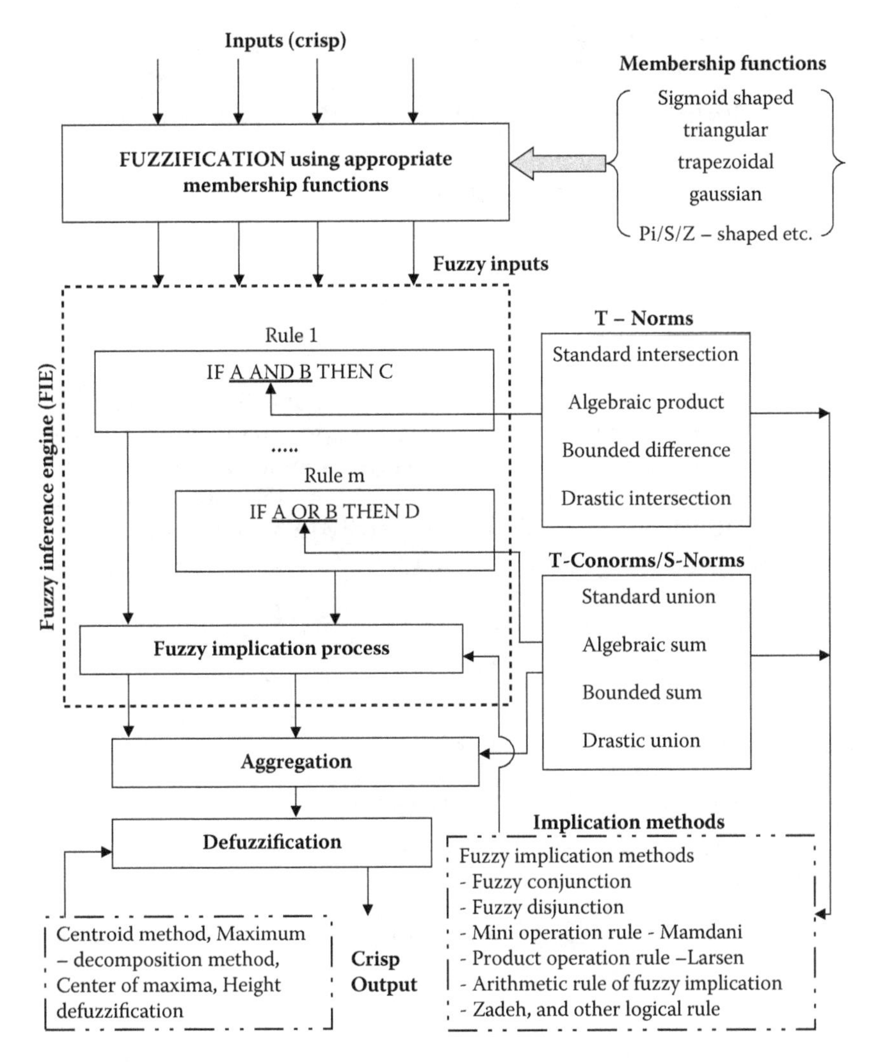

FIGURE 6.14
Comprehensive fuzzy implication process. (Adapted from Raol, J. R., and J. Singh. *Flight Mechanics Modeling and Analysis*. Boca Raton, FL: CRC Press.)

The T-norm operator is used (1) to combine the clauses in the antecedent portion of a given rule (e.g., *if* u_1 is A_1 *and* u_2 is A_2); and (2) to map the input fuzzy sets into output fuzzy sets. The T-norms normally used as fuzzy intersections are given by [3,8]

$$\text{Standard intersection (SI): } T_{SI}(x, y) = \min(x, y) \quad (6.17)$$

$$\text{Algebraic product (AP): } T_{\text{AP}}(x,y) = x \bullet y \quad \text{(by Zadeh)} \tag{6.18}$$

$$\text{Bounded difference/product (BD/BP): } T_{\text{BD|BP}}(x,y) = \max(0, x+y-1) \tag{6.19}$$

Drastic intersection/product (DI/DP):

$$T_{\text{DI||DP}}(x,y) = \begin{cases} x & \text{when} \quad y=1 \\ y & \text{when} \quad x=1 \\ 0 & \text{otherwise} \end{cases} \tag{6.20}$$

Here, $x = \mu_A(u)$, $y = \mu_B(u)$, and $u \in U$. We assume that the fuzzy sets (A and B) are normalized, with membership grade values ranging between 0 and 1. The definition of T-norms should satisfy certain axioms [7,8], when x, y, and z are in the range [0,1] for the entire membership grade. The fuzzy-set intersections that satisfy these axiomatic skeletons are bounded by the following inequality:

$$T_{\text{DI}}(x,y) \le T(x,y) \le T_{\text{SI}}(x,y) \tag{6.21}$$

The T-norm operations of Equations 6.17 through 6.20 as well as many fuzzy-set operations and operators are illustrated next, with some specific membership functional shapes and forms [2–8].

EXAMPLE 6.1

Consider the fuzzy sets A and B, defined by the membership functions "trimf" and "trapmf" (inbuilt MATLAB® functions), respectively, as

$$\mu_A(u) = trimf(u,[a,b,c]) \tag{6.22}$$

$$\mu_B(u) = trapmf(u,[d,e,f,g]) \tag{6.23}$$

In the present case, we have $a = 3$, $b = 6$, $c = 8$, $d = 1$, $e = 5$, $f = 7$, and $g = 8$. The time history of the discrete input u to the fuzzy sets A and B is given as $u = 0,1,2,3,4,5,6,7,8,9,10$. The fuzzified values of the input u, passed through membership functions $\mu_A(u) = trimf(u,[a,b,c])$ (Figure 6.15) and $\mu_B(u) = trapmf(u,[d,e,f,g])$ (Figure 6.16) are given as:

$$\begin{aligned} x = \mu_A(u) = \{&0/0 + 0/1 + 0.0/2 + 0.0/3 + 0.33/4 \\ &+ 0.667/5 + 1/6 + 0.5/7 + 0/8 + 0/9 + 0/10\} \\ y = \mu_B(u) = \{&0/0 + 0/1 + 0.25/2 + 0.5/3 + 0.75/4 \\ &+ 1.000/5 + 1/6 + 1.0/7 + 0/8 + 0/9 + 0/10\} \end{aligned}$$

Thus, we can compute the values of the various T-norms.

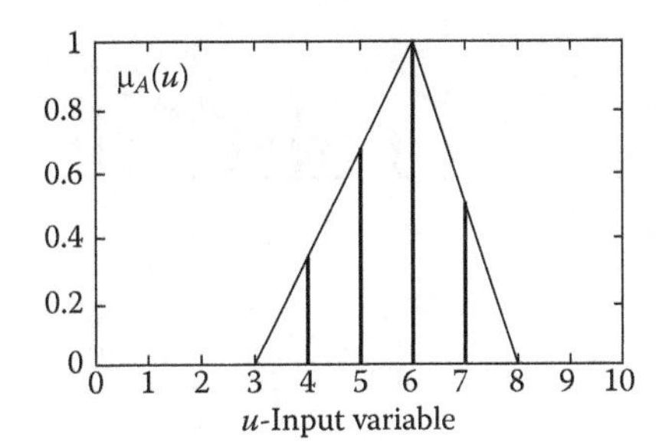

FIGURE 6.15
Triangular membership function.

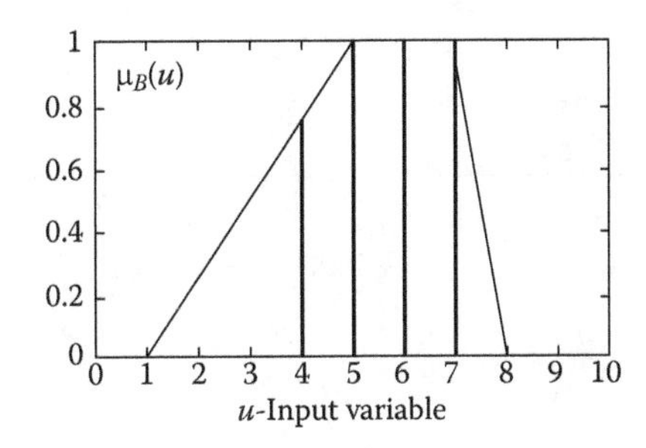

FIGURE 6.16
Trapezoidal membership function.

Solution 6.1

The computed values (with the pictorial representation in Figure 6.17) for T-norms are as follows:

1. SI

$$T1 = T(x,y) = min(x,y)$$
$$= \{0/0 + 0/1 + 0/2 + 0/3 + 0.33/4 + 0.667/5 + 1.0/6 + 0.5/7 + 0/8 + 0/9 + 0/10\}$$

2. AP

$$T2 = T(x,y) = x \bullet y$$
$$= \{0/0 + 0/1 + 0/2 + 0/3 + 0.25/4 + 0.667/5 + 1.0/6 + 0.5/7 + 0/8 + 0/9 + 0/10\}$$

3. BD or BP

$$T3 = T(x,y) = \max(0, x + y - 1)$$
$$= \{0/0 + 0/1 + 0/2 + 0/3 + 0.0833/4 + 0.667/5 + 1.0/6 + 0.5/7 + 0/8 + 0/9 + 0/10\}$$

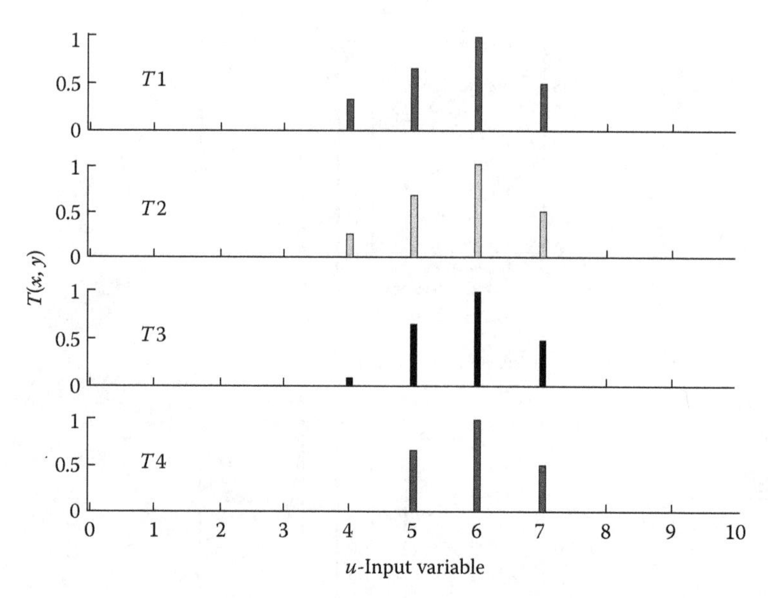

FIGURE 6.17
Pictorial representation of T-norms.

4. DI or DP

$$T(x,y) = \begin{cases} x & \text{when} \quad y = 1 \\ y & \text{when} \quad x = 1 \\ 0 & \text{otherwise} \end{cases}$$

$$T4 = \{0/0 + 0/1 + 0/2 + 0/3 + 0/4 + 0.667/5 \\ +1.0/6 + 0.5/7 + 0/8 + 0/9 + 0/10\}$$

The above results are provided in a tabular form in Table 6.1.

6.1.4.2 Fuzzy Implication Process Using T-norm

Before using the T-norm in FIPs [2–4], let us understand the basics of FIP. Consider two fuzzy sets A and B that belong to the UODs U and V, respectively; then, the fuzzy implication is defined as

$$\text{Rule: IF } A \text{ THEN } B = A \rightarrow B \equiv AXB \tag{6.24}$$

In Equation 6.24, AXB is the Cartesian product (CP) of the two fuzzy sets A and B. The CP is an essential operation of all FIEs and signifies the fuzzy relationship. Suppose u and v are the elements in the UODs of U and V, respectively, for fuzzy sets A and B; then, their fuzzy relation is defined as

$$R = \{((u,v), \mu_R(u,v)) \mid (u,v) \in UXV\} \tag{6.25}$$

EXAMPLE 6.2

Assume $U = V = \{1,2,3,4,5\}$; then, the fuzzy relation such that u and v are approximately equal [2,3,8] can be expressed as

$$\mu_R(u,v) = \begin{cases} 1 & |u-v| = 0 \\ 0.8 & |u-v| = 1 \\ 0.3 & |u-v| = 2 \\ 0 & \text{otherwise} \end{cases}$$

The convenient representation of the fuzzy relation as a Sagittal diagram [2] and a relational matrix (2D-membership array) are given in Figures 6.18 and 6.19. Unlike in the crisp-set values, in this example, the approximate equals can have various grades of being equal to each other, the extreme being the exact value 1 of the membership function, at which point the values of u and v are exactly the same. This is due to the fact that the value belongs to the fuzzy membership by a stochastic degree. As the difference between u and v increases, the membership-function's value of "belonging to the being-equal set" decreases. When the difference is very large, this value approaches zero.

EXAMPLE 6.3

To understand FIP using a different T-norm definition, consider A and B as two discrete fuzzy sets defined as

$$A = \{\mu_A(u)/u\} = (1/1 + 0.7/2 + 0.2/3 + 0.1/4) \; \forall \; u \in U \tag{6.26}$$

TABLE 6.1
The Results of the Four T-norms

u	$T1$	$T2$	$T3$	$T4$
0	0	0	0	0
1	0	0	0	0
2	0	0	0	0
3	0	0	0	0
4	0.33	0.25	0.0833	0
5	0.667	0.667	0.667	0.667
6	1	1	1	1
7	0.5	0.5	0.5	0.5
8	0	0	0	0
9	0	0	0	0
10	0	0	0	0

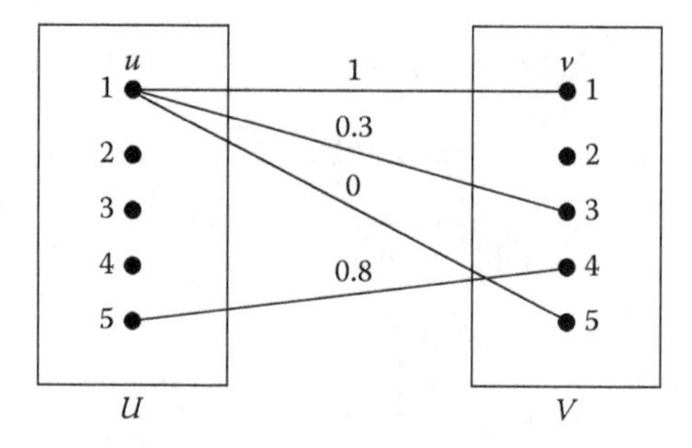

FIGURE 6.18
Sagittal diagram (with some connectivities).

$$M_R = \begin{bmatrix} 1 & 0.8 & 0.3 & 0 & 0 \\ 0.8 & 1 & 0.8 & 0.3 & 0 \\ 0.3 & 0.8 & 1 & 0.8 & 0.3 \\ 0 & 0.3 & 0.8 & 1 & 0.8 \\ 0 & 0 & 0.3 & 0.8 & 1 \end{bmatrix}$$

FIGURE 6.19
Membership matrix (with all connectivities).

$$B = \{\mu_B(v)/v\} = (0.8/1 + 0.6/2 + 0.4/3 + 0.2/4)\ \forall\, v \in V \tag{6.27}$$

The CP of A and B can be obtained using a fuzzy conjunction (FC), which is usually defined by T-norms, i.e.:

$$\begin{aligned} A \to B &= AXB \\ &= \sum\nolimits_{UXV} \mu_A(u) * \mu_B(v)/(u,v) \end{aligned} \tag{6.28}$$

Here, $u \in U$ and $v \in V$, and * is an operator representing a T-norm. Compute the FC using different T-norm operators.

Solution 6.3

For SI we have

$$\begin{aligned} A \to B &= AXB \\ &= \sum\nolimits_{UXV} (\mu_A(u) \cap \mu_B(v))/(u,v) \\ &= \sum\nolimits_{UXV} \min(\mu_A(u), \mu_B(v))/(u,v) \end{aligned} \tag{6.29}$$

It should be noted that Equation 6.29 uses the basic definition of SI given in Equation 6.17. Using Equations 6.26 and 6.27 in Equation 6.29 results in the following outcomes:

$$A \to B = \begin{Bmatrix} \min(1,0.8)/(1,1), \min(1,0.6)/(1,2), \min(1,0.4)/(1,3), \min(1,0.2)/(1,4), \\ \min(0.7,0.8)/(2,1), \min(0.7,0.6)/(2,2), \min(0.7,0.4)/(2,3), \min(0.7,0.2)/(2,4), \\ \min(0.2,0.8)/(3,1), \min(0.2,0.6)/(3,2), \min(0.2,0.4)/(3,3), \min(0.2,0,2)/(3,3), \\ \min(0.1,0.8)/(4,1), \min(0.1,0.6)/(4,2), \min(0.1,0.4)/(4,3), \min(0.1,0.2)/(4,4) \end{Bmatrix}$$

$$= \begin{Bmatrix} 0.8/(1,1) + 0.6/(1,2) + 0.4/(1,3) + 0.2/(1,4) + \\ 0.7/(2,1) + 0.6/(2,2) + 0.4/(2,3) + 0.2/(2,4) + \\ 0.2/(3,1) + 0.2/(3,2) + 0.2/(3,3) + 0.2/(3,4) + \\ 0.1/(4,1) + 0.1/(4,2) + 0.1(4,3) + 0.1/(4,4) \end{Bmatrix}$$

This can be conveniently represented using a relational matrix, as shown in Table 6.2 and Figure 6.20 [2]. It should be noted that the Min-Operation Rule of Fuzzy Implication (MORFI; Mamdani) also uses the same definition as mentioned in Equation 6.29.

TABLE 6.2

The Relational Matrix R_1^T

u/v	**1**	**2**	**3**	**4**
1	0.8	0.6	0.4	0.2
2	0.7	0.6	0.4	0.2
3	0.2	0.2	0.2	0.2
4	0.1	0.1	0.1	0.1

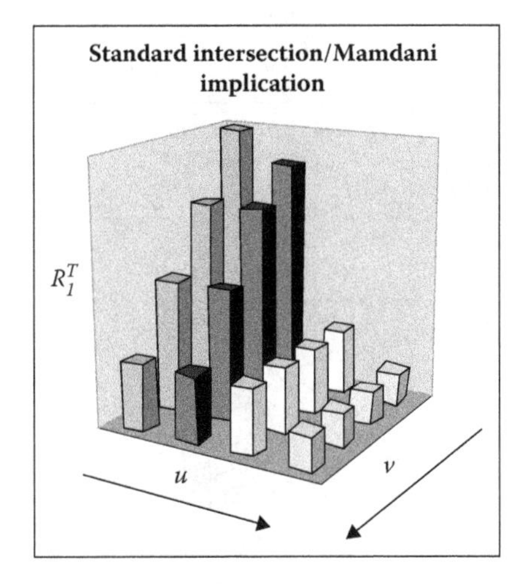

FIGURE 6.20
Graphical representation of the relational matrix R_1^T.

For AP, we have

$$
\begin{aligned}
A \to B &= AXB \\
&= \sum\nolimits_{UXV} (\mu_A(u) \bullet \mu_B(v))/(u,v)
\end{aligned} \tag{6.30}
$$

It should be noted that Equation 6.30 uses the basic definition of AP as given in Equation 6.18 for the • function. Using Equations 6.26 and 6.27 in Equation 6.30 results in

$$
A \to B = \begin{Bmatrix} 0.8/(1,1) + 0.6/(1,2) + 0.4/(1,3) + 0.2/(1,4) + \\ 0.56/(2,1) + 0.42/(2,2) + 0.28/(2,3) + 0.14/(2,4) + \\ 0.16/(3,1) + 0.12/(3,2) + 0.08/(3,3) + 0.04/(3,4) + \\ 0.08/(4,1) + 0.06/(4,2) + 0.04/(4,3) + 0.02/(4,4) \end{Bmatrix}
$$

This can be conveniently represented using a relational matrix, shown in Table 6.3 and Figure 6.21 [2]. It should be noted that the Product Operation Rule of Fuzzy Implication (PORFI; Larsen) also uses the same definition as mentioned in Equation 6.30.

TABLE 6.3

The Relational Matrix R_2^T

u/v	1	2	3	4
1	0.8	0.6	0.4	0.2
2	0.56	0.42	0.28	0.14
3	0.16	0.12	0.08	0.04
4	0.08	0.06	0.04	0.02

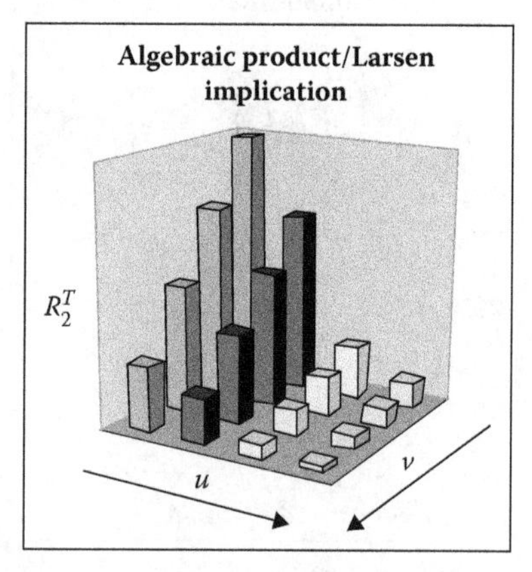

FIGURE 6.21
Graphical representation of the relational matrix R_2^T.

For BD, we have

$$\begin{aligned} A \to B &= AXB \\ &= \sum_{UXV} \max(0, \mu_A(u) + \mu_B(v) - 1)/(u, v) \end{aligned} \tag{6.31}$$

Using Equations 6.26 and 6.27 in Equation 6.31 results in

$$A \to B = \begin{Bmatrix} 0.8/(1,1) + 0.6/(1,2) + 0.4/(1,3) + 0.2/(1,4) + \\ 0.5/(2,1) + 0.3/(2,2) + 0.1/(2,3) + 0/(2,4) + \\ 0/(3,1) + 0/(3,2) + 0/(3,3) + 0/(3,4) + \\ 0/(4,1) + 0/(4,2) + 0/(4,3) + 0/(4,4) \end{Bmatrix}$$

It should be noted that Equation 6.31 uses the basic definition of BD as given in Equation 6.19 and, in addition, can be conveniently represented using a relational matrix [2], as shown in Table 6.4 and Figure 6.22.

TABLE 6.4

The Relational Matrix R_3^T

u/v	**1**	**2**	**3**	**4**
1	0.8	0.6	0.4	0.2
2	0.5	0.3	0.1	0.0
3	0.0	0.0	0.0	0.0
4	0.0	0.0	0.0	0.0

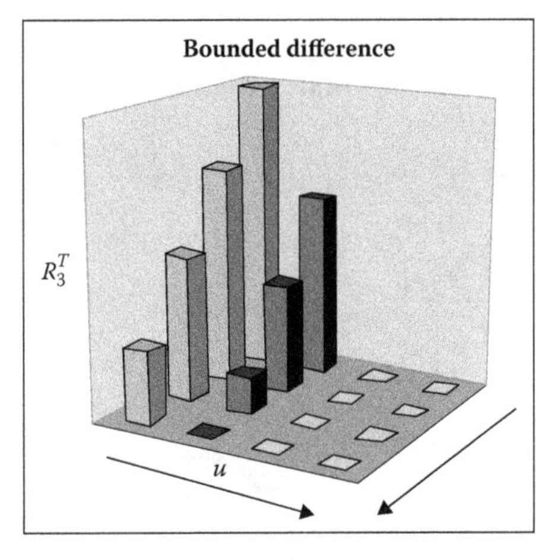

FIGURE 6.22
Graphical representation of the relational matrix R_3^T.

For DI, we have

$$A \to B = AXB$$
$$= \begin{cases} \mu_A(u)/(u,v) & \text{if} \quad \mu_B(v) = 1 \\ \mu_B(v)/(u,v) & \text{if} \quad \mu_A(u) = 1 \\ 0/(u,v) & \text{otherwise} \end{cases} \tag{6.32}$$

Using Equations 6.26 and 6.27 in Equation 6.32 results in

$$A \to B = \begin{cases} 0.8/(1,1) + 0.6/(1,2) + 0.4/(1,3) + 0.2/(1,4) + \\ 0/(2,1) + 0/(2,2) + 0/(2,3) + 0/(2,4) + \\ 0/(3,1) + 0/(3,2) + 0/(3,3) + 0/(3,4) + \\ 0/(4,1) + 0/(4,2) + 0/(4,3) + 0/(4,4) \end{cases}$$

It should be noted that Equation 6.32 uses the basic definition of DI as given in Equation 6.20; moreover, it can be conveniently represented with a relational matrix, as shown in Table 6.5 and Figure 6.23.

TABLE 6.5
The Relational Matrix R_4^T

u/v	1	2	3	4
1	0.8	0.6	0.4	0.2
2	0.0	0.0	0.0	0.0
3	0.0	0.0	0.0	0.0
4	0.0	0.0	0.0	0.0

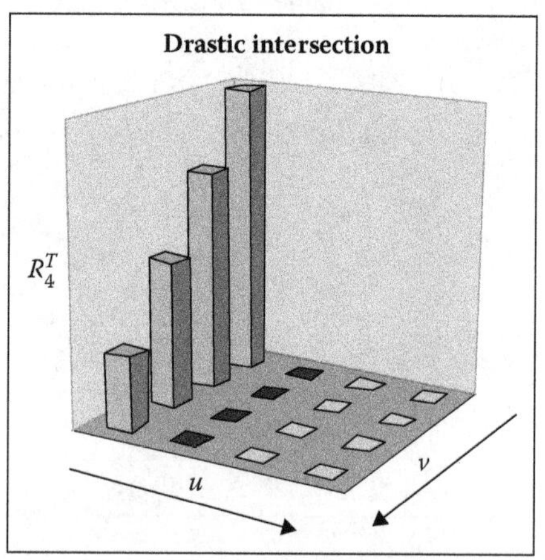

FIGURE 6.23
Graphical representation of the relational matrix R_4^T.

6.1.4.3 Triangular Conorm or S-norm

In FL, the operator corresponding to the operator *or* of Boolean logic is *max* (see Equation 6.12). The other possible *or* is defined by Equation 6.15. It is called a triangular conorm (T-conorm) or S-norm and is defined as follows: The union of two fuzzy sets A and B is specified by a binary operation on the unit interval, i.e., a function of the form

$$S: [0,1]X[0,1] \rightarrow [0,1]$$

or more specifically,

$$\mu_{A \cup B}(u) = S(\mu_A(u), \mu_B(u)) \tag{6.33}$$

This fuzzy operator is used (1) to combine the clauses in the antecedent part of a given rule, such as "*if* u_1 is A_1 *or* u_2 is A_2" (see the S-norm block in Figure 6.14); and (2) in FIPs. The T-conorms used as fuzzy unions [3,8] are

$$\text{Standard union (SU): } S_{\text{SU}}(x,y) = \max(x,y) \tag{6.34}$$

$$\text{Algebraic sum (AS): } S_{\text{AS}}(x,y) = x + y - x \bullet y \text{ (Zadeh)} \tag{6.35}$$

$$\text{Bounded sum (BS): } S_{\text{BS}}(x,y) = \min(1, x+y) \tag{6.36}$$

$$\text{Drastic union (DU): } S_{\text{DU}}(x,y) = \begin{cases} x & \text{when} \quad y = 0 \\ y & \text{when} \quad x = 0 \\ 1 & \text{otherwise} \end{cases} \tag{6.37}$$

$$\text{Disjoint sum: } S_{\text{DS}}(x,y) = \max\{\min(x, 1-y), \min(1-x, y)\} \tag{6.38}$$

Here, $x = \mu_A(u)$, $y = \mu_B(u)$, and $u \in U$. The fuzzy set unions that satisfy certain axioms are bounded by the following inequality:

$$S_{\text{SU}}(x,y) \leq S(x,y) \leq S_{\text{DU}}(x,y) \tag{6.39}$$

EXAMPLE 6.4

Consider Example 6.3. Compute the S-norms.

SOLUTION 6.4

For triangular S-norms (conorms), the computed values (as shown in Figure 6.24) are as follows:

1. SU

$$S(x,y) = \max(x,y)$$
$$S1 = \{0/0 + 0/1 + 0.25/2 + 0.5/3 + 0.75/4 + 1.0/5 + 1.0/6 + 1.0/7 + 0/8 + 0/9 + 0/10\}$$

2. AS

$$S(x,y) = x + y - x \bullet y$$
$$S2 = \{0/0 + 0/1 + 0.25/2 + 0.5/3 + 0.833/4 + 1.0/5 + 1.0/6 + 1.0/7 + 0/8 + 0/9 + 0/10\}$$

3. BS

$$S(x,y) = \min(1, x+y)$$
$$S3 = \{0/0 + 0/1 + 0.25/2 + 0.5/3 + 1.0/4 + 1.0/5 + 1.0/6 + 1.0/7 + 0/8 + 0/9 + 0/10\}$$

4. DU

$$S(x,y) = \begin{cases} x & \text{when} \quad y = 0 \\ y & \text{when} \quad x = 0 \\ 1 & \text{otherwise} \end{cases}$$
$$S4 = \{0/0 + 0/1 + 0.25/2 + 0.5/3 + 1.0/4 + 1.0/5 + 1.0/6 + 1.0/7 + 0/8 + 0/9 + 0/10\}$$

Table 6.6 shows the above results.

6.1.4.4 Fuzzy Inference Process Using S-norm

The CP of A and B can be obtained using a fuzzy disjunction (FD), which is usually defined by T-conorms or S-norms, as follows:

$$\begin{aligned} A \to B &= AXB \\ &= \sum_{UXV} \mu_A(u) \dot{+} \mu_B(v)/(u,v) \end{aligned} \tag{6.40}$$

Here, $u \in U$ and $v \in V$, and $\dot{+}$ is an operator representing a T-conorm or S-norm.

EXAMPLE 6.5

For evaluation and comparison of S-norms, consider Example 6.2 (Equations 6.26 and 6.27). Compute the FD using different T-conorm operators.

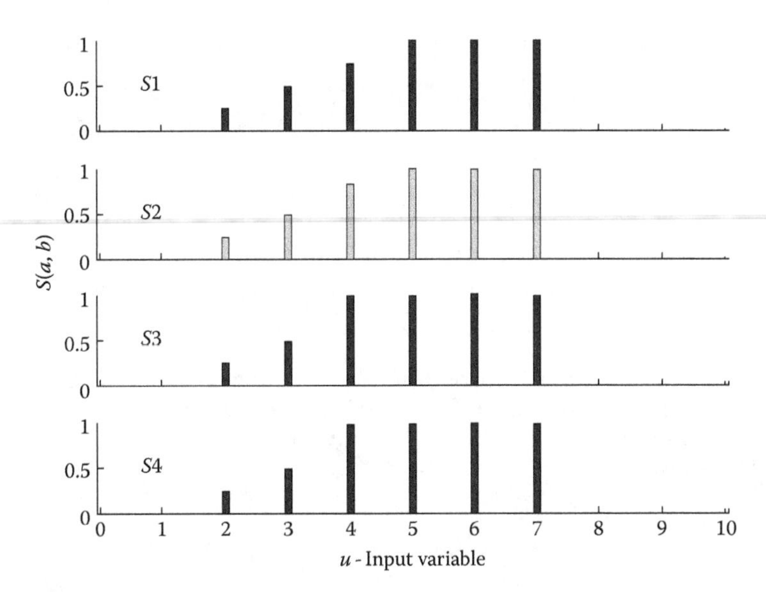

FIGURE 6.24
Pictorial representation of T-conorms or S-norms.

TABLE 6.6
The Results of the Four S-Norms

u	*S1*	*S2*	*S3*	*S4*
0	0	0	0	0
1	0	0	0	0
2	0.25	0.25	0.25	0.25
3	0.5	0.5	0.5	0.5
4	0.75	0.833	1	1
5	1	1	1	1
6	1	1	1	1
7	1	1	1	1
8	0	0	0	0
9	0	0	0	0
10	0	0	0	0

Solution 6.5

For SU, we have

$$\begin{aligned} A \rightarrow B &= AXB \\ &= \sum_{UXV} \mu_A(u) \cup \mu_B(v)/(u,v) \\ & \sum_{UXV} \max(\mu_A(u), \mu_B(v))/(u,v) \end{aligned} \tag{6.41}$$

Using Equations 2.26 and 2.27 in Equation 2.41 yields the following result:

$$A \rightarrow B = \begin{Bmatrix} \max(1,0.8)/(1,1), \max(1,0.6)/(1,2), \max(1,0.4)/(1,3), \max(1,0.2)/(1,4), \\ \max(0.7,0.8)/(2,1), \max(0.7,0.6)/(2,2), \max(0.7,0.4)/(2,3), \max(0.7,0.2)/(2,4), \\ \max(0.2,0.8)/(3,1), \max(0.2,0.6)/(3,2), \max(0.2,0.4)/(3,3), \max(0.2,0,2)/(3,3), \\ \max(0.1,0.8)/(4,1), \max(0.1,0.6)/(4,2), \max(0.1,0.4)/(4,3), \max(0.1,0.2)/(4,4) \end{Bmatrix}$$

$$= \begin{Bmatrix} 1/(1,1) + 1/(1,2) + 1/(1,3) + 1/(1,4) + 0.8/(2,1) + 0.7/(2,2) \\ +0.7/(2,3) + 0.7/(2,4) + 0.8/(3,1) + 0.6/(3,2) + 0.4/(3,3) + 0.2/(3,4) \\ +0.8/(4,1) + 0.6/(4,2) + 0.4/(4,3) + 0.2/(4,4) \end{Bmatrix}$$

It should be noted that Equation 6.41 uses the basic definition of SU given in Equation 6.34 and can also be conveniently represented with a relational matrix [2], as presented in Table 6.7, and Figure 6.25.

For AS, we have

$$\begin{aligned} A \rightarrow B &= AXB \\ &= \sum_{UXV} (\mu_A(u) + \mu_B(v) - \mu_A(u) \bullet \mu_B(v))/(u,v) \end{aligned} \tag{6.42}$$

TABLE 6.7

The Relational Matrix R_1^S

u/v	1	2	3	4
1	1	1	1	1
2	0.8	0.7	0.7	0.7
3	0.8	0.6	0.4	0.2
4	0.8	0.6	0.4	0.2

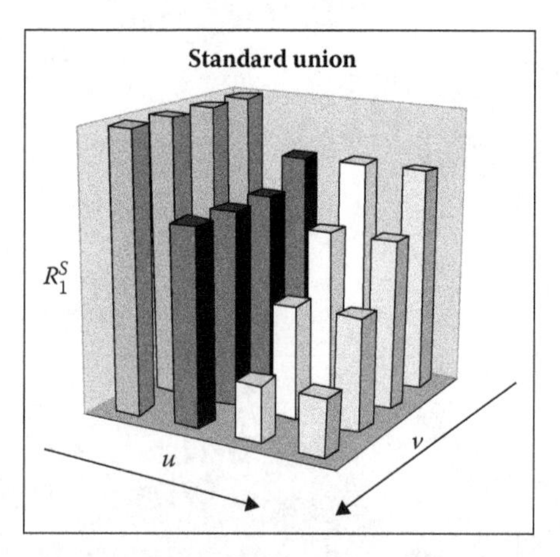

FIGURE 6.25
Graphical representation of the relational matrix R_1^S.

Using Equations 6.26 and 6.29 in Equation 6.42 provides the following result:

$$A \rightarrow B = \begin{Bmatrix} 1/(1,1)+1/(1,2)+1/(1,3)+1/(1,4)+0.94/(2,1) \\ +0.88/(2,2)+0.82/(2,3)+0.76/(2,4)+0.84/(3,1) \\ +0.68/(3,2)+0.52/(3,3)+0.36/(3,4)+0.82/(4,1) \\ +0.64/(4,2)+0.46/(4,3)+0.28/(4,4) \end{Bmatrix}$$

It should be noted that Equation 6.42 uses the basic definition of AS given in Equation 6.35 and can be conveniently represented using a relational matrix as shown in Table 6.8 and Figure 6.26.

For BS, we have

$$\begin{aligned} A \rightarrow B &= AXB \\ &= \sum\nolimits_{UXV} \min(1, \mu_A(u) + \mu_B(v))/(u,v) \end{aligned} \qquad (6.43)$$

TABLE 6.8
The Relational Matrix R_2^S

u/v	1	2	3	4
1	1	1	1	1
2	0.94	0.88	0.82	0.76
3	0.84	0.68	0.52	0.36
4	0.82	0.64	0.46	0.28

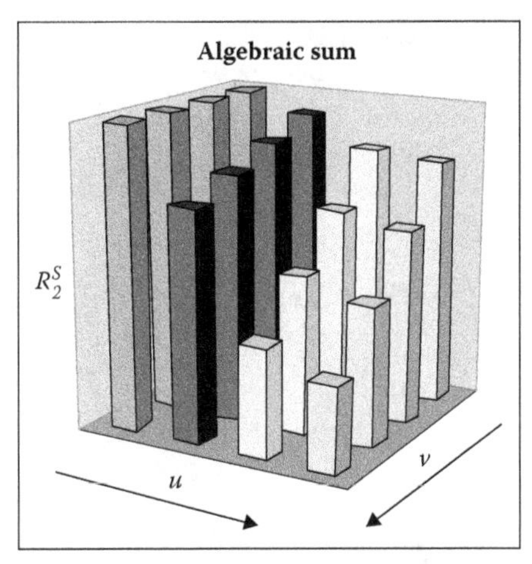

FIGURE 6.26
Graphical representation of the relational matrix R_2^S.

Using Equations 6.26 and 6.27 in Equation 6.43, yields the following outcome:

$$A \to B = \begin{Bmatrix} 1/(1,1)+1/(1,2)+1/(1,3)+1/(1,4) \\ +1/(2,1)+1/(2,2)+1/(2,3)+0.9/(2,4) \\ +1/(3,1)+0.8/(3,2)+0.6/(3,3)+0.4/(3,4) \\ +0.9/(4,1)+0.7/(4,2)+0.5/(4,3)+0.3/(4,4) \end{Bmatrix}$$

It should be noted that Equation 6.43 uses the basic definition of BS given in Equation 6.36 and can be conveniently represented using a relational matrix, as shown in Table 6.9 and Figure 6.27.

For DU, we have

$$A \to B = AXB \\ = \begin{cases} \mu_A(u)/(u,v) & \text{if} \quad \mu_B(v)=0 \\ \mu_B(v)/(u,v) & \text{if} \quad \mu_A(u)=0 \\ 1/(u,v) & \text{otherwise} \end{cases} \tag{6.44}$$

TABLE 6.9

The Relational Matrix R_3^S

u/*v*	**1**	**2**	**3**	**4**
1	1	1	1	1
2	1	1	1	0.9
3	1	0.8	0.6	0.4
4	0.9	0.7	0.5	0.3

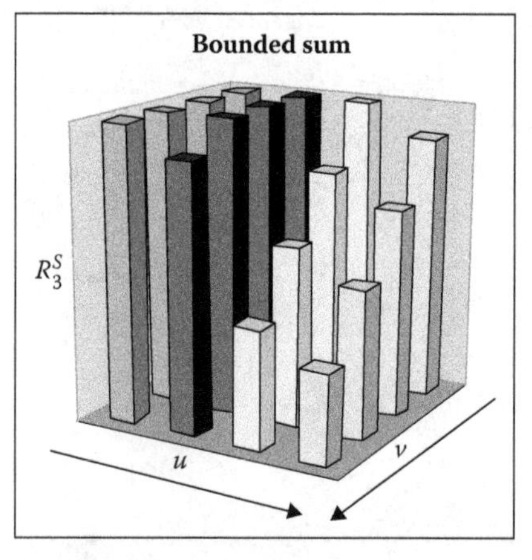

FIGURE 6.27
Graphical representation of relational matrix R_3^S.

Using Equations 6.26 and 6.27 in Equation 6.44 yields the following result:

$$A \to B = \begin{Bmatrix} 1/(1,1)+1/(1,2)+1/(1,3)+1/(1,4) \\ +1/(2,1)+1/(2,2)+1/(2,3)+1/(2,4) \\ +1/(3,1)+1/(3,2)+1/(3,3)+1/(3,4) \\ +1/(4,1)+1/(4,2)+1/(4,3)+1/(4,4) \end{Bmatrix}$$

It should be noted that Equation 6.44 uses the basic definition of DU given in Equation 6.37 and can be conveniently represented using a relational matrix as shown in Table 6.10 and Figure 6.28.

6.1.4.4.1 *Fuzzy Complements*

Consider a fuzzy set A with a membership grade of $\mu_A(u)$ for crisp input u in the UOD U. The standard definition of a fuzzy complement is given by

$$c(\mu_A(u)) = \mu_{\bar{A}}(u) = 1 - \mu_A(u) \tag{6.45}$$

TABLE 6.10
Relational Matrix R_4^S

u/v	1	2	3	4
1	1	1	1	1
2	1	1	1	1
3	1	1	1	1
4	1	1	1	1

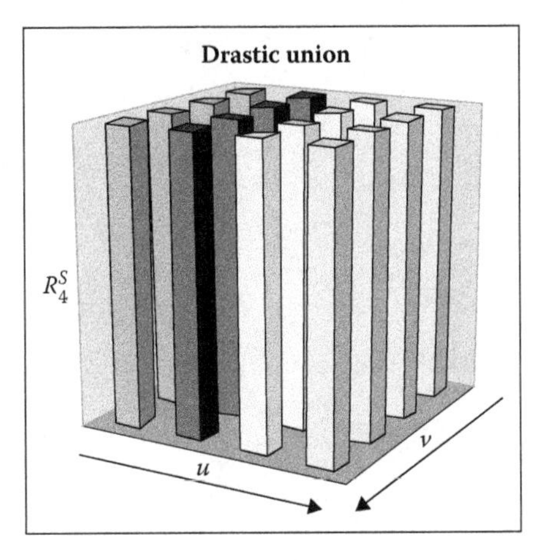

FIGURE 6.28
Graphical representation of the relational matrix R_4^S.

Similar to the above definition of standard fuzzy complement (SFC) any other function must satisfy the following two axioms to be considered a fuzzy complement [3,8]:

Axiom c1: Boundary condition: $c(0) = 1$ and $c(1) = 0$

Axiom c2: Monotonic nonincreasing: if $a < b$, then $c(a) \geq c(b)$

Here, $a = \mu_A(u)$ and $b = \mu_A(v)$, for some $u, v \in U$ in fuzzy set A. The above axioms are called the *axiomatic skeleton* for fuzzy complements. The other axioms, which help in creating a subclass of fuzzy complements from the general class, which satisfies above two axioms c1 and c2, are as follows:

Axiom c3: c is a continuous function

Axiom c4: c is involutive, i.e., $c(c(a)) = a$

A few possible relationships and connectivities between the FL operators (FLORs) and implication functions are depicted in Figure 6.29.

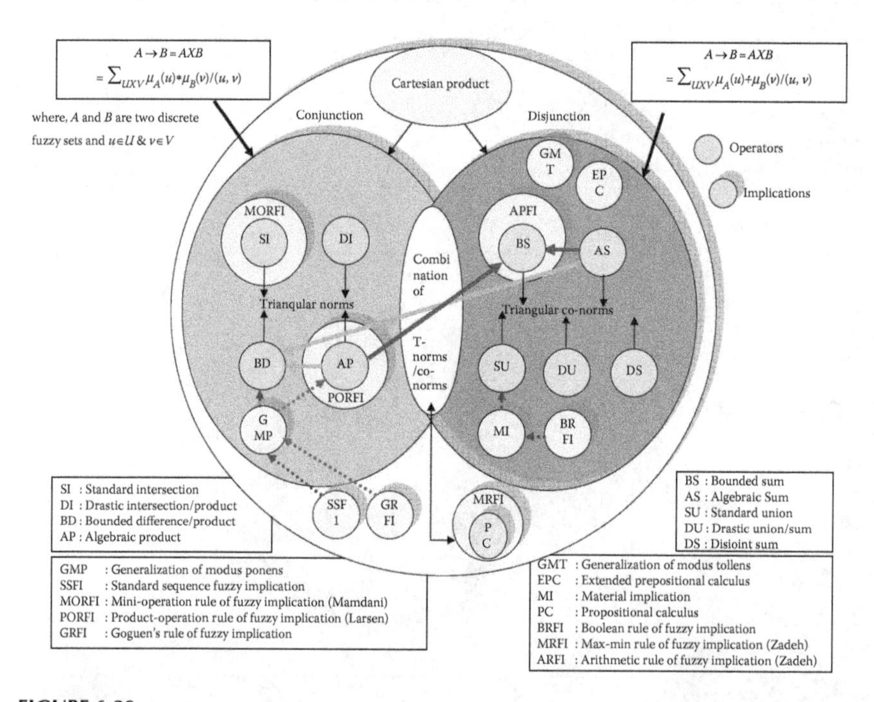

FIGURE 6.29
Depiction of the interconnectivity of fuzzy operators and implication functions.

6.1.5 Relationships between Fuzzy Logic Operators

Some of the interconnectivities and relationships between the T-norm and the S-norm operators have been defined. The various relations between the FLORs [6] are given as follows:

FLOR 1

$$\begin{aligned} T_{BD}(x,y) &= \max(0, x+y-1) \\ T_{BD}(x,y) &= \max(0, S_{AS}(x,y)+x\bullet y-1), \text{ since } S_{AS}(x,y)=x+y-xy \\ T_{BD}(x,y) &= \max(0, S_{AS}(x,y)+T_{AP}(x,y)-1), \text{ since } T_{AP}(x,y)=xy \end{aligned} \tag{6.46}$$

FLOR 2

$$\begin{aligned} S_{BS}(x,y) &= \min(1, x+y) \\ S_{BS}(x,y) &= \min(1, S_{AS}(x,y)+xy) \\ S_{BS}(x,y) &= \min(1, S_{AS}(x,y)+T_{AP}(x,y)) \end{aligned} \tag{6.47}$$

FLOR 3

$$T_{AP}(x,y) = x+y-S_{AS}(x,y), \text{ since } S_{AS}(x,y)=x+y-xy \tag{6.48}$$

FLOR 4

$$T_{SI}(x,y) = x+y-S_{SU}(x,y) \tag{6.49}$$

Assume, $x = 0.2$ & $y = 0.8$; then,

$$\begin{aligned} S_{SU}(x,y) &= \max(x,y) = \max(0.2, 0.8) = 0.8 \\ \text{RHS} &= 0.2+0.8-0.8 = 0.2 \\ \text{LHS} &= T_{SI}(x,y) = \min(x,y) = \min(0.2, 0.8) = 0.2 \\ \text{LHS} &= \text{RHS} \end{aligned}$$

$$T_{SI}(x,y) = T_{AP}(x,y)+S_{AS}(x,y)-S_{SU}(x,y) \text{ (by putting FLOR 3 in FLOR 4)}$$

FLOR 5

$$T_{BD}(x,y) = x+y-S_{BS}(x,y) \tag{6.50}$$

Figure 6.29 shows a pictorial view of these relationships and interconnectivities, in addition to the fuzzy operators shown in Figure 6.14. The arrows originating from the S-norm AS and the T-norm AP and subsequently ending at the T-norm BD indicate the fuzzy relation FLOR 1, i.e., Equation 6.46. Similarly, arrows originating from the S-norm AS and the T-norm AP and ending at the S-norm BS indicate the fuzzy relation FLOR 2, i.e., Equation 6.47. Figure 6.29 shows that the fuzzy implication functions MORFI, PORFI, and ARFI (discussed in Section 6.2) apply the operators SI, AP, and

BS, respectively (as given in Equations 6.17, 6.18, and 6.36). The Max-Min Rule of Fuzzy Implication (MRFI) is derived from a propositional calculus (PC) implication, which in turn uses operators from both the T-norm and S-norm. Similarly, the Boolean Rule of Fuzzy Implication (BRFI), derived from a material implication (MI), uses the operator SU (given in Equation 6.34) from the S-norm. Standard-Sequence Fuzzy Implication (SSFI) and Goguen's Rule of Fuzzy Implication (GRFI), derived from generalization of *modus ponens* (GMP), use the operators BD (given in Equation 6.19) and AP (given in Equation 6.18), respectively. The foregoing observations are very important, because normally one would not expect any direct or obvious correspondence between the conjunction and disjunction definitions of a CP. The above analysis shows that these definitions are related and, hence, enhance the power of the inference process.

6.1.6 Sup (max)–Star (T-norm) Composition

The fuzzy operators T-norm and S-norm can also be used in rule optimization for combining two rules. This operation is known as "Sup (max)–star (T-norm) composition." Consider fuzzy sets *A*, *B*, and *C*, with elements *u*, *v*, and *w* in the UOD *U*, *V*, and *W*, respectively. Consider the following sequential fuzzy conditional rules:

$$R = \text{IF } A \text{ THEN } B \text{ (rule 1); } S = \text{IF } B \text{ THEN } C \text{ (rule 2)}$$

Here, *R* and *S* are the fuzzy relations in *UXV* and *VXW*, respectively. It is possible to combine two rules into a single rule by absorbing the intermediate result *B* and find the relationship between the antecedent and the ultimate consequent directly, i.e.,

$$RoS = \text{IF } A \text{ THEN } C \text{ (new rule)}$$

The composition of *R* and *S* is a fuzzy relation denoted by *RoS* (where *o* implies *rule composition*), i.e., a fuzzy relation defined on *UXW*. Figure 6.30 shows a pictorial representation of the fuzzy relations *R* and *S* [2]. In general, a fuzzy composition is defined as

$$RoS = \{[(u,w), \sup_v(\mu_R(u,v) * \mu_S(v,w))], u \in U, v \in V, w \in W\} \quad (6.51)$$

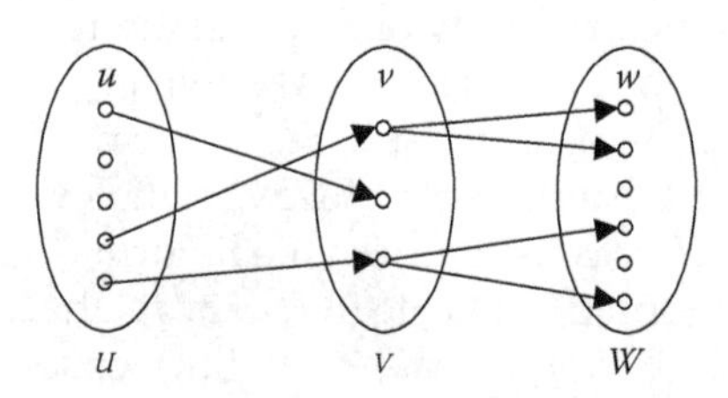

FIGURE 6.30
Sup-star composition (with some connectivities).

Here, "sup" means supremum and is denoted as ∪, i.e., union or max (SU); and * represents the T-norms of either (1) SI, (2) AP, (3) BD, or (4) DI. There are different methods to compute the composition of fuzzy relations.

6.1.6.1 Maximum–Minimum Composition (Mamdani)

Max–min composition is defined as follows:

$$\begin{aligned}\mu_{RoS}(u,w) &= \max_v \min(\mu_R(u,v),\mu_S(v,w)) \\ &= \cup_v(\mu_R(u,v)\cap\mu_S(v,w))\end{aligned} \tag{6.52}$$

If fuzzy relations R and S are given by relational matrices [2] (Tables 6.11 and 6.12), then

	0.1	0.2	0.0	1.0	(R: first row all columns)
	0.9	0.2	0.8	0.4	(S: first column all rows)
min	0.1	0.2	0.0	0.4	($\mu_R(u,v)\cap\mu_S(v,w)$)
max	0.4				($\cup_v$)

After completing the operations for all rows of R with all columns of S, RoS can be represented as shown in Table 6.13 (based on the relational matrices for R and S). It should be noted that in Equation 6.52 (for *min*), the basic definition of SI is used, i.e., Equation 6.17.

TABLE 6.11

Relational Matrix R

R	a	b	c	d
1	0.1	0.2	0.0	1.0
2	0.3	0.3	0.0	0.2
3	0.8	0.9	1.0	0.4

TABLE 6.12

Relational Matrix S

S	α	β	γ
a	0.9	0.0	0.3
b	0.2	1.0	0.8
c	0.8	0.0	0.7
d	0.4	0.2	0.3

TABLE 6.13

RoS Results (Method 1)

RoS	α	β	γ
1	0.4	0.2	0.3
2	0.3	0.3	0.3
3	0.8	0.9	0.8

TABLE 6.14

RoS Results (Method 2)

RoS	α	β	γ
1	0.4	0.2	0.3
2	0.27	0.3	0.24
3	0.8	0.9	0.72

6.1.6.2 Maximum Product Composition (Larsen)

Maximum product composition can be defined as

$$\mu_{RoS}(u,w) = \max_v \underbrace{(\mu_R(u,v) \bullet \mu_S(v,w))}_{\text{algebraic product}} \tag{6.53}$$

Consider the same example as used in Section 6.1.6.1. The max-product composition operation for the first row (all columns) of the fuzzy relation R with the first column (and all rows) of the fuzzy relation S can be given as

	0.1	0.2	0.0	1.0	(R: first row all columns)
	0.9	0.2	0.8	0.4	(S: first column all rows)
•	0.09	0.04	0.0	0.4	$(\mu_R(u,v) \bullet \mu_S(v,w))$
max	0.4				$(\cup_v)$

After completing the operations for all rows of R with all columns of S, RoS can be represented as in Table 6.14. It should be noted that in Equation 6.53 (for •) the basic definition of AP, given in Equation 6.18, is used.

6.1.7 Interpretation of the Connective "*and*"

The connective *and* is usually implemented as FC in Cartesian space.

If (A AND B), *then* C; here, the antecedent is interpreted as a fuzzy set in the product space, say UXV, with the membership function given by [2]

$$\mu_{AXB}(u,v) = \min\{\mu_A(u), \mu_B(v)\} \text{ (same as Equation 6.17)}$$

or

$$\mu_{AXB}(u,v) = \mu_A(u) \bullet \mu_B(v) \quad \text{(same as Equation 6.18)}$$

Here, U and V are the UODs associated with A and B, respectively, and $u \in U, v \in V$.

6.1.8 Defuzzification

The defuzzification process is used to obtain a crisp value from, or using the fuzzy outputs from, an inference engine. Consider the discrete aggregated fuzzy output set B (Figure 6.31), defined as

$$v = \left[0\ 1\ 2\ 3\ 4\ 5\ 6\ 7\ 8\ 9\ 10\ 11\ 12\ 12.6\ 13.6\ 14.6\ 15.6\right] \tag{6.54}$$

$$\mu_B(v) = \left[0\ 0.2\ 0.4\ 0.4\ 0.4\ 0.6\ 0.8\ 0.8\ 0.8\ 0.6\ 0.6\ 0.6\ 0.6\ 0.6\ 0.4\ 0.2\ 0\right] \tag{6.55}$$

The different methods for obtaining the defuzzified output are discussed next.

6.1.8.1 Centroid Method, or Center of Gravity or Center of Area

The defuzzifier determines the center of gravity (centroid), v', of a fuzzy set B and uses this value as the output of the FLS. For a continuous aggregated fuzzy set, the centroid is given by

$$v' = \frac{\int_S v\mu_B(v)dv}{\int_S \mu_B(v)dv}, \quad \text{where } S \text{ denotes the support of } \mu_B(v) \tag{6.56}$$

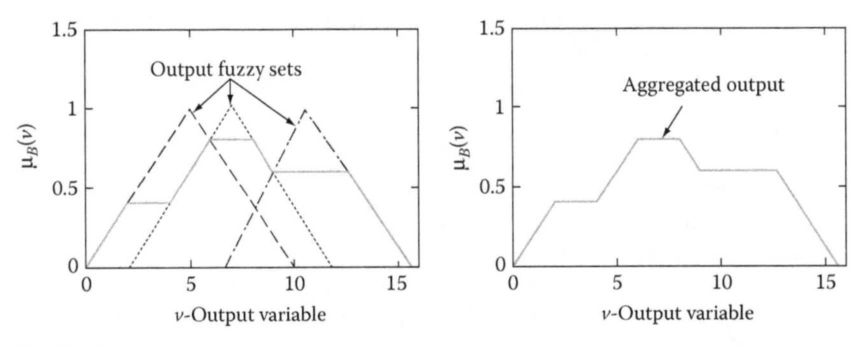

FIGURE 6.31
Output fuzzy sets (left) and aggregated sets (right).

In discrete fuzzy sets, the centroid can be presented as

$$v' = \frac{\sum_{i=1}^{n} v(i)\mu_B(v(i))}{\sum_{i=1}^{n} \mu_B(v(i))} \tag{6.57}$$

Substituting the numerical values from Equations 6.54 and 6.55 in Equation 6.57, we obtain the following results:

$$\begin{aligned}\sum_{i=1}^{n} v(i)\mu_B(v(i)) &= 0\times 0+1\times 0.2+2\times 0.4+3\times 0.4+4\times 0.4+5\times 0.6+6\times 0.8\\ &\quad +7\times 0.8+8\times 0.8+9\times 0.6+10\times 0.6+11\times 0.6+12\times 0.6\\ &\quad +12.6\times 0.6+13.6\times 0.4+14.6\times 0.2+15.6\times 0\\ &= 64.72\end{aligned}$$

$$\begin{aligned}\sum_{i=1}^{n} \mu_B(v(i)) &= 0+0.2+0.4+0.4+0.4+0.6+0.8+0.8+0.8+0.6+0.6+0.6\\ &\quad +0.6+0.6+0.4+0.2+0\\ &= 8\end{aligned}$$

$$v' = \frac{64.72}{8} = 8.09$$

6.1.8.2 Maximum Decomposition Method

In the maximum decomposition method, the defuzzifier examines the aggregated fuzzy set and chooses the output v, for which $\mu_B(v)$ is the maximum. In the example discussed above, finding the defuzzified value for the discrete-output of the fuzzy set B (defined by Equations 6.54 and 6.55) is not advisable because the maximum value of $\mu_B(v) = 0.8$ (Equation 6.55) is at more than one point, i.e., at $v = 6, 7, 8$.

6.1.8.3 Center of Maxima or Mean of Maximum

In a multimode fuzzy region in which more than one rule fires, the center-of-maxima technique finds the highest plateau and then the next highest plateau. The midpoint between the centers of these plateaus is selected as the output of the FLS. In this case, the defuzzified value for the discrete-output fuzzy set B (defined by Equations 6.54 and 6.55) can be computed as follows: since the maximum value of $\mu_B(v) = 0.8$ is at $v = 6, 7, 8$, the defuzzified value, computed as the mean of maximum, is equal to $(6+7+8)/3=7$.

6.1.8.4 Smallest of Maximum

The smallest of maximum is the smallest value of v at which the aggregated-output fuzzy set has the maximum membership grade. Because the maximum value of $\mu_B(v)=0.8$, according to Equation 6.55, is at $v=6, 7, 8$, the defuzzified value computed as the smallest of maximum is equal to min $(v = 6, 7, 8) = 6$.

6.1.8.5 Largest of Maximum

The largest of maximum is the largest value of v at which the aggregated-output fuzzy set has the maximum membership grade. Since the maximum value of $\mu_B(v)=0.8$ is at $v=6, 7, 8$, the defuzzified value computed as the largest of maximum is equal to $\max(v = 6, 7, 8) = 8$.

6.1.8.6 Height Defuzzification

In height defuzzification, the defuzzifier first evaluates $\mu_{Bi}(v_i')$ at v_i' and then computes the output of the FLS given by

$$v' = \frac{\sum_{i=1}^{m} v_i' \mu_{B_i}(v_i')}{\sum_{i=1}^{m} \mu_{B_i}(v_i')} \tag{6.58}$$

where m represents the number of output fuzzy sets B_i obtained after fuzzy implication, and v_i' represents the centroid of the fuzzy region i. In the example above (and also Equations 6.54 and 6.55), the number of output fuzzy sets is one, i.e., $m = 1$. The defuzzified output after applying Equation 6.58 with $m = 1$ yields the same value as that computed in Section 6.1.9.1 and equal to 8.09.

6.1.9 Steps of the Fuzzy Inference Process

Consider the ith fuzzy rule (with more than one part in the antecedent) for a MISO system defined as

$$R^i\text{: IF } u \text{ is } T_u^i \quad \text{AND} \quad v \text{ is } T_v^i \quad \text{THEN} \quad w \text{ is } T_w^i \tag{6.59}$$

Here, u, v, and w are the fuzzy or linguistic variables, whereas T_u, T_v, and T_w are their linguistic values (*low*, *high*, *large*, and so on). To obtain the crisp output using an FIS, the following steps are performed:

Step 1: Fuzzify the inputs u and v using membership functions ($\mu^i(u)$, and $\mu^i(v)$) for the ith rule. Essentially, this involves defining the appropriate FMFs.

Step 2: Because the antecedent part of every rule has more than one clause, a FLOR is used to resolve the antecedent to a single number between 0 and 1, which gives the degree of support (or firing strength) for the ith rule. The firing strength can be expressed by

$$\alpha^i = \mu^i(u) \times \mu^i(v) \tag{6.60}$$

Here, the symbol * represents a T-norm. The most popular T-norms used are SI and AP, represented below:

$$\alpha^i = \min(\mu^i(u), \mu^i(v)) \quad \text{or} \quad \alpha^i = \mu^i(u) \bullet \mu^i(v) \tag{6.61}$$

Step 3: Apply an implication method to shape the consequent part (the output fuzzy set) based on the antecedent. The input to the implication process is a single number (α) determined by the antecedent, and the output is a fuzzy set. The most commonly used methods are MORFI (Mamdani) and PORFI (described in Section 6.2)

$$\mu^i(w)' = \min(\alpha^i, \mu^i(w)) \tag{6.62}$$

$$\mu^i(w)' = \alpha^i \bullet \mu^i(w) \tag{6.63}$$

Step 4: Because more than one rule (i.e., more than one output fuzzy set) can be fired simultaneously, it is essential to combine the corresponding output fuzzy sets into a single composite fuzzy set; this process of combining is known as aggregation. The inputs to the aggregation process are outputs of the implication process, and the output of the aggregation process is a single fuzzy set that represents the output variable. The order in which rules are fired does not affect the aggregation process. The most commonly used aggregation method is the max (SU) method. Suppose that rules 3 and 4 are fired at the same time; then, the composite output fuzzy set could be expressed as shown below:

$$\mu(w) = \max(\mu^3(w)', \mu^4(w)') \tag{6.64}$$

Keep in mind that Equation 6.64 represents the final output membership curve or function.

Step 5: To determine the crisp value of the output variable w, use the defuzzification process. The input to this process is the output from the aggregation process (Equation 6.64), and the output is a single crisp number.

6.2 Fuzzy Implication Functions

Because of its approximate reasoning capability, FL can become an ideal tool to develop various applications that require logical reasoning to model imprecisely defined events. The core of an FL-based system is the FIS, in which the fuzzy *if...then* rules are processed using fuzzy implication methods (FIMs) to finally obtain the output as fuzzy sets. An FIM plays a crucial role in the successful designing of the FIS and FLS. It becomes necessary to select an appropriate FIM from the existing methods. If any new implication method is considered, it should satisfy some of the intuitive criteria of GMP and generalized *modus tollens* (GMT; discussed in Section 6.3) so that this FIM can be fitted into the logical development of any system using FL.

6.2.1 Fuzzy Implication Methods

In any FIS, fuzzy implication facilitates mapping between input and output fuzzy sets so that fuzzified inputs can be mapped to desirable output fuzzy sets. Essentially, a fuzzy *if...then* rule, provided by the domain expert, is interpreted as a fuzzy implication. Consider a simple rule such as

$$\text{IF } u \text{ is } A, \text{ THEN } v \text{ is } B \tag{6.65}$$

Here, "*if* u is A" is known as the antecedent or premise, and the clause "*then* v is B" is the consequent part of the fuzzy rule. The crisp variable u, fuzzified by the set A in UOD U, is an input to the inference engine, whereas the crisp variable v, represented by the set B in UOD V, is an output from the inference engine. The fuzzified output of the inference engine, computed using sup-star composition, is given by the following formula:

$$B = RoA \tag{6.66}$$

Here, o is a compositional operator, and R is a fuzzy relation in the product space $U \times V$. Equation 6.66, in the form of the membership function, is

$$\mu_B(v) = \mu_R(u,v) \; o \; \mu_A(u) \tag{6.67}$$

A fuzzy implication, denoted by $\mu_{A \to B}(u,v)$, is also a type of relation that assists in the mapping between input and output. Hence, Equation 6.67 can be rewritten as follows:

$$\mu_B(v) = \mu_{A \to B}(u,v) \; o \; \mu_A(u) \tag{6.68}$$

There are seven standard methods of interpreting the fuzzy *if…then* rule to define an FIP, as listed below [3,8]:

1. The FC approach:

$$\mu_{A \to B}(u,v) = \mu_A(u) * \mu_B(v) \tag{6.69}$$

   Here, the symbol * represents a T-norm operator. This operator is a general symbol for the fuzzy "AND" operation.
2. The FD approach:

$$\mu_{A \to B}(u,v) = \mu_A(u) \dotplus \mu_B(v) \tag{6.70}$$

   Here, the symbol "$\dotplus$" represents an S-norm operator. This operator is the general symbol of the fuzzy "OR" operation.
3. An MI-based approach:

$$\mu_{A \to B}(u,v) = \mu_{\bar{A}}(u) \dotplus \mu_B(v) \tag{6.71}$$

   Here, $\mu_{\bar{A}}(u)$ is a *fuzzy complement* of $\mu_A(u)$. The symbol of the S-norm operator is used here.
4. The PC approach:

$$\mu_{A \to B}(u,v) = \mu_{\bar{A}}(u) \dotplus \mu_A(u) * \mu_B(v) \tag{6.72}$$

   PC applies the two operators "AND" and "OR", i.e., it uses both the T- and S-norms.
5. The extended PC (EPC) approach:

$$\mu_{A \to B}(u,v) = \mu_{\bar{A}}(u) x \mu_{\bar{B}}(v) \dotplus \mu_B(v) \tag{6.73}$$

   The EPC utilizes the complement and the S-norm.

6. The generalization of *modus ponens* approach:

$$\mu_{A\to B}(u,v)=\sup\{c\in[0,1],\mu_A(u)*c\le\mu_B(v)\} \tag{6.74}$$

7. The generalization of *modus tollens* approach:

$$\mu_{A\to B}(u,v)=\inf\{c\in[0,1],\mu_B(v)\dot{+}c\le\mu_A(u)\} \tag{6.75}$$

By applying different combinations of T-norms and S-norms (as discussed in Section 6.1), a variety of approaches can be used to interpret the fuzzy *if…then* rules, i.e., numerous different fuzzy implications can be derived. However, not all fuzzy implications or interpretations completely satisfy the intuitive criteria of GMP and GMT as discussed in Section 6.3. The most commonly used specific fuzzy implications that satisfy one or more of these intuitive criteria are given below:

1. MORFI (Mamdani) is derived by applying the SI operator of T-norms in Equation 6.69:

$$R_{\text{MORFI}}=\mu_{A\to B}(u,v)=\min(\mu_A(u),\mu_B(v)) \tag{6.76}$$

2. PORFI (Larsen) is derived by applying the AP operator of T-norms in Equation 6.69:

$$R_{\text{PORFI}}=\mu_{A\to B}(u,v)=\mu_A(u)\mu_B(v) \tag{6.77}$$

3. The arithmetic rule of fuzzy implication (ARFI; Zadeh/Lukasiewicz) is derived by using the BS operator of S-norms and the complement operator in Equation 6.71 to obtain the following:

$$\begin{aligned} R_{\text{ARFI}}=\mu_{A\to B}(u,v) &= \mu_{\bar{A}}(u)\dot{+}\mu_B(v)\\ &= \min(1,\mu_{\bar{A}}(u)+\mu_B(v))\\ &= \min(1,1-\mu_A(u)+\mu_B(v)) \end{aligned} \tag{6.78}$$

4. The max-min rule of fuzzy implication (MRFI; Zadeh) is derived by using the SI operator of T-norms, the SU operator of S-norms, and the fuzzy-complement operator in Equation 6.72:

$$\begin{aligned} R_{\text{MRFI}}=\mu_{A\to B}(u,v) &= \mu_{\bar{A}}(u)\dot{+}\mu_A(u)*\mu_B(v)\\ &= \max(\mu_{\bar{A}}(u),\mu_A(u)*\mu_B(v))\\ &= \max(1-\mu_A(u),\mu_A(u)*\mu_B(v))\\ &= \max(1-\mu_A(u),\min(\mu_A(u),\mu_B(v))) \end{aligned} \tag{6.79}$$

5. SSFI is derived by using the BD operator of T-norms in Equation 6.74:

$$
\begin{aligned}
R_{\text{SSFI}} = \mu_{A\to B}(u,v) &= \sup\{c \in [0,1], \mu_A(u) * c \le \mu_B(v)\} \\
&= \sup\{c \in [0,1], \max(0, \mu_A(u) + c - 1) \le \mu_B(v)\} \\
&= \begin{cases} 1 \text{ if } \mu_A(u) \le \mu_B(v) \\ 0 \text{ if } \mu_A(u) > \mu_B(v) \end{cases}
\end{aligned}
\tag{6.80}
$$

6. BRFI is derived by using the SU operator of S-norms and the fuzzy-complement operator in Equation 6.71:

$$
\begin{aligned}
R_{\text{BRFI}} = \mu_{A\to B}(u,v) &= \mu_{\bar{A}}(u) \dot{+} \mu_B(v) \\
&= \max(\mu_{\bar{A}}(u), \mu_B(v)) \\
&= \max(1 - \mu_A(u), \mu_B(v))
\end{aligned}
\tag{6.81}
$$

7. GRFI is derived by using the AP operator of T-norms in Equation 6.74:

$$
\begin{aligned}
R_{\text{GRFI}} = \mu_{A\to B}(u,v) &= \sup\{c \in [0,1], \mu_A(u) * c \le \mu_B(v)\} \\
&= \sup\{c \in [0,1], \mu_A(u)c \le \mu_B(v)\} \\
&= \begin{cases} 1 & \text{if } \mu_A(u) \le \mu_B(v) \\ \dfrac{\mu_B(v)}{\mu_A(u)} & \text{if } \mu_A(u) > \mu_B(v) \end{cases}
\end{aligned}
\tag{6.82}
$$

Apart from the above-mentioned seven methods, there are a few more ways to perform fuzzy implication operations that require a combination of both the T-norms and S-norms.

EXAMPLE 6.6

Consider Example 6.3 (Equations 6.26 and 6.27) to understand several methods of FIP.

SOLUTION 6.6

For an ARFI, we have

$$
\begin{aligned}
A \to B &= AXB \\
&= \sum_{UXV} (1 \cap (1 - \mu_A(u) + \mu_B(v)))/(u,v) \\
&= \sum_{UXV} \min(1, 1 - \mu_A(u) + \mu_B(v))/(u,v)
\end{aligned}
\tag{6.83}
$$

From the above, we can see that Equation 6.83 partially uses T-norm and S-norm operators. Using Equations 6.26 and 6.27 in Equation 6.83 yields the following result:

$$A \to B = \begin{Bmatrix} 0.8/(1,1) + 0.6/(1,2) + 0.4/(1,3) + 0.2/(1,4) \\ +1/(2,1) + 0.9/(2,2) + 0.7/(2,3) + 0.5/(2,4) \\ +1/(3,1) + 1/(3,2) + 1/(3,3) + 1/(3,4) \\ +1/(4,1) + 1/(4,2) + 1/(4,3) + 1/(4,4) \end{Bmatrix}$$

The above results are represented as the relational matrix [2] shown in Table 6.15 and Figure 6.32.

For the MRFI (Zadeh), we have

$$\begin{aligned} A \to B &= AXB \\ &= \sum_{UXV} ((\mu_A(u) \cap \mu_B(v)) \cup (1 - \mu_A(u)))/(u,v) \\ &= \sum_{UXV} \max(\min(\mu_A(u), \mu_B(v)), \mu_{\bar{A}}(u))/(u,v) \end{aligned} \tag{6.84}$$

TABLE 6.15

Relational Matrix R_1^{TS}

u/v	**1**	**2**	**3**	**4**
1	0.8	0.6	0.4	0.2
2	1	0.9	0.7	0.5
3	1	1	1	1
4	1	1	1	1

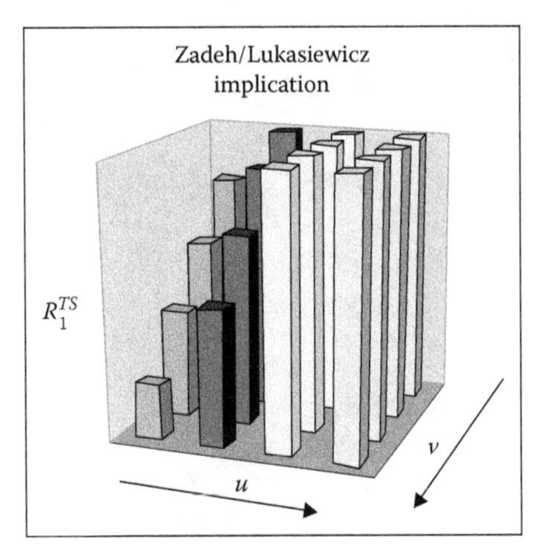

FIGURE 6.32
Graphical representation of the relational matrix R_1^{TS}.

Using Equations 6.26 and 6.27 in Equation 6.84 presents the following result:

$$A \to B = \begin{Bmatrix} 0.8/(1,1)+0.6/(1,2)+0.4/(1,3)+0.2/(1,4) \\ +0.7/(2,1)+0.6/(2,2)+0.4/(2,3)+0.3/(2,4) \\ +0.8/(3,1)+0.8/(3,2)+0.8/(3,3)+0.8/(3,4) \\ +0.9/(4,1)+0.9/(4,2)+0.9/(4,3)+0.9/(4,4) \end{Bmatrix}$$

The relational matrix is depicted in Table 6.16 and Figure 6.33.

For an SSFI, we have

$$\begin{aligned} A \to B &= AXB \\ &= \sum_{UXV} (\mu_A(u) > \mu_B(v))/(u,v) \end{aligned} \tag{6.85}$$

Here,

$$\mu_A(u) > \mu_B(v) = \begin{cases} 1 \text{ if } \mu_A(u) \le \mu_B(v) \\ 0 \text{ if } \mu_A(u) > \mu_B(v) \end{cases}$$

TABLE 6.16

Relational Matrix R_2^{TS}

u/v	**1**	**2**	**3**	**4**
1	0.8	0.6	0.4	0.2
2	0.7	0.6	0.4	0.3
3	0.8	0.8	0.8	0.8
4	0.9	0.9	0.9	0.9

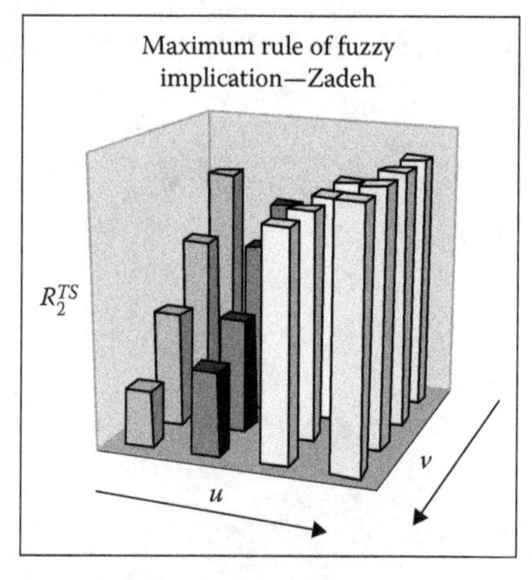

FIGURE 6.33
Graphical representation of the relational matrix R_2^{TS}.

Using Equations 6.26 and 6.27 in Equation 6.85 yields the following result:

$$A \to B = \begin{Bmatrix} 0/(1,1)+0/(1,2)+0/(1,3)+0/(1,4) \\ +1/(2,1)+0/(2,2)+0/(2,3)+0/(2,4) \\ +1/(3,1)+1/(3,2)+1/(3,3)+1/(3,4) \\ +1/(4,1)+1/(4,2)+1/(4,3)+1/(4,4) \end{Bmatrix}$$

This can be conveniently represented using a relational matrix, as shown in Table 6.17 and Figure 6.34.

For a BRFI, we have

$$\begin{aligned} A \to B &= AXB \\ &= \sum_{UXV} ((1-\mu_A(u)) \cup \mu_B(v))/(u,v) \\ &= \sum_{UXV} (\mu_{\bar{A}}(u) \cup \mu_B(v))/(u,v) \\ &= \sum_{UXV} \max(\mu_{\bar{A}}(u), \mu_B(v))/(u,v) \end{aligned} \tag{6.86}$$

TABLE 6.17

Relational Matrix R_3^{TS}

u/v	**1**	**2**	**3**	**4**
1	0	0	0	0
2	1	0	0	0
3	1	1	1	1
4	1	1	1	1

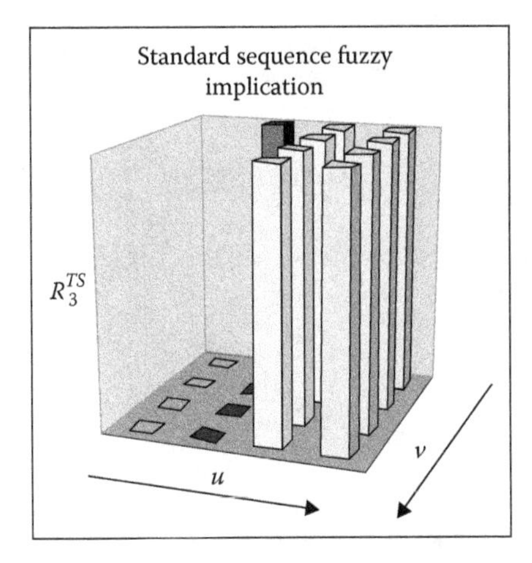

FIGURE 6.34
Graphical representation of the relational matrix R_3^{TS}.

The complement of Equation 6.26 is $\mu_{\bar{A}}(u)/u = (0/1 + 0.3/2 + 0.8/3 + 0.9/4)$. Using $\mu_{\bar{A}}(u)/u$ and Equation 6.27 in Equation 6.86 results in the following Boolean fuzzy implication:

$$A \to B = \begin{Bmatrix} 0.8/(1,1) + 0.6/(1,2) + 0.4/(1,3) + 0.2/(1,4) \\ +0.8/(2,1) + 0.6/(2,2) + 0.4/(2,3) + 0.3/(2,4) \\ +0.8/(3,1) + 0.8/(3,2) + 0.8/(3,3) + 0.8/(3,4) \\ +0.9/(4,1) + 0.9(4,2) + 0.9(4,3) + 0.9/(4,4) \end{Bmatrix}$$

This is conveniently represented with a relational matrix, as shown in Table 6.18 and Figure 6.35.

For a GRFI, we have

$$\begin{aligned} A \to B &= AXB \\ &= \sum\nolimits_{UXV} (\mu_A(u) >> \mu_B(v))/(u,v) \end{aligned} \tag{6.87}$$

TABLE 6.18
Relational Matrix R_4^{TS}

u/v	**1**	**2**	**3**	**4**
1	0.8	0.6	0.4	0.2
2	0.8	0.6	0.4	0.3
3	0.8	0.8	0.8	0.8
4	0.9	0.9	0.9	0.9

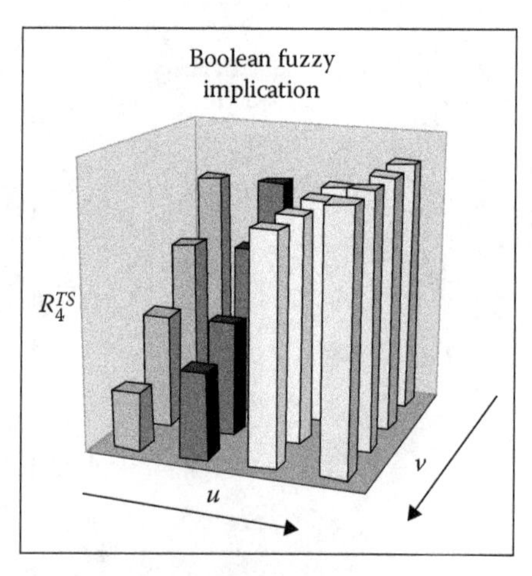

FIGURE 6.35
Graphical representation of the relational matrix R_4^{TS}.

Here,
$$\mu_A(u) >> \mu_B(v) = \begin{cases} 1 & \text{if } \mu_A(u) \le \mu_B(v) \\ \dfrac{\mu_B(u)}{\mu_A(v)} & \text{if } \mu_A(u) > \mu_B(v) \end{cases}$$

Using Equations 6.26 and 6.27 in Equation 6.87 yields the following result:

$$A \to B = \begin{Bmatrix} 0.8/(1,1) + 0.6/(1,2) + 0.4/(1,3) + 0.2/(1,4) \\ +1/(2,1) + 0.85/(2,2) + 0.57/(2,3) + 0.28/(2,4) \\ +1/(3,1) + 1/(3,2) + 1/(3,3) + 1/(3,4) + 1/(4,1) \\ +1/(4,2) + 1/(4,3) + 1/(4,4) \end{Bmatrix}$$

This is represented using a relational matrix, as shown in Table 6.19 and Figure 6.36.

TABLE 6.19

Relational Matrix R_5^{TS}

u/*v*	**1**	**2**	**3**	**4**
1	0.8	0.6	0.4	0.2
2	1	0.85	0.57	0.28
3	1	1	1	1
4	1	1	1	1

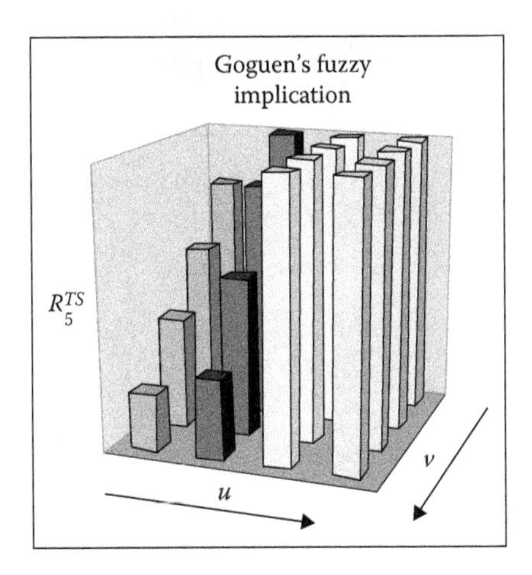

FIGURE 6.36
Graphical representation of the relational matrix R_5^{TS}.

6.2.2 Comparative Evaluation of the Various Fuzzy Implication Methods s with Numerical Data

In Example 6.7, we evaluate several FIMs using numerical data by applying MATLAB.

EXAMPLE 6.7

Compute the norms of various fuzzy implication functions of the previous sections and compare the results.

SOLUTION 6.7

To compare various FIMs, the norms of the relational matrices are computed using MATLAB. Table 6.20 shows the values of the norms of R_1^T, R_2^T, R_3^T, R_4^T, R_1^S, R_2^S, R_3^S, R_4^S, R_1^{TS}, R_2^{TS}, R_3^{TS}, R_4^{TS}, and R_5^{TS} relational matrices for the same example using Equations 6.26 and 6.27. We can see from the column under $\|\boldsymbol{R}^T\|$ that the SI is relatively stronger than the results of other T-norms. Similarly, the DU under column $\|R^S\|$ is relatively stronger than the other S-norms. We can also see that the norms under the column $\|R^S\|$ are relatively stronger compared to the norms under column $\|R^T\|$. Most of the norms under column $\|R^S\|$ fall between the norms under columns $\|R^T\|$ and $\|R^S\|$, because the implication methods under it partially apply the T-norm and S-norm operators.

MORFI and PORFI have found extensive applications in practical control engineering due to their computational simplicity [2].

TABLE 6.20

Norms of Several Relational Matrices

Implication Methods	$\|\boldsymbol{R}^T\|$	Implication Methods	$\|\boldsymbol{R}^T\|$	Implication Methods	$\|\boldsymbol{R}^T\|$
Standard intersection–Mamdani implication	1.555	Standard union	2.869	Arithmetic rule of fuzzy implication–Zadeh/Lukasiewicz implication	3.391
Algebraic product–Larsen implication	1.359	Algebraic sum	3.109	Maximum rule of fuzzy implication–Zadeh	2.807
Bounded difference	1.236	Bounded sum	3.370	Standard sequence fuzzy implication	2.877
Drastic intersection	1.095	Drastic union	4.000	Boolean fuzzy implication	2.828
				Goguen's fuzzy implication	3.310

$\| \;\|$ indicates the norm of relational matrix.

6.2.3 Properties of Fuzzy If-Then Rule Interpretations

GMP and GMT are two ideal inference rules for our day-to-day reasoning and thought processes. To compute the consequences when the FIMs are applied in the fuzzy inference process (so that they can be compared with the consequences of GMP), the following formula is used:

$$B' = RoA' \tag{6.88}$$

Here, o, known as the compositional operator, is represented using the sup-star notation, with "sup" as supremum and "star" as the T-norm operator. In the present case, SU or the "max" operator for "sup" and SI or the "min" operator for "star" have been used. In terms of the membership function, Equation 6.88 is rewritten as

$$\mu_{B'}(v) = \sup_{u\in U}\{\mu_{A\to B}(u,v) * \mu_{A'}(u)\}$$

or,

$$y = \mu_{B'}(v) = \sup_{u\in U}\{\min[\mu_{A\to B}(u,v), \mu_{A'}(u)]\} \tag{6.89}$$

Here, $\mu_{A\to B}(u,v)$ is an FIM, and $x = \mu_{A'}(u)$ is premise 1 of the GMP rule (see Table 6.21), containing any one of the following:

$$\mu_{A'}(u) = \mu_A(u),\ \mu_{A'}(u) = \mu^2{}_A(u),\ \mu_{A'}(u) = \sqrt{\mu_A(u)},\quad \text{or}\quad \mu_{A'}(u) = 1 - \mu_A(u)$$

Similarly, to compute the consequences in a GMT (so that they can be compared with the consequences of GMP) when FIMs are applied in the fuzzy inference process, the following formula is used:

$$A' = RoB' \tag{6.90}$$

TABLE 6.21

GMP Intuitive Criteria—A Forward Chain-Inference Rule

GMP Criteria	*u* is *A′* (premise 1)	*v* is *B′* (consequence)
C1	*u* is *A*	*v* is *B*
C2-1	*u* is very *A*	*v* is very *B*
C2-2	*u* is very *A*	*v* is *B*
C3-1	*u* is more or less *A*	*v* is more or less *B*
C3-2	*u* is more or less *A*	*v* is *B*
C4-1	*u* is not *A*	*v* is unknown
C4-2	*u* is not *A*	*v* is not *B*

TABLE 6.22
GMT Intuitive Criteria—A Backward-Chain Inference Rule

GMT Criteria	v is B' (premise 1)	u is A' (consequence)
C5	v is not B	u is not A
C6	v is not very B	u is not very A
C7	v is not more or less B	u is not more or less A
C8-1	v is B	u is unknown
C8-2	v is B	u is A

Equation 6.90, in terms of the membership function, is rewritten as

$$\mu_{A'}(u) = \sup_{v \in V} \{\min[\mu_{A \to B}(u, v), \mu_{B'}(v)]\} \tag{6.91}$$

Here, $\mu_{B'}(v)$ is the premise 1 of the GMT (see Table 6.22) containing any one of the following:

$$\mu_{B'}(v) = 1 - \mu_B(v),\ \mu_{B'}(v) = 1 - \mu^2{}_B(v),\ \mu_{B'}(v) = 1 - \sqrt{\mu_B(v)},\ \text{or}\ \mu_{B'}(v) = \mu_B(v)$$

6.3 Forward- and Backward-Chain Logic Criteria

The intuitive criteria of GMP and GMT for both forward and reverse logic are described in this section. These are two important rules that can also be used in FL for approximate reasoning or inference [2,8].

6.3.1 Generalization of *Modus Ponens* Rule

This is a forward-driven inference rule defined by adopting the following *modus operandi* [8]:

$$\begin{aligned} &\text{Premise 1} &&: u \text{ is } A' \\ &\text{Premise 2} &&: \text{IF } u \text{ is } A \text{ THEN } v \text{ IS } B \\ &\text{Consequence} &&: v \text{ is } B' \end{aligned}$$

Here, A' and A are input fuzzy sets, B' and B are output fuzzy sets, and u and v are the variables corresponding to the input and output fuzzy sets, respectively. The fuzzy set A' of premise 1 can have the following values: A, *very* A, *more or less* A, and *not* A. The linguistic values "very" and "more or less" are hedges and are defined in terms of their membership

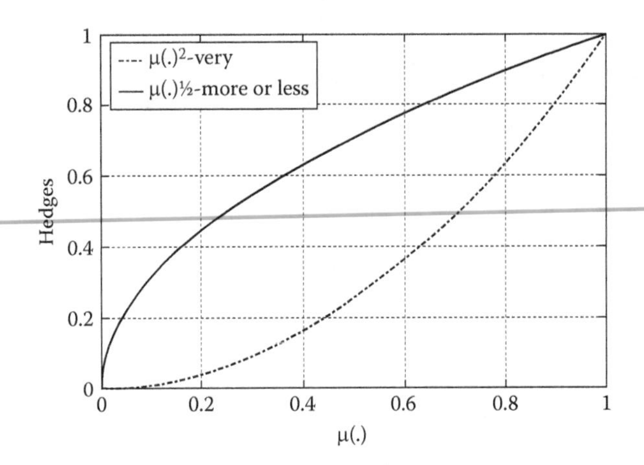

FIGURE 6.37
Variables, with hedges "very" and "more or less." (From Kashyap, S. K., J. R. Raol, and A. V. Patel. 2008. In *Foundations of Generic Optimization Vol. 2*, pp. 313–386, ed. R. Lowen and A. Verschoren. New York: Springer. With permission.)

grade as $\mu(.)^2$ and $\mu(.)^{1/2}$, respectively. (.) denotes the fuzzy sets A or B. Figure 6.37 shows the profiles of these hedges. The criteria of GMP, relating premise 1 and the consequence for any given premise 2, are provided in Table 6.21 [8]. There are seven criteria under GMP, and each can be related to our everyday reasoning. If a fundamental relation between "u is A" and "v is B" is not strong in premise 2, then the satisfaction of the criteria C2-2 and C3-2 is allowed.

6.3.2 Generalization of *Modus Tollens* Rule

This is a backwards goal-driven inference rule, defined by the following procedure [8]:

$$\begin{array}{lll} \text{Premise 1} & : & v \text{ is } B' \\ \text{Premise 2} & : & \text{IF } u \text{ is } A \text{ THEN } v \text{ IS } B \\ \text{Consequence} & : & u \text{ is } A' \end{array}$$

Fuzzy set B' of premise 1 could have the following values: *not* B, not *very* B, not *more or less* B, and B. The linguistic values such as "not very" and "not more or less" are known as hedges and are defined in terms of

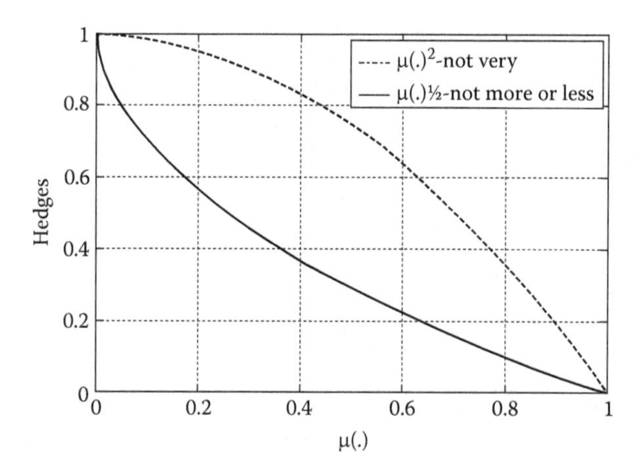

FIGURE 6.38
Variables, with hedges "not very" and "not more or less." (From Kashyap, S. K., J. R. Raol, and A. V. Patel. 2008. In *Foundations of Generic Optimization Vol. 2*, pp. 313–386, ed. R. Lowen and A. Verschoren. New York: Springer. With permission.)

their membership grade as $1-\mu(.)^2$ and $1-\mu(.)^{1/2}$, respectively; Figure 6.38 shows these profiles. The criteria of GMT, relating premise 1 and the consequence for any given premise 2, are presented in Table 6.22 [8].

6.4 Tool for the Evaluation of Fuzzy Implication Functions

In this section, a methodology to determine whether any of the existing FIMs satisfies a given set of intuitive criteria of GMP and GMT is described. MATLAB and graphics have been used to develop a user-interactive package for evaluating implication methods with respect to these criteria [5,7]. In this section, only the main procedural steps are given and thereafter illustrated for only one FIM—MORFI. The details of the evaluation of all seven FIMs against all the criteria of GMP and GMT using this tool are described in [5,7].

6.4.1 Study of Criteria Satisfaction Using MATLAB® Graphics

This tool helps in visualizing the results both analytically and numerically, in addition to facilitating the use of plots. Table 6.23 shows the menu panels for the toolbox (based on figures 3–5 of [5]). The first menu panel

TABLE 6.23

Menu Panel Ideas for Selection of FIM, Premise 1 of GMP, and Premise 1 of GMT Criteria

MENU	MENU	MENU
Selection of FIM	Selection of premise 1 of GMP criteria	Selection of premise 1 of GMT criteria
MORFI	C1: x is A	C5: y is not B
PORFI	C2-1/C2-2: x is very A	C6: y is not very B
ARFI	C3-1/C-2: x is more or less A	C7: y is not more or less B
MRFI	C4-1/C4-2: x is not A	C8-1/C8-2: y is B
BRFI	Exit	Exit
GRFI		
Exit		

(column 1 of Table 6.23) helps the user to select a particular FIM for evaluation. The next two columns help select premise 1 from the GMP and GMT criteria, respectively, to be applied to the chosen FIM. The procedural steps to establish the satisfaction or otherwise of GMP and GMT criteria using MATLAB graphics are described next [5,7].

First, generate 2D plots of the selected implication method. Let the fuzzy input set A and output set B have the following membership grades:

$$\mu_A(u) = \begin{bmatrix} 0 & 0.05 & 0.1 & 0.15 & ,..., & 1 \end{bmatrix} \tag{6.92}$$

$$\mu_B(v) = \begin{bmatrix} 0 & 0.05 & 0.1 & 0.15 & ,..., & 1 \end{bmatrix} \tag{6.93}$$

The plots are generated by taking one value of Equation 6.92 at a time for the entire $\mu_A(u)$ of Equation 9.92 and applying each value to the selected implication methods of Equations 6.77 through 6.83. In the plots, the x-axis is $\mu_A(u)$, and the y-axis is $\mu_{A \to B}(u, v)$ for each value of $\mu_B(v)$. Figure 6.39 shows the 2D plot of a MORFI FIM. Coding with symbols indicates the values of the FIM method computed by varying fuzzy set μ_B between 0 and 1, with a fixed interval of 0.05. It is important to realize that there could be infinite possible values if the interval of μ_B is reduced to a very small value. In procedural steps 2 and 3, the GMP and GMT criteria are applied to such FIMs, and the consequences are realized visually and studied analytically.

One by one, premise 1 of all GMP criteria (C1 to C4-2) are applied to the chosen FIM. This is illustrated here only for the MORFI FIM.

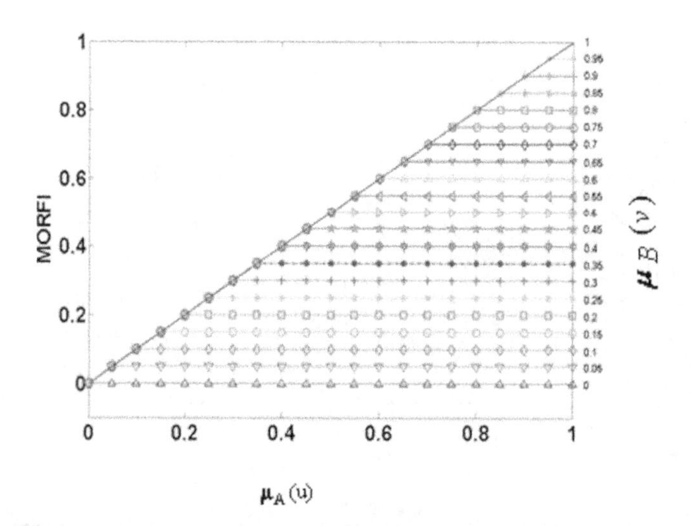

FIGURE 6.39
2D plot for min-operation rule of fuzzy implication. (From Kashyap, S. K., J. R. Raol, and A. V. Patel. 2008. In *Foundations of Generic Optimization Vol. 2*, pp. 313–386, ed. R. Lowen and A. Verschoren. New York: Springer. With permission.)

MORFI: C1 [5,7]: $\mu_{A'}(u) = \mu_A(u)$ is applied to the right-hand side of Equation 6.89 to obtain the consequence $\mu_{B'}(v)$. We start with an attempt to interpret the *min* operation of Equation 6.89 by considering Figure 6.39, the 2D view of the FIM $\mu_{A\to B}(u,v)$, and premise 1 $\mu_{A'}(u)$ (Table 6.21). Figures 6.40 and 6.41 (for only one value of $\mu_B(v)$) show this particular superimposition. We can see that $\mu_{A'}(u)$ is always larger than or equal to $\mu_{A\to B}(u,v)$ for any value of $\mu_A(u)$. This means that the outcome of the *min* operation is $\mu_{A\to B}(u,v)$ (Figure 6.39). Furthermore, from Figures 6.40 and 6.41, we can see that $\mu_{A\to B}(u,v) = \min(\mu_A(u), \mu_B(v))$ converges to $\mu_B(v)$, also the max. value of $\mu_{A\to B}(u,v)$, for $\mu_A(u) \geq \mu_B(v)$. Hence, the supremum of $\mu_{A\to B}(u,v)$ is $\mu_B(v)$, i.e., $\mu_{B'}(v) = \mu_B(v)$. We can therefore infer that MORFI satisfies the intuitive criterion C1 of GMP. This is also proven by an analytical method as follows:

$$\begin{aligned}
\mu_{B'}(v) &= \sup_{u\in U}\{\min[\min\{\mu_A(u),\mu_B(v)\},\mu_A(u)]\} \\
&= \sup_{u\in U}\begin{Bmatrix} y1 = \min\{\mu_A(u),\mu_A(u)\};\ \text{for } \mu_A(u)\leq\mu_B(v) \\ y2 = \min\{\mu_B(v),\mu_A(u)\};\ \text{for } \mu_A(u)>\mu_B(v) \end{Bmatrix} \\
&= \sup_{u\in U}\begin{Bmatrix} y1 = \mu_A(u);\ \text{for } \mu_A(u)\leq\mu_B(v) \\ y2 = \mu_B(v)\ \ \text{for } \mu_A(u)>\mu_B(v) \end{Bmatrix}
\end{aligned} \tag{6.94}$$

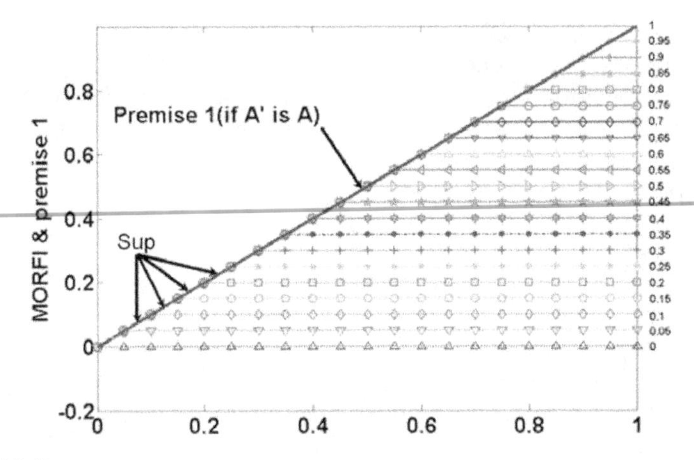

FIGURE 6.40
Superimposition of min-operation rule of fuzzy implication and premise 1 of C1. (From Kashyap, S. K., J. R. Raol, and A. V. Patel. 2008. In *Foundations of Generic Optimization Vol. 2*, pp. 313–386, ed. R. Lowen and A. Verschoren. New York: Springer. With permission.)

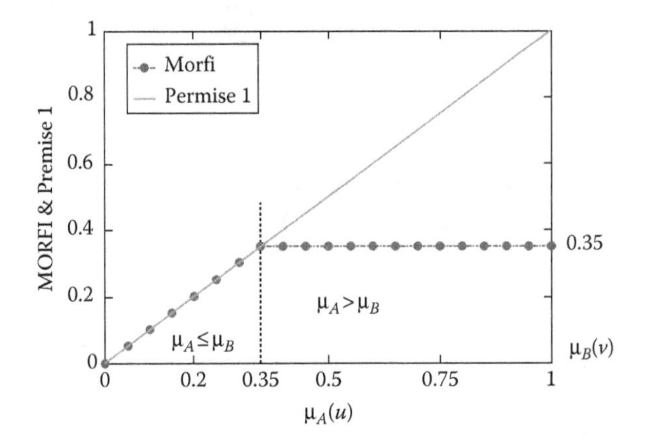

FIGURE 6.41
MORFI and premise 1 of C1 for $\mu_B = 0.35$. (From Kashyap, S. K, Raol, J. R., and Patel, A. V. In *Foundations of Generic Optimization Vol. 2*, pp. 313–386, ed. R. Lowen and A. Verschoren. New York: Springer. With permission.)

The outcome of the *min* operation between $\mu_{A \to B}(u,v)$ and $\mu_{A'}(u)$ consists of $y1$ and $y2$; the outcome begins with $y1$, which increases to a maximum value of $\mu_B(v)$ with an increase in $\mu_A(u)$ from zero to $\mu_B(v)$, and $y2$ begins from the maximum value of $y1$ and remains constant at that value in spite of further increases in $\mu_A(u)$. Hence, it is observed that the supremum is $y2$, i.e., $\mu_{B'}(v) = \mu_B(v)$.

MORFI: C2-1/C2-2 [5,7]: $\mu_{A'}(u) = \mu^2_A(u)$ is applied to the RHS of Equation 6.89 to obtain the consequence $\mu_{B'}(v)$. Figures 6.42 and 6.43 illustrate the superimposition of $\mu_{A\to B}(u,v)$ and $\mu_{A'}(u)$. The area below the intersection point of $\mu_{A\to B}(u,v)$ and $\mu_{A'}(u)$ corresponds to the *min* operation of Equation 6.89, and the supremum of the resultant area is those intersection points having values equal to $\mu_B(v)$. We can therefore infer that MORFI satisfies the intuitive criterion C2-2 (not C2-1) of GMP. The analytical process is as follows:

$$\mu_{B'}(v) = \sup_{u\in U}\{\min[\min\{\mu_A(u),\mu_B(v)\},\mu^2_A(u)]\}$$

$$= \sup_{u\in U}\begin{Bmatrix} y1 = \min\{\mu_A(u),\mu^2_A(u)\};\text{ for } \mu_A(u)\le\mu_B(v) \\ y2 = \min\{\mu_B(v),\mu^2_A(u)\};\text{ for } \mu_A(u)>\mu_B(v)\end{Bmatrix} \tag{6.95}$$

$$= \sup_{u\in U}\left\{\begin{matrix} y1 = \mu^2_A(u); \quad \text{since } \mu^2_A(u)\le\mu_A(u);\ \text{for}\ \ \mu_A(u)\le\mu_B(v) \\ \begin{Bmatrix} y21 = \mu^2_A(u);\ \text{for}\ \ \mu_A(u)\le\sqrt{\mu_B(v)} \\ y22 = \mu_B(v);\ \text{for}\ \ \mu_A(u)>\sqrt{\mu_B(v)}\end{Bmatrix};\ \text{for}\ \ \mu_A(u)>\mu_B(v)\end{matrix}\right\}$$

The outcome of the *min* operation between $\mu_{A\to B}(u,v)$ and $\mu_{A'}(u)$ consists of $y1$, $y21$, and $y22$. Because $\sqrt{\mu_B(v)} > \mu_B(v)$, $y1$ and $y21$ can be treated as one, having a value $\mu^2_A(u)$ for the value of $\mu_A(u)\le\sqrt{\mu_B(v)}$, the outcome begins with $y1/y21$, which increases to a maximum value

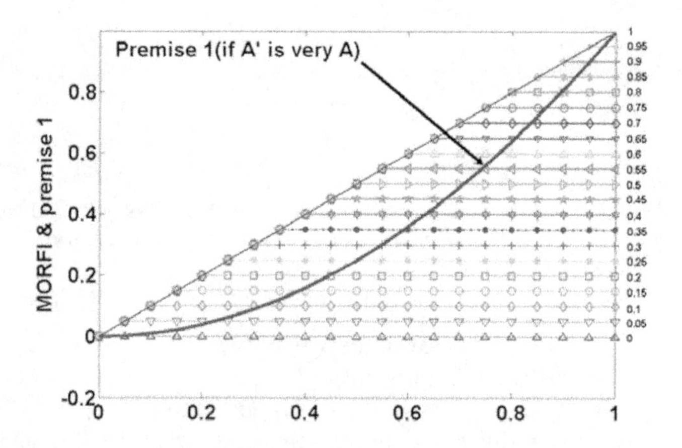

FIGURE 6.42
Superimposition of min-operation rule of fuzzy implication and premise 1 of C2-1/C2-2. (From Kashyap, S. K., J. R. Raol, and A. V. Patel. 2008. In *Foundations of Generic Optimization Vol. 2*, pp. 313–386, ed. R. Lowen and A. Verschoren. New York: Springer. With permission.)

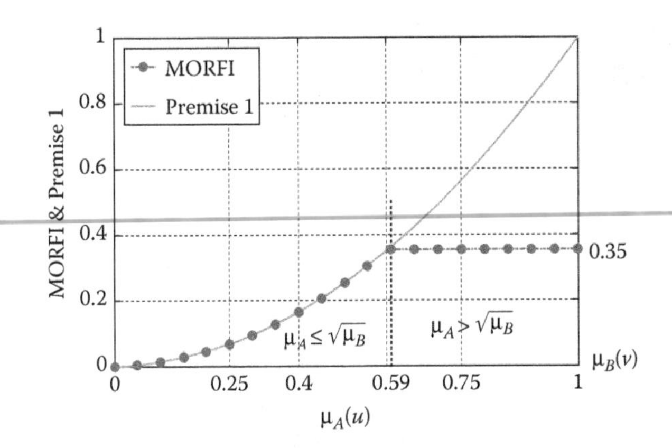

FIGURE 6.43
min-operation rule of fuzzy implication and premise 1 of C2-1/C2-2 for $\mu_B = 0.35$. (From Kashyap, S. K., J. R. Raol, and A. V. Patel. 2008. In *Foundations of Generic Optimization Vol. 2*, pp. 313–386, ed. R. Lowen and A. Verschoren. New York: Springer. With permission.)

of $\mu_B(v)$ with an increase in $\mu_A(u)$ from zero to $\sqrt{\mu_B(v)}$. The function $y22$ begins from the maximum value of $y1/y21$ and remains constant at that value in spite of any further increase in $\mu_A(u)$. Thus, we can see that the supremum is fixed at $y22$, i.e., $\mu_{B'}(v) = \mu_B(v)$.

Again, premise 1 of all GMT criteria, i.e., C5 to C8-2, are applied to the FIM one by one. The relational matrix of the FIM should be transposed, as shown in Figure 6.44. The x-axis now represents the fuzzy set $\mu_B(v)$ (in case of GMP, it is $\mu_A(u)$) and the y-axis represents the implication $\mu_{A\to B}(u,v)$ computed for each value of the fuzzy set $\mu_A(u)$. Transposition of R is required because the inference rule of GMT is a backward, goal-driven rule.

MORFI: C8-1/8-2 [5,7]: $\mu_{B'}(v) = \mu_B(v)$ is applied to the RHS of Equation 6.91 to obtain the consequence $\mu_{A'}(u)$. We can see from Figure 6.45 that $\mu_{B'}(v)$ is always larger than or equal to $\mu_{A\to B}(u,v)$ for any value of $\mu_B(v)$. This means that the outcome of the *min* operation is $\mu_{A\to B}(u,v)$ itself (Figure 6.44). We can also see that $\mu_{A\to B}(u,v) = \min(\mu_A(u), \mu_B(v))$ converges to $\mu_A(u)$, also the max value of $\mu_{A\to B}(u,v)$, for $\mu_B(v) \ge \mu_A(u)$. The supremum of $\mu_{A\to B}(u,v)$ is $\mu_A(u)$, i.e., $\mu_{A'}(u) = \mu_A(u)$. MORFI therefore satisfies the intuitive criterion C8-2 (not C8-1) of GMT. The analytical proof is given next.

$$\mu_{A'}(u) = \sup_{v\in V}\{\min[\min(\mu_A(u), \mu_B(v)), \mu_B(v)]\}$$

$$= \sup_{v\in V}\begin{cases} y1 = \min[\mu_B(v), \mu_B(v)] = \mu_B(v); & \text{for } \mu_B(v) \le \mu_A(u) \\ y2 = \min[\mu_A(u), \mu_B(v)] = \mu_A(u); & \text{for } \mu_B(v) > \mu_A(u) \end{cases} \quad (6.96)$$

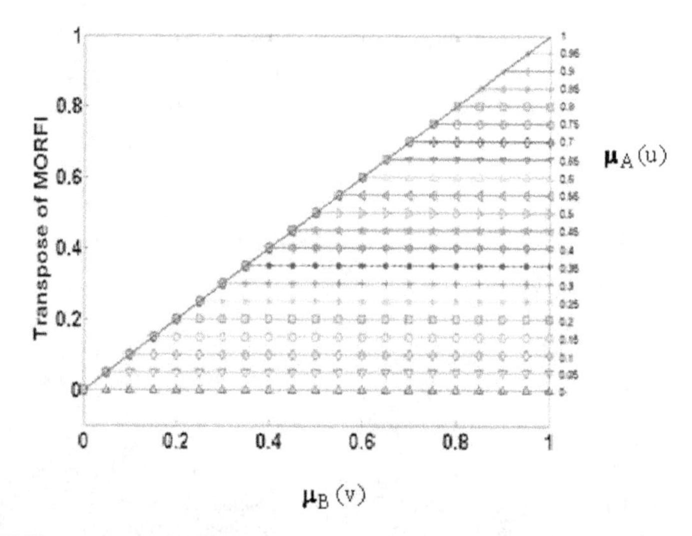

FIGURE 6.44
A 2D plot of the min-operation rule of fuzzy implication-transpose. (From Kashyap, S. K., J. R. Raol, and A. V. Patel. 2008. In *Foundations of Generic Optimization Vol. 2,* pp. 313–386, ed. R. Lowen and A. Verschoren. New York: Springer. With permission.)

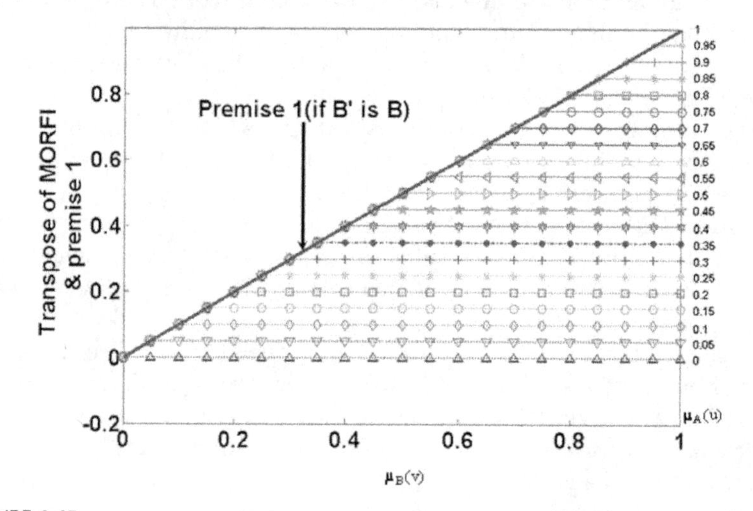

FIGURE 6.45
Superimposition of min-operation rule of fuzzy implication and premise 1 of C8-1/C8-2. (From Kashyap, S. K., J. R. Raol, and A. V. Patel. 2008. In *Foundations of Generic Optimization Vol. 2,* pp. 313–386, ed. R. Lowen and A. Verschoren. New York: Springer. With permission.)

The outcome of the *min* operation between $\mu_{A\to B}(u,v)$ and $\mu_{B'}(v)$, for some fixed value of $\mu_A(u)$, consists of $y1$, having value $\mu_B(v)$ when $\mu_B(v) \le \mu_A(u)$, which increases up to a value of $\mu_A(u)$; then, $y2$ becomes equal to the fixed value of $\mu_A(u)$ when $\mu_B(v) > \mu_A(u)$. We can conclude that the supremum of $y1$ and $y2$ will be the curve $\mu_A(u)$, i.e., $\mu_{A'}(u) = \mu_A(u)$.

6.5 Development of New Implication Functions

The various fuzzy implication functions and FIMs (Section 6.2) use FLORs, such as the T-norm and/or the S-norm. There could be a possibility of the existence of more such implication methods, which can be obtained using various unexplored combinations of fuzzy operators in Equations 6.69 through 6.75. Based on this fact, an effort was undertaken in a study by Kashyap [7] to derive a few new FIMs using MI, PC, and fuzzy operators. After the derivations, the consequences of these new implication methods were tested with those of the GMP and GMT criteria using the MATLAB graphics-based tool presented in Section 6.4. It is possible that many such new fuzzy implication functions might not satisfy all the GMP and GMT criteria; however, the approach described in Section 6.4 and here provides a way to arrive at new FIMs that might be useful in certain applications of FL, particularly in the analysis and design of control and AI systems. This tool is flexible, and therefore new avenues can be explored based on the intuitive experience of the user or designer and his or her special needs for the control-design and MSDF/AI processes. The tool and the derivation of the new FIM may be a possibility in the search for new FIMs; however, the study does not guarantee that the new FIM would be superior to existing ones. The results of this section should not be considered final, but rather an indication of a new direction toward FIM goals. Based on the existing FIM and the MATLAB-graphic user interface (GUI) tool for the evaluation of FIM, the new implication methods (proposed after graphical visualization and study of associated analytical derivations) are listed below:

- The PCBSAP rule of fuzzy implication is derived by applying the AP operator of T-norms and the BS operator of S-norms in Equation 6.73 as follows:

$$\begin{aligned} R_{\text{PCBSAP}} = \mu_{A\to B}(u,v) &= \mu_{\bar{A}}(u) \dot{+} \mu_A(u) * \mu_B(v) \\ &= \min(1, \mu_{\bar{A}}(u) + \mu_A(u)\mu_B(v)) \\ &= \min(1, 1 - \mu_A(u) + \mu_A(u)\mu_B(v)) \end{aligned} \tag{6.97}$$

- The PCSUAP rule of fuzzy implication is derived by applying the AP operator of T-norms and the SU operator of S-norms in Equation 6.72 as shown below:

$$\begin{aligned} R_{\text{PCSUAP}} = \mu_{A\to B}(u,v) &= \mu_{\bar{A}}(u) \dot{+} \mu_A(u) * \mu_B(v) \\ &= \max(\mu_{\bar{A}}(u), \mu_A(u)\mu_B(v)) \\ &= \max(1 - \mu_A(u), \mu_A(u)\mu_B(v)) \end{aligned} \tag{6.98}$$

- The PCBSSI rule of fuzzy implication is derived by applying the SI operator of T-norms and the BS operator of S-norms in Equation 6.72.

$$\begin{aligned} R_{\text{PCBSSI}} = \mu_{A\to B}(u,v) &= \mu_{\bar{A}}(u) \dotplus \mu_A(u) * \mu_B(v) \\ &= \mu_{\bar{A}}(u) \dotplus \min(\mu_A(u), \mu_B(v)) \\ &= \min(1, 1-\mu_A(u) + \min(\mu_A(u), \mu_B(v))) \end{aligned} \tag{6.99}$$

- The PCBSBP rule of fuzzy implication is derived by applying the BP operator of T-norms and the BS operator of S-norms in Equation 6.72 as shown underneath:

$$\begin{aligned} R_{\text{PCBSBP}} &= \mu_{A\to B}(u,v) = \mu_{\bar{A}}(u) \dotplus \mu_A(u) * \mu_B(v) \\ &= \min(1, \mu_{\bar{A}}(u) + \max(0, \mu_A(u) + \mu_B(v) - 1)) \\ &= \min(1, 1-\mu_{\text{A}}(u) + \max(0, \mu_A(u) + \mu_B(v) - 1)) \end{aligned} \tag{6.100}$$

- The PCSUBP rule of fuzzy implication is derived by applying the BP operator of T-norms and the SU operator of S-norms in Equation 6.72 as follows:

$$\begin{aligned} R_{\text{PCSUBP}} &= \mu_{A\to B}(u,v) = \mu_{\bar{A}}(u) \dotplus \mu_A(u) * \mu_B(v) \\ &= \max(\mu_{\bar{A}}(u), \max(0, \mu_{\text{A}}(u) + \mu_B(v) - 1)) \\ &= \max(1-\mu_A(u), \max(0, \mu_A(u) + \mu_B(v) - 1)) \end{aligned} \tag{6.101}$$

- The PCASBP rule of fuzzy implication is derived by applying the BP operator of T-norms and the AS operator of S-norms in Equation 6.72 to yield the following:

$$\begin{aligned} R_{\text{PCASBP}} &= \mu_{A\to B}(u,v) = \mu_{\bar{A}}(u) \dotplus \mu_A(u) * \mu_B(v) \\ &= 1-\mu_A(u) + \max(0, \mu_A(u) + \mu_B(v) - 1) \\ &\quad -(1-\mu_A(u))\max(0, \mu_A(u) + \mu_B(v) - 1) \end{aligned} \tag{6.102}$$

- The PCASAP rule of fuzzy implication is derived by applying the AP operator of T-norms and the AS operator of S-norms in Equation 6.72 as represented below:

$$\begin{aligned} R_{\text{PCASAP}} = \mu_{A\to B}(u,v) &= \mu_{\bar{A}}(u) \dotplus \mu_A(u) * \mu_B(v) \\ &= 1-\mu_A(u) + \mu_A(u)\mu_B(v) - (1-\mu_A(u))\mu_A(u)\mu_B(v) \\ &= 1-\mu_A(u) + \cancel{\mu_A(u)\mu_B(v)} - \cancel{\mu_A(u)\mu_B(v)} + \mu^2{}_A(u)\mu_B(v) \\ &= 1-\mu_A(u)(1-\mu_A(u)\mu_B(v)) \end{aligned} \tag{6.103}$$

- The PCASSI rule of fuzzy implication is derived by applying the SI operator of T-norms and the AS operator of S-norms in Equation 6.72.

$$
\begin{aligned}
R_{\text{PCASAP}}\mu_{A\to B}(u,v) &= \mu_{\bar{A}}(u) \dot{+} \mu_A(u) * \mu_B(v) \\
&= 1-\mu_A(u)+\min(\mu_A(u),\mu_B(v)) \\
&= 1-\mu_A(u)+\min(\mu_A(u),\mu_B(v)) \\
&\quad -(1-\mu_A(u))\min(\mu_A(u),\mu(v)) \\
&= 1-\mu_A(u)+\min(\mu_A(u),\mu_B(v)) \\
&\quad -\min(\mu_A(u),\mu_B(v))+\mu_A(u)\min(\mu_A(u),\mu_B(v)) \\
&= 1-\mu_A(u)(1-\min(\mu_A(u),\mu_B(v)))
\end{aligned}
\tag{6.104}
$$

- The MIAS rule of fuzzy implication is derived by applying the AS operator of S-norms in Equation 6.71 as depicted below:

$$
\begin{aligned}
R_{\text{MIAS}} = \mu_{A\to B}(u,v) &= \mu_{\bar{A}}(u) \dot{+} \mu_B(v) \\
&= 1-\mu_A(u)+\mu_B(v)-(1-\mu_A(u))\mu_B(v) \\
&= 1-\mu_A(u)+\mu_B(v)-\mu_B(v)+\mu_A(u)\mu_B(v) \\
&= 1-\mu_A(u)(1-\mu_B(v))
\end{aligned}
\tag{6.105}
$$

The procedural steps used to generate the 2D plots of new implication methods are the same as those discussed in Section 6.4. Some modification is carried out in the existing MATLAB graphics tool by adding new cases to existing cases and writing the mathematical equations pertaining to any of the new implication methods, to derive these implication methods and conduct their "satisfaction" studies by checking against the intuitive criteria of the GMP and GMT rules. At this point, it is important to mention that the various implication methods discussed in Section 6.4 and this section might not be applicable to all types of applications, and it is up to any domain expert to select an appropriate method suitable for a particular application. The user can cut down the effort level by first considering those implication methods that satisfy the maximum number of intuitive criteria of GMP and GMT. For the purpose of illustration, only one new FIM (from the list above) is considered, and the discussion is centered on the criteria that are satisfied by the selected FIM; all new FIMs do not satisfy all the criteria, and the discussions related to this aspect are not repeated in the present book. The complete details can be found in works by Kashyap [7]. The purpose here is to illustrate how a new FIM can be evolved and evaluated using the same MATLAB GUI tool, as explained in

Section 6.4. Whether the new FIM is very useful or not is a separate question. However, applications of some of the new FIMs have been validated and are presented in Section 8.4.

6.5.1 Study of Criteria Satisfaction by New Implication Function Using MATLAB and GUI Tools

Let us begin with the implication method PCBSAP. A 2D plot of this method is shown in Figure 6.46. The details of the graphical and analytical developments related to certain new fuzzy implication functions that do not satisfy certain GMP and GMT criteria, except the one below to establish the procedure, are not presented here.

PCBSAP: C1: $\mu_{A'}(u) = \mu_A(u)$ is applied to the RHS of Equation 6.89 to derive the consequence $\mu_{B'}(v)$. Figures 6.47 and 6.48 illustrate the superimposed plots of $\mu_{A \to B}(u, v)$ and $\mu_{A'}(u)$. We can see that the area below the intersection point of $\mu_{A \to B}(u, v)$ and $\mu_{A'}(u)$ corresponds to the *min* operation of Equation 6.89. The supremum of the resultant area is the intersection points having values equal to $1/2 - \mu_B(v)$. Therefore, we can conclude that PCBSAP does not satisfy the intuitive criterion C1 of GMP.

Further analysis is elaborated next:

$$
\begin{aligned}
\mu_{B'}(v) &= \sup_{u \in U}\{\min[\min\{1, 1 - \mu_A(u) + \mu_A(u)\mu_B(v)\}, \mu_A(u)]\} \\
&= \sup_{u \in U}\{\min[1 - \mu_A(u) + \mu_A(u)\mu_B(v), \mu_A(u)]\} \\
&= \sup_{u \in U}\begin{Bmatrix} y1 = \mu_A(u); & \text{for } \mu_A(u) \le \mu^{\min}{}_A(u) \\ y2 = 1 - \mu_A(u) + \mu_A(u)\mu_B(v); & \text{for } \mu_A(u) > \mu^{\min}{}_A(u) \end{Bmatrix}
\end{aligned} \tag{6.106}
$$

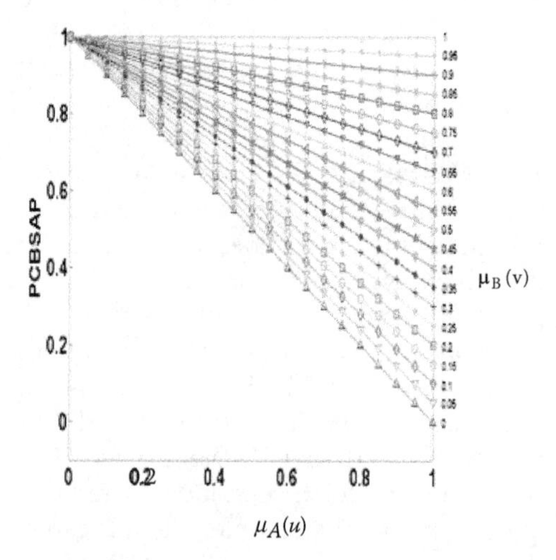

FIGURE 6.46
2D plots of PCBSAP.

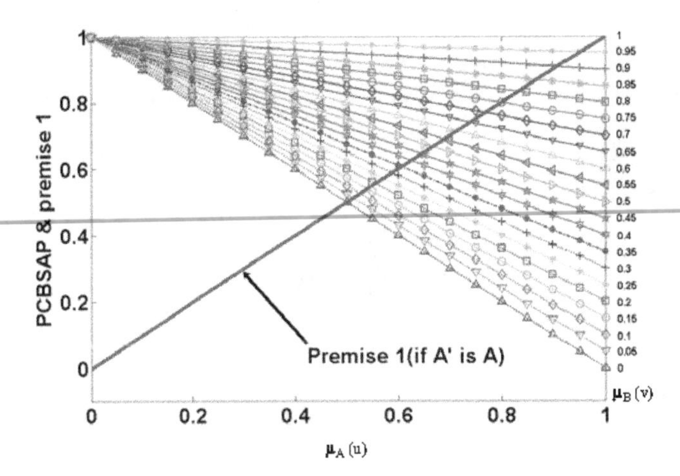

FIGURE 6.47
Superimposed plots of PCBSAP and premise 1 of C1.

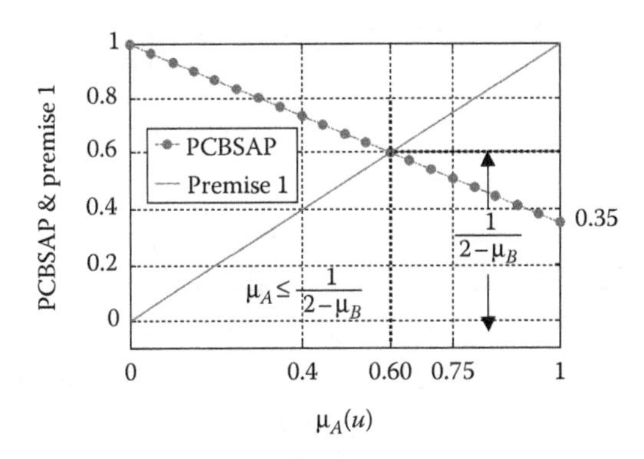

FIGURE 6.48
PCBSAP and premise 1 of C1 for $1/2-\mu_B$.

Here, $\mu_A^{\min}(u) = \dfrac{1}{2-\mu_B(v)}$ is obtained by solving the expression $1-\mu_A(u)+\mu_A(u)\mu_B(v)=\mu_A(u)$. We can see from the above equation that the outcome of the *min* operation between $\mu_{A\to B}(u,v)$ and $\mu_{A'}(u)$ is either $1-\mu_A(u)+\mu_A(u)\mu_B(v)$ or $\mu_{A'}(u)$. Moreover, we can see from the nature of the equations that $\mu_{A'}(u)$ increases with increase in $\mu_A(u)$, whereas $1-\mu_A(u)+\mu_A(u)\mu_B(v)$ decreases; hence, the supremum of the *min* operation is the point of intersection of $y1$ and $y2$, i.e., $1-\mu_A(u)+\mu_A(u)\mu_B(v)=\mu_A(u)$ or $\mu_{B'}(v) = \dfrac{1}{2-\mu_B(v)}$.

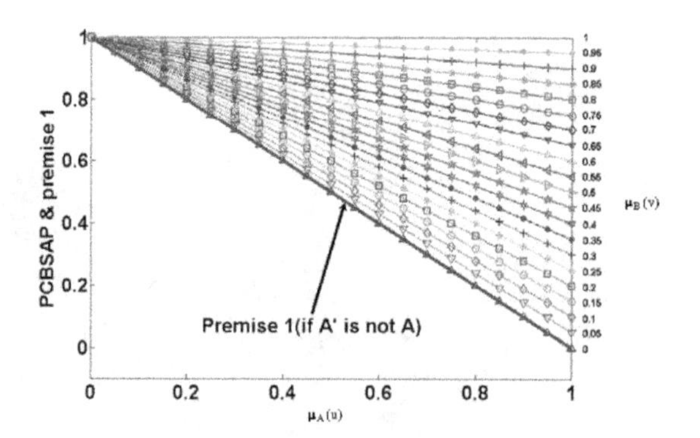

FIGURE 6.49
Superimposed plots of PCBSAP and premise 1 of C4-1/C4-2.

PCBSAP: C4-1/C4-2: $\mu_{A'}(u) = 1 - \mu_A(u)$ is applied to the RHS of Equation 6.89 to obtain the consequence $\mu_{B'}(v)$. We can see from Figure 6.49 that $\mu_{A\to B}(u,v)$ is always greater than or equal to $\mu_{A'}(u)$; thus, the outcome of the "min" operation is always $\mu_{A'}(u)$, i.e., $1 - \mu_A(u)$, for any value of $\mu_B(v)$. Because the supremum of $1 - \mu_A(u)$ is always unity, $\mu_{B'}(v) = 1$. Thus, PCBSAP satisfies the C4-1 (not C4-2) criterion of GMP. The analytical process is elaborated below:

$$\begin{aligned}
\mu_{B'}(v) &= \sup_{u\in U}\{\min[\min\{1, 1-\mu_A(u)+\mu_A(u)\mu_B(v)\}, 1-\mu_A(u)]\} \\
&= \sup_{u\in U}\{\min[1-\mu_A(u)+\mu_A(u)\mu_B(v), 1-\mu_A(u)]\} \\
&= \sup_{u\in U}\begin{Bmatrix} y1 = 1-\mu_A(u); & \text{for} \quad \mu_A(u)\mu_B(v) \ge 0 \\ y2 = 1-\mu_A(u)+\mu_A(u)\mu_B(v); & \text{for} \quad \mu_A(u)\mu_B(v) < 0 \end{Bmatrix}
\end{aligned} \tag{6.107}$$

We can see from Equation 6.107 that $y2$ does not exist because of the nonvalid condition $\mu_A(u)\mu_B(v) < 0$; therefore, the consequence $\mu_{B'}(v)$ would be the supremum of $y1$, and the solution would be unity only.

PCBSAP: C8-1/C8-2: $\mu_{B'}(v) = \mu_B(v)$ is applied to the RHS of Equation 4.21 to obtain the consequence $\mu_{A'}(u)$. Figure 6.50 illustrates the superimposed plots of $\mu_{A\to B}(u,v)$ and $\mu_{B'}(v)$. From the figure, we can see that $\mu_{B'}(v)$ is always equal to or less than the implication $\mu_{A\to B}(u,v)$ for any value of $\mu_A(u)$; therefore, the outcome of the *min* operation results in $\mu_{B'}(v)$ itself. Hence, the supremum of $\mu_{B'}(v)$ is unity only because the maximum value of $\mu_{B'}(v) = \mu_B(v)$ is unity. Therefore, we can conclude that PCBSAP satisfies the intuitive criterion C8-1 (not C8-2) of GMT. The analytical process is explained below:

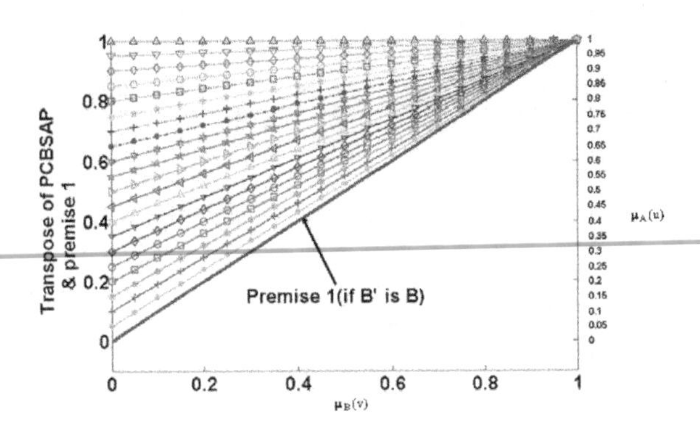

FIGURE 6.50
Superimposed plots of PCBSAP and premise 1 of C8-1/C8-2.

$$\begin{aligned}\mu_{A'}(u) &= \sup_{v\in V}\{\min[\min(1,1-\mu_A(u)+\mu_A(u)\mu_B(v)),\mu_B(v)]\}\\ &= \sup_{v\in V}\{\min[1-\mu_A(u)+\mu_A(u)\mu_B(v),\mu_B(v)]\}\\ &= \sup_{v\in V}\begin{Bmatrix} y1 = 1-\mu_A(u)+\mu_A(u)\mu_B(v); & \text{for} \quad \mu_B(v)>1 \\ y2=\mu_B(v); & \text{for} \quad \mu_B(v)\le 1\end{Bmatrix}\end{aligned} \tag{6.108}$$

we can see that $y1$ is not valid because $\mu_A(u) > 1$ is not possible. Hence, the outcome of the *min* operation between $\mu_{A\to B}(u,v)$ and $\mu_{B'}(v)$ is always $\mu_B(v)$; thus, $\mu_{A'}(u) = 1$.

Tables 6.24 and 6.25 summarize the results of all investigations of the various new implication methods in comparison with the intuitive criteria of GMP and GMT. Not all the results are described in the previous section. We can see that the implication methods, such as PCBSAP, PCBSSI, PCBSBP, and MIAS, satisfy exactly the same intuitive criteria of GMP and GMT, the total number of satisfied criteria being two. Other methods satisfy only one intuitive criterion of GMP and GMT. The logical explanation of these observations can be found in the study by Kashyap [7]. The use of a few existing and some new FIMs in the decision making process is described in Chapter 8.

6.6 Fuzzy Logic Algorithms and Final Composition Operations

An FL-based system can model any continuous function or system. The quality of approximation depends on the quality of rules that are formed by an expert. Fuzzy engineering is a function approximation using fuzzy

TABLE 6.24

Comparison of True and Computed Consequences of GMP (New Fuzzy Implication Methods)

		FIM							
		PCBSAP		PCSUAP		PCBSSI		PCBSBP	
Criteria	T	Computed	SF	Computed	SF	Computed	SF	Computed	SF
C1	μ_B	$\frac{1}{2-\mu_B(v)}$	N	$0.5 \cup \mu_B(v)$	N	$\frac{1+\mu_B(v)}{2}$	N	$0.5 \cup \mu_B(v)$	N
C2-1	μ_B^2	Refer Equation 5.11 of [7]	N	$\frac{3-\sqrt{5}}{2} \cup \mu_B(v)$	N	$\frac{3+4\mu_B(v)-\sqrt{5+4\mu_B(v)}}{2}$	N	$\frac{3-\sqrt{5}}{2} \cup \mu_B(v)$	N
C2-2	μ_B		N		N		N		N
C3-1	$\sqrt{\mu_B}$	Refer Equation 5.12 of [7]	N	$\frac{\sqrt{5}-1}{2} \cup \mu_B(v)$	N	$\frac{\sqrt{5+4\mu_B(v)}-1}{2}$	N	$\frac{\sqrt{5}-1}{2} \cup \mu_B(v)$	N
C3-2	μ_B		N		N		N		N
C4-1	1	1	Y	1	Y	1	Y	1	Y
C4-2	μ_B		N		N		N		N

Continued

TABLE 6.24 *(Continued)*

Criteria	T	FIM					
		PCSUBP		PCASBP		PCASAP	
		Computed	SF	Computed	SF	Computed	SF
C1	μ_B	$0.5 \cup \mu_B(v)$	N	$0.5 \cup \mu_B(v)$	N	Refer Equation 5.34 of [7] $\mu_{B'}(v) = \sup_{u \in U}\left\{\min\left[1-\mu_A(u)(1-\mu_A(u)\mu_B(v)), \mu_A(u)\right]\right\}$ $= \sup_{u \in U}\begin{Bmatrix} y1 = \mu_A(u); & \text{for } \mu_A(u) \le \mu_A^{\min}(u) \\ y2 = 1-\mu_A(u)(1-\mu_A(u)\mu_B(v)); & \text{for } \mu_A(u) > \mu_A^{\min}(u) \end{Bmatrix}$ $= \sup_{u \in U}\begin{Bmatrix} y1 = \mu_A(u); & \text{for } \mu_A(u) \le \mu_A^{\min}(u) \\ y2 = 1-\mu_A(u)+\mu_A^2(u)\mu_B(v); & \text{for } \mu_A(u) > \mu_A^{\min}(u) \end{Bmatrix}$	N
C2-1	μ_B^2		N		N	Refer Equation 5.35 of [7] $\mu_B'(v) = \sup_{u \in U}\left\{\min\left[1-\mu_A(u)(1-\mu_A(u)\mu_B(v)), \mu_A^2(u)\right]\right\}$	N
C2-2	μ_B	$\frac{3-\sqrt{5}}{2} \cup \mu_B(v)$	N	$\frac{3-\sqrt{5}}{2} \cup \mu_B(v)$	N	$= \sup_{u \in U}\begin{Bmatrix} y1 = \mu_A^2(u); & \text{for } \mu_A(u) \le \mu_A^{\min}(u) \\ y2 = 1-\mu_A(u)(1-\mu_A(u)\mu_B(v)); & \text{for } \mu_A(u) > \mu_A^{\min}(u) \end{Bmatrix}$ $= \sup_{u \in U}\begin{Bmatrix} y1 = \mu_A^2(u); & \text{for } \mu_A(u) \le \mu_A^{\min}(u) \\ y2 = 1-\mu_A(u)+\mu_A^2(u)\mu_B(v); & \text{for } \mu_A(u) > \mu_A^{\min}(u) \end{Bmatrix}$	N
C3-1	$\sqrt{\mu_B}$		N		N	Refer Equation 5.36 of [7] $\mu_B'(v) = \sup_{u \in U}\left\{\min\left[1-\mu_A(u)(1-\mu_A(u)\mu_B(v)), \sqrt{\mu_A(u)}\right]\right\}$	N
C3-2	μ_B	$\frac{\sqrt{5}-1}{2} \cup \mu_B(v)$		$\frac{\sqrt{5}-1}{2} \cup \mu_B(v)$	N	$= \sup_{u \in U}\begin{Bmatrix} y1 = \sqrt{\mu_A(u)}; & \text{for } \mu_A(u) \le \mu_A^{\min}(u) \\ y2 = 1-\mu_A(u)(1-\mu_A(u)\mu_B(v)); & \text{for } \mu_A(u) > \mu_A^{\min}(u) \end{Bmatrix}$ $= \sup_{u \in U}\begin{Bmatrix} y1 = \sqrt{\mu_A(u)}; & \text{for } \mu_A(u) \le \mu_A^{\min}(u) \\ y2 = 1-\mu_A(u)+\mu_A^2(u)\mu_B(v); & \text{for } \mu_A(u) > \mu_A^{\min}(u) \end{Bmatrix}$	N
C4-1	1	1	Y	1	Y	1	Y
C4-2	μ_B		N		N		N

Continued

TABLE 6.24 *(Continued)*

Criteria	T	FIM			
		PCASSI		MIAS	
		Computed	SF	Computed	SF
C1	μ_B	$0.5 \cup \dfrac{1}{2-\mu_B(v)}$	N	$\dfrac{1}{2-\mu_B(v)}$	N
C2-1	μ_B^2	Refer Equation 5.39 of [7]	N		N
C2-2	μ_B	$\sup_{u\in U} = \left\{\begin{matrix} \left\{\begin{matrix} y11=\mu_A^2(u); & \text{for } \mu_A(u)\le 1 \\ y12=1-\mu_A(u)+\mu_A^2(u); & \text{for } \mu_A(u)>1 \end{matrix}\right\}; & \text{for } \mu_A(u)\le\mu_B(v) \\ \left\{\begin{matrix} y21=\mu_A^2(u); & \text{for } \mu_A(u)\le\mu_A^{\min}(u) \\ y22=1-\mu_A(u)+\mu_A(u)\mu_B(v); & \text{for } \mu_A(u)>\mu_A^{\min}(u) \end{matrix}\right\}; & \text{for } \mu_A(u)>\mu_A(v) \end{matrix}\right\}$	N	Refer Equation 5.43 of [7]	N
C3-1	$\sqrt{\mu_B}$	Refer Equation 5.40 of [7]	N		N
C3-2	μ_B	$\mu_{B'}(v)=\sup_{u\in U}\left\{\min\left[1-\mu_A(u)+\mu_A(u)\min(\mu_A(u),\mu_B(v)),\sqrt{\mu_A(u)}\right]\right\}$ $=\sup_{u\in U}\left\{\begin{matrix} y1=\min\left[1-\mu_A(u)+\mu_A^2(u),\sqrt{\mu_A(u)}\right]; & \text{for } \mu_A(u)\le\mu_B(v) \\ y2=\min\left[1-\mu_A(u)+\mu_A(u)\mu_B(v),\sqrt{\mu_A(u)}\right]; & \text{for } \mu_A(u)>\mu_B(v) \end{matrix}\right\}$ $=\sup_{u\in U}\left\{\begin{matrix} \left\{\begin{matrix} y11=\sqrt{\mu_A(u)}; & \text{for } \mu_A(u)\le 0.56 \\ y12=1-\mu_A(u)+\mu_A^2(u); & \text{for } \mu_A(u)>0.56 \end{matrix}\right\}; & \text{for } \mu_A(u)\le\mu_B(v) \\ \left\{\begin{matrix} y21=\sqrt{\mu_A(u)}; & \text{for } \mu_A(u)\le\mu_A^{\min}(u) \\ y22=1-\mu_A(u)+\mu_A(u)\mu_B(v); & \text{for } \mu_A(u)>\mu_A^{\min}(u) \end{matrix}\right\}; & \text{for } \mu_A(u)>\mu_B(v) \end{matrix}\right\}$	N	Refer Equation 5.44 of [7]	N
C4-1	1	1	Y	1	Y
C4-2	μ_B		N		N

T = true consequence of GMP; SF = satisfaction flag: "Y" if computed consequence matches with true one and "N" if computed consequence does not match.

TABLE 6.25

Comparison of True and Computed Consequences of GMT (New Fuzzy Implication Methods)

		FIM					
		PCBSAP		PCSUAP		PCBSSI	
Criteria	T	Computed	SF	Computed	SF	Computed	SF
C5	$\mu_{\bar{A}}$	$0.5 \cup \frac{1}{1+\mu_A(u)}$	N	$1-\mu_A(u) \cup \frac{\mu_A(u)}{1+\mu_A(u)}$	N	$0.5 \cup 1-\frac{\mu_A(u)}{2}$	N
C6	$1-\mu_A^2$	Refer Equation 5.47 of [7] $\mu_{A'}(u) = \sup_{v \in V} \left\{ \min\left[\min\left(1, 1-\mu_A(u)+\mu_A(u)\mu_B(v)\right), 1-\mu_B^2(v) \right] \right\}$ $= \sup_{v \in V} \left\{ \min\left[1-\mu_A(u)+\mu_A(u)\mu_B(v), 1-\mu_B^2(v) \right] \right\}$ $= \sup_{v \in V} \begin{Bmatrix} y1 = 1-\mu_A(u)+\mu_A(u)\mu_B(v);\ \text{for}\ \mu_B(v) \le \mu_B^{\min}(v) \\ y2 = 1-\mu_B^2(v);\ \text{for}\ \mu_B(v) > \mu_B^{\min}(v) \end{Bmatrix}$	N	Refer Equation 5.51 of [7]	N	Refer Equation 5.55 of [7]	N
C7	$1-\sqrt{\mu_A}$	Refer Equation 5.48 of [7] $\mu_{A'}(u) = \sup_V \left\{ \min\left[\min\left(1, 1-\mu_A(u)+\mu_A(u)\mu_B(v)\right), 1-\sqrt{\mu_B(v)} \right] \right\}$ $= \sup_{v \in V} \left\{ \min\left[1-\mu_A(u)+\mu_A(u)\mu_B(v), 1-\sqrt{\mu_B(v)} \right] \right\}$ $= \sup_{v \in V} \begin{Bmatrix} y1 = 1-\mu_A(u)+\mu_A(u)\mu_B(v);\ \text{for}\ \mu_B(v) \le \mu_B^{\min}(v) \\ y2 = 1-\sqrt{\mu_B(v)};\ \text{for}\ \mu_B(v) > \mu_B^{\min}(v) \end{Bmatrix}$	N	Refer Equation 5.52 of [7]	N	Refer Equation 5.56 of [7]	N
C8-1	1	1	Y	$1-\mu_A(u) \cup \mu_A(u)$	N	1	Y
C8-2	μ_A		N		N		N

Continued

TABLE 6.25 *(Continued)*

Criteria	T	FIM							
		PCBSBP		PCSUBP		PCASBP		PCASAP	
		Computed	SF	Computed	SF	Computed	SF	Computed	SF
C5	$\mu_{\bar{A}}$	$0.5 \cup 1-\mu_A(u)$	N	$1-\mu_A(u) \cup \frac{\mu_A(u)}{2}$	N	$1-\mu_A(u) \cup \frac{\mu_A^2(u)-\mu_A(u)+1}{1+\mu_A(u)}$	N	$\frac{\mu_A^2(u)-\mu_A(u)+1}{1+\mu_A^2(u)}$	N
C6	$1-\mu_A^2$	$\frac{\sqrt{5}-1}{2} \cup 1-\mu_A(u)$	N	Refer Equation 5.63 of [7]	N	Refer Equation 5.67 of [7]	N	Refer Equation 5.71 of [7]	N
C7	$1-\sqrt{\mu_A}$	$\frac{3-\sqrt{5}}{2} \cup 1-\mu_A(u)$	N	Refer Equation 5.64 of [7]	N	Refer Equation 5.68 of [7]	N	$\frac{1+2\mu_A^2(u)-\sqrt{1+4\mu_A^3(u)}}{2\mu_A^2(u)}$	N
C8-1	1	1	Y	$1-\mu_A(u) \cup \mu_A(u)$	N	$\mu_A^2(u)+\mu_A(u)-2\mu_A(u)+1$	N	$\frac{1}{1+\mu_A(u)}$	N
C8-2	μ_A		N		N		N		N

Continued

TABLE 6.25 *(Continued)*

		FIM			
		MIAS		PCASSI	
Criteria	T	Computed	SF	Computed	SF
C5	$\mu_{\bar{A}}$	$\frac{1}{1+\mu_A(u)}$	N	$0.5 \cup \frac{1}{1+\mu_A(u)}$	N
C6	$1-\mu_A^2$	Refer Equation 5.75 of [7]	N	Refer Equation 5.79 of [7]	N
				$\mu_{A'}(u) = \sup_{v \in V}\left\{\min\left[1-\mu_A(u)+\mu_A(u)\mu_B(v), 1-\mu_B^2(v)\right]\right\}$ $= \sup_{v \in V}\begin{Bmatrix} y1 = 1-\mu_A(u)+\mu_A(u)\mu_B(v);\ \text{for}\ \mu_B(v) \le \mu_B^{\min}(v) \\ y2 = 1-\mu_B^2(v);\ \text{for}\ \mu_B(v) > \mu_B^{\min}(v) \end{Bmatrix}$	
C7	$1-\sqrt{\mu_A}$	Refer Equation 5.76 of [7]	N	Refer Equation 5.78 of [7]	N
		$\mu_{A'}(u) = \sup_{v \in V}\left\{\min\left[1-\mu_A(u)\left(1-\min\left(\mu_A(u),\mu_B(v)\right)\right), 1-\sqrt{\mu_B(v)}\right]\right\}$ $= \sup_{v \in V}\begin{Bmatrix} y1 = \min\left[1-\mu_A(u)+\mu_A(u)\mu_B(v), 1-\sqrt{\mu_B(v)}\right] & \text{for}\ \mu_B(v) \le \mu_A(u) \\ y2 = \min\left[1-\mu_A(u)+\mu_A^2(u), 1-\sqrt{\mu_B(v)}\right] & \text{for}\ \mu_B(v) > \mu_A(u) \end{Bmatrix}$ $= \sup_{v \in V}\begin{Bmatrix} \begin{Bmatrix} y11 = 1-\mu_A(u)+\mu_A(u)\mu_B(v) & \text{for}\ \mu_B(v) \le \mu_B^{\min}(v) \\ y12 = 1-\sqrt{\mu_B(v)} & \text{for}\ \mu_B(v) > \mu_B^{\min}(v) \end{Bmatrix} & \text{for}\ \mu_B(v) \le \mu_A(u) \\ \begin{Bmatrix} y21 = 1-\mu_A(u)+\mu_A^2(u) & \text{for}\ \mu_B(v) \le \left(\mu_A(u)\left(1-\mu_A(u)\right)\right)^2 \\ y22 = 1-\sqrt{\mu_B(v)} & \text{for}\ \mu_B(v) > \left(\mu_A(u)\left(1-\mu_A(u)\right)\right)^2 \end{Bmatrix} & \text{for}\ \mu_B(v) > \mu_A(u) \end{Bmatrix}$			
C8-1	1	$1-\mu_A(u)+\mu_A^2(u)$	N	1	Y
C8-2	μ_A		N		N

T = true consequence of GMP; SF = satisfaction flag: "Y" if computed consequence matches with true one and "N" if computed consequence does not match.

systems, and it is based on mathematics of function approximation (FA) and statistical learning theory. The basic unit of a fuzzy algorithm is the "*if...then*" rule, for example: "*If* the water in the washing machine is dirty, *then* add more detergent powder." Thus, a fuzzy system is a set of if–then rules, which maps the input sets, like "dirty water," to output sets like "more detergent powder." Overlapping rules are used to define polynomials and richer functions. A set of such possible rules would be as follows [2–4]:

Rule 1: *If* the air is cold, *then* set the air conditioning motor speed to stop.

Rule 2: *If* the air is cool, *then* set the motor speed to slow.

Rule 3: *If* the air is just right, *then* set the motor speed to medium.

Rule 4: *If* the air is warm, *then* set the motor speed to fast.

Rule 5: *If* the air is hot, *then* set the motor speed to blast.

This set provides the first-cut fuzzy system, and more rules can either be presumed and added by experts or added by learning new rules adaptively from training datasets obtained from the system. ANNs can be used to learn the rules from the data. Fuzzy engineering mainly deals with first defining these rules, then tuning such rules, and finally adding new rules or pruning old rules. In an additive fuzzy system, each input partially fires all rules in parallel, and the system acts an associative processor as it computes the output, F(x). The FLS then combines the partially fired *then* fuzzy sets into a sum and converts this sum to a scalar or vector output. Thus, a match-and-sum fuzzy approximation can be viewed as a generalized AI expert system or a neural-like fuzzy associative memory. The additive fuzzy systems belong to the proven universal approximators for rules that use fuzzy sets of any shape and are computationally simple.

A fuzzy variable's values can be considered labels of fuzzy sets, as provided below: temperature→fuzzy variable→linguistic values, such as low, medium, normal, high, very high, and so forth. This leads to membership values on the UOD. The dependence of a linguistic variable on another variable is described using a fuzzy conditional statement, as shown below:

R: if S1 (is true), then S2 (is true). Or S1→S2; more specifically: (1) *if* the load is small, *then* torque is very high; and (2) *if* the error is negative and large, *then* the output is negative and large. A composite conditional statement in an FA would be as follows:

R1: if S1, then (if S2, then S3), which is equivalent to the following statement:

R1: if S1, then R2; and R2: if S2, then S3.

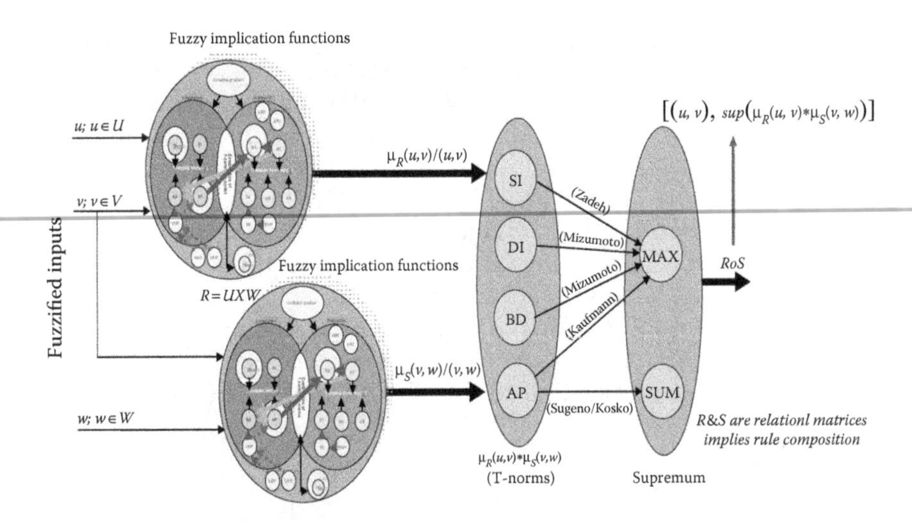

FIGURE 6.51
Fuzzy implication functions—aggregation process.

An FA is formed by combining 2 or 3 fuzzy conditional statements: *if* the speed error is negative large, *then* (*if* the change in speed error is *not* (negative large *or* negative medium), *then* the change in fuel is positive large),..., *or*,..., *or*,..., and so on.

The composite picture of the fuzzy implication functions and the aggregation process is depicted in Figure 6.51. The composite operation can be realized using various combinations of T-norms and supremum operators. Composite operations using the MAX-SI or MAX-AP combinations are easy to implement compared with the other possible combinations. The accuracy of the final result depends mostly on the composite operation and on the fuzzy implication functions. Hence, it may be possible to obtain a better compositional output by logically selecting the different fuzzy implication functions and combining them with the appropriate T-norms and supremum operators. This is a topic requiring further study.

6.7 Fuzzy Logic and Fuzzy Integrals in Multiple Network Fusion

The problem in multiple networks (NWs) is how to combine the results from those NWs to obtain the best estimate of the optimal result [10,11]. In fusion methods, the difference in the performance of each NW is considered when combining the NWs. The concept is based on FL, which essentially uses the so-called fuzzy integral (FI). This fusion method combines the outputs of separate NWs with the importance of each NW. This importance is subjectively assigned as the nature of FL. Assume a 2-layered (2-L) neural NW classifier that has outputs as estimates of Bayesian *a posteriori* probabilities [10]. The 2-L NW has N_i neurons in the input layer, N_h neurons in the hidden layer, and N_o neurons in the output layer. In essence, N_i is the number of input features and N_o is the number of classes; N_h is selected appropriately. The operation of this 2-L NW is a nonlinear decision making process. Given an unknown input x and the class set $\Omega = (\omega_1, \ldots, \omega_{No})$, each output neuron determines the probability $P(\omega_i \mid X)$ of x belonging to this class, as shown in the following equation:

$$P(\omega_i \mid X) \cong f\left[\sum_{k=1}^{Nh} w_{ik} f\left(\sum_{j=1}^{Ni} w_{kj} x_j\right)\right] \tag{6.109}$$

Here, the w's are the respective weights, and $f = \dfrac{1}{1+e^{-x}}$ is the sigmoid function. The neural NW with the maximum value is selected as the corresponding class. The classification of an input X is based on a set of real-valued measurements $P(\omega_i \mid X)$, $1 \le i \le No$. This represents the probability that X originates from each of the N_0 classes under the given condition of the input X. Each NW estimates a set of approximations of those true values, as shown below:

$$P_k(\omega_i \mid X), \quad 1 \le i \le N_o; \quad 1 \le k \le n \tag{6.110}$$

The conventional approach is to use the formula for averages, as follows:

$$P(\omega_i \mid X) = \frac{1}{n}\sum_{k=1}^{n} P_k(\omega_i \mid X); \quad 1 \le i \le N_o \tag{6.111}$$

The above formula can be thought of as an averaged Bayes' classifier. One can use the weighted average as depicted below:

$$P(\omega_i \mid X) = \sum_{k=1}^{n} r_k P_k(\omega_i \mid X); \quad 1 \le i \le N_o \tag{6.112}$$

where $\sum_{k=1}^{n} r_k = 1$

The FI is a nonlinear functional, and it is defined with reference to a fuzzy measure. The fuzzy measure is represented as follows [10]:

g: $2^y \rightarrow [0,1]$ is a fuzzy measure for the following conditions

$$(a)\quad g(\varphi) = 0;\; g(y) = 1$$
$$(b)\quad g(A) \le g(B)\ \ if\ \ A \subset B \qquad (6.113)$$
$$(c)\quad \text{if}\left[A_i\right]_{i=1} \text{ is an increasing sequence of measureable sets}$$

Then, we have

$$\lim_{i\to\infty} g(A_i) = g(\lim_{i\to\infty} A_i) \qquad (6.114)$$

The FI is now defined as follows:

Let y be a finite set and h: $y \rightarrow \{0, 1\}$ be a fuzzy subset of the set y. The FI over y of the function h is defined by

$$\begin{aligned} h(y) \circ g(.) &= \max_{E\subseteq y} \{\min[\min_{y\in E} h(y), g(E)]\} \\ &= \max_{\alpha\in[0,1]}[\min(\alpha, g(F_\alpha))] \end{aligned} \qquad (6.115)$$

Here, $F_\alpha = \{y \mid h(y) \ge \alpha\}$. In the above definition, we have the following interpretations: (1) $h(y)$ is the degree to which h is satisfied by y; (2) min(y) measures the degree to which h is satisfied by all the elements in E; (3) g is a measure of the degree to which the subset of the objects, E, satisfies the concept that is measured by g; and (4) the *min* signifies the degree to which E satisfies the measure g and min $h(y)$. The biggest of these terms is taken by the max operation. Thus, the FI signifies the maximal grade of agreement between the objective evidence and the expectation. Therefore, an FI can be used to arrive at a consensus in classification problems [10].

7

Decision Fusion

The objective of decision fusion is to arrive at one final decision or action from an entire surveillance volume at any instant of time using outputs from different levels—for example, level 1, object refinement (OR), and level 2, situation refinement (SR)—of any multisensor data fusion (MSDF) system, especially in defense systems (see the JDL model in Section 2.1). The procedure is also applicable to other civilian data fusion (DF) systems.

7.1 Symbol- or Decision-Level Fusion

Symbol-level fusion (used synonymously for decision fusion) represents high-level information (higher than both the kinematic-level and image-level fusion processes), wherein the symbol represents a decision. A symbol can represent an input in the form of a decision; fusion describes both a logical and a statistical inference. In the case of symbolic data, inference methods from artificial intelligence can be used as computational mechanisms, such as fuzzy logic (FL), since the fusion of symbolic information would require reasoning and inference in the presence of uncertainty.

There are generally three domains of objects which can provide information [12]: (1) the concrete domain of physical objects with physical characteristics, physical states, activities and tangible entities (e.g., an aircraft or a robot); (2) the abstract domain—the "mind-thought" process, which covers the domain of the mind of all living beings and all intangible entities; and (3) the symbol domain—where the characteristics of concrete and abstract entities are transformed into representations of common systems of symbols (for communication and possible dynamic interactions). Interestingly, all three domains can be found in multiple interacting systems, for example, a group of mobile robots, which complicates decision making in terms of coordination and team autonomy. The sensing and data processing part of the robotic system builds the world model of the robot's environment (the robot's "mind"—in fact the "mind" constructs an internal model of the outside world), which contains the physical objects and symbols (e.g., sign boards, traffic signs, and so on). The decision process consists of (1) decoding information, interpretations, and associations by using previous experiences; and (2) perception of interpreted and associated sensory impressions that would lead to meaning (this is also called *new information*). The sensing process precedes the decision process, which is followed by the behavioral process. The sensing and decision processes together are called the *information process cycle*, with appropriate feedback where applicable. According to one theory of intelligence [13,14], the unknowable external world, noumena, is distinguished from the perceptual stimuli resulting from that world, phenomena. The noumena, which are not directly knowable (e.g., fire, river), are the sources of the perceptual stimuli. Many worldly phenomena represent a partial projection (a small cross-section, features, and so on) of noumena. Humans organize perceived phenomena into schemata, which represent individual phenomena and abstractions. This is a systematic procedure involving some regularity or pattern in these phenomena. For example, the properties measured by the sensors are the phenomena, and the vector of these properties is a form of schemata. The schemata, also used in the theory of genetic algorithms, can include temporal sequences and images. The reasoning process about the world requires an abstraction from the phenomena: (1) categories of objects, (2) relations between objects, (3) actions, and (4) events, many of which can be represented by labels, that is, symbols. Thus, a symbol is a sign that represents a thing.

The symbol states of information consists of (1) a set of symbolic representation-like text and sound in a language and pictures, and (2) information products and outcomes such as documents and speeches [12]. This symbol-level information can be used for encoding and decoding our thought processes. Symbols—such as signs, letters, sounds in and of languages, pictures, maps, and objects representing something—can form information products and outcomes such as data, messages, facts, reports, books, intelligent information, speeches, models, simulation outputs, computer programs, media products, financial products (e.g., coins and currency notes), ethnic and religious symbols, national and political symbols (e.g., emblems, flags, and election, party, or union symbols or logos). In effect, information process refinement (PR) starts from the signs, progresses to data to symbols, facts, and ideas, and then to knowledge and wisdom. Symbols represent input in the form of a decision, where as fusion describes a logical and statistical inference. The significant advantage of symbol-level fusion is an increase in the truth value. This type of fusion can also be considered decision fusion.

The main decision-level fusion approaches are identity- and knowledge-based methods [15,16]. In the identity-based approach, maximum *a posteriori* (MAP), maximum likelihood (ML), and Dempster–Shafer (D–S) methods are used. In the knowledge-based approach, the methods used are logic templates, syntax rule, neural network (NW), and FL methods. Many of these approaches are also applicable to feature-level fusion (for image level or even for extracting the patterns or features from speech signals). In feature-level fusion, the object is the characters' space. The object in decision fusion is decision action space. Decision fusion depends mainly upon external knowledge, and hence more on inference from the external knowledge [17]. Interestingly, the results obtained and fused from decision fusion can be used to classify images, detect changes, and detect and recognize targets.

The principles of symbolic fusion are as follows:

1. Primitives in the world model should be expressed as a set of properties; the schema is such a representation, where the properties are the symbolic labels or numerical measures.
2. The observation model should be expressed in a common coordinate system, that is, the information should be properly associated; it could be on the basis of spatial or temporal coordinates, or on the basis of some relation between properties.
3. The observation and model should be expressed in a common vocabulary, or "context," which is a collection of symbols and relations used to describe a situation; knowledge of the "context" provides a set of symbols and relations and leads to a process of prediction and verification.
4. The properties should incorporate an explicit representation of uncertainty, this being precision and confidence.
5. The primitives should be accompanied by a confidence factor determined by probabilistic technique or FL in the framework of possibility theory.

In symbolic form or fusion, the predict, match, and update cycles, like the cycles for kinematic fusion (using Kalman filter [KF]), can be defined for a symbolic description composed of schema. In the prediction stage, *a priori* information of "context" is applied to predict the evolution of schemes in the model, as well as the existence and location of new schema [13]. The prediction stage thus selects the perceptual actions that detect the expected phenomenon. The match stage associates new perceptual phenomenon with the predictions from the internal model. The primary method is the spatial location, which could be an association based on similar properties. Then, the update stage combines the prediction and observation stages to construct an internal model that can be thought of as a "short term memory" with a certain quantity of information.

In Chapters 7 and 8, the use of FL is regarded as helping a decision process either in KF or situation assessment (SA).

7.2 Soft Decisions in Kalman Filtering

The KF has been used as one of the most promising algorithms for recursive estimation of states of linear and nonlinear systems. The accuracy of the filter is based on (1) the accuracy of the mathematical model of the actual dynamic system and measurement device, due to the random uncertainties; and (2) its tuning parameters Q (process noise covariance matrix) and R (measurement noise covariance matrix). In some applications, we may encounter a modeling error, that is, when the true models are not accurately known or they are difficult to realize or implement, we have to use approximate representations. Modeling errors are often compensated for by process noise-tuning parameters (of Q) that are selected on a trial-and-error basis. The final solutions obtained through this approach may not provide optimal filter performance. Although there are a few adaptive filters that can be used for such purposes, these are computationally very demanding and time varying, especially in the case of the extended KF (EKF).

The gain of a KF determines how much weighting should be given to the present measurement (in fact to the residuals): (1) If the measurement data are highly contaminated with noise, then less weight is automatically assigned to that data, and the filter depends on the model of the target (i.e., state propagation, see Equations 2.23 and 2.24); and (2) if the measurement data are less noisy, then more weight is assigned to that data, and the estimated state is a combination of state-predicted (through a target model) and observation data. Thus, the KF has an inherent decision-making capability that helps in soft switching between the process model and measurement model (via appropriate weight assignments, Kalman gain, and P, Q, and R). This decision in soft switching is based on Kalman gain that depends on the relative value of tuning parameters and matrices Q and R.

Measurement noise variance reflects the noise level in measurement data; higher R (in terms of the norm of the matrix) means the data is very noisy, and *vice versa*. In cases of very high R, the filter does not trust the measurement data, and therefore assigns low weight to the correction part of the state update through Kalman gain (Equation 2.20). This can also be interpreted as that the filter relies mostly on the process model. Similarly, for very low R, the filter relies mostly on the measurement model. Table 7.1 summarizes the soft decisions in a KF [7].

TABLE 7.1

Soft Decision Making in a Kalman Filter

Tuning of Q and R Parameters and Matrices	Kalman Gain	Soft Decisions
High R or low Q	Low	Less faith on measurements and more faith on the predicted states
Low R or high Q	High	Enough or more faith on measurements and less faith on the predictions
Moderate R or Q	Moderate	Moderate faith on measurements
High initial P	High	Less faith in initial states
Low initial P	Low	High or enough initial faith on states
Moderate initial P	Moderate	Moderate initial faith on states

7.3 Fuzzy Logic–Based Kalman Filter and Fusion Filters

FL assists in modeling conditions that are inherently imprecisely defined and FL-based methods, in the form of approximate reasoning, which provide decision support and expert systems (ES) with good reasoning capabilities (this is called an FL-type 1 system). FL can also be used for tuning KFs. Algorithms can be developed by considering a combination of FL and KF [18–20]. The proper combination of FL- and KF-based approaches can be used to obtain improved accuracy and performance in tracking and in MSDF systems. In such systems, FL can aid soft decision making in the filtering process by using fuzzy *if…then* rules for making a judgment on the use of, for example, residuals in navigating the prediction or for filtering in the direction of achieving accurate results in either tracking process, feature selection, detection, matching, or MSDF.

In this section, two schemes based on KF and fuzzy Kalman filter (FKF) are studied for target-tracking applications and their performances evaluated. The concept of FL is extended to state-vector level DF for similar sensors. The performances of FL-based fusion methods are compared with the conventional fusion method, also called state-vector fusion (SVF), to track a maneuvering target. The FL concept is combined with KF at the measurement update level. The equations for the FKF are the same as those for KF except for the following equation [18]:

$$\hat{X}(k+1, k+1) = \tilde{X}(k+1, k) + KB(k+1) \tag{7.1}$$

Here, $B(k + 1)$ is regarded as an output of the FL-based process variable (FLPV) and is generally a nonlinear function of the innovations $\mathbf{e}$ of the KF. It is assumed that positions in x–y axes measurements of the target are available. The FLPV vector consists of the modified innovation sequence for x and y axes:

$$B(k+1) = [b_x(b_x(k+1)b_y(k+1)] \tag{7.2}$$

To determine the FLPV vector, the innovation vector $\mathbf{e}$ is first separated into its x and y components, $\mathbf{e}_x$ and $\mathbf{e}_y$. The target motion in each axis is assumed to be independent. The FLPV vector for the x direction is developed and then generalized to include y direction. This vector consists of two inputs, $\mathbf{e}_x$ and $\dot{\mathbf{e}}_x$, and single output $b_x(k + 1)$, where $\dot{\mathbf{e}}_x$ is computed by

$$\dot{\mathbf{e}}_x = \frac{\{\mathbf{e}_x(k+1) - \mathbf{e}_x(k)\}}{T} \tag{7.3}$$

Here, T is the sampling interval in seconds; the expression of Equation 7.3 can be extended to y direction, and even to z direction if required.

7.3.1 Fuzzy Logic–Based Process and Design

FLP is obtained via a fuzzy inference system (FIS); see Figures 6.13 and 6.14. Sections 6.1.4 and 6.1.9 list the steps to build an FIS. The antecedent membership functions that define the fuzzy values for inputs $\mathbf{e}_x$ and $\dot{\mathbf{e}}_x$ and the membership function for the output b_x needed to develop the FLP are shown in Figure 7.1 [20]. The linguistic variables or labels to define membership functions are large negative (LN), medium negative (MN), small negative (SN), zero error (ZE), small positive (SP), medium positive (MP), and large positive (LP). The rules for the inference in FIS are generally created based on the experience and intuition of the domain expert, one such rule being [9,18,20]:

$$\text{IF } \mathbf{e}_x \text{ is LP AND } \dot{\mathbf{e}}_x \text{ is LP THEN } b_x \text{ is LP} \tag{7.4}$$

Having $\mathbf{e}_x$ and $\dot{\mathbf{e}}_x$ with large positive values indicates an increase in the innovation sequence at faster rate. Then, the future value of $\mathbf{e}_x$ (and hence $\dot{\mathbf{e}}_x$) can be reduced by increasing the present value of $b_x(\approx Z - H\tilde{X})$ by a large magnitude. This will generate 49 rules to implement FLP. The output b_x at any instant of time can be computed using (1) the inputs $\mathbf{e}_x$ and $\dot{\mathbf{e}}_x$, (2) input membership functions, (3) the 49 rules [9,18,20], (4) FIS, (5) the

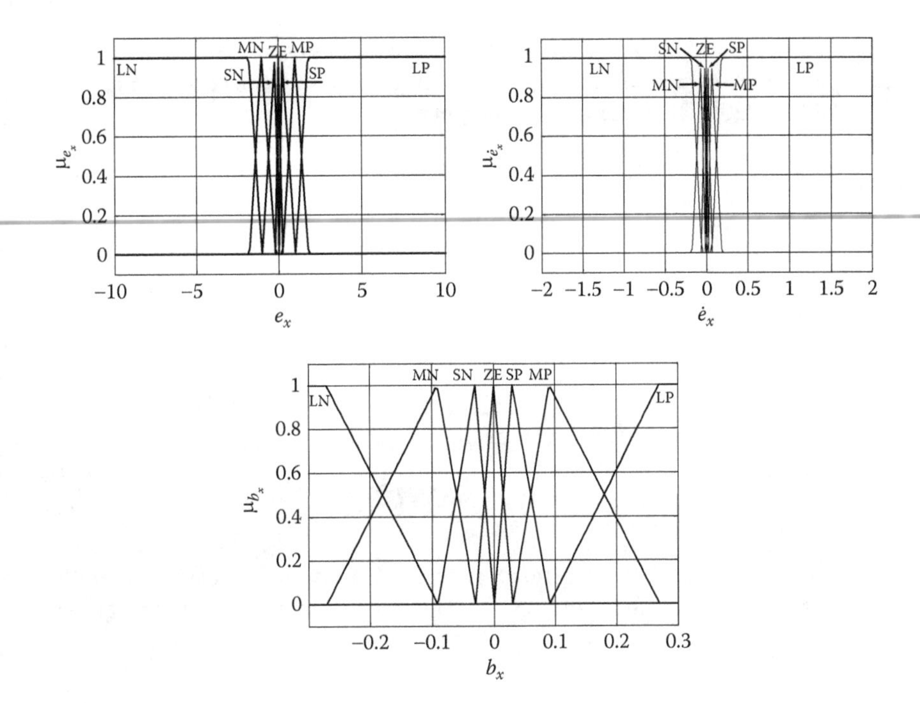

FIGURE 7.1
Fuzzy membership functions for error input and its finite difference and for the output fuzzy logic–based process variable.

TABLE 7.2
Features of FIS for FKF

Fuzzy Implication Methods	Mamdani
AND operator	Minimum
Fuzzy implication	Minimum
Aggregation	Maximum
Defuzzification	Centroid

aggregator, and (6) defuzzification. These steps are given in Chapter 6, Section 6.1.9. The properties and features of the fuzzy operators and fuzzy implication methods used are given in Table 7.2.

7.3.2 Comparison of Kalman Filter and Fuzzy Kalman Filter

The simulated data for the x-axis position are generated using the constant acceleration model (CAM) with process noise increment and with $T = 0.1$ s., the total number of scans being $N = 100$. The simulation uses parameter values and related information such as (1) initial states of the target: $(x, \dot{x}, \ddot{x})$ are 0 m, 100 m/s, and 0 m/s^2, respectively, and (2) process noise variance ($Q = 0.0001$) [20]. The CAM model is given as

$$F = \begin{bmatrix} 1 & T & T^2/2 \\ 0 & 1 & T \\ 0 & 0 & 1 \end{bmatrix} \tag{7.5}$$

$$G = \begin{bmatrix} T^3/6 & T^2/2 & T \end{bmatrix} \tag{7.6}$$

The target state equation is given as

$$X(k+1) = FX(k) + Gw(k) \tag{7.7}$$

where k is the scan number and w is white Gaussian process noise with zero mean and covariance matrix Q. The measurement equation is given as

$$Z_m(k) = HX(k) + v(k) \tag{7.8}$$

$$H = [1 \quad 0 \quad 0] \tag{7.9}$$

where v is white Gaussian measurement noise with zero mean and covariance matrix Q ($R = \sigma^2$; σ is the standard deviation of noise with a value of 10 meters). The initial conditions, F, G, H, Q, and R, for both filters, KF and FKF, are the same. The initial state vector $\hat{\mathbf{X}}(0/0)$ is close to the true initial states. The KF and FKF algorithms were coded in MATLAB®. The results for both filters are compared in terms of true and estimated states, i.e. states errors with bounds at every scan number. FKF performs much better than KF [7,20]. The consistency checks on these filters were performed using the normalized cost function (CF), computed using the formula:

$$\text{CF} = \frac{1}{N}\sum_{k=1}^{N} \mathbf{e}(k)S(k)^{-1}\mathbf{e}(k)^T \tag{7.10}$$

where $\mathbf{e}$ is the innovation sequence vector, and S is the innovation covariance matrix. The filter performance is deemed consistent if its normalized CF. Equation 7.10 is equal to the dimension of the vector of observables. The CF for KF (= 2.94) was found to be very close to the theoretical value of 3, whereas for FKF its CF value of 2.85 was slightly different from the theoretical value 3. However, it is still comparable with the KF value and is not much different from the theoretical number. Hence, the FKF can be treated as an approximately consistent filter. The performance of both filters in terms of states errors was also evaluated, with the FKF showing better performance than KF. The procedure of this section validated the application of the FKF for target tracking and shows it is comparable to or has a better performance than the KF. The results [7,20] are not given here, however, further applications of both filters are evaluated and some results are given in Section 7.3.3.

7.3.3 Comparison of Kalman Filter and Fuzzy Kalman Filter for Maneuvering Target Tracking

To use FKF to track a maneuvering target, a redesign of the FLPV to capture the various possible maneuver modes of the target is required. This involves [9,20]: (1) proper selection of membership functions of input and output (I/O); (2) tuning of the membership functions; (3) selection of fuzzy operators (e.g., T-norm and S-norm; see Section 6.2); and (4) selection of fuzzy implication, aggregation, and defuzzification techniques. Here we use MATLAB-based functions such as "genfis1()" to create the initial FLP vector and "anfis()" to tune it. The required training and check data are obtained from true and measured target positions. Figure 7.2 depicts the procedure to obtain a tuned FLP vector [20].

7.3.3.1 Training Set and Check-Set Data

The target states are simulated using a 3-DOF kinematic model with process noise acceleration increments and additional accelerations. Measurement data are obtained with a sampling interval of 1 second, and a total of 150 scans are generated. The data simulation is done with: (1) initial states $(x, \dot{x}, \ddot{x}, y, \dot{y}, \ddot{y})$ of target as (100, 30, 0, 100, 20, 0); (2) process noise variance

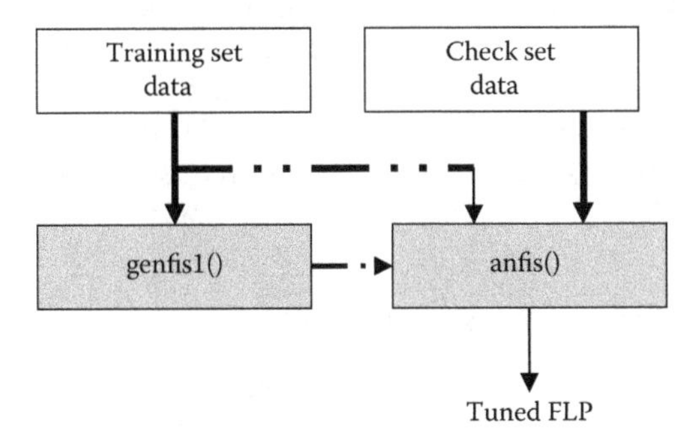

FIGURE 7.2
Procedure for tuning the fuzzy logic–based process.

$Q = 0.1$ (it is assumed that $Q_{xx} = Q_{yy} = Q$); and (3) measurement noise variance ($R = 25$). We also assume that $R_{xx} = R_{yy} = R$. The target has an additional acceleration of (x_{acc}, y_{acc}) at scans 25 and 100 and an acceleration of ($-x_{acc}$, $-y_{acc}$) at scans 50 and 75. Data simulation is carried out with process noise vector **w** (a 2 × 1 vector) modified to include these additional accelerations at the specified scan points, in order to induce a specific maneuver [9,20]:

$$\left.\begin{aligned} w(1) &= guass() \times \sqrt{Q_{xx}} + x_{acc} \\ w(2) &= guass() \times \sqrt{Q_{yy}} + y_{acc} \end{aligned}\right\} \tag{7.11}$$

$$\left.\begin{aligned} w(1) &= guass() \times \sqrt{Q_{xx}} - x_{acc} \\ w(2) &= guass() \times \sqrt{Q_{yy}} - y_{acc} \end{aligned}\right\} \tag{7.12}$$

At the other scan points, the vector **w** is simply defined without these additional accelerations terms. Acceleration $x_{acc} = -9 \times 9.8$ m/s^2 and $y_{acc} = 9 \times 9.8$ m/s^2 are used in the above equations and the function gauss() is used to generate Gaussian random numbers with mean 0 and variance 1. First, the initial FLPV is created for the x axis and tuned using inputs u_x^1, u_x^2, and output o_x is obtained using:

$$u_x^1(k) = z_x(k) - x(k) \tag{7.13}$$

$$u_x^2(k) = \frac{u_x^1(k) - u_x^1(k-1)}{T} \tag{7.14}$$

$$\text{output}_x(k) = m u_x^1(k) \tag{7.15}$$

Here, x and z_x are the true and measured target x position, respectively. m is the unknown parameter and is 2 for the present case. The first half of the total simulated data is taken for training and the remaining half is taken as the check-set data. The same procedure is followed to get the tuned FLPV for the y axis; then the trained FLP vector is plugged in to the FKF and its performance is compared with the KF for the two cases discussed next.

7.3.3.2 Mild and Evasive Maneuver Data

The mild maneuver (MM) data are generated with minor modifications in the acceleration injection points, with a total of 17 scans generated. Accelerations are injected at scans 8 ($x_{acc} = 6$ m/s^2 and $y_{acc} = -6$ m/s^2) and 15 ($x_{acc} = -6$ m/s^2 and $y_{acc} = 6$ m/s^2) only [20]. The evasive maneuver (EM)

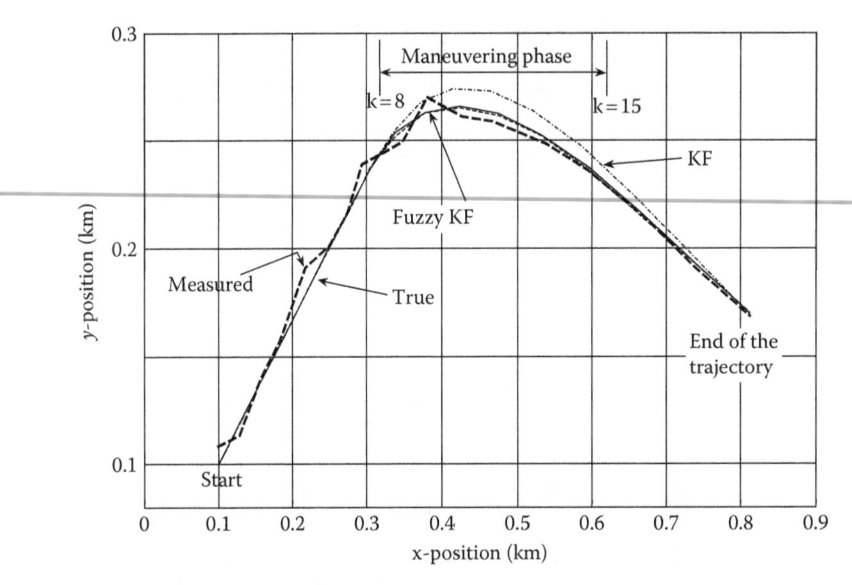

FIGURE 7.3
True, measured, and estimated x–y positions of the target for a mild maneuver.

data are generated with the same points for acceleration injection but with a maneuver magnitude of $40 \times 9.8\ \text{m/s}^2$ (instead of $9 \times 9.8\ \text{m/s}^2$). The results are obtained for 100 Monte Carlo simulation runs with the initial state vectors of KF and FKF kept close to the initial true states. The initial state-error covariance matrices for both the filters are unity values. The measured, true, and estimated x–y target positions for MM data are shown in Figure 7.3 [20]. The estimated trajectories using KF and FKF compare reasonably well with the true ones. Some discrepancies exist in the maneuvering phase of flight, where FKF exhibits better performance than KF. Similar observations can be made for the case of EM data (the detailed results are not shown here). However, Figure 7.4 shows the comparison of root sum square position error (RSSPE), root sum square velocity error (RSSVE), and root sum square average error (RSSAE) for both the filters; these errors are found to be somewhat large for the KF compared to those for FKF [20]. In the previous sections, the feasibility of fuzzy logic–based KF has been established for target tracking applications.

7.3.4 Fuzzy Logic–Based Sensor Data Fusion

SVF is generally used for the integration of estimated states, weighted with predicted state-error covariance matrices as follows:

$$\hat{X}_f^{SV}(k) = \hat{X}_1^{KF}(k) + \hat{P}_1^{KF}(k)\left(\hat{P}_1^{KF}(k) + \hat{P}_2^{KF}(k)\right)^{-1}\left(\hat{X}_2^{KF}(k) - \hat{X}_1^{KF}(k)\right) \tag{7.16}$$

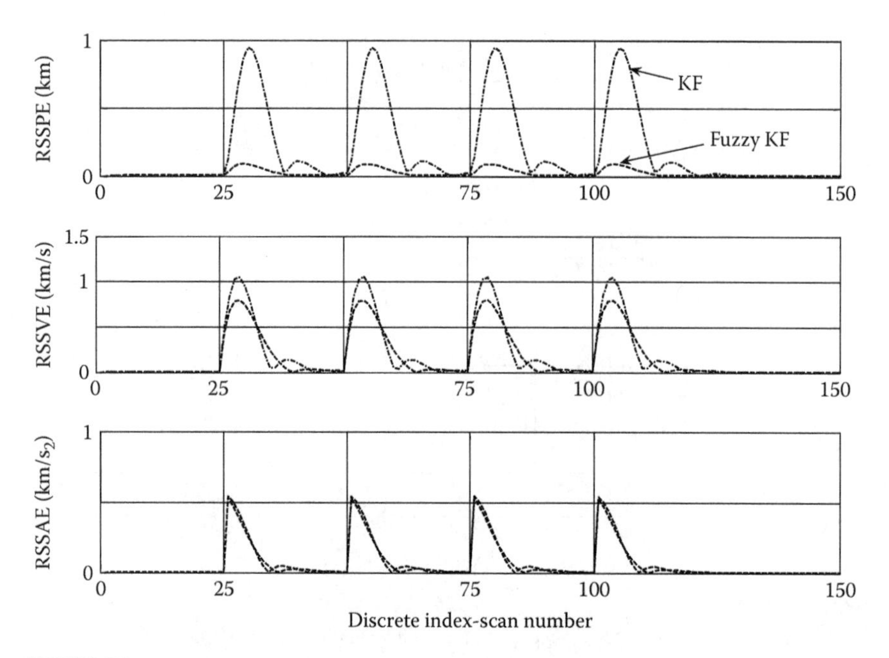

FIGURE 7.4
Root sum square position error, root sum square velocity error, and root sum square average error plots for Kalman filter and fuzzy Kalman filter for evasive maneuver data.

$$\hat{P}_f^{SV}(k) = \hat{P}_1^{KF}(k) - \hat{P}_1^{KF}(k)\left(\hat{P}_1^{KF}(k) + \hat{P}_2^{KF}(k)\right)^{-1} \hat{P}_1^{KF}(k) \tag{7.17}$$

where $\hat{X}_1^{KF}$ and $\hat{X}_2^{KF}$ are the estimated states obtained using the basic KF for sensor 1 and sensor 2, respectively, and $\hat{P}_1^{KF}$ and $\hat{P}_2^{KF}$ are the associated state-error covariance matrices. Next, we will study different ways to perform fusion using fuzzy logic–based KF schemes.

7.3.4.1 Kalman Filter Fuzzification

In Kalman filter fuzzification (KFF), the original data from each sensor are processed by a respective KF to estimate the states of a target (position, velocity, and acceleration). The error signal for each channel is generated by taking the difference of the measured and estimated positions of the target for that particular channel. The average estimation error is computed by:

$$\overline{e}_{\text{idn}}^{KF}(k) = \frac{e_{x_{\text{idn}}}^{KF}(k) + e_{y_{\text{idn}}}^{KF}(k) + e_{z_{\text{idn}}}^{KF}(k)}{M} \tag{7.18}$$

$M = 3$ for the total number of measurement channels, and idn = 1, 2 for the sensor identity number. The error signals are generated by:

$$\left.\begin{aligned} e_{x_{\text{idn}}}^{KF}(k) &= x_{m_{\text{idn}}}(k) - \hat{x}_{\text{idn}}^{KF}(k) \\ e_{y_{\text{idn}}}^{KF}(k) &= y_{m_{\text{idn}}}(k) - \hat{y}_{\text{idn}}^{KF}(k) \\ e_{z_{\text{idn}}}^{KF}(k) &= z_{m_{\text{idn}}}(k) - \hat{z}_{\text{idn}}^{KF}(k) \end{aligned}\right\} \tag{7.19}$$

where $x_{m_{\text{idn}}}, y_{m_{\text{idn}}}$, and $z_{m_{\text{idn}}}$ are the target position measurements in the x, y, and z axes and $\hat{x}_{\text{idn}}^{KF}, \hat{y}_{\text{idn}}^{KF}$, and $\hat{z}_{\text{idn}}^{KF}$ are the corresponding estimated positions from the KF. The fused states are given by [19,20]:

$$\hat{X}_f^{KFF}(k) = w_1(k)\hat{X}_1^{KF}(k) + w_2(k)\hat{X}_2^{KF}(k) \tag{7.20}$$

where $\{w_1, w_2\}$ are the weights generated by the FIS for sensor 1 and sensor 2, and the normalized values of the error signals $\bar{e}_1$ and $\bar{e}_2$ are the inputs to the FIS (associated with each sensor). The weights w_1 and w_2 are obtained as follows:

1. Fuzzification: These normalized error signals are fuzzified to values in the interval of [0, 1] using corresponding membership functions labeled by linguistic variables. The membership functions for both error signals are kept the same and the variables have the attributes ZE, SP, MP, LP, and very large positive (VLP). Figure 7.5 shows the membership functions for error signals $\bar{e}_{\text{idn}}$ and weights w_{idn} [20].
2. Rule generation and FIS process: The rules are created based on magnitude of error signals reflecting the uncertainty in sensor measurements. Some rules for sensor 1 and sensor 2 are given as follows [19,20]:

   Sensor 1:

   If $\bar{e}_1$ is LP AND $\bar{e}_2$ is VLP Then w_1 is MP
   If $\bar{e}_1$ is ZE AND $\bar{e}_2$ is MP Then w_1 is LP

   Sensor 2:

   If $\bar{e}_1$ is ZE AND $\bar{e}_2$ is VLP Then w_2 is ZE

   If $\bar{e}_1$ is ZE AND $\bar{e}_2$ is ZE Then w_2 is MP

   Table 7.3 gives the fuzzy rule base for the outputs w_1 and w_2, for sensor 1 and 2, respectively.

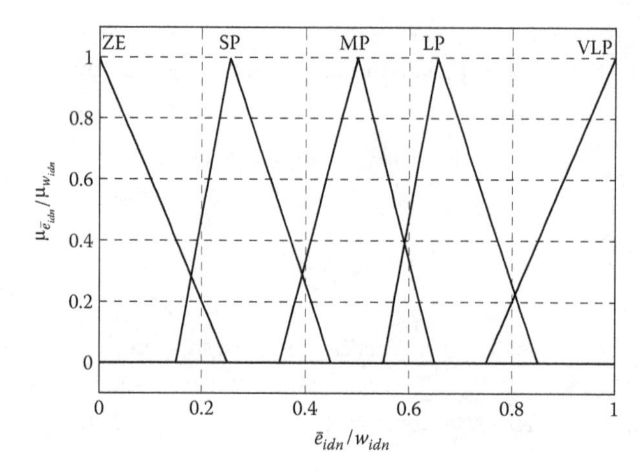

FIGURE 7.5
Error signals—weights w_{sid} membership functions for Kalman filter and fuzzy Kalman filter fuzzification.

TABLE 7.3
Fuzzy Rule Base of 25 Rules for Sensors 1 and 2

$\bar{e}_1$	$\bar{e}_2$ For Sensor 1					$\bar{e}_2$ For Sensor 2				
	ZE	**SP**	**MP**	**LP**	**VLP**	**ZE**	**SP**	**MP**	**LP**	**VLP**
ZE	MP	MP	LP	LP	VLP	MP	MP	SP	SP	ZE
SP	MP	MP	MP	LP	LP	MP	MP	MP	SP	SP
MP	SP	MP	MP	MP	LP	LP	MP	MP	MP	SP
LP	ZE	SP	SP	MP	MP	VLP	LP	LP	MP	MP
VLP	ZE	ZE	SP	MP	MP	VLP	VLP	LP	MP	MP

3. Defuzzification: The crisp values of w_1 and w_2, obtained by defuzzifying the aggregated output fuzzy sets (using the center of area [COA] method) are used in the fusion specified by Equation 7.20.

7.3.4.2 Fuzzy Kalman Filter Fuzzification

An alternative architecture is shown in Figure 7.6 [20]. The basic steps to compute weights are the same as for KFF, but with the following changes: (1) in Equations 7.18 and 7.19, superscript "KF" is replaced with "FKF" meaning that state estimation is performed using FKF instead of KF; and (2) the fused states are obtained by:

$$\hat{X}_f^{\text{FKFF}}(k) = \hat{X}_1^{\text{FKF}}(k) + w_1(k)\left(w_1(k) + w_2(k)\right)^{-1}\left(\hat{X}_2^{\text{FKF}}(k) - \hat{X}_1^{\text{FKF}}(k)\right) \tag{7.21}$$

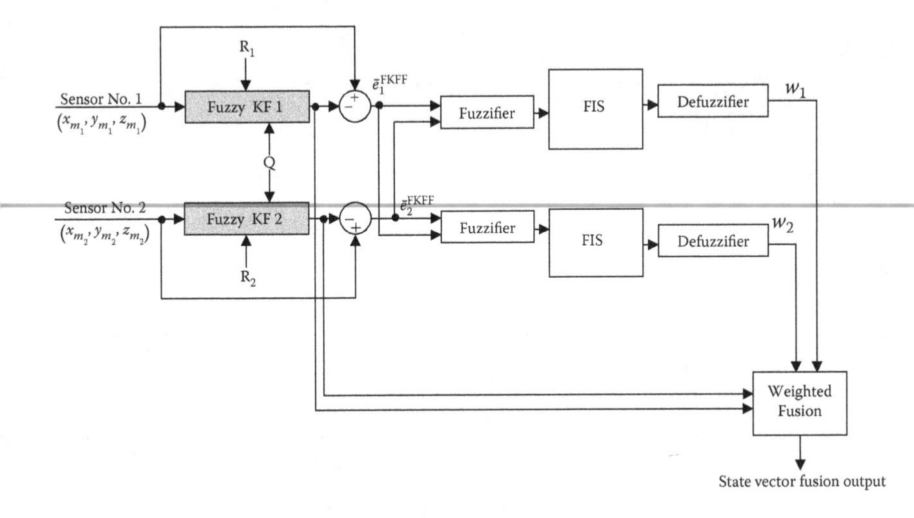

FIGURE 7.6
An alternative scheme for fusion using fuzzy logic.

Here the values of the weights might be different from the previous ones. This new SVF equation is obtained from Equation 7.16 by replacing $\hat{P}_1^{KF}$ and $\hat{P}_2^{KF}$ with w_1 and w_2, respectively.

7.3.4.3 Numerical Simulation Results

The trained FLP vector is obtained as shown earlier (as used in FKF). To compare the performance of the fusion algorithms (SVF, KFF, and FKFF), another set of data is generated by modifying the acceleration injection points used to generate training set and check-set data, with a total of 25 scans generated. The accelerations are injected at scan 8 as $x_{acc} = 6\ m/s^2$ and $y_{acc} = -6\ m/s^2$ and at scan 15 as $x_{acc} = -6\ m/s^2$ and $y_{acc} = 6\ m/s^2$, and the measurements for the two sensors are generated with SNR = 10 for sensor 1 and SNR = 20 for sensor 2. The data for each sensor are processed by KF and FKF for 100 Monte Carlo simulation runs, and their initial states (80% of true initial state) and error covariance matrices are kept the same. The performance of these filters is compared in terms of RSSPE, RSSVE, and RSSAE, however, Figure 7.7 illustrates velocity error comparisons for these three schemes [20]. From these and related plots not shown here [20], the following observations are made: (1) FKFF performs better than SVF and KFF; and (2) during the maneuver, the FKFF has fewer state errors compared to other fusion methods. The two weights sum to approximately one for KFF and FKFF methods, as expected.

An extension of FKF is illustrated in Example 8.3 in Chapter 8.

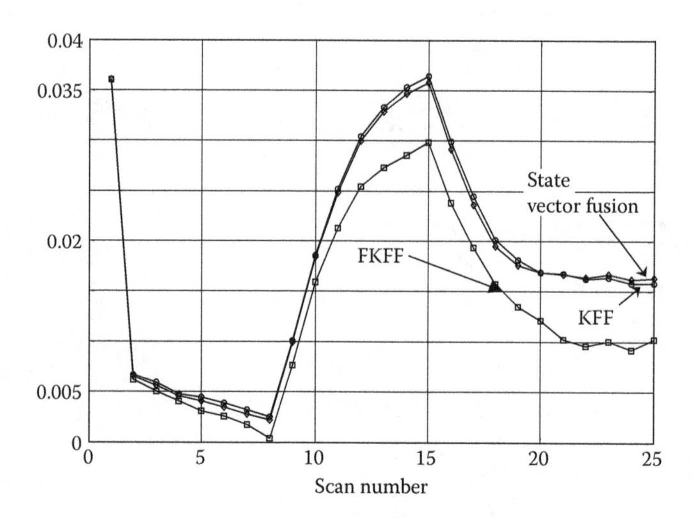

FIGURE 7.7
Root sum square velocity errors for the three fusion approaches: state-vector fusion, Kalman filter, and fuzzy Kalman filter fuzzification.

7.4 Fuzzy Logic in Decision Fusion

As we have seen in Chapter 2, level 2 fusion is also known as SR. It forms SA by relating the objects to the existing SA or by relating object assessments mutually. SR helps develop a description of the current relationships among objects and events in the context of the environment. It can also assess the meaning of level 1 estimated state and/or identity results in the context of background or supporting data. SA can be divided into three stages: (1) perception and event detection (PED), (2) current SA, and (3) near-future situation prediction. The PED reduces the workload of overall SA process by detecting changes in existing SA; that is, if a situation is already assessed and nothing changes, the situation does not need any more evaluation until new events occur. If a new event occurs, then the current situation is assessed along with the prediction in order to determine what could happen in the near future. The various levels of SA are modeled based on abstraction and reasoning. As the steps in the MSDF ladder increase, the intensity of abstraction increases.

SA aids in decision making; for example, for a pilot of a fighter aircraft [21]. The various decisions and actions that a pilot must make are (1) avoid collision with any nearby flying objects; (2) access the intentions of enemy aircraft; and (3) communicate with nearby friendly aircraft. The pilot's inherent capability for making various decisions and taking action works well when the number of nearby flying objects is small. In a complex scenario, it is difficult for a pilot to make a quick and accurate decision. Here, a mathematical model of the SR algorithm is required that can aid the pilot in quick and accurate decision making, allowing the pilot to concentrate more on flying his or her aircraft.

Let us consider the OR as a numerical procedure that enables us to assign properties to the objects of interest, for example, a missile with certain properties, such as acceleration, velocity, and position. The OR also uses rules based on geometry and kinematics and presumes that a certain object cannot change its properties by breaking these rules, while also remaining the same object; meaning that if the properties change, then it may be a different object.

A missile has the property of targeting an aircraft, whereas an aircraft has the property of being targeted by a missile. With these properties, one can conclude the relationship between the objects but cannot determine whether the missile is going to target an aircraft now. It is possible that the missile is not interested in targeting an aircraft at the present time, but may at a later time. We could assume that the world is the totality of certain observed facts and these facts are the application of relations to objects. Figure 7.8 illustrates object and situation assessments for a typical battlefield scenario [21,22]. The objects are the missile, fighter aircraft, and tank. The properties of these objects are the positions, velocities, directions, and identities computed through the numerical procedures of level 1 fusion. Using SA, we can then conclude that the missile is approaching an aircraft and a tank pointing towards the missile at any given instant of time.

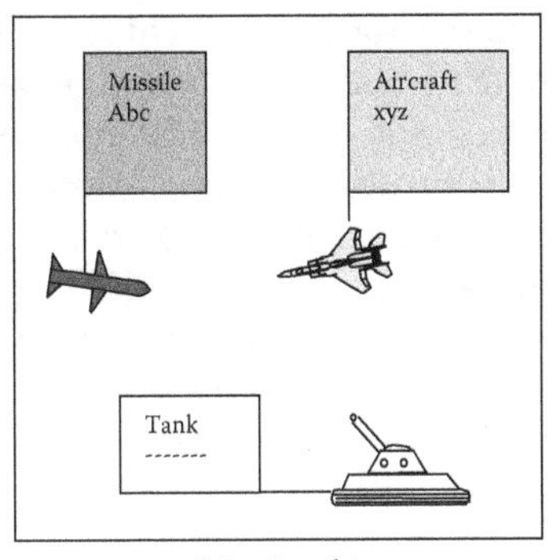

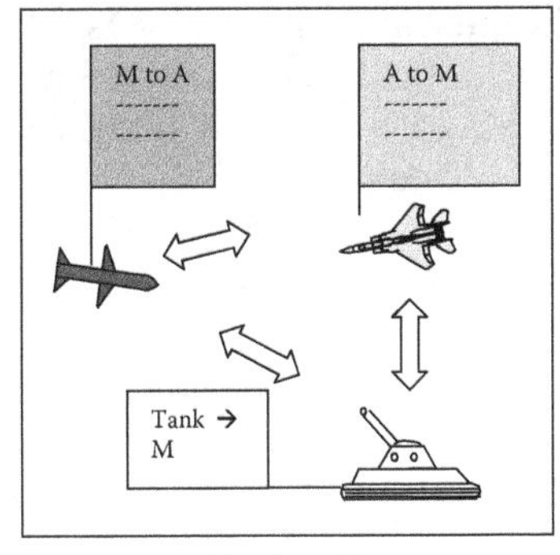

SA - Level 2

M to A : Missile approaching aircraft
A to M : Aircraft approaching missile
Tank → to M : Tank pointing missile

FIGURE 7.8
Object assessment and situation assessment.

7.4.1 Methods Available to Perform Situation Assessments

The most commonly used techniques in SA systems are high-level classification methods; the basic requirement is that they should be able to handle uncertainty with an ease in modeling situations. The common methods for assessing situations are (1) artificial neural networks (ANN)s—an ANN would have been trained in some situations; then it would be given the data of the current situation, which might have just occurred. The ANN predicts the situation, and if there is a close match, further decisions are made, and so on; (2) forward-chain ES (FCES; classical); (3) Bayesian NW (BNW), also called belief NW (more appropriate for D–S networks; causal net or inference net); (4) FL—FL/FMF/FIS; and (5) hybrid methods—FL and BNW, ANN, FL, and so on. An ANN can easily recognize situations by learning from training data. The NW must be sufficiently trained; a lack of sufficient training data is generally a problem. The FCES requires modeling by an expert and cannot update its knowledge automatically (i.e., it cannot adapt itself from the data), as the system contains only the knowledge of its designer.

BNW can be modeled as an ES and is also able to update beliefs (probabilities for BNW and "masses" for D–S methods). It has the ability to investigate hypotheses of the future. To make the system handy, the nodes in the BNW are often discrete and an expert can easily enter estimates of the probabilities for one situation leading to another. This will lead to a "quite good" NW. However, the system is difficult to use in real time. Also, BNW needs continuous input data, that is, the data must be classified first. The FLES can represent human knowledge and experience in the form of fuzzy rules. These rules can be tuned adaptively or new rules can be created dynamically using sets of I/O data and ANN. This learning method generates optimal fuzzy inference (FI engine of knowledge base) rapidly and with good accuracy as required for time-critical missions. The DF systems produced through partially processed sensory information and intelligence would have uncertain, incomplete and inaccurate information due to limited sensor capabilities. The FL can handle this for tasks such as (1) SA and decision making through modeling the entity, and (2) models and the associated fuzzy sets (and possibility theory). Hybrid methods utilize the FL approach to classify the continuous input that goes to BNW as discrete input and make the net continuous.

7.4.2 Comparison between Bayesian Network and Fuzzy Logic

Using FL, numerical data can be classified into fuzzy sets of discrete variables. In the classification of numerical data that measures the temperature of a certain material, a normal practice is to assign a grade to the membership function, that is, very cold, quite hot, and so on. The problem arises when assigning a grade to temperatures such as 19° and 21°C; the difference between the temperatures is not very big, but when classified using hard boundaries, 19°C is treated as cold and 21°C is treated as hot. Trying to model a system using hard boundaries could result in erroneous outputs. Classification can be more precise using fuzzy boundaries. A temperature of 19°C could be classified as cold with a membership value of 0.6 and as hot with a membership value of 0.4. In a similar manner, a temperature of 21°C can be classified as both hot and cold. The importance of fuzzy classification becomes more pronounced when dealing with noisy signals.

7.4.2.1 Situation Assessment Using Fuzzy Logic

When the level of uncertainties increases, situation assessment requires exact reasoning with an upward climb in the DF levels. Using only numerical procedures, it is very hard to model the uncertainties so that situations can be assessed as accurately as possible. An application of FL at the higher levels of DF could be a good choice for precise decision making. If one of the outputs of SA is "Aircraft is nonfriendly and targeting tank," then this event can be interpreted as if the situation assessor makes the decision that "Aircraft is nonfriendly and targeting tank." Fusion comes into the picture when there is more than one such decision for the same object of interest seen by multiple sensors of different types and accuracies. The different accuracy levels of each sensor dictate different confidence levels while making decisions. To have an accurate decision, it is essential to fuse the decisions (outputs form situation assessor) using the FL approach. This is a decision fusion paradigm and not a direct DF.

Consider a scenario of an unknown aircraft seen by two sensors of different types. The first sensor provides identity information and the second sensor measures the direction of the moving aircraft. The goal of FLSA is to make a decision about whether the behavior of the unknown aircraft is hostile or friendly [21,22]. The two inputs are "direction" and "identity" and the single output is "behavior" in FIS, i.e., the situation assessor. The inputs are fuzzified using corresponding membership functions. The input direction and identity have two membership functions, each with linguistic labels {departing, approaching} and {friend, foe}, respectively, and the output has membership functions with linguistic labels {friendly, hostile}. Inputs are aggregated using following inference rules [21–23]:

Rule 1: IF aircraft is departing OR identity is friend THEN behavior is friendly

Rule 2: IF aircraft is approaching OR identity is foe THEN behavior is hostile

In the FL "AND" is min(A, B) and OR is max(A, B), with A and B as two fuzzy sets representing input direction and identity, respectively. There is a possibility that both rules will be fired simultaneously, and then there will be output fuzzy set for each rule. These outputs are combined using an aggregation process. The present case uses max(C1, C2), where C1 and C2 are fuzzy output sets. The final crisp output is produced by applying the defuzzification process (e.g., COA has been used) to obtain the resultant output fuzzy set. For a given crisp input, the corresponding membership value for each membership function is given as input 1 (direction) with membership grade (e.g., {0.2, 0.8}), and input 2 (identity) with membership grade (e.g., {0.3, 0.7}). Both rules are fired concurrently. In the first rule, the inference is OR giving the combined membership a value of max(0.2,0.3) = 0.3. The membership function of that behavior is friendly and is cut at membership value 0.3. In the second rule, the membership function of that behavior is hostile and is cut at membership value min(0.8,0.7) = 0.7. Then the truncated output fuzzy sets are combined and COA is calculated to determine the total hostility of an aircraft, as shown Figure 7.9 [21].

7.4.3 Level-3 Threat Refinement and Level-4 Process Refinement

In level 3, fusion projects the current situation into the future to draw inferences about enemy threats, as well about as enemy vulnerabilities and opportunities for operations. This requires information from level 1 and level 2 so that quantitative estimates of an object's behavior can be determined and an expected course of action can be assessed. The threat refinement (TR) aspects are the identification of possible capabilities, intent of hostile objects, and the expected outcome. The PR is a metaprocess, a process concerned with other processes. Level 4 performs four key functions:

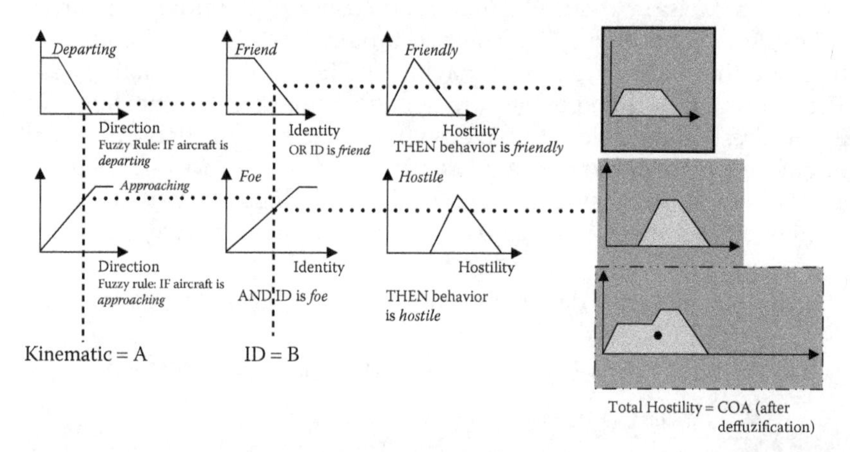

FIGURE 7.9
Fuzzy logic–based situation assessment to determine total hostility.

(1) monitors the DF process performance to provide information about real-time control and long-term performance; (2) identifies the information needed to improve the multilevel fusion product; (3) determines the source-specific requirements to collect relevant information; and (4) allocates and directs the sources to achieve mission goals. The latter function may be outside the domain of specific DF functions.

7.4.4 Fuzzy Logic–Based Decision Fusion Systems

An FL-based architecture for decision fusion systems (DFS) is shown in Figure 7.10. A typical scenario is obtained by defining the number of targets, target types, the identity number of each target, flight plan of each target through kinematic simulation, number of sensors and specification of each sensor in terms of field of view, probability of detection, sampling interval, measurement frame and its accuracies, and so on. The measurements for this scenario are generated using various sensors, for example, radar warning receiver (RWR), radio detection and ranging (RADAR), infrared search and track (IRST), forward looking infrared receiver (FLIR), and identification friend or foe (IFF). There is a separate block called *OA* and *SA* for each sensor. The purpose of the OA and SA blocks is to assess

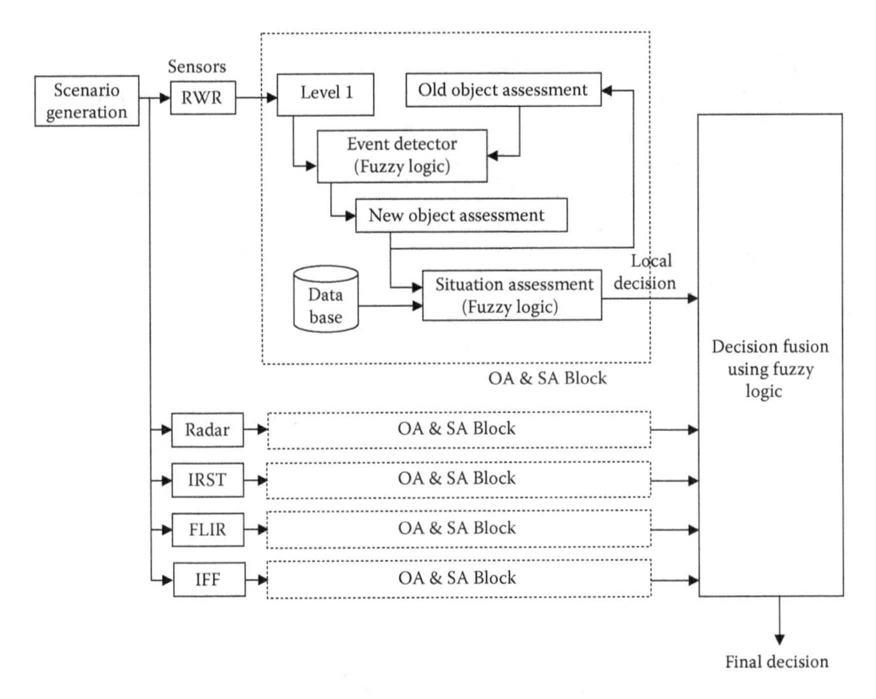

FIGURE 7.10
A fuzzy logic–based simple decision fusion system.

the current situation (e.g., the battlefield) by processing the measurements from each sensor at the local level. The outputs from the OA and SA blocks are fused using FL to make a final decision for various purposes. For example, the decision could be that "a particular fighter aircraft is a threat to us, destroy the aircraft" or it could be that "although the fighter is an enemy aircraft, it is moving away from us, wait for some time."

7.4.4.1 Various Attributes and Aspects of Fuzzy Logic–Based Decision Fusion Systems

All the algorithms can be developed in MATLAB on the Windows XP platform. The measurement data should be obtained by modeling each sensor as realistically as possible by taking all the available information such as field of view, tracking accuracy, probability of detection, false alarm density, and so on. Algorithms for level 1 fusion would be for gating, data association, KF, interacting multiple models, SVF (state-vector fusion) and measurement data level fusion (MDLF). The output of level 1 forms an OA consisting of information such as (1) track number, (2) track class, i.e. friend (.), neutral (.), and foe (.), (3) track type, i.e. fighter (.), bomber (.), transport (.), airborne warning and air control systems (.), commercial (.), and (4) track kinematics, i.e. position, speed, and covariance matrix. Here, (.) indicates the assigned membership grade value. The classification of an object helps to ease the decision making. If the aircraft is classified as *friend*, then the system does not need to know what type of target it is. On other hand, if it is classified as a foe, it is essential to know the type of the aircraft. A "foe" fighter is more harmful than a "friend" fighter; an approaching transport aircraft may be regarded as a "friendly" aircraft, and the receding fighter aircraft can be ignored.

The main purpose of an event detector (ED) is to compare the current output (say, at kth scan) from level 1 fusion with stored level 1 output of the previous event (say at k–1th scan). If an ED finds significant changes in outputs, then it assumes that a new event has occurred; for example, if an object changes it speed and bearing, or a new object with a different identity and class enters into the surveillance volume. An ED reduces the workload of the situation assessor of DFS by detecting changes in existing SA; that is, if a situation is already assessed and nothing changes, the situation does not need any further evaluation until new events occur. An ED can be realized using the FL approach. The outputs of level 1 fusion (containing object attributes) at two successive scans are fuzzified through an appropriate membership function and are represented by graded membership values. The way to conclude if an event has occurred within the attributes is by checking the statistical significance.

The database could contain flight corridors of friendly objects (path or place), information about terrain, and so on. Database information and outputs from the current OA can be used to assess the current situation (e.g., the battlefield). These could also store information about a situation that could be used to solve a specific problem or subsets of a problem associated with SA. The database would be more "intelligent" by using past experiences. It could then be used to make an accurate decision based on a particular situation that has occurred.

SA helps evaluate a situation, for example, a battlefield, by comparing the information stored in the database with the current OA. Experienced decision makers rely on SA for most of their decisions, that is, they select actions that have worked well in earlier, similar situations. Because of the presence or lack of certain essential characteristics, they can relate the current situation to past situations and to the actions that have worked well in past cases. SA creates relevant relations between the objects in the environment. It is essential to understand the outcome of SA or ontology for relations. The most common relations put into practice by the situation assessor are [21] as follows:

1. Pair: Two or more objects flying in a specific pattern, for example, formation flight of fighter aircraft. The fuzzy rules could be as follows [7,21]:
   - IF two aircraft have the same *bearing*, *elevation*, and *speed* THEN they have the same kinematics.
   - IF two aircraft have the same *kinematics*, *identity*, and *class* and are at a short *distance* from each other THEN they form a relation *pair.*
2. Along: An object flying along a static object, for example, civilian aircraft flying along an air lane. The fuzzy rules could be [7, 21] as follows:
   - IF an aircraft has the same *bearing* as an air lane, and IF it is *close* to the air lane THEN the aircraft is flying *along* the air lane.
   - IF an aircraft *class* is civilian THEN there is a higher possibility that the aircraft is flying *along* the air lane.
3. Attacking: An object attacking other dynamic or static objects, for example, a fighter aircraft attacking another fighter aircraft or a bomber attacking a place. The fuzzy rules could be as follows [7,21]:
   - IF an aircraft has high *speed,* a close *distance* to another aircraft, and a *bearing* towards it THEN the aircraft is trying to *close* in on the other one.
   - IF an aircraft is *closing* in on another, and has a different *identity,* and is a fighter aircraft THEN the aircraft is *attacking* the other one.

The above rules were created by a group of highly qualified and experienced individuals working in a relevant technical domain. The output for each sensor from the OA and SA blocks assists in local decision making with a certain degree of accuracy. A unified decision with a higher accuracy and robustness is obtained by fusing local decisions using an FL approach.

In Section 7.5 and in Chapter 8, some examples illustrate the application of FL for the development of a decision support system (DSS) that could be used to aid a fighter aircraft pilot in decision making in an air combat (AC) scenario (e.g., air-to-air, air-to-ground, ground-to-air, and so on.). This similar analysis procedure is equally applicable to other civilian systems, with different specifics. FL can be applied to decision fusion, a methodology that helps make a certain decision based on processing a certain scenario. For example, the pilot of a combat aircraft needs to make various tactical decisions based on what he or she observes from the sensors and surroundings. The decisions made by a pilot are based on his or her past experiences as they relate to sets of realistic scenarios. However, the pilot's response depends upon the amount of dynamic mental memory he or she has left (like random access memory in a computer!). Naturally, if there is a likelihood of complex scenarios, then the response from the pilot may be slow, which in turn will be reflected in his or her decision making ability. Due to the limited memory factor, it is therefore necessary to have a replica (in terms of mathematical model, that is, decision support system) of a pilot's mental model (PMM) for inference or decision making.

7.5 Fuzzy Logic Bayesian Network for Situation Assessment

Decision making in an AC is a complex task. Threat assessment (TA) and SA are the main components in this process. The AC operators of military airborne platforms depend on observed data from multiple sensors and sources to achieve their mission. The operators combine these data manually to produce a coherent air surveillance picture that portrays tracks of airborne targets and their classifications. In many cases, this air surveillance picture is analyzed manually (mentally) to determine the behavior of each target with respect to the owner's ship and other targets in the region and assess the intent and/or threat posed and any impact these might have on the planned mission [21,22]. As the number of targets grows or the situation escalates, there is a potential for an overload in the volume of available data and an overworking of the operators. Therefore it is desirable to assist the operators by automating some of the SA and TA processes.

The main problem in such decision making in any AC task is that of uncertainty. This can be handled via FL, belief functions, or ANNs. The probabilistic approach is based on rigorous theory, but it requires a vast amount of storage and computational manipulation, making it computationally burdensome. An alternative is the BNW (see Sections 2.3 and 2.5) since the Bayesian approach has many feasible features [24,25]. If the BNW is integrated with the FL, which assigns the data to discrete sets, this hybrid approach would be able to handle many of the requirements of SA. The design and implementation of an expert system, called the "intelligent system" for SA in air combat, or ISAC, as an aid to pilots engaged in SA tasks is discussed next.

7.5.1 Description of Situation Assessment in Air Combat

The ISAC is a pilot-in-the-loop (PIL) real-time simulator which consists of (1) an integration of airborne sensor models (ISM), (2) an interactive graphical user interface (GUI), i.e. an exercise controller (EC) for AC scenario generation and platform models, (3) PMMs of concurrent BNWs, (4) data processing algorithms, and (5) a graphical display [22]. The schematic of the ISAC is shown in Figure 7.11.

7.5.1.1 Exercise Controller

The EC developed in the C++ module consists of platform models of fighter, bomber, missile, rotorcraft, and transport aircraft. ECs are used to create any

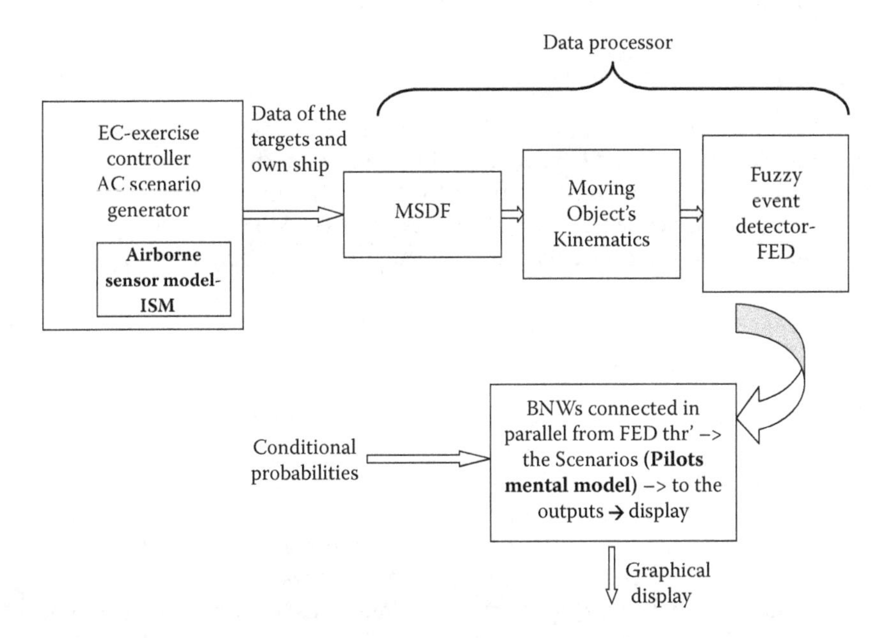

FIGURE 7.11
The ISAC simulator schematic with fuzzy logic Bayesian networks.

typical air-to-air combat scenarios, which can have a maximum of six targets (excluding the owner's ship). ECs have a simple user interface consisting of a display area, status area, and menu area. The status area has important parameters such as speed, course, bearing, coordinates, and radar status. The targets are represented by prespecified shapes and colors, which distinguish the platform types they represent. The class and identification of all targets should be specified for the simulation. Each target is controlled either by predefined trajectories or by user interaction in real time.

7.5.1.2 Integrated Sensor Model

The integrated sensor model is a MATLAB-SIMULINK–based module that has functional models of different sensors: (1) Doppler radar, (2) IRST, (3) RWR, and (4) electro-optical tracking system.

7.5.1.3 Data Processor

The data processor (DP) module consists of (1) MSDF, (2) relative kinematics data (RKD), and (3) fuzzy event detector (FED), and combines and classifies data received from multiple sensors. EKF is used to estimate the states of the targets using fused measurements from multiple sensors. FED classifies the RKD data into qualitative forms or events: the speed is "low," "medium," or "high." The MATLAB FL toolbox is used to design appropriate membership functions for data classification.

7.5.1.4 Pilot Mental Model

The PMM emulates the pilot's information processing, SA, and decision-making functions based on information received from the DP. Agents based on BNW technology are used to assess the occurrence of different situations in AC scenarios using the HUGIN C++ API software tool [26].

Finally, the *graphical display* provides the updated probabilities of all the agents.

7.5.2 Bayesian Mental Model

A mental model of the SA requires (1) the capability to quantitatively represent key SA concepts such as situations, events, and the PMM; (2) a mechanism to reflect both diagnostic and inferential reasoning; and (3) the ability to deal with various levels and types of uncertainties [21–25]. BNWs are ideal tools for meeting such requirements. BNWs are directed acyclic graphs in which the nodes represent probabilistic variables whose probability distribution is denoted as a belief value (more so for the D–S belief NWs), and in which links represent informal or causal dependencies among the variables. The weights in the BNWs are the conditional probabilities that are attached to each cluster of parent–child nodes in the NW [25]. The three agents based on BNW are used to assess the occurrence of different situations in the AC task with their tasks as [21,22] (1) pair agent—two or more targets are in formation (e.g., a pair of aircraft); (2) along agent—aircraft flying along an air lane; and (3) attack agent—one target attacking another target (e.g., fighter attacking the owner's ship). The process consists of (1) a BNW to represent the PMM, and (2) a belief update algorithm to reflect the propagation.

7.5.2.1 Pair Agent Bayesian Network

Figure 7.12 depicts these three BNW models [21,22]. The pair agent BNW (PAN) computes the updated probabilities when the speed, elevation, and course, and the ID, distance, and class are the independent or information nodes. The inputs "distance," "course," "elevation," and "speed" have three states: small, medium, and large. The "identification" has three states: friend, unknown, and foe. The "class" has four states: fighter, bomber, transport, and missile. The node "kinematics" has two states: same and different. The pair node is the hypothesis node and has two states: yes and no. The rules of the pair agent are [21,22] (1) if two aircraft have the same course, elevation, and speed, then they have the same kinematics; and (2) if two aircraft have the same kinematics, the same ID, the same class, and are at a short distance from each other, then they form a pair.

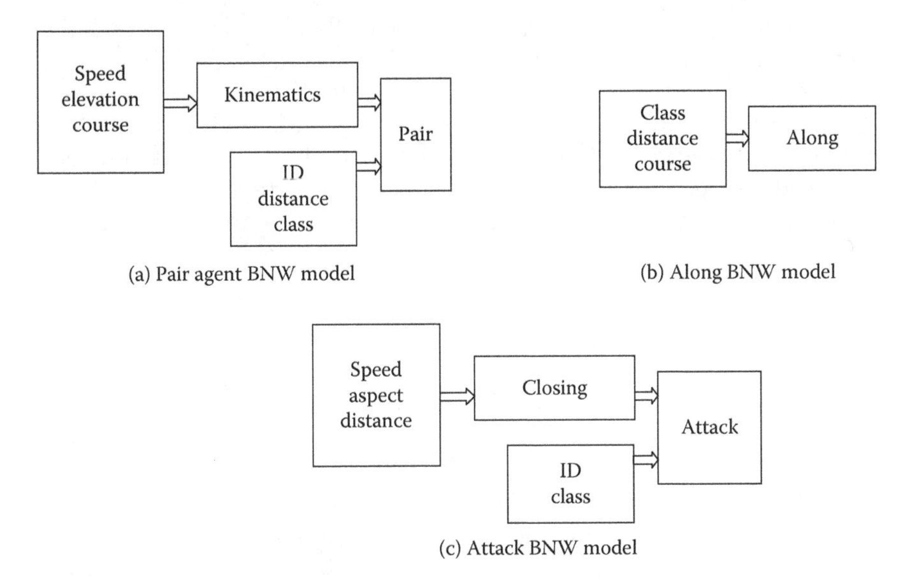

FIGURE 7.12
Bayesian network models for situation assessment.

7.5.2.2 Along Agent Bayesian Network

The along agent BNW (AAN) model computes the relationship between the air lane and the aircraft. The "distance" and "course" have three states: small, medium, and large. The "class" has four states: fighter, bomber, transport, and missile. The AAN node is the hypothesis node and has two states: yes and no. The rules are (1) if an aircraft has the same course as an air lane, and if it is close to the air lane, then the aircraft is flying along the air lane; and (2) if an aircraft is in transport there is a higher possibility that the aircraft is flying along the air lane.

7.5.2.3 Attack Agent Bayesian Network

The attack agent BNW (AtAN) model computes the attacking probabilities. The states of "identification," "class," "distance," and "speed" are the same as in the pair agent. The "aspect" node has three states: small, medium, and high. The closing node has two states: yes and no. The attack node is the hypothesis node and has two states: yes and no. The rules are (1) if an aircraft has high speed, a close distance to another aircraft, and is heading towards the other aircraft, it is trying to close in on the other; and (2) if an aircraft is closing in on another, has a different ID, and is a fighter aircraft, then the aircraft is attacking the other.

7.5.3 Results and Discussions

The scenario consists of six targets (five aircraft and one missile) with the scenario data shown in Table 7.4. Figure 7.13 shows AtAN probabilities—it detected four active relations between the owner's ship and the targets 1(foe), 2(foe), 4(unknown), and 5(foe)—with target 5 having the highest probability because the target is a missile and closes in on the owner's ship from behind [22]. Targets 3 and 6 have lowest probability since both are friends. The PAN detected a pair between targets 1 and 2 which lasted

TABLE 7.4

Data on Scenario

Target No.	Class	ID
1	Fighter	Foe
2	Fighter	Foe
3	Fighter	Friend
4	Fighter	Unknown
5	Missile	Foe
6	Transport	Friend

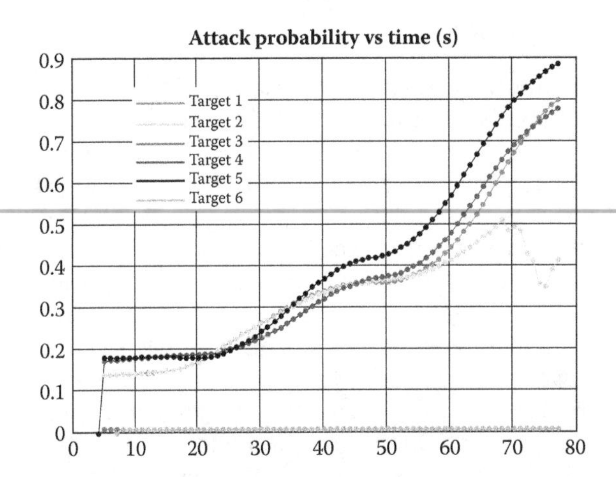

FIGURE 7.13
Attack agent Bayesian network probabilities for an air combat scenario.

for 20 seconds (results not shown), and it did not detect a pair between targets 3 and 4 because their identifications were different, as shown in Table 7.4. AAN found the relationship between target 6 and the air lane. The air lane is a virtual object and was inserted in the database before the simulation. Thus, the FL–BNW combination using several agents can accomplish the assigned job with fairly good precision.

7.6 Fuzzy Logic–Based Decision Fusion in a Biometric System

It is very important to safeguard proper access to computers, communication networks, and private information. In this context, user authentication relies on tokens and passwords that may be lost or forgotten [11]. This problem can be overcome by the use of biometric authentication, which verifies the user's identity based on his or her facial features, voice, and fingerprints. All of that information could be used and fused for final decision making. This authentication should be transparent to human–computer interaction (HCI) to maximize its usability. While none of the biometrics guarantee absolute reliability on their own, they can reinforce one another when used jointly and fused to maximize verification performance, where decisions based on individual biometrics are fused. Speaker verification, face identification, and finger print verification systems have been developed and tested, and a fusion approach based on FL in a biometric verification decision was developed [11].

7.6.1 Fusion in Biometric Systems

One speech utterance, one fingerprint image, and one face image for every subject were randomly grouped (480 data groups from the true claimant and 7200 from the imposters). The verification results of individual biometrics by means of majority votes were fused, and a marked improvement of 48% was recorded, relative to speaker verification only.

Then, fusion by weighted average scores was tried. The verification scores obtained from the spoken utterances, facial images, and fingerprint images were scaled to the same range of values by minimum–maximum normalization. A fixed weight was assigned to each biometric element. These weights were normalized and then used in the linear combination of the verification scores to obtain the fusion score, or the weighted scores of the verification scores. The weights were assigned by threefold cross-validation. To find the values that gave the best performance, the verification set was divided into three equal portions and each portion was used in turn for testing while the other two were used for optimizing the weights, which were varied within the [0, 1] range in steps of 0.1. The equal error rates of the three testing blocks were then averaged. An improvement of 52% was recorded, relative to fusion by majority voting [11].

7.6.2 Fuzzy Logic Fusion

In a biometric system, there could be various uncertainties: (1) the light is either too bright or too dark, the facial image is at an angle, or the expression is different from the original registered image for the face identification process; (2) the fingerprint image might be off-center, have faded fingerprints, or be smudged; and (3) the speaker's utterances might be drowned in noise or his or her voice characteristics and/or style might have changed. Since all these uncertainties cannot be precisely quantified or modeled, one can use the FL to process the imprecise information.

For FIS there were (1) six inputs (two for the face and four for the finger print), and (2) two output variables (one face and one finger print). The fuzzy sets for the two output variables were represented by triangular membership functions and were defined as low, medium, and high output weightings for each biometric entity. For defuzzification, the COA method was used (see Chapter 6). The input variables' fuzzy sets were either linear or Gaussian combination membership functions, $f(x)$. The least favored external condition for each input fuzzy variable was represented by the set $1 - f(x)$. The six input variables are

1. "Face-finding confidence" with discrete levels set as 0, 2.5, 5, 7.5, and 10. The higher input levels signify higher confidence in face detection. Here a triangular membership function was used to seek a high level of confidence in finding a face.

2. "Illumination" determines the average intensity of the face image. High or low corresponds to bright or dark environments; here a Gaussian shape was used.
3. "CorePosX" is for the *x* coordinate of the fingerprint image core; a centrally placed *f*(*x*) was used and high or low values implied an off-centered image.
4. "CorePosY" for the *y* coordinate.
5. "Darkness"; large values imply darker images, low values were favored.
6. "Low clarity" is for light pixels with intensities > 110 and < 160; large values imply faded images, hence low values were preferred.

The shapes of *f*(*x*) for these fuzzy variables were chosen accordingly.

Two groups of fuzzy *if … then* rules were used. One group was related to the output variable "face" for weighting the inputs "face-finding confidence" and "illumination." The second group was for the output "fingerprint," based on the values of the inputs "CorePosX," "CorePosY," "darkness," and "low clarity." The fuzzy rules assure the following basic properties: (1) if all the input variables are favorable, then set the output variable to high; (2) if any one condition is not favorable, then set the output variable to medium; and (3) if there are multiple unfavorable conditions, then set the output variable to low. The verification set was divided into three segments: one portion was used for testing and the remaining two were used for optimizing the weights. The weights were normalized and the weighted average was determined for the verification results. Because of the use of the FL-based approach in decision making, a further improvement of 19% was achieved [11].

8

Performance Evaluation of Fuzzy Logic–Based Decision Systems

A systematic approach should be followed to find out if any of the existing implication methods discussed in Chapter 6 satisfy a given set of intuitive criteria of *generalized modus ponens* (GMP) and of *generalized modus tollens* (GMT). MATLAB® with graphics is used to develop a user-interactive package to evaluate implication methods with respect to these criteria. The graphical method of investigation is much quicker and requires less effort from the user compared to the analytical method. Also, the analytical method seeks a diagnosis of the various curves involved in finding consequences (i.e., the nature of curves with respect to the variation of fuzzy sets $\mu_A(u)$ and $\mu_B(v)$) when the intuitive criteria of GMP and GMT are applied to various implication methods.

8.1 Evaluation of Existing Fuzzy Implication Functions

Tables 8.1 and 8.2 summarize the results of various implication methods (similar to [8]) tested against the intuitive criteria of GMP and GMT using the new tools discussed in Section 6.4 [5,7]. Fuzzy implication methods (FIMs) such as the mini-operation rule of fuzzy implication (MORFI) and the product-operation rule of fuzzy implication (PORFI) satisfy exactly the same intuitive criteria of GMP and GMT—the total number of satisfied intuitive criteria is four. A similar observation can be made for implication methods such as the arithmetic rule of fuzzy implication (ARFI), the Boolean rule of fuzzy implication (BRFI), and Goguen's rule of fuzzy implication (GRFI). These methods satisfy only two intuitive criteria of GMP and GMT. With these criteria, the max-min rule of fuzzy implication (MRFI) has the minimum satisfactory number, equal to one. The logical explanation here is that the corresponding curve profiles of implication methods (with respect to the shape of the envelope), such as MORFI and PORFI, are similar, with both starting from the same origin and ending with the membership grade $\mu_B(v)$. Similarly, implication methods such as ARFI, BRFI, and GRFI have similar curve profiles, with each starting with

TABLE 8.1

Comparison of True and Computed Consequences of GMP

		IM											
		MORFI		PORFI		ARFI		MRFI		BRFI		GRFI	
Criteria	T	Computed	SF	Computed	SF	Computed	SF	Computed	SF	Computed	SF	Computed	SF
C1	μ_B	μ_B	Y	μ_B	Y	$\frac{1+\mu_B}{2}$	N	$0.5 \cup \mu_B$	N	$0.5 \cup \mu_B$	N	$\sqrt{\mu_B}$	N
C2-1	μ_B^2	μ_B	N	μ_B	N	$\frac{3+2\mu_B-\sqrt{5+4\mu_B}}{2}$	N	$\frac{3-\sqrt{5}}{2} \cup \mu_B$	N	$\frac{3-\sqrt{5}}{2} \cup \mu_B$	N	$(\mu_B)^{2/3}$	N
C2-2	μ_B		Y		Y		N		N		N		N
C3-1	$\sqrt{\mu_B}$	μ_B	N	μ_B	N	$\frac{\sqrt{5+4\mu_B}-1}{2}$	N	$\frac{\sqrt{5}-1}{2} \cup \mu_B$	N	$\frac{\sqrt{5}-1}{2} \cup \mu_B$	N	$(\mu_B)^{2/3}$	N
C3-2	μ_B		Y		Y		N		N		N		N
C4-1	1	$0.5 \cap \mu_B$	N	$\frac{\mu_B}{1+\mu_B}$	N	1	Y	1	Y	1	Y	1	Y
C4-2	$\mu_{\bar{B}}$		N		N		N		N		N		N

T = true consequence of GMP; SF = satisfaction flag; "Y" if computed consequence matches with the true one/"N" if computed consequence does not match with the true one.

TABLE 8.2

Comparison of True and Computed Consequences of GMT

		IM											
		MORFI		PORFI		ARFI		MRFI		BRFI		GRFI	
Criteria	T	Computed	SF	Computed	SF	Computed	SF	Computed	SF	Computed	SF	Computed	SF
C5	$\mu_{\bar{A}}$	$0.5 \cap \mu_A$	N	$\frac{\mu_A}{1+\mu_A}$	N	$1-\frac{\mu_A}{2}$	N	$0.5 \cup 1-\mu_A$	N	$0.5 \cup 1-\mu_A$	N	$\frac{1}{1+\mu_A}$	N
C6	$1-\mu_A^2$	$\frac{\sqrt{5}-1}{2} \cap \mu_A$	N	$\frac{\mu_A \sqrt{\mu^2+4}-\mu_A^2}{2}$	N	$\frac{1-2\mu_A+\sqrt{1+4\mu_A}}{2}$	N	$\frac{\sqrt{5}-1}{2} \cup \mu_A$	N	$\frac{\sqrt{5}-1}{2} \cup \mu_A$	N	$\frac{\sqrt{1+4\mu_A^2}-1}{2\mu_A^2}$	N
C7	$1-\sqrt{\mu_A}$	$\frac{3-\sqrt{5}}{2} \cap \mu_A$	N	$\frac{2\mu_A+1-\sqrt{4\mu_A+1}}{2}$	N	$\frac{3-\sqrt{1+4\mu_A}}{2}$	N	$\frac{3-\sqrt{5}}{2} \cup \mu_A$	N	$\frac{3-\sqrt{5}}{2} \cup \mu_A$	N	$\frac{2+\mu_A-\sqrt{\mu_A^2+4\mu_A}}{2}$	N
C8-1	1	μ_A	N	μ_A	N	1	Y	$\mu_A \cup 1-\mu_A$	N	1	Y	1	Y
C8-2	μ_A		Y		Y		N		N		N		N

unity and finally converging to $\mu_B(v)$. MRFI has a unique curve profile that does not match any other methods, and this separates MRFI from the existing implication methods. Finally, we can conclude that the similarity in curve profiles of these methods (MORFI and PORFI in one group and ARFI, BRFI, and GRFI in another group) may lead to an equal number of satisfied intuitive criteria for GMP and GMT.

8.2 Decision Fusion System 1—Formation Flight

In this section, fuzzy logic–based decision software (FLDS) residing in its own ship platform is used to decide whether two enemy fighter aircrafts have formation flight during the course of an AC scenario. Kinematic data using point mass models are generated for two aircrafts of the same class and identity in the pitch plane (with no motion in the *x-y* plane) using MATLAB, with the following parameters: (1) the initial state of aircraft 1 is $X_1 = [x \ \dot{x} \ z \ \dot{z}]$ = [0 m 166 m/s 1000 m 0 m/s]; (2) the initial state of aircraft 2 is $X_2 = [x \ \dot{x} \ z \ \dot{z}]$ = [0 m 166 m/s 990 m 0 m/s]; (3) the sensor update rate is 1 Hz; (4) the total simulation time is 30 seconds; (5) the aircraft motion has constant velocity; and (6) the kinematic model is $X_i(k+1) = FX_i(k) + Gw_i(k)$, where k(=1, 2,..., 30) is the scan or the index number, i(=1, 2) is the aircraft number, F is the state transition matrix, G is the process noise gain matrix, and w is the white Gaussian process noise with covariance matrix Q = 0.1*eye(4, 4):

$$F = \begin{bmatrix} 1 & T & 0 & 0 \\ 0 & 1 & 0 & 0 \\ 0 & 0 & 1 & T \\ 0 & 0 & 0 & 1 \end{bmatrix} \quad \text{and} \quad G = \begin{bmatrix} \frac{T^2}{2} & 0 & 0 & 0 \\ 0 & T & 0 & 0 \\ 0 & 0 & \frac{T^2}{2} & 0 \\ 0 & 0 & 0 & T \end{bmatrix}$$

The aircraft maintain a formation flight from $t=0$ to 5 s, then split apart at $t = 5$ s, and remain in that mode for up to $t=10$ s. From the 10th second to the 15th second, they fly with constant separation and start approaching each other from the 15th second. From the 20th second, they again form a pair, and stay in formation flight for another 10 seconds. Figure 8.1 shows the trajectories in the pitch plane, and the elevation angles of the two aircraft as seen from the owner's ship. The performance of FLDS depends upon the proper selection of membership functions of fuzzy sets, fuzzy

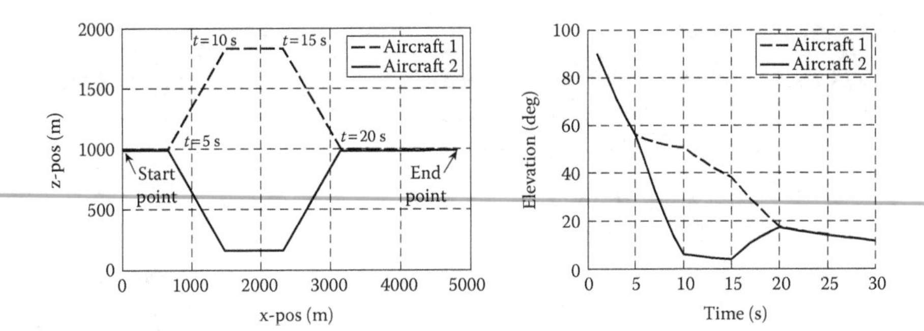

FIGURE 8.1
Trajectory and elevation angles of two aircraft.

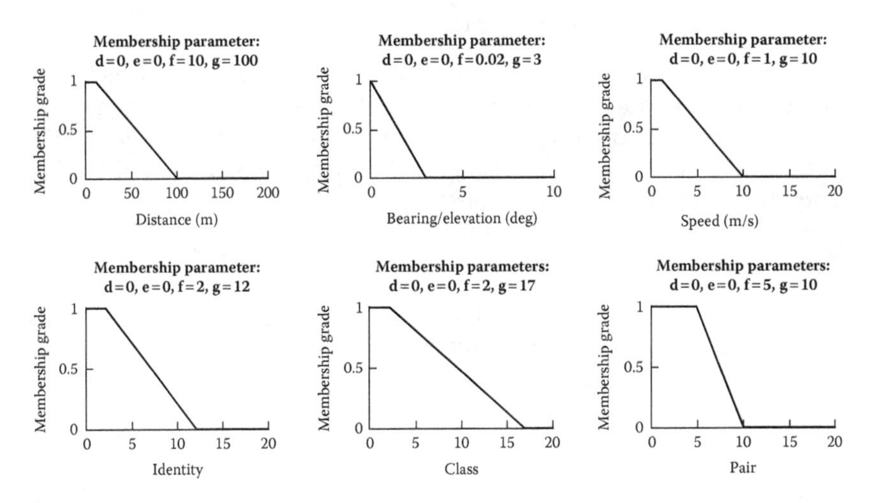

FIGURE 8.2
Input and output membership functions for fuzzy logic–based decision software.

rules, FIMs, aggregation methods, and defuzzification techniques, which are described next.

8.2.1 Membership Functions

The inputs to FLDS are numerical differences (absolute values) of the aircraft's bearing, elevation, separation distance along the z-axis, speed, identity, and class. For each input and output, there is a membership function, which fuzzifies the data between 0 and 1. The trapezoidal-shaped membership functions are chosen as shown in Figure 8.2. Note that the limits (d, e, f, and g) of these functions are provided for the sake of concept proving based on the designers' or authors' intuition. In practice, these limits should be provided by an expert in the relevant domain.

8.2.2 Fuzzy Rules and the Fuzzy Implication Method

Fuzzy rules provide a way to map the inputs to outputs and are processed using FIM in a fuzzy inference engine (FIE). The rules used to decide whether two aircraft form a pair are as follows:

Rule 1: If two aircraft have the same bearing, elevation, and speed, then they have same kinematics.

Rule 2: If two aircraft have the same kinematics, identity, class, and are at a short distance from each other, then they form a pair. The fuzzy rules are processed by FIM, and the PORFI or Larsen implication is used (see Chapter 6).

8.2.3 Aggregation and Defuzzification Method

An aggregation method is used to combine the output fuzzy sets (each of which is made due to a rule being triggered) to get a single fuzzy set. Here, the bounded sum (BS) operator of a T-conorm or S-norm is used in the aggregation process. The aggregated output fuzzy set is defuzzified using the center of area (COA) method.

8.2.4 Fuzzy Logic–Based Decision Software Realization

FLDS is implemented in a MATLAB or Simulink® environment using MATLAB's fuzzy logic toolbox. Figures 8.3 through 8.5 show the schematics

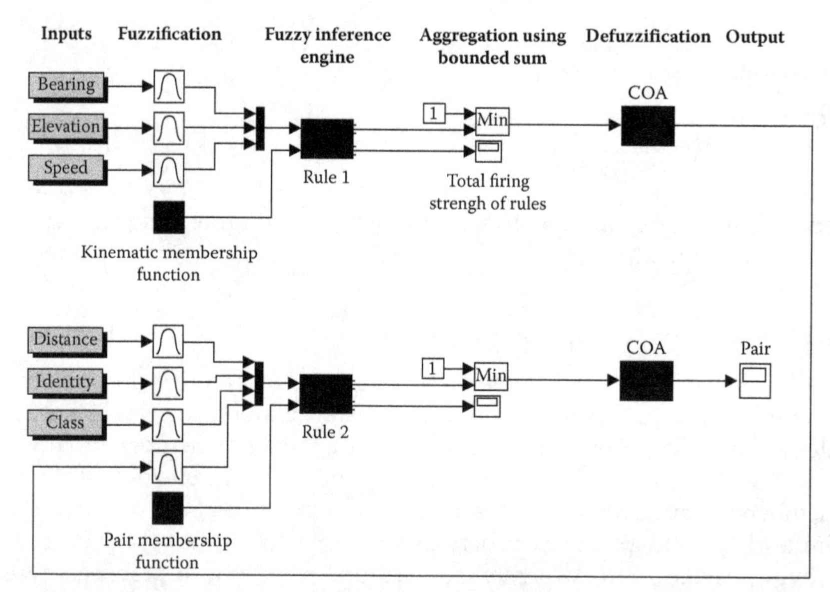

FIGURE 8.3
Fuzzy logic–based decision software in MATLAB and Simulink.

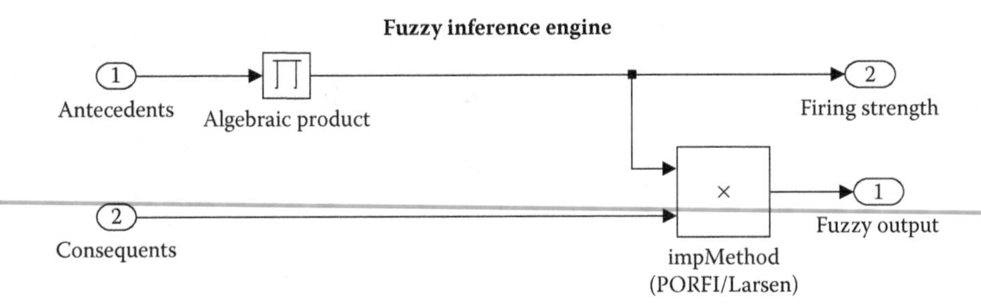

FIGURE 8.4
Fuzzy inference engine subblock for rule 1 and rule 2.

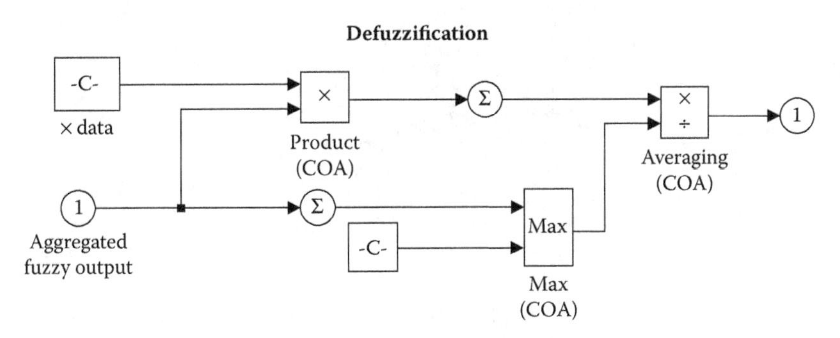

FIGURE 8.5
Defuzzification subblock.

of the FLDS system, a subblock of FIE/S for rule 1 and rule 2, and a subblock for defuzzification. After running FLDS, the results at various stages (fuzzified inputs elevation, distance, final output, and so on) of FLDS are stored. Figure 8.6 illustrates the fuzzified values of inputs, distance and elevation, and notes that during the formation flight (0–5 seconds and 20–30 seconds) the membership grades of both the inputs are fairly high, and during the nonformation flight (5–20 seconds) the membership grade is nearly zero. There are other inputs such as speed, bearing, aircraft identity, and class that are used to decide whether both the aircrafts form a pair. Figure 8.6 also shows the defuzzified output pair of FLDS, and we can see that FLDS is able to correctly detect the aircraft pair and split periods; it assigns a large weight during the pair period and zero weight during the split period.

8.3 Decision Fusion System 2—Air Lane

Here, we describe the steps required to develop FLDS to decide if a particular aircraft is flying along the air lane. A sensitivity study is carried out

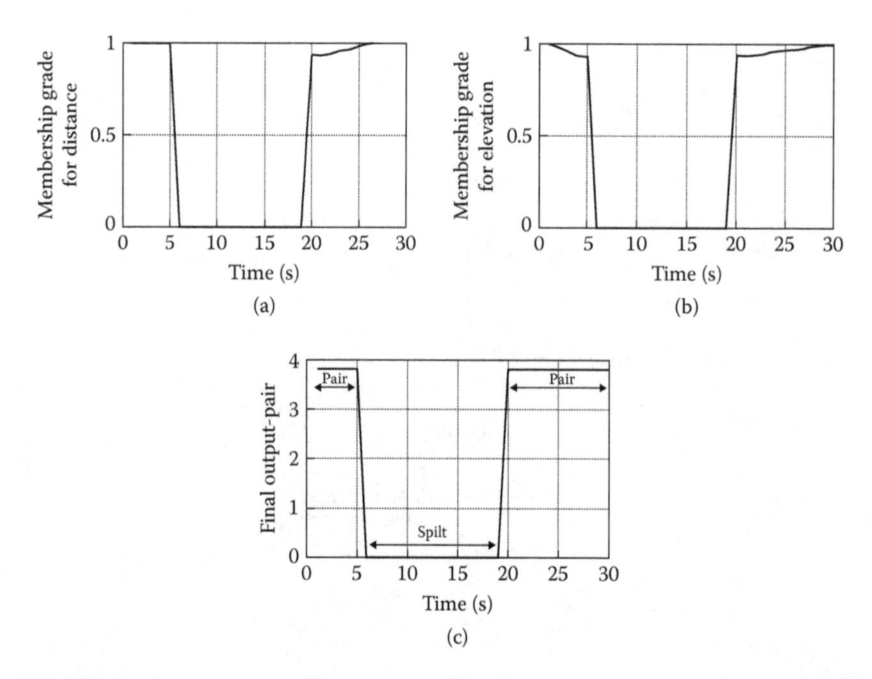

FIGURE 8.6
Input and output of the fuzzy logic–based decision software: (a) distance fuzzified, (b) elevation fuzzified, and (c) the output of FLDS.

to check the performance of FLDS with respect to the fuzzy aggregation operator. Simulated data for scenario generation are obtained with the following parameters: (1) the initial state of the aircraft is $X = [x \ \dot{x} \ y \ \dot{y}] =$ [2990 m 0 m/s 0 m 332 m/s]; (2) the air lane is located along the y-axis at x = 3000 m; (3) the sensor update rate is 1 Hz; (4) the simulation time is 30 seconds; (5) the aircraft motion has constant velocity; and (6) the kinematic model is $X(k+1) = FX(k) + Gw(k)$, where $k(= 1, 2, \ldots, 30)$ is the scan number, F is the state transition matrix (same as in Section 8.2), G is the process noise gain matrix, and w is the white Gaussian process noise with process covariance Q: 1.0 $eye(4,4)$. Figure 8.7 depicts the aircraft and air lane locations in the yaw plane and the bearings of the aircraft and air lanes at various data points as seen from the origin point (0, 0). Other aspects related to the fuzzy inference system (FIS) are discussed next.

8.3.1 Membership Functions

The inputs are fuzzified values of the distance (which is the absolute separation between the aircraft and the air lane along the y-axis), the absolute value of bearing difference between them, and the class of the aircraft. We assume that the aircraft is civilian. The membership functions used

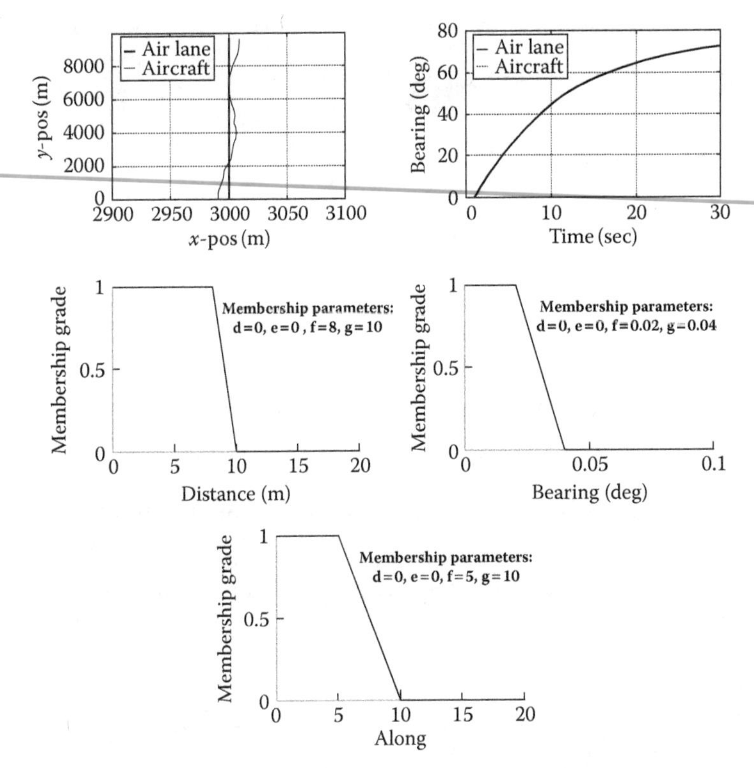

FIGURE 8.7
Input (distance, bearing) and output (along) membership functions.

to fuzzify the inputs and outputs are of a trapezoidal shape as shown in Figure 8.7.

8.3.2 Fuzzy Rules and Other Methods

The rules used to decide that a particular aircraft flies along an air lane or not are as follows:

Rule 1: If the aircraft has the same bearing as the air lane and if it is close to the air lane, then the aircraft is flying along the air lane.

Rule 2: If the aircraft is civil, then there is a high possibility that the aircraft is flying along the air lane.

PORFI is used for this example. The bounded sum (BS), algebraic sum (AS), and standard union (SU) operators of the T-conorm and S-norm are used one-by-one in the aggregation process. The aggregated output fuzzy set is defuzzified using the COA method.

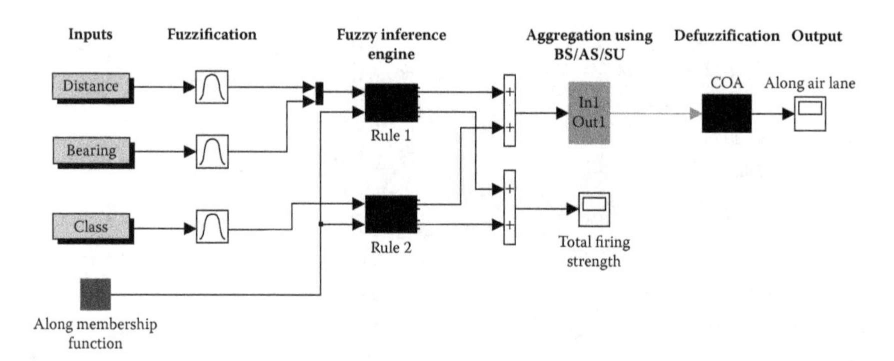

FIGURE 8.8
Fuzzy logic–based decision software for the air lane example using MATLAB and Simulink.

8.3.3 Fuzzy Logic–Based Decision Software Realization for System 2

FLDS is realized using MATLAB or Simulink, as shown in Figure 8.8. Some important subblocks were already given in Section 8.2, and an additional subblock is "aggregation using BS/AS/SU" only. Inside this subblock are the different aggregation methods that a user can select. The results at each stage are stored after the execution of this FLDS for each aggregation method. This includes fuzzified inputs such as bearing and distance and the final output. Figure 8.9 illustrates the fuzzified inputs, such as distance and bearing, and the comparison of the final outputs obtained from different aggregation methods. We observe that (1) if the SU operator is used for aggregation, a constant output is observed regardless of whether the aircraft is along the air lane; (2) for the AS operator, a smooth transition takes place between 0 and 1, which means a hard decision is made by the FLDS about the aircraft, i.e., whether it is along the air lane; (3) for the BS operator, a nonsmooth transition is observed between 0 and 1, which seems to be intuitively correct, and which gives a level of confidence while making a decision about an aircraft, i.e., whether it is flying along the air lane (the larger the final output, the higher the confidence); and (4) the use of BS in the aggregation process provides better results than other methods such as SU and AS.

8.4 Evaluation of Some New Fuzzy Implication Functions

Both existing and new fuzzy implication functions and their derivations, as well as an evaluation of their performance with respect to intuitive criteria of GMP and GMT, were discussed in Chapter 6. The use of the existing fuzzy implication function, called PORFI, was demonstrated in Sections

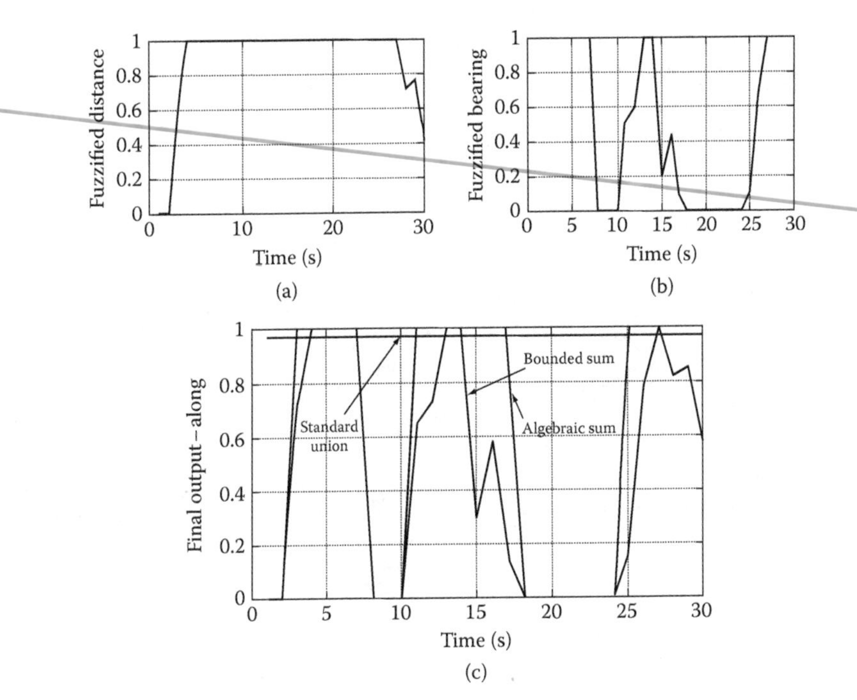

FIGURE 8.9
Fuzzy logic–based decision software for the air lane decision: (a) distance fuzzified, (b) bearing fuzzified, (c) final output.

8.2 and 8.3. Now, the same case studies are revisited, but one of the new implication functions, called PCBSAP, is used to process the fuzzy rules.

To compare the performance of Simulink-based FLDS applied to different fuzzy implication functions, some changes have been made. Figure 8.10 depicts the modified subblock. A similar change has been made in the Simulink subblock (FIE/S) for the case study in Section 8.3.

The embedded MATLAB-based fuzzy implication method (EMFIM) shown in Figure 8.10 is used to define various FIMs (existing and new ones). These FIMs are selected using the front-end menu developed using MATLAB graphics, via the "selection flag" in the figure. Figure 8.11 shows a comparison of the final outputs of FLDS obtained for implication functions PORFI (existing FIM) and PCBSAP (a new FIM). The outputs are comparable for up to 20 seconds of the total simulation time. During this period, the two aircraft form a pair and then spilt. After the 20th second, a mismatch between the outputs is observed. In Figure 8.11 we see that the FLDS output obtained for PORFI quickly converges to the max value of unity, whereas the FLDS output for PCBSAP takes more time before converging to the unity value. The output for PCBSAP is more realistic because it matches the way the data simulation was carried out.

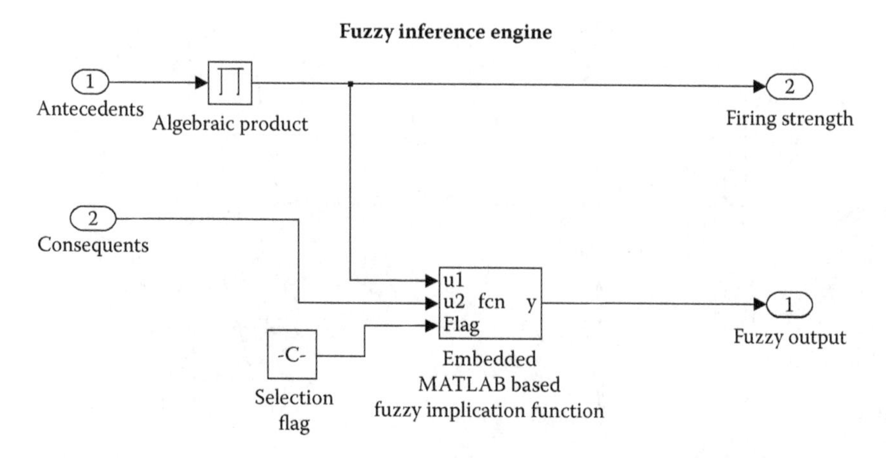

FIGURE 8.10
Modified subblock of fuzzy logic–based decision software for rule 1 and rule 2 to study a new fuzzy implication method.

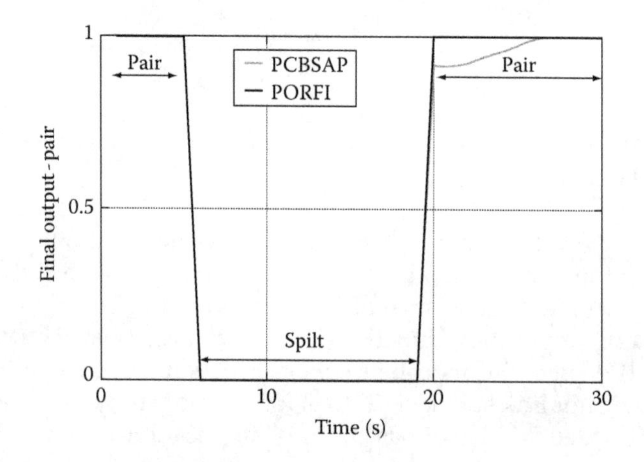

FIGURE 8.11
The outputs of fuzzy logic–based decision software using different fuzzy implication methods—one existing and one new for formation flight.

We can also test the new FIM by applying it to FLDS for identifying whether a particular aircraft is flying along the air lane. Section 8.3 concluded that the results obtained using the aggregation operator bounded sum matched perfectly with the way the data simulation was carried out. Based on this fact, a new implication function is applied with a combination of "bounded sum" operators only, and the results obtained are compared with those obtained when the existing implication function

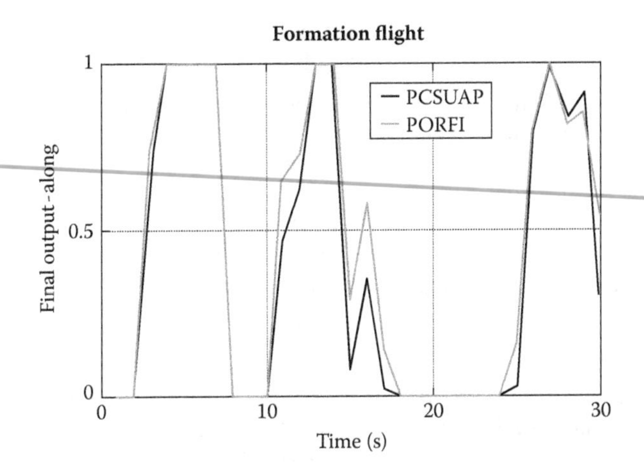

FIGURE 8.12
Performance of fuzzy logic–based decision software for two fuzzy implication methods—one existing and one new for the air lane case.

PORFI was used. Figure 8.12 compares only the final outputs (along the lane) for another new FIM called propositional calculus standard union algebraic product (PCSUAP) and the existing PORFI. Figure 8.12 shows that the outputs of FLDS for these two implication functions are fairly comparable, except a slight lag or lead observed in the output of FLDS for PCSUAP when compared with the output obtained for PORFI. The peaks of the outputs still match perfectly.

The above examples establish that a newly developed FIM (which were also validated with GMP and GMT criteria in Section 6.5) could be used for decision making. They give almost identical or somewhat better performances than some of the existing FIM. However, more rigorous study is required to evaluate all the newly developed FIMs and their applicability to general control systems as well as in DF systems, and the analyses of the new FIMs reported in this volume are not exhaustive.

8.5 Illustrative Examples

Although the examples illustrated in this section may not have direct applications to the fusion or decision processes, they can serve to strengthen the base for the application of fuzzy logic to decision fusion and related DF processes. These examples are mainly centered on the topics of Chapters 6 and 7. Using the results of these examples, one can build other decision-level fusion processes.

EXAMPLE 8.1

Using Equation 6.7 and $u = -15:0.1:15$:

(a) Simulate three trapezoidal functions (corresponding to a rule fired) having the following parameters:
Rule 1: $d = -11$, $e = -9$, $f = -2$ and $g = 1$
Rule 2: $d = -6$, $e = -4$, $f = 1$ and $g = 4$
Rule 3: $d = 1$, $e = 2$, $f = 7$ and $g = 8$

(b) Combine the fuzzified outputs using the following equation:

$$y_0 = \max(0.5y_2, \max(0.8y_1, 0.05y_3))$$

Here, y_1, y_2, y_3 are the outputs for rule 1, rule 2, and rule 3, respectively.

(c) Defuzzify the aggregated output (y_0) using the following methods, and compare the methods: (1) centroid, (2) bisector, (3) middle of maximum, (4) smallest of maximum, and (5) largest of maximum.

SOLUTION 8.1

The trapezoidal membership function is defined in Equation 6.7.

(a) Using the equation of the membership function, the fuzzy membership function for each rule is simulated using the MATLAB code ch8_ex1.m (see Figure 8.13). As we know, any fuzzy rule can be defined as follows: IF x1 is A AND x2 is B THEN y1 is C, where the part before THEN is known as the antecedent and the part after is known as the consequent. The antecedent in the above rule are combined using the fuzzy logic AND operator (T-norm).

(b) The outputs of the antecedent part of rule 1, rule 2, and rule 3 are 0.8, 0.5, and 0.05, respectively, and are represented by dotted horizontal lines in Figure 8.13. The fuzzy output of each rule is obtained using a fuzzy implication function known as Larsen algebraic product, e.g., for rule 1, 0.8^*y_1. To get a single fuzzy output, the outputs of the three rules are combined using the equation $\max(0.5y_2, \max(0.8y_1, 0.05y_3))$ (see the bottom plot of Figure 8.13).

(c) To get a crisp output, the output obtained after aggregation operation is defuzzified. Figure 8.14 shows (with "dot") the defuzzified outputs corresponding to these methods.

EXAMPLE 8.2

An aircraft can become unstable during a flight if there is a loss of control surface effectiveness due to damaged control surfaces. To make the aircraft resistant to a loss of control surface effectiveness, there should be some mechanism to estimate parameters representing the amount of loss of control surface effectiveness. An extended Kalman filter (EKF) can be used to estimate the parameters of a control distribution matrix (B) as augmented states of the system. In this method, the true parameters, estimated parameters, and feedback gain (computed using linear quadratic regulator [LQR] algorithm under healthy conditions) are used to compute the feedback gain using a pseudo-inverse technique to reconfigure the impaired aircraft [27–29]. The limitation

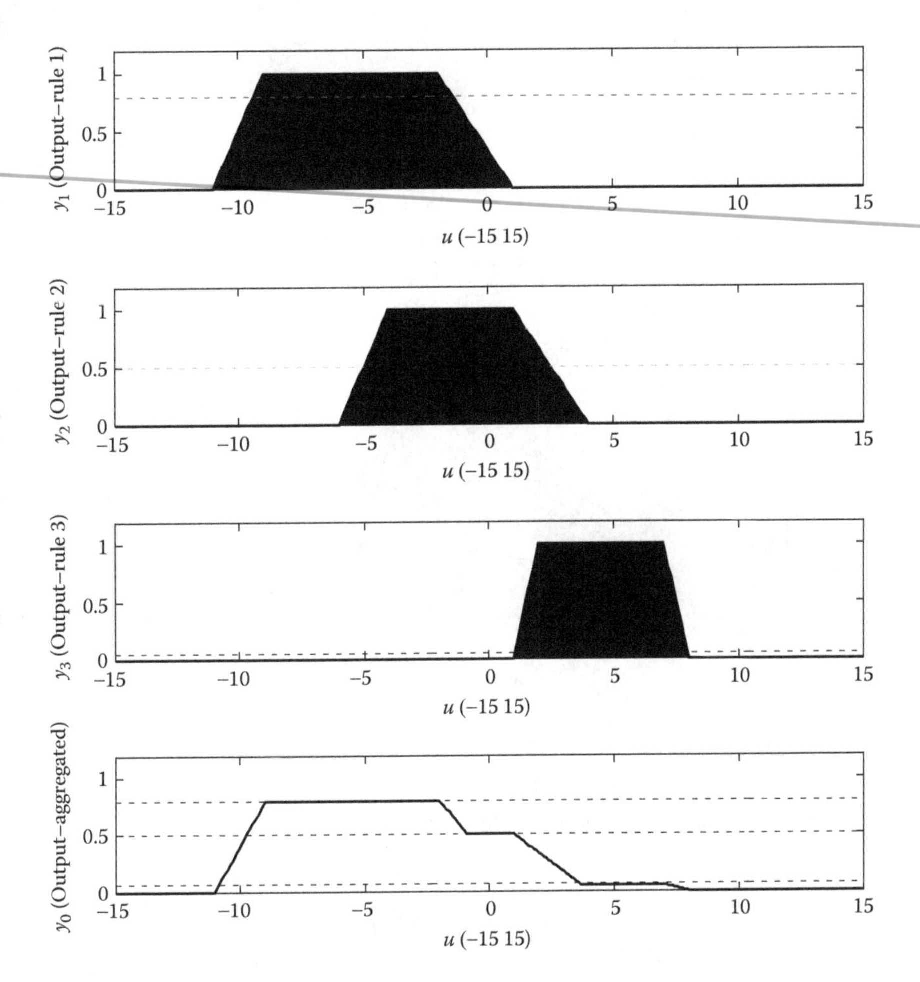

FIGURE 8.13
Membership functions for Example 8.1.

of this approach is that the model of the system must be accurately known, which is not always the case. Sometimes models are complexly mathematically formulated and may be highly nonlinear and time-varying. In such a situation, an adaptive neuro fuzzy inference system (ANFIS) can be employed [30], wherein fuzzy logic is used to predict or estimate the system states, used in control law to achieve the desired system response. In this scheme, sensor data at the present instant are smoothed by averaging the previous samples over a selected window length using a sliding window technique. ANFIS can be tested offline using inputs as errors between the nominal states and faulty states of the aircraft; it is then used to estimate the factor of effectiveness and hence the elements of control distribution matrix under faulty conditions. The reconfiguration is then carried out by computing new feedback gain using pseudo-inverse technique.

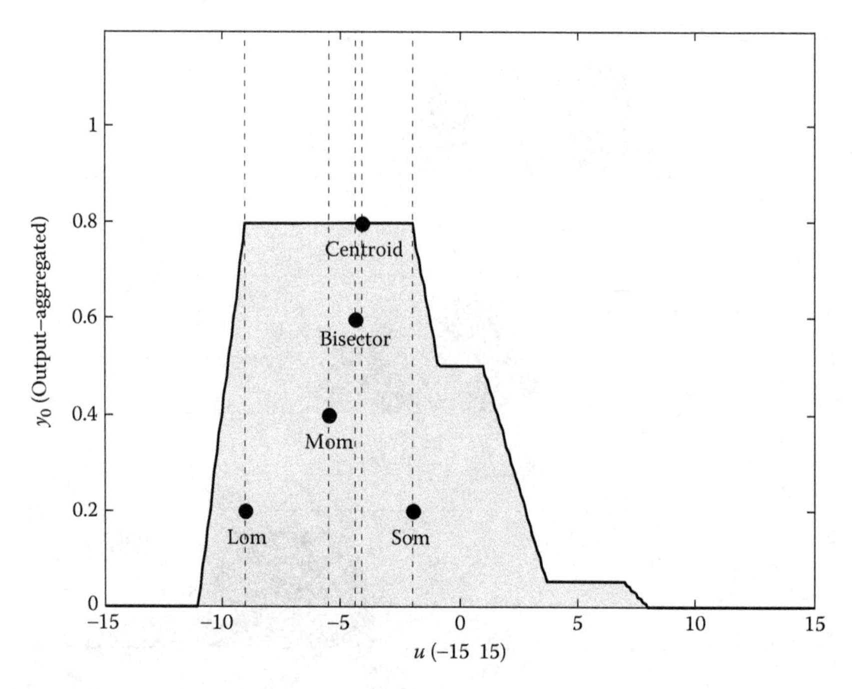

FIGURE 8.14
Defuzzified outputs (dots) for Example 8.2.

We can simulate the longitudinal dynamics of a Delta-4 aircraft [31] (Tables 8.3 through 8.5) using the following state-space matrices:

$$A = \begin{bmatrix} -0.033 & 0.0001 & 0.0 & -9.81 \\ 0.168 & -0.367 & 260 & 0.0 \\ 0.005 & -0.0064 & -0.55 & 0.0 \\ 0.0 & 0.0 & 1.0 & 0.0 \end{bmatrix}; \quad B = \begin{bmatrix} 0.45 \\ -5.18 \\ -0.91 \\ 0.00 \end{bmatrix}$$

$$\dot{x} = Ax + Bu_c = Ax + B(-Kx) = (A - BK)x$$

Here, the state vector is given by $x = [u, w, q, \theta]^T$ and $K = [0.0887\ \ 0.0046\ \ -0.8968\ \ -2.106]$ is the feedback gain computed using the LQR method. We can introduce the control surface fault and develop solutions using the EKF and ANFIS methods (using MATLAB).

Solution 8.2

The control surface fault is introduced by multiplying "B" vector with factor 0.8, i.e., a 20% loss of the control surface effectiveness. Run the ch8_ex2 (.m program from the directory in Section 8.5, the MATLAB SW provided). Figure 8.15

TABLE 8.3

Specifications for Delta-4 Aircraft

Wing area (m²)		576	
Aspect ratio		7.75	
Chord $\bar{c}$ (m)		9.17	
Total related thrust (kN)		730	
c.g.		0.3 $\bar{c}$	
Pilot's location	l_{x_p} (m)	25	
(relative to c.g.)	l_{z_p} (m)	+2.5	
Weight (kg)		Approach: 264,000	All other flight conditions: 300,000
Inertia (kg m²)	I_{xx}	2.6×10^7	3.77×10^7
	I_{yy}	4.25×10^7	4.31×10^7
	I_{zz}	6.37×10^7	7.62×10^7
	I_{xz}	3.4×10^6	3.35×10^6

Source: McLean, D. 1990. *Automatic flight control systems*. London: Prentice Hall International.

TABLE 8.4

Selected Flight Conditions

	Flight Conditions			
Parameter	**1**	**2**	**3**	**4**
Height (m)	Sea level	6100	6100	12200
Mach no.	0.22	0.6	0.8	0.875
U_0(m/s)	75	190	253	260
$\bar{q}$ (Nm²)	3460	11730	20900	10100
α_0 (degrees)	+2.7	+2.2	+0.1	+4.9
γ_0 (degrees)	0	0	0	0

Source: McLean, D. 1990. *Automatic flight control systems*. London: Prentice Hall International. With permission.

shows the actual and estimated control matrix elements obtained through EKF and ANFIS schemes using noisy measured data. For offline training in an ANFIS scheme, the state error is computed using the filtered measured data of the impaired aircraft and the nominal states. Filtering is carried out using the sliding window technique with a window length of 20 sampling intervals. The estimated parameters are close to the true values for both schemes. The delay in estimation is noted in both schemes, but the estimated values settle to the true values somewhat earlier in the ANFIS scheme. Figure 8.16 shows the error (a true state-estimated state) for cases with and without reconfiguration. We can interpret that

TABLE 8.5

Stability Derivatives (Longitudinal Axis)

Stability Derivatives	Flight Conditions			
	1	2	3	4
X_u	−0.02	−0.003	−0.02	−0.033
X_w	0.1	0.04	0.02	0.0
X_{δ_E}	0.14	0.26	0.0	0.45
$X_{\delta_{th}}$	0.17×10^{-4}	0.45×10^{-4}	0.45×10^{-4}	0.45×10^{-4}
Z_u	−0.23	−0.08	−0.01	−0.17
Z_w	−0.634	−0.618	−0.925	−0.387
$Z_{\delta E}$	−2.9	−6.83	−9.51	−5.18
$Z_{\delta_{th}}$	0.06×10^{-5}	0.05×10^{-5}	0.05×10^{-5}	0.05×10^{-5}
M_u	-2.55×10^{-5}	3.28×10^{-4}	14.21×10^{-4}	54.79×10^{-4}
M_w	−0.005	−0.007	−0.0011	−0.006
$M_{\dot{w}}$	−0.003	−0.001	−0.001	−0.0005
M_q	−0.61	−0.77	−1.02	−0.55
M_{δ_E}	−0.64	−1.25	−1.51	−0.91
$M_{\delta_{th}}$	1.44×10^{-5}	1.42×10^{-5}	1.42×10^{-5}	1.42×10^{-5}

the reconfigured aircraft states converge to the true states as desired and that the error is less compared to the error under no reconfiguration.

EXAMPLE 8.3

Apply the concept of fuzzy logic–based Kalman filter (FKF) of Section 7.3 for measurement fusion (that is the data level fusion, see Section 3.1) and compare the results with KF fusion (KFF), FKF fusion (FKFF), and state-vector fusion (SVF), using the same numerical simulation.

SOLUTION 8.3

The concept of FKF is extended to measurement fusion, and the algorithm is called FKF measurement fusion (FKMF). The basic equations of FKMF are the same as those of FKF with the following additional changes:

- Measurement noise covariance matrix, $R = \begin{bmatrix} R_1 & 0 & 0 & 0 \\ 0 & R_2 & 0 & 0 \\ 0 & 0 & . & 0 \\ 0 & 0 & 0 & R_m \end{bmatrix}$,

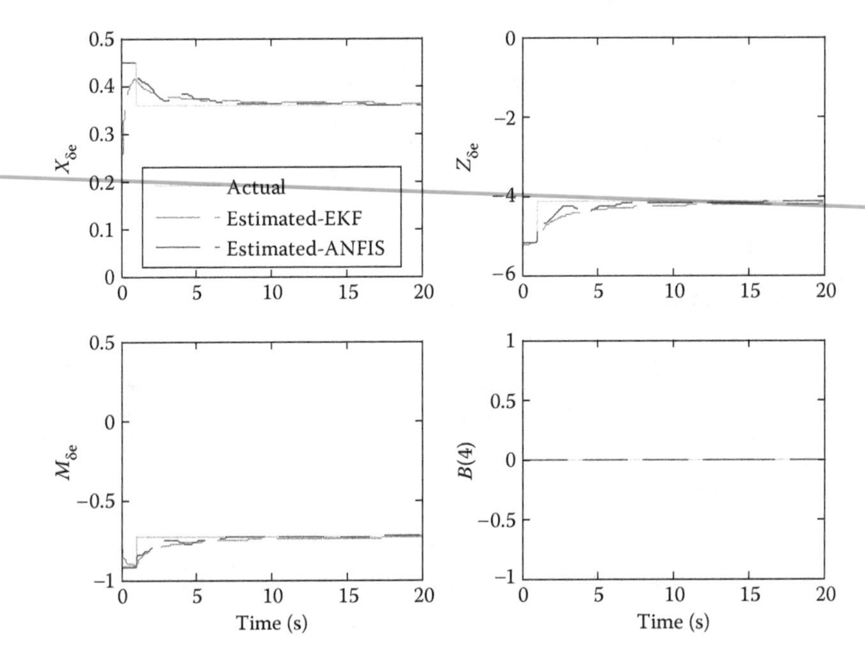

FIGURE 8.15
Estimated and true values of the control distribution matrix.

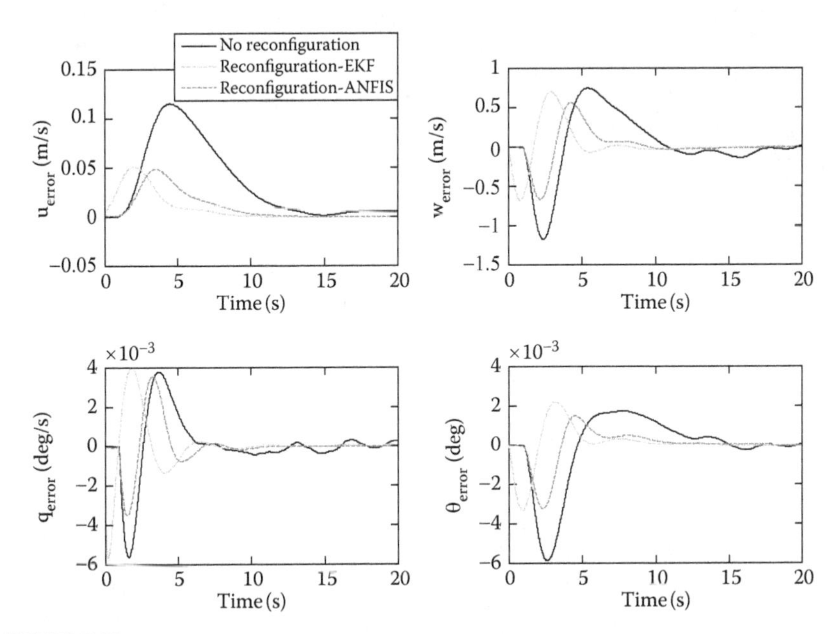

FIGURE 8.16
Comparison of state errors.

where R_m is measurement noise covariance matrix of m-th sensor. In the present case, the number of sensors $m = 2$.

- For $m = 2$, the observation matrix, $H = \begin{bmatrix} H_1 \\ H_2 \end{bmatrix}$ and concatenated

  sensors measurements $Z_m = \begin{bmatrix} Z_{m_1} \\ Z_{m_2} \end{bmatrix}$

The FLP in Equation 7.2 is computed sequentially and separately for sensor 1 and sensor 2 and then concatenated as a final fused FLP:

$$B^m(k+1) = \left[b_x^m\ (k+1)\ \ b_y^m\ (k+1) \right],\quad m = 1 \text{ to } 2$$

The final FLP is given by

$$B(k+1) = \left[b_x^1(k+1)\ \ b_y^1(k+1)\ \ b_x^2(k+1)\ \ b_y^2(k+1) \right]^T$$

Run the ch8_ex3 (.m program from the directory in Section 8.5, the MATLAB SW provided). Figure 8.17 shows the comparison of true, measured (sensor 1 and sensor 2), and estimated x–y trajectory of the target using fusion algorithms—SVF, KFF, FKFF, and FKMF. Figures 8.18 to 8.20 show the comparison of the root sum square position, velocity, and acceleration errors for these algorithms. It is clear that there are fewer overall errors for FKMF compared to other fusion algorithms. Since only position measurements are used in these algorithms, the merit of FKMF over other algorithms is not clearly visible in the case of errors in target accelerations.

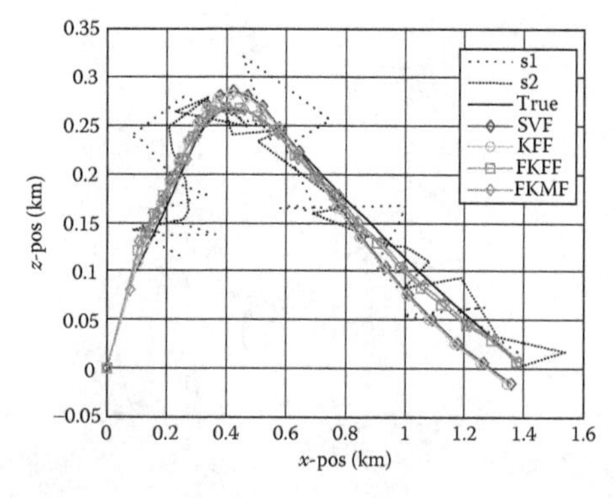

FIGURE 8.17
True, measured, and estimated x–y positions.

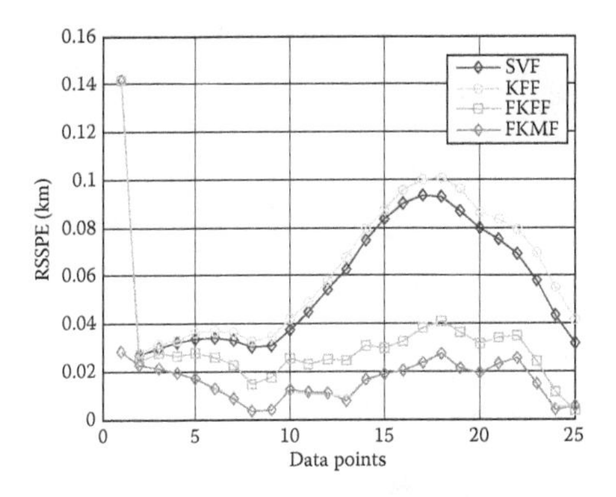

FIGURE 8.18
Root sum square position errors for four fusion approaches: state-vector fusion, Kalman filter fusion, fuzzy logic–based Kalman filter fusion, and fuzzy logic–based Kalman filter measurement fusion.

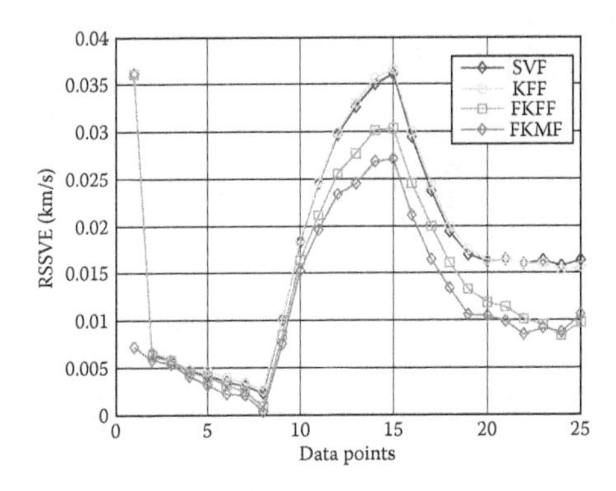

FIGURE 8.19
Root sum square velocity errors for four fusion approaches: state-vector fusion, Kalman filter fusion, fuzzy logic–based Kalman filter fusion, and fuzzy logic–based Kalman filter measurement fusion.

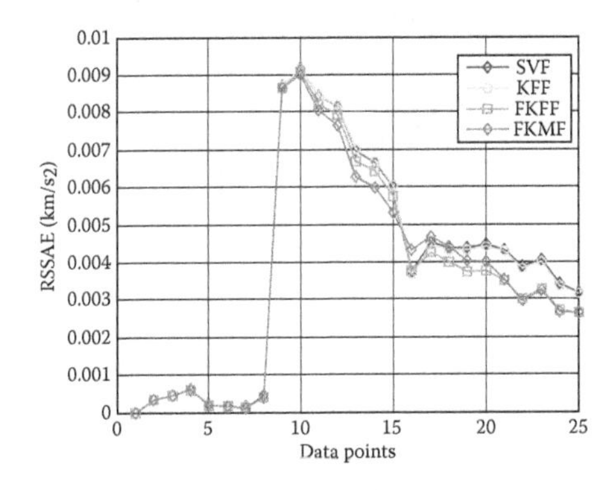

FIGURE 8.20
Root sum square acceleration errors for four fusion approaches: state-vector fusion, Kalman filter fusion, fuzzy logic–based Kalman filter fusion, and fuzzy logic–based Kalman filter measurement fusion.

Epilogue

Fuzzy logic and systems (FL/S) have been prevalent for four decades and have applications in many engineering systems and household appliances. The FL/S discussed in this book are called Type-1 FL/S—Type-0 is the conventional deterministic systems (like defuzzified output of the fuzzy inference system [FIS]/fuzzy control system) and Type-2 FL/S are very recent ideas, and are meant to handle uncertainty in rule-based systems and related developments [32]. Type-1 fuzzy sets are certain, whereas Type-2 FL/S can handle rule uncertainties and even measurement uncertainties. In probability theory, the variance provides a measure of dispersion about the mean [33]. Type-2 FL/S provides a similar measure of dispersion, and linguistic and random uncertainties have a definite place in Type-2 FL/S. This new dispersion in Type-2 FL/S can be thought of as being related to a linguistic confidence interval. A Type-2 FL/S has more degrees of freedom in design than does a Type-1 FL/S and is expected to give a better performance than a Type-1 FL/S. Therefore, new theoretical results and new applications, e.g., data fusion and decision fusion based on Type-2 FL/S, are required.

Exercises

II.1. The fuzzy logic membership function (FLMF) concept is used in KF in Section 7.3 to obtain fuzzy logic (FL)-based KF, via Equations 7.1 to 7.4. This mainly applies to residuals in the KF. Can you think of another approach to using the FLMF concept in KF?

II.2. What is the real reason for checking the consistency of filter performance?

II.3. In a crisp representation of the numbers and their sums, differences and equality are governed by fixed definitions. What are the rules for fuzzy members?

II.4. Is it possible to consider a very old man a young boy, and a very young boy an old man? If so, how?

II.5. Why are a domain expert's knowledge and experience that are needed to build FL-based systems?

II.6. Why is FL very suitable for decision making and in decision support systems (DM/DSS)?

II.7. Can crisp logic be considered a special case of FL? If so, why?

II.8. Why is defuzzification required in FL/S?

II.9. What is the type in an inner FL system and an output FL system?

II.10. How can FL/S be used for system identification or estimation?

II.11. In Table 6.20, why does the norm of the relational matrices decrease from the standard intersection to the drastic intersection, while the norm increases from the standard union to the drastic union?

II.12. What are the most common forms of fuzzy membership functions?

II.13. What are the two methods of obtaining fuzzy rules (rule bases)?

II.14. What are the features of the two methods in Exercise II.13?

II.15. How can system identification and parameter estimation methods be useful in Exercise II.14?

II.16. When does the conventional control system start to become an intelligent system?

II.17. What are other types of logic in the decision making process?

II.18. What are the features and operators/rules used in these other logics?

II.19. What is the difference between vagueness and uncertainty?

II.20. What is a feedback fuzzy system (FFS)?

II.21. What is the use of FFS?

II.22. Draw an overall block diagram with appropriate elements for modeling a nonlinear system (NLS) with FFS. (Hint: See Figure A.1.)

II.23. What is the meaning of "noise" in the case of the NLS/FFS modeling exercise?

II.24. In what way can the FFS be adjusted for modeling?

II.25. What does "pruning the fuzzy rules" mean and why is it necessary?

II.26. Why do the min and max operators in FL/S have more than one "internal" definition, unlike crisp logic, AND, OR, and so on where the results are always unique?

II.27. What is the relationship of decision fusion to data fusion?

II.28. Draw a block diagram for centralized fusion architecture in the case of decision fusion from the classifier's point of view.

II.29. How do the decision fusion methods of voting, Bayesian, and Dempster–Shafer compare?

II.30. How do static and dynamic decision making and fusion work?

II.31. It is often said that sensor/decision fusion results are better than would have been possible with only an individual sensor. What is the meaning of "better" here?

II.32. Is the data/decision fusion process the process of "reduction?" How and why?

II.33. Can fuzzy set theory be considered a method of decision fusion?

II.34. Can you visualize the fusion of FL, artificial neural networks (ANNs), and genetic algorithms in the context of data fusion itself?

II.35. Changes in the target aspect with respect to the sensor (such as radar or radio frequency-based seekers used in homing missiles to intercept air-breathing targets) can cause the apparent center of sensor reflections to move from one point to another. This causes noisy or jittered angular measurements. This form of noise is called angle noise, angle scintillations, angle fluctuations, or glint noise [34,35]. Analysis of the probability density function of glint noise reveals that it has a non-Gaussian, heavy-tailed (due to the spiky pattern of glint noise) distribution. The glint noise can be treated as a mixture of Gaussian noise with moderate variance and a heavy-tailed Laplacian noise with large variance:

$$p_{glint}(w) = (1-\varepsilon)p_G(w) + \varepsilon p_L(w)$$

Here, ε is the "glint probability" and subscripts G and L stand for Gaussian and Laplacian distributions. The probability density functions (pdf's) of these noises with zero mean are

$$p_L(w) \triangleq \frac{1}{2\eta} e^{-|w|/\eta}$$

$$p_G(w) \triangleq \frac{1}{\sqrt{2\pi}\sigma_G} e^{-w^2/2\sigma_G^2}$$

Here, σ_G^2 is the variance of Gaussian noise and η is related to Laplacian variance σ_L^2 as $2\eta^2 = \sigma_L^2$. Within the context of target-track estimation using KF, how would you represent or model glint noise and use it in KF with FL or fuzzy rules? Give the important steps.

II.36. ANNs are biologically inspired, with abilities to learn and adapt (see Figure A.6) due to (1) perception, (2) multilayer feed-forward networks, (3) recurrent networks, and (4) a radial basis function (RBF) neural network. The RBF network is characterized by radial basis functions. An RBF network has good local interpolation, global generalization ability, and is more biologically plausible because of its finite response. Also, it can initially have zero neurons in a hidden layer, and online neurons can be created and pruned. One such RBF network is known as minimum resource allocation network (MRAN) [35,36]. MRAN has been used in several applications varying from function approximation to nonlinear system identification and its application in flight control. A topology of RBF neural networks with Gaussian functions (as a type of radial basis function), which can be easily used for data fusion (Exercise III.7), is shown below [36,37]. The RBF neural network response at each of its output node is given by

$$y_k = b_k + \sum_{j=1}^{n} \alpha_{kj}\phi_j(U);\; k = 1,\ldots,p$$

$$\phi_j(U) = \exp\left(\frac{-\|U - \mu_j\|}{\sigma_j^2}\right)$$

where $U = [u_1,\ldots,u_m]$ is the input vector to network and $\|.\|$ denotes the Euclidean norm. Simulate the following nonlinear function having six Gaussian functions using MRAN [35,36]:

$$y(x) = \exp\left[-\frac{(x_1-0.3)^2+(x_2-0.2)^2}{0.01}\right] + \exp\left[-\frac{(x_1-0.7)^2+(x_2-0.2)^2}{0.01}\right]$$
$$+\exp\left[-\frac{(x_1-0.3)^2+(x_2-0.8)^2}{0.01}\right] + \exp\left[-\frac{(x_1-0.7)^2+(x_2-0.8)^2}{0.01}\right]$$
$$+\exp\left[-\frac{(x_1-0.1)^2+(x_2-0.5)^2}{0.02}\right] + \exp\left[-\frac{(x_1-0.9)^2+(x_2-0.5)^2}{0.02}\right]$$

II.37. Why does the FL–based KF (FLKF) generally not meet the filter-consistency check as accurately as the standard KF?

II.38. Why is decision making in KF called soft decision making?

II.39. What type of decision making is used in FLKF?

II.40. What kind of fusion processes can be used for data that are not homogeneous quantities?

II.41. What is the major difference between the forward- and backward-chaining inference rules?

II.42. Is it possible to combine FL and the Bayesian method for data fusion or decision fusion?

II.43. Can you visualize a decision fusion strategy?

II.44. Is it possible to combine FL, a genetic algorithm, and ANN to form a decision fusion system?

II.45. Is it possible to use FL for image fusion?

II.46. What are the strong and weak aspects of FL for fusion?

II.47. In what way can the combination of FL, ANN, and KF be used to track a maneuvering target?

II.48. How one can apply the FL, FMF, or FIS process to image processing or fusion?

II.49. Is there any mapping involved in the use of FL/FIS to design a fusion system based on FL? Is this mapping linear or nonlinear?

II.50. What does FL represent in the case of noisy sensors used for tracking/data fusion, since the noise is already represented by the Gaussian probability density function (pdf) with often a zero mean and a given standard deviation?

II.51. In Chapter 6, while discussing FL operators and their interconnectivity, we came across conjunction and disjunction Cartesian products. Can you discuss these in terms of the fusion process?

II.52. How can FL be used for data classification? What are the different clustering techniques available for classification?

References

1. Zadeh, L. A. 1965. Fuzzy sets. *Inf Control* 8:338–53.
2. King, R. E. 1999. *Computational intelligence in control engineering*. New York: Marcel Dekker.
3. Passino, K. M., and S. Yurkovich. 1998. *Fuzzy control*. Menlo Park, CA: Addison-Wesley Longman.
4. Kosko, B. 1997. *Fuzzy engineering*. Upper Saddle River, NJ: Prentice Hall.
5. Kashyap, S. K., J. R. Raol, and A. V. Patel. 2008. Evaluation of fuzzy implications and intuitive criteria of GMP and GMT. In *Foundations of generic optimization, Vol. 2: Applications of fuzzy control, genetic algorithms and neural networks (Mathematical modelling: Theory and applications)*, pp. 313–386, ed. R. Lowen and A. Verschoren. New York: Springer.

6. Kashyap, S. K., and J. R. Raol. 2007. Interpretation and unification of fuzzy set operations and implications. *J Syst Soc India* 16(1):26–33.
7. Kashyap, S. K. 2008. Decision fusion using fuzzy logic. Doctoral thesis, University of Mysore, Mysore, India.
8. Li-Xin, W. 1994. *Adaptive fuzzy systems and control, design and stability analysis*. Englewood Cliffs, NJ: Prentice Hall.
9. Raol, J. R., and J. Singh. 2009. *Flight mechanics modeling and analysis*. Boca Raton, FL: CRC Press.
10. Cho, S. B., and J. H. Kim. 1995. Multiple network fusion using fuzzy logic. *IEEE Trans Neural Netw* 6(2):497–501.
11. Lau, C. W., B. Ma, H. M. Meng, Y. S. Moon, and Y. Yam. 2004. Fuzzy logic decision fusion in a multimodal biometric system. http://www.se.cuhk.edu.hk/hccl/publications/pub/lau_icslp2004.pdf (accessed December 2008).
12. Wik, M. W. 2006. A three processes effects model based on the meaning of information. 11th ICCRTS Coalition Command and Control in the Networked Era, September 26–28, 2006. http://www.dodccrp.org/events/11th_ICCRTS/html/presentations/018.pdf (accessed December 2008).
13. Crowley, J. L., and Y. Demazeu. 1993. Principles and techniques of sensor data fusion. LIFIA (IMAG), France. http://www-prima.imag.fr/Prima/Homepages/jlc/papers/SigProc-Fusion.pdf (accessed July 2008).
14. Kant, E. 1958. *Critique of pure reason*, translated by N. Kemp Smith. New York: Random House (original work published in 1781).
15 Abidi, M. A., and R. C. Gonzalez, eds. 1992. *Data fusion in robotics and machine intelligence*. USA: Academic Press.
16. Hall, D. L. 1992. *Mathematical techniques in multi-sensor data fusion*. Norwood, MA: Artech House.
17. Chao, W., Q. Jishuang, and L. Zhi. Data fusion, the core technology for future on-board data processing system. Institute of Remote Sensing Applications, Chinese Academy of Sciences, Beijing, China.
18. Klein, L. A. 2004. *Sensor and data fusion: A tool for information assessment and decision making*. Washington, DC: SPIE Press.
19. Egfin Nirmala, D., V. Vaidehi, S. Indira Gandhi. 2005. Data fusion using fuzzy logic for multi target tracking. In *Proceedings of International Radar Symposium* India, pp. 75–80.
20. Kashyap, S. K., and J. R. Raol. 2008. Fuzzy logic applications for filtering and fusion for target tracking. *Def Sci J* 58:120–35.
21. Ivansson, J. 2002. Situation assessment in a stochastic environment using Bayesian networks. Master's thesis, Division of Automatic Control, Department of Electrical Engineering, Linkoping University.
22. Narayana Rao, P., S. K. Kashyap, and G. Girija. 2008. Situation assessment in air combat: A fuzzy-Bayesian hybrid approach. In *Proceedings of the International Conference on Aerospace Science and Technology*, NAL, Bangalore, India.
23. Endsley, M. R. 1993. A survey of situation awareness requirements in air-to-air combat fighters. *Int J Aviat Psychol* 3:157–68.
24. Neapolitan, R. E. 1990. *Probabilistic reasoning in expert systems*. New York: John Wiley & Sons.

25. Pearl, J. 1988. *Probabilistic reasoning in intelligent systems: Networks of plausible inference.* San Francisco, CA: Morgan Kaufmann Publishers, Inc.
26. HUGIN Expert, http//: www.HUGIN.dk (accessed December 2008).
27. Hajiyev, Ch. M., and F. Caliskan. 2001. Integrated sensor/actuator FDI and reconfigurable control for fault tolerant flight control system design. *Aeronaut J* 105:525–33.
28. Hajiyev, C., and F. Caliskan. 2003. *Fault diagnosis and reconfiguration in flight control systems.* Boston: Kluwer Academic Publishers.
29. Eva Wu, N., Y. Zhang, and K. Zhou. 2000. Detection, estimation, and accommodation of loss of control effectiveness. *Int J Adapt Control Process* 14:775–95.
30. Savanur, S. R., S. K. Kashyap, and J. R. Raol. 2008. Adaptive Neuro-Fuzzy based control surface fault detection and reconfiguration. In *Proceedings of the International Conference on Aerospace Science and Technology*, Bangalore, India.
31. McLean, D. 1990. *Automatic flight control systems.* London: Prentice Hall International.
32. Mendel, J. M. Why we need Type-2 fuzzy logic system. http://www.informit.com/articles/article.aspx?p=21312-32k (accessed January 2009).
33. Raol, J. R., G. Girija, and J. Singh. 2004. *Modelling and parameter estimation of dynamic systems, IEE Control Engineering. Series Book.* Vol. 65. London: IEE.
34. Hewer, G. A., R. D. Martin, and J. Zeh. 1987. Robust preprocessing for Kalman filtering of glint noise. *IEEE Trans Aerosp Electron Syst* 23:120–8.
35. Wu, W. 1993. Target tracking with glint noise. *IEEE Trans Aerosp Electron Syst* 29:174–85.
36. Sundararajan, N., P. Saratchandran, and Y. W. Lu. 2001. *Radial basis function neural networks with sequential learning—MRAN and its applications, Progress in neural processing 11.* Singapore: World Scientific Publishing.
37. Lu, Y., N. Sundararajan, and P. Saratchandran. 1997. A sequential learning scheme for function approximation using minimal radial basis function neural networks. *Neural Comput* 9:461–78.

Part III

Pixel- and Feature-Level Image Fusion

J. R. Raol and V. P. S. Naidu

9

Introduction

We work with visual images every day of our lives. We recognize static and moving objects naturally and promptly. We teach our children by showing them line diagrams, pictures, and images. We give these images names and ask our children to recognize the objects and give their names. Our daily lives are a world of pictures, 2D and 3D images, and sounds. Our biological neural networks instantaneously recognize these images and, when required, quickly perform fusion of the images. One of our basic needs in evolution has been the recognition, registration, and fusion of the images we encounter daily, even every second or millisecond. Our imaging system is based on the ability to see light-illuminated objects. However, in the technological world many other methods of imaging are available: laser, nuclear magnetic resonance, and infrared (IR). We live in an environment of images and sounds. We have an uncanny ability to recognize and relate human faces—in far away places, we often say that the face of a person is similar to the face of someone whom we have seen earlier, and we can even see the faces and features of persons and objects (including animals, birds, and insects) in the clouds, and profiles in the silhouettes of mountains!

Image fusion is the process of combining information from two or more sensed or acquired images into a single composite image that is more informative and more suitable for visual perception and computer or visual processing. The objective is to reduce uncertainty, minimize redundancy in the output, and maximize relevant information pertaining to an application or a task. For example, if a visual image is fused with a thermal image, a target that is warmer or colder than its background can be easily identified, even when its color and spatial details are similar to those of its background. In image fusion, the image data appear in the form of arrays of numbers, which represent brightness (intensity), color, temperature, distance, and other scene properties. These data could be 2D or 3D. The 3D data are essentially the volumetric images and/or video sequences in the form of spatial-temporal volumes. The approaches for image fusion are (1) hierarchical image decomposition, (2) neural networks in fusion of visible and IR images, (3) laser detection and ranging (LADAR) and passive IR images for target segmentation, (4) discrete wavelet transforms (DWTs), (5) principal component analysis (PCA), and (6) principal components substitution (PCS).

Image fusion can be categorized as of the following [1–4]:

1. Low-level: Pixel-level fusion is done in the spatial or transform domain (TD). TD algorithms globally create the fused image. The algorithms that work in the spatial domain have the ability to focus on desired image areas, limiting changes in other areas. Multiresolution analysis can use filters with increasing spatial extent to generate a sequence of images (pyramids) from each image, separating the information observed at different resolutions. Then, at each position in the transformed image, the value in the pyramid showing the highest saliency is taken. An inverse transform of the composite image is used to create the fused image. Intensity gradients can be used as saliency measures. Dual-tree complex wavelet transform can outperform most other grayscale image fusion methods.
2. Mid-level: Feature-level fusion algorithms typically segment the images into regions and fuse the regions using their various properties. The multiscale edge representations using wavelet transform can also be used for image fusion.
3. High-level: Symbol-level fusion algorithms combine image descriptions in the form of relational graphs. Since decision fusion and symbol-level fusion are very closely related, fuzzy logic theory can also be used for this level of fusion.

In pixel-level fusion, there are three broad approaches: (1) color transformation (CT), (2) statistical and numerical (SN), and (3) multiresolution method (MRM). In the CT method, there is an advantage in the possibility of presenting the data in different color channels, including hue (H, dominant wavelength), intensity (I), and saturation (S; called HIS approach). The red, green, and blue (RGB) image is divided into spatial (I) and spectral (H and S) information. There are two kinds of possible transformations: (1) transform the three images (channels) presented in RGB into I, H, and S directly; and (2) separate the color channels into average brightness representing the surface roughness (intensity), dominant wavelength (hue), and purity (saturation). The HIS approach has become standard procedure in image analysis. It has the ability to enhance the color of highly correlated data and can fuse disparate data.

In the SN method, the PCA utilizes all channels' data as input and combines the multisensor data into one major part. The PCS regards one channel's data as a major part at first, and then substitutes it with the other channel's data if it has a better fused result. These methods do not preserve the spectral characteristics of the source image well, since the fused image features will be altered to a great extent.

The MRM method is preferable (as are the pyramid-based and wavelet-based methods). The Gaussian pyramid and enhanced Laplacian (ELP) decomposition can only decompose and interpolate images with a decimation factor of 2, resulting in many restrictions. Laplacian pyramids correspond to a band-pass representation, while Gaussian pyramids correspond to a low-pass representation. Pyramids are an example of a wavelet representation of an image and correspond to the decomposition of an image into spatial/frequency bands. The wavelets separate information into frequency bands; in the case of images we present high-frequency information, such as texture and so forth, in a finely sampled grid [5]. Coarse information can be represented in a coarser grid where the lower sampling rate is acceptable. Coarse features can be detected in the coarse grid using a small template size (this is referred to as *multiresolution* or *multiscale resolution*). Wavelets correspond to a mixture of spatial and frequency domains; the position information at each level is known to the accuracy of that grid resolution and contributing frequencies are bandwidth limited at each grid resolution. Since wavelet coefficients that have large absolute values contain information about the salient features of images, such as edges and lines, a good rule is to take the maximum absolute values of the corresponding wavelet coefficients. The maximum absolute value, within a window, is used as an activity measure of the central pixel of the window. In the wavelet fusion method, the transformed images are combined in the transform domain using a defined fusion rule and then transformed back to the spatial domain to obtain the fused image. The wavelet package (WP)–based method uses a WP to further decompose multitemporal images at low- or high-frequency parts. At the same level, it utilizes a threshold and weight algorithm to fuse the corresponding low-frequency parts, and at the same time applies a Lis high-pass filter to fuse the high frequency parts. The fused image is then restored by inverse DWT process. The WT package fusion method [6] decomposes the image recursively at first, which means decomposing the low-frequency part of the previous level. Let the grayscale image after decomposition by WT be Io. Then

Io = base image (approximations) + vertical + horizontal + diagonal details

The base image is decomposed at the second level and so on, thus, the nth decomposition will comprise $3n + 1$ subimage sequences. The $3n + 1$ subimage sequences of the nth level will be fused by applying different rules to low- and high-frequency parts. Then the inverse WT is taken to restore the fused image. The WP method decomposes every part of the previous level recursively, to either low frequency or high frequency; thus there will be 4-power n subimage sequences at the nth decomposition, which provides the possibility of utilizing much more flexible fusion rules to acquire a better-quality fused result. The general WT method always fuses the low-frequency part of the decomposed images. The WT package-based method decomposes both the low-frequency and high-frequency parts recursively at each level, and then fuses the different images' corresponding parts at the same level: the low-frequency parts by threshold and weight and high-frequency parts by high-pass filtering. The low-frequency subimages are fused using the following algorithm [5,6]:

```
if abs (IM1(i,j)-IM2(i,j)) < threshold1
        IM1(i,j) = w1 * IM1(i,j) + w2 * IM2(i,j);
elseif abs(IM1(i,j)-IM2(i,j)) ≥ threshold1 AND abs(IM1(i,j)
        -IM2(I,j)) < threshold2
        IM1(i,j) = w3 * IM1(i,j) + w4 * IM2(i,j);
If abs (IM1(i,j)-IM2(i,j)) ≥ threshold2
        IM1(i,j) = IM2(i,j);
end
```

Threshold1 and threshold2 are thresholds of pixels' values, and *w*s are fusion weights among the pairs of pixel sequences. Each pair of *w*s would satisfy $w1 + w2 = 1$. Finally, all of the fused subimage sequences are restored recursively. The fusion rules that can be used are (1) maximum selection—use the coefficients in each subband with the largest magnitude; (2) weighted average—normalized correlation between two images' subband over a small local area; and (3) window-based verification (WBV) scheme—a binary decision map to choose between each pair of coefficients using a majority filter.

In Part III, we will discuss some image registration methods, image-level fusion methods, and performance evaluation aspects.

10

Pixel- and Feature-Level Image Fusion Concepts and Algorithms

The data or information fusion process can take place at different levels, such as the signal, pixel, feature, and symbolic levels. Signal-level processing and data fusion have been discussed in Part I. In Part II, we discussed some aspects of decision fusion based on fuzzy logic. In the pixel-level fusion process, a composite image is built from several input images based on their respective pixels (picture elements). One of the applications is the fusion of forward-looking infrared (FLIR) and low-light visible images (LLTV) obtained by an airborne sensor system to aid the pilot navigate in poor weather conditions and darkness. In pixel-level fusion, some basic aspects of the fusion result are as follows [7–10]: (1) the data fusion (DF) process should carry over all the useful and relevant information from the input images to the composite image, to the greatest extent possible; (2) the DF scheme should not introduce any additional inconsistencies not originally present in the input images, which would distract the observer or other subsequent processing stages; and (3) the DF process should be shift and rotational invariant. The fusion results should not depend on the location or orientation of an object from the input images. Additionally, there should be (1) temporal stability, that is, the gray level changes in the fused image sequence should only be caused by gray level changes in the input sequences, and not by the fusion process; and (2) temporal consistency, that is, gray level changes occurring in the input image sequences should be present in the fused image sequence without any delay or contrast change.

10.1 Image Registration

In the image registration process, two (or more) sensed images of the same scene are superimposed [1–4]. These images might have been taken at different times, from different viewpoints, or from different sensors. The main idea is to geometrically align the sensed image and the reference image. Certain differences in images of the same scene might arise due to differing imaging conditions. Image registration is required for several applications: (1) multispectral classification, (2) environmental monitoring, (3) change detection, (4) image mosaic, (5) weather forecasting, (6) creating super-resolution images, (7) integrating information into geographical information systems, (8) combining computer tomography and nuclear magnetic resonance (NMR) data, (9) map updating (cartography), (10) target localization, and (11) automatic quality control. The image registration applications can be divided into four groups [3]:

1. Multiview analysis: Images of the same scene are acquired from different viewpoints in order to obtain a larger 2D or 3D representation of the observed scene. The examples are image mosaics of the surveyed area and shape recovery from stereo.
2. Multitemporal analysis: Images of the same scene are acquired at different times in order to find and evaluate changes in the scene that appeared in the time between the acquisition of the images. This includes monitoring global land usage, landscape planning, automatic change detection for security monitoring, motion tracking, monitoring of healing therapy, and monitoring tumor growth.
3. Multimodal analysis: Images of the same scene are acquired by different sensors in order to integrate the information obtained from the different sources to obtain a more complex and detailed scene representation. Examples include the fusion of information from sensors with different characteristics, such as panchromatic images, images offering a better spatial resolution, color/multispectral images with a better spectral resolution, radar images independent of cloud cover and solar illumination. Also included are combinations of sensors recording the anatomical body structure, such as magnetic resonance image (MRI), ultrasound or CT scans, and sensors monitoring functional and metabolic body activities, like positron emission tomography (PET), single photon emission computed tomography (SPECT), and magnetic resonance spectroscopy (MRS).
4. Scene to model registration: Images of a scene and a model of the scene are registered. The model could be maps or digital elevation models in a geographic information system (GIS), another scene with similar content, or an average specimen. The aim is to localize the acquired images in the scene or model and compare them. Some examples are the registration of aerial or satellite data onto maps or other GIS layers, target template matching with real-time images, automatic quality inspection, comparison of the

patient's image with digital anatomical atlases, and specimen classification.

Low-level fusion algorithms assume there is a correspondence between pixels in the input images. If the camera's intrinsic and extrinsic parameters (IEPs) do not change and only the environmental parameters change, then the acquired images will be spatially registered. When the IEPs are different, the registration of images is required. Mid-level fusion algorithms assume that the correspondence between the features in the images is known. High-level fusion algorithms require correspondence between image descriptions for comparison and fusion.

Although the most common method of image registration is to manually align the images (which is time consuming and inaccurate), there are two main approaches: (1) area-based matching (ABM), and (2) feature-based matching (FBM). In the registration method, it is important to find enough accurate control input pairs and a procedure to interpolate the image. Any image registration method should take into account the assumed type of geometric deformation between the images, radiometric deformations, noise corruption, required registration accuracy, and application-dependent data characteristics. The main registration methods follow these steps:

1. Feature detection: Salient objects such as regions, edges, contours, and corners are detected. These features are represented by their center of gravity (CG) and end points.
2. Feature matching: The correspondence between the observed image and the reference image is ascertained. Various feature descriptors and spatial relationships are used for this purpose.
3. Transform model estimation: The parameters of the mapping functions are determined based on the feature correspondence.
4. Image resampling and transformation: The image is transformed using mapping functions.

10.1.1 Area-Based Matching

The detected features in the two images (the reference image and the sensed image) can be matched using the image intensity values in their close neighborhoods, spatial distribution, or the symbolic representation for feature matching.

ABM is also called the *correlation-like* or template-matching method [3]. This method merges feature detection with matching and does not attempt to detect salient objects. Windows of a predetermined size are used for correspondence estimation. In this classical method, a small rectangular or circular window of pixels in the image is compared statistically with windows of the same size as the reference image. The centers of the matched windows are treated as control inputs. These are used to solve for mapping function parameters between the reference image and the sensed image. The cross-correlation methods use matching of the intensities directly, and do not bother with the structural analysis. The normalized cross-correlation (NCC) and least-squares (LS) techniques are also widely used for this purpose. NCC is based on the maximum value of the correlation coefficient between the reference image and the sensed image. LS is based on minimizing the differences in the gray values between the reference image and the sensed image. The area correlation in the spatial domain can be used to match the feature points. These points are extracted by Gabor wavelet decomposition.

10.1.1.1 Correlation Method

The correlation method is based on normalized correlation, defined as follows:

$$CC(i,j) = \frac{\Sigma\left(w - E(w)\right)\left(I(i,j) - E\left(I(i,j)\right)\right)}{\sqrt{\Sigma\left(w - E(w)\right)^2 \Sigma\left(I(i,j) - E\left(I(i,j)\right)\right)^2}} \tag{10.1}$$

The CC value is computed for window pairs from the sensed and reference images. The maximum of CC is sought. For subpixel accuracy, the interpolation of CC values is used. Correlation-based methods are still being used because of their ease in the hardware implementation and hence are useful in real-time applications. Feature points can be detected using a wavelet transform (WT) algorithm, and then the cross-correlation method can be used to match the detected points across the images. Often, ABM is used to find the similarity of the features in the reference and sensed images.

10.1.1.2 Fourier Method

The Fourier representation of the images in the frequency domain is used in this method. The Fourier method computes the cross-power spectrum of the sensed and reference images, seeking the location of the peak in its inverse. This method is very robust in the presence of the correlated noise and the frequency dependent noise. This ABM can be effectively implemented in the fast Fourier transform (FFT) domain. Some of the FFT can be used to achieve invariance to translation, rotation, and scale. These ABM algorithms are easy to implement because of their simple mathematical models.

10.1.1.3 Mutual Information Method

The mutual information (MI) method is a leading method in multimodal registration, especially in medical imaging. For example, a comparison of the anatomical and functional images of the body of the patient would lead to a useful diagnosis. Another useful application is the area of remote sensing (discussed in Section 2.4). The main idea is to maximize the MI. Here, the various possibilities include the use of joint entropy, the MI, and the normalized MI. In this approach, all of the image data and image intensities are used. An optimization method might be required for maximization of the MI.

10.1.2 Feature-Based Methods

FBMs use the image features (the salient structures in the image), derived by a feature extraction algorithm [1–3]. Usually, the features include edges, contours, surfaces, corners, line intersections, and points, for example: (1) significant regions such as forests, fields, lakes, and ponds; (2) significant lines in region boundaries, coastlines, roads, and rivers; and (3) significant points in region corners, line intersections, and points on curves with high curvatures. The features should be distinct and should be spread all over the image. First, these features should be efficiently detectable in both of the images that are being compared, and they should also be stable at fixed positions, otherwise a comparison cannot be correctly made. The number of common elements in the detected sets of the features should be very high. FBMs do not work directly with the values of image intensity. Statistical features such as moment invariants or centroids and higher level structural and syntactic descriptions are also used. FBMs can be used for feature detection and feature matching. The various approaches in FBMs are as follows:

1. Use of the corresponding centers of gravity of regions as corresponding control points (CPs) to estimate the registration parameters; these centers of gravity should be invariant with the rotation, scaling, and skewing of the corresponding images. They should also be stable, even in the presence of noise and variation in gray levels. The region features are detected using segmentation methods.
2. Use of structural similarity detection techniques.
3. Use of the affine moment invariants principle, a segmentation technique, in registering an image with an affine geometric distortion.
4. Use of contour-based methods that use regional boundaries and other strong edges as matching primitives.

5. Use of a combined invariant moment shape descriptor and improved chain-code matching to establish correspondence between potentially matched regions detected in two images.
6. Use of line segments as primitive in a registration process. Line features are object contours, coastal lines, and roads. The correspondence of lines is expressed by pairs of line ends.
7. Use of wavelet decomposition for generating the pyramidal structure, because of its multiresolution characteristics.
8. Use of Gabor WT decomposition to extract feature points.
9. Use of local modulus maxima of the WT to find feature points and the cross-correlation method to build correspondence between these feature points.
10. Use of maximum of WT coefficients to form the basic features of a correlation-based automatic registration algorithm.

To apply FBMs for feature matching, it is assumed that the two sets of features in the two images under the test are detected. The goal is to find pairwise relationships between these features using spatial correspondences or other descriptions of the detected features.

10.1.2.1 Spatial Relation

Information about the distances between the CPs (such as line endings, distinctive points, and centers of gravity) and their spatial distributions is used in this method of feature matching. The methods are (1) graph matching—a number of features are evaluated, and after a particular transformation they fall within a given range next to the features in the reference image; the transformation parameters with the highest scores are used; (2) clustering method—using the points connected by abstract edges and line segments, the cluster is detected; its centroid represents the most probable vector of matching parameters; and (3) chamfer matching—line features in the images are matched using the minimization of the generalized distance between them.

10.1.2.2 Invariant Descriptors

A correspondence between the features from the two images is determined using invariant descriptors of these features [3]. The descriptors should be invariant, unique, stable, and independent. Often a trade-off amongst these requirements might be necessary. The most invariant descriptors from the two images are paired. The simplest descriptor is the image intensity function itself. For matching CC, any distance or MI criteria can be used.

A wild forest, for example, can be described by elongation parameters, compactness, number of holes, and so on. A detected feature can be represented by its signature, for example, the longest structure and the angles between the remaining structures. The closed-boundary regions can also be used as features. The invariant descriptors can be defined as radial-shaped vectors, polygons, moment-based, perimeters, areas, compactness, extracted contours and slopes of the tangents in the contours, histograms of line ratios, histograms of angle differences, ellepticity, angle, thinness, and so on.

10.1.2.3 Relaxation Technique

This technique is part of solution to the consistent labeling problem. The idea is to label each feature from the sensed image with the label of a feature from the reference image. This method can be enhanced by a descriptor approach using corner sharpness, contrast, slope, and so on.

10.1.2.4 Pyramids and Wavelets

It is often necessary to attempt to reduce the computational burden associated with the registration techniques and algorithms. We can use the subwindow to minimize the computational cost. We can start with a coarser-resolution grid and then proceed to higher-resolution grids, or use the sparse regular grid of windows and perform cross-correlation and related methods. The foregoing are the simple examples of the pyramidal techniques [3–6]. The point is to start with the coarser grids or windows and proceed step by step to finer grids in a hierarchical manner for feature matching. We can use Gaussian pyramids, simple averaging, or wavelets for obtaining a coarse resolution in the first place. Then, systematically and gradually, the estimates of the parameters of the mapping functions or the correspondence are improved, while leading to a finer resolution, thereby monitoring and controlling the computational cost, desired resolution, and accuracy of feature matching. The major aspect of the initial feature matching would have already been obtained using the coarse-resolution grids, and then the finer matching can be carried out, if required, for the finer details. However, the consistency should be checked to avoid false matching, using the coarser grid or window. We can use concepts of median pyramid and averaging pyramid based on the process applied to the number of pixels. Wavelet decomposition can also be used in the pyramidal approach using the orthogonal or biorthogonal wavelets for multiresolution analysis. The main idea is to use the wavelet coefficients of the two images and seek correspondences between the features from the two images in terms of the important coefficients.

10.1.3 Transform Model

Once feature correspondence is established, a mapping function is constructed that transforms the sensed image to lay it over the reference image. The mapping functions could be (1) similarity transform, (2) affine, (3) perspective projection, or (4) elastic transform. The controlled points (CPs) between the sensed and the reference image should correspond closely.

10.1.3.1 Global and Local Models

A model (similarity transformation) that preserves the shape is given as [3]

$$\begin{aligned} u &= s\left(x\cos(\phi) - y\sin(\phi)\right) + t_x \\ v &= s\left(x\sin(\phi) + y\cos(\phi)\right) + t_y \end{aligned} \tag{10.2}$$

where s is the scaling parameter, ϕ is the rotation parameter, and t is the translation parameter. Another more general model is given as

$$\begin{aligned} u &= a_0 + a_1 x + a_2 y \\ v &= b_0 + b_1 x + b_2 y \end{aligned} \tag{10.3}$$

This model can be defined by three noncollinear CPs. It preserves the straight line parallelism. Other models are given in [3]. The parameters of these mapping functions and models are determined using LS methods (see the Appendix). The errors at the CPs should be minimized.

It is important for various types of images to handle the local situations. The LS method would average out such local aspects. Hence, locally sensitive registration methods are required. The weighted LS method can be used (see Section 2.2), since we can use proper weighting parameters to capture and retain local variations in the features.

10.1.3.2 Radial Basis Functions

This type of model is given (for u and v) as

$$u = a_0 + a_1 x + a_2 y + \sum_{j=1}^{n} c_j h(x, x_j) \tag{10.4}$$

The term under the summation sign is the radially symmetric function. This transformation model is a combination of the global model and the radial basis function (RBF). The main point here is that the function depends on the radius (the distance) of the point from the CPs, and not on the location or the particular position. Any RBF can be used.

10.1.3.3 Elastic Registration

The idea in elastic registration is not to use any parametric mapping function. In this approach, the images are regarded as pieces of rubber or elastic sheets [3]. Here, the external and internal forces to stretch the image or make it stiff are brought into play. Image registration is performed by achieving the minimum energy state in an iterative manner. However, some similarity functions are needed to define the forces. These functions again depend on the intensity values and the correspondence of the boundary structures. We can use the fluid registration approach or the diffusion-based registration approach. Due to the possibility of various types of elastic models, there is an immense scope of development of new and efficient image registration methods within the precincts of the elastic registration methods.

10.1.4 Resampling and Transformation

For the purpose of transformation, each pixel from the sensed image is transformed using the estimated mapping functions. Another approach is to determine the registered image data from the sensed image data using the coordinates of the target pixel and the inverse of the estimated mapping function. The coordinate system used is the same as that of the reference image. The image interpolation that takes place in the sensed image on the regular grid is realized via convolution of the image with an interpolation kernel. For interpolation, we can use any of the following functions: (1) bilinear, (2) nearest neighbor, (3) splines, (4) Gaussians, and (5) sinc functions.

10.1.5 Image Registration Accuracy

There are several types of errors that can occur during the process of image registration [3]: (1) localization errors, (2) matching errors, and (3) alignment errors. Localization errors occur due to the displacement of the CP coordinates which in turn is due to their inaccurate detection. These errors can be reduced using an optimal feature detection algorithm. Matching errors indicate the number of false matches in the correspondence of the CPs of both the images. We can use the consistency check or the cross-validation to identify the false matches. Alignment errors are the difference between the mapping model and the actual between-image geometric distortion. Alignment errors can be assessed using another comparative method. Visual assessment by a domain expert is still used as a complementary method to objective error evaluation methods.

10.2 Segmentation, Centroid Detection, and Target Tracking with Image Data

Tracking moving objects or targets using image data involves processing images and producing an estimate of the object's current position and velocity vectors at each step. There is additional uncertainty regarding the origin of the received data, which may or may not include measurements from the targets. This may be due to random clutter (false alarms). This data association problem was addressed in Chapter 3. In this section, the focus is on the use of the algorithm for tracking with segmentation from imaging sensors [7,11]. This centroid detection and tracking (CDT) algorithm is independent of the size of targets and is less sensitive to the intensity of the targets. The typical characteristics of the target obtained by motion-recognition or object/pattern-recognition are used to associate images with the target being tracked. The motion recognition characteristics location, velocity, and acceleration (as a total state vector) are generated using data from successive frames (interscan level), whereas the object/pattern-recognition characteristics, geometric structure such as shape and size, and energy levels (different gray levels in the image) in one or more spectral bands are obtained using image data at the intrascan level. The CDT algorithm combines both object- and motion-recognition methods for target tracking from imaging sensors. The properties of the image considered are the intensity and size of the cluster. The pixel intensity is discretized into layers of gray level intensities with sufficient target pixel intensities within the limits of certain target layers. The CDT algorithm involves the conversion of the data from one image into a binary image by applying upper and lower threshold limits for the target layers. The binary image is then converted into clusters using the nearest neighbor criterion. For the known target size, the information is used to set limits for removing those clusters, which differ sufficiently from the size of the target cluster. This reduces computations. Then the centroid of the clusters is computed and used for tracking the target (see Figure 10.1). The associated methods described are given next.

10.2.1 Image Noise

Noise can be additive, and the measured image $x(i, j)$ is the sum of the true image $s(i, j)$ and the noise-image $v(i, j)$.

$$x(i, j) = s(i, j) + v(i, j) \tag{10.5}$$

The noise $v(i, j)$ is zero mean and white and is described by variance σ^2. Salt-and-pepper (SP) noise is referred to as *intensity spikes* (speckle or data drop-out noise) and is caused by errors in the image data transmission,

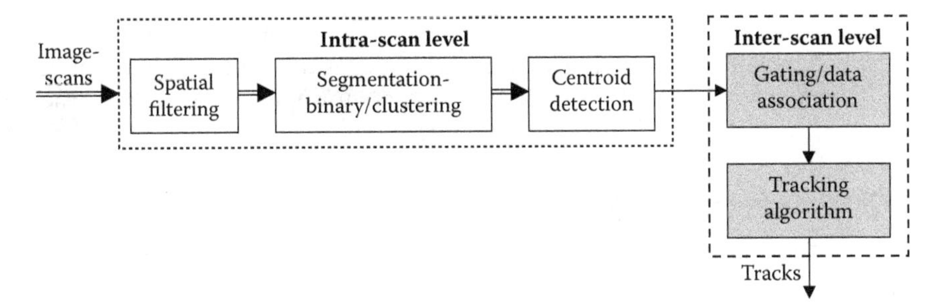

FIGURE 10.1
Centroid detection and tracking algorithm steps.

malfunctioning pixel elements in the camera or sensors, faulty memory locations, or timing errors in the digitization. The noise-corrupted pixels are set alternatively to zero or maximum value, giving the image an SP-like appearance. Uncorrupted pixels remain unchanged.

$$x(i,j) = \begin{cases} 0 & \text{rand} < 0.5d \\ 255 & 0.5d \leq \text{rand} < d \\ s(i,j) & \text{rand} \geq d \end{cases} \tag{10.6}$$

Here, rand is the uniformly distributed random numbers in the interval zero to one, and d is the noise density and it is real positive. Gaussian noise is electronic noise in the image acquisition system and can be generated with zero mean Gaussian distribution (randn) with its standard deviation as σ.

$$v(i,j) = \text{randn}\ \sigma \tag{10.7}$$

$$x(i,j) = s(i,j) + v(i,j) \tag{10.8}$$

10.2.1.1 Spatial Filter

The images acquired through sensors are generally contaminated by various types of noise processes and sources: sensor problems in a camera, detector sensitivity variation, environmental effects, the discrete nature of radiation, dust on the optics, quantization errors, or transmission errors. Noise can be handled in two ways: spatial-domain filtering (SDF) or frequency-domain filtering (FDF). SDF refers to the image plan itself and is based on the direct manipulation of pixels in the image. FDF is based on modifying the Fourier transform of an image. In SD filters, a mask (a template, kernel, window, or filter) is a small 2D (of 3×3 dimension) array or image in which the coefficients determine the process, such as image smoothing or sharpening [4]. The two primary categories of spatial filters for noise removal from the image are mean filters (linear) and order (nonlinear) filters.

10.2.1.2 Linear Spatial Filters

Linear filtering is achieved by local convolution with size $n \times n$ kernel mask and by a series of shift-multiply-sum operators with the $n \times n$ kernel. In this process (with a smoothing kernel mask), the brightness value of each pixel is replaced by a weighted average of the pixel brightness values in a neighborhood surrounding the pixel. The weighting factors for various pixels in the neighborhood are determined by the corresponding values in the mask. If the brightness value of any neighborhood pixel is unusual due to the influence of noise, then the brightness averaging will tend to reduce the effect of the noise by distributing it among the neighboring pixels.

$$y(i,j) = \sum_{k=i-w}^{i+w} \sum_{l=j-w}^{j+w} x(k,l)h(i-k,j-l) \tag{10.9}$$

Here, x is the input image, h is the filter function/impulse response/convolution mask, y is the output image, (i, j) is the index of the image pixel, $i,j = 1,2,\ldots,N$, $w = (n-1)/2$, N is the size of the input image, and $n \times n$ is the order of the filter or mask (an odd number). The mean filter is the simplest linear spatial filter. The coefficients in the kernel are nonnegative and equal. Masks of different sizes can be obtained as

$$h_k = \frac{\text{ones}(k,k)}{k^2} \tag{10.10}$$

Here, ones(k,k) is a $k \times k$ matrix having all elements as unity.

10.2.1.3 Nonlinear Spatial Filters

Nonlinear filters are data dependent. Order filters (OF) are a kind of nonlinear filter. OFs are based on order statistics and are implemented by arranging the neighborhood pixels in order from the smallest to largest gray-level values. The most useful OF is the median filter, which selects the middle pixel value from the order set. The operation is performed by applying a sliding window similar to a convolution operation in a linear SDF. The median is computed by sorting all the pixel values from the surrounding neighborhood into numerical order and then replacing the pixel being considered with the middle pixel value. The unimportant or nonrepresentative pixel in a neighborhood will not significantly affect the median value.

10.2.2 Metrics for Performance Evaluation

The main idea is to compute a single number that reflects the quality of the smoothed, filtered, and fused image.

10.2.2.1 Mean Square Error

The mean square error (MSE) of the true and filtered images will be nearly zero when the corresponding true and filtered image pixels are truly alike. The filtering algorithm that gives the minimum MSE value is preferable here.

$$\text{MSE} = \frac{1}{N^2}\sum_{i=1}^{N}\sum_{j=1}^{N}\left[s(i,j) - y(i,j)\right]^2 \tag{10.11}$$

Here, s is the true image, y is the filtered image, N is the size of the image, and (i, j) is the index of the pixel.

10.2.2.2 Root Mean Square Error

Root mean square error (RMSE) will increase when the corresponding filtered image pixels deviate from the true image pixels. The filtering algorithm that gives the minimum value is preferable here.

$$\text{RMSE} = \sqrt{\frac{1}{N^2}\sum_{i=1}^{N}\sum_{j=1}^{N}\left[s(i,j) - y(i,j)\right]^2} \tag{10.12}$$

10.2.2.3 Mean Absolute Error

Here again, a minimum mean absolute error (MAE) is sought for a good match.

$$\text{MAE} = \frac{1}{N^2}\sum_{i=1}^{N}\sum_{j=1}^{N}\left|s(i,j) - y(i,j)\right| \tag{10.13}$$

10.2.2.4 Percentage Fit Error

The filtering algorithm that gives the minimum percentage fit error (PFE) is preferable here.

$$\text{PFE} = \frac{\text{norm}(s - y)}{\text{norm}(s)} \times 100 \tag{10.14}$$

10.2.2.5 Signal-to-Noise Ratio

The impact of noise on the image can be described by the signal-to-noise ratio (SNR). A comparison between two values for differently filtered images gives a comparison of the quality. The SNR will be infinity when both true and filtered images are truly alike. The filtered algorithm that gives the high SNR value is preferable here.

$$\mathrm{SNR} = 10\log_{10}\left[\frac{\sum_{i=1}^{N}\sum_{j=1}^{N} s(i,j)^2}{\sum_{i=1}^{N}\sum_{j=1}^{N}\left[s(i,j)-y(i,j)\right]^2}\right] \tag{10.15}$$

10.2.2.6 Peak Signal-to-Noise Ratio

The peak signal-to-noise ratio (PSNR) value will be infinity when both true and filtered images are alike. The filtered algorithm that gives the high PSNR value is preferable here.

$$\mathrm{PSNR}_{af} = 10\log_{10}\left[\frac{I_{max}^2}{\sum_{i=1}^{N}\sum_{j=1}^{N}[s(i,j)-y(i,j)]^2}\right] \tag{10.16}$$

Here, I_{max} is the maximum pixel intensity value.

The performance of the mean and median filter to remove the SP noise with density 0.03 and Gaussian noise with mean zero and standard deviation $\sigma = 1.4$ is shown for simulated data in Figure 10.2. The performance of both types of filters to remove the salt-and-pepper (SAP) noise and Gaussian noise is found to be satisfactory.

10.2.3 Segmentation and Centroid Detection Techniques

This section deals with image segmentation and CDT aspects.

10.2.3.1 Segmentation

Segmentation decomposes the image into two different regions: texture segmentation and particle segmentation. In texture segmentation, an image is partitioned into different regions (microimages) and each region is defined by a set of feature characteristics. In particle segmentation, an image is partitioned into object regions and background regions. Here, segmentation refers to the task of extracting objects or particles of interest

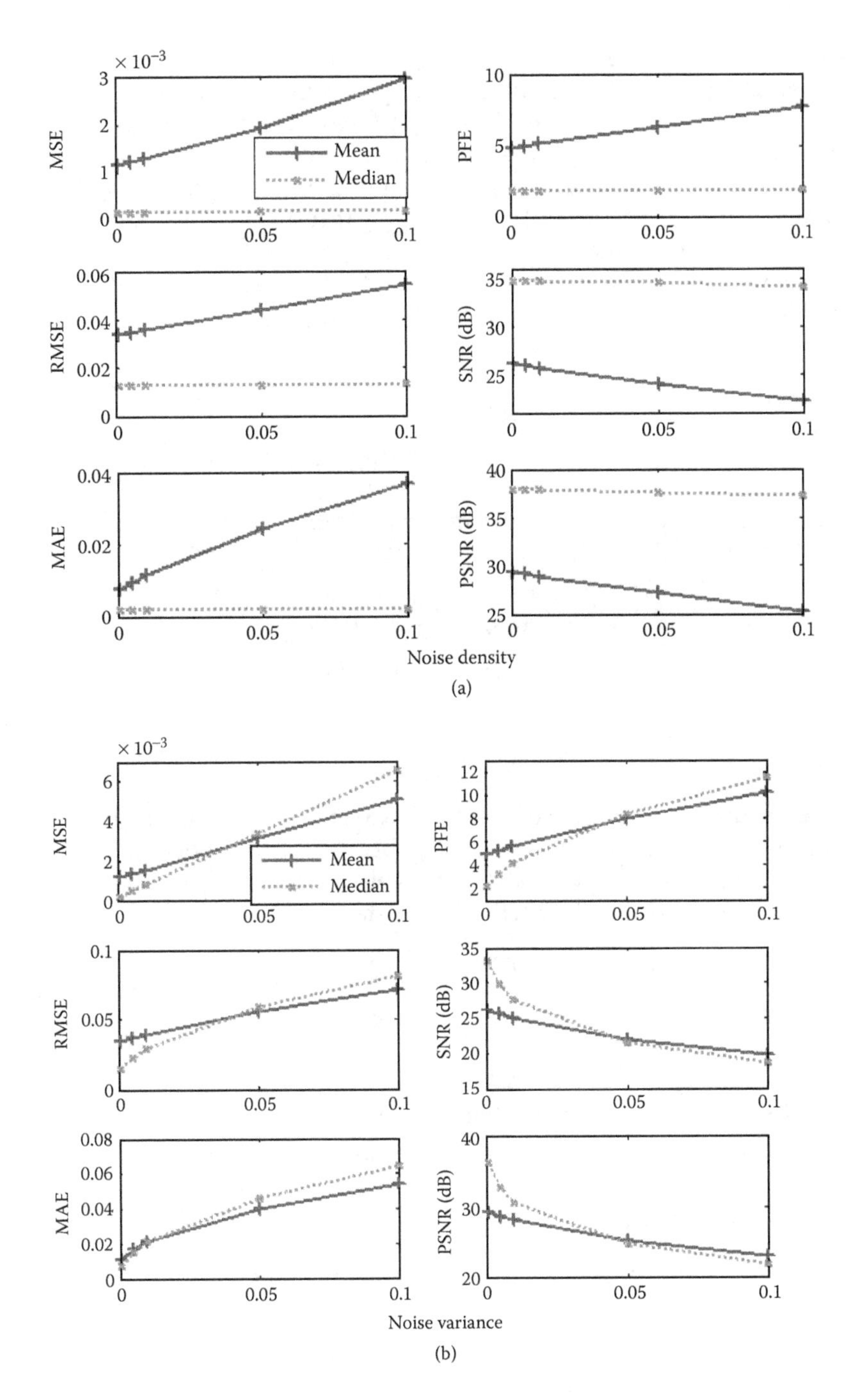

FIGURE 10.2
Performance of the noise filters for (a) salt-and-pepper noise and (b) Gaussian noise.

as precisely as possible from the image. In the CDT algorithm, particle segmentation is used to separate the target from background, when the target is not fully visible [8]. The pixel intensities are discretized into 256 gray levels, and particle segmentation is carried out in two steps: (1) the gray-level image is transformed into a binary image using lower and upper threshold limits of the target, the thresholds being determined using pixel intensity histograms from the target and its environment; and (2) these detected pixels are then grouped into clusters using the nearest neighbor data association method [8,9]. The gray image $\mathrm{Im}(i, j)$ is converted into a binary image with intensity $\beta(i, j)$:

$$\beta(i,j) = \begin{cases} 1 & I_L \leq \mathrm{Im}(i,j) \leq I_U \\ 0 & \text{otherwise} \end{cases} \tag{10.17}$$

where I_L and I_U are the lower and upper limits of the target intensity. The detection probability of the pixel is defined as

$$\begin{aligned} P\{\beta(i,j)=1\} &= p(i,j) \\ P\{\beta(i,j)=0\} &= 1-p(i,j) \end{aligned} \tag{10.18}$$

where $p(i,j) = \frac{1}{\sigma\sqrt{2\pi}} \int_{I_L}^{I_U} e^{\frac{-(x-\mu)^2}{2\sigma^2}} dx$ and the gray image $I(i,j)$ is assumed to have Gaussian distribution with mean μ and variance σ^2. The binary image is grouped into clusters using nearest neighbor data association [8,9]. A pixel belongs to the cluster only if the distance between this pixel and at least one other pixel of the cluster is less than the distance d_p, given by [11]

$$\sqrt{\frac{1}{p_t}} < d_p < \sqrt{\frac{1}{p_v}} \tag{10.19}$$

where p_t and p_v are detection probabilities of target and noise pixels, respectively. d_p affects the size, shape, and number of clusters. d_p should be close to $\sqrt{1/p_t}$ to minimize the gaps in the target image [8,9].

10.2.3.2 Centroid Detection

The centroid of the cluster is computed using the following equation:

$$(x_c, y_c) = \frac{1}{\sum_{i=1}^{n}\sum_{j=1}^{m} I_{ij}} \left(\sum_{i=1}^{n}\sum_{j=1}^{m} iI(i,j), \sum_{i=1}^{n}\sum_{j=1}^{m} jI(i,j) \right) \tag{10.20}$$

where $[x_c, y_c]$ is the centroid of the cluster, I_{ij} is the intensity of the $(i, j)th$ pixel and n and m are the dimensions of the cluster.

10.2.4 Data Generation and Results

The model of the FLIR sensor for the generation of synthetic images [7,9,11] is described next. We consider a 2D array to be

$$m = m_\xi \times m_\eta \tag{10.21}$$

pixels. Here, each pixel is expressed by a single index $i = 1, \cdots, m$ with the intensity I of pixel i given as

$$I_i = s_i + n_i \tag{10.22}$$

where s_i is the target intensity and n_i is the noise in pixel i, (the noise being Gaussian with zero mean and covariance σ^2). The total target-related intensity is given by

$$s = \sum_{i=1}^{m} s_i \tag{10.23}$$

For the number of pixels m_s covered by the target, the average target intensity over its extent is given by

$$\mu_s = \frac{s}{m_s} \tag{10.24}$$

The average pixel SNR is

$$r' = \frac{\mu_s}{\sigma} \tag{10.25}$$

To simulate the motion of the target in the frame, kinematic models of target motion are used.

The state model used to describe the constant velocity target motion is given by

$$X(k+1) = \begin{bmatrix} 1 & T & 0 & 0 \\ 0 & 1 & 0 & 0 \\ 0 & 0 & 1 & T \\ 0 & 0 & 0 & 1 \end{bmatrix} X(k) + \begin{bmatrix} \frac{T^2}{2} & 0 \\ T & 0 \\ 0 & \frac{T^2}{2} \\ 0 & T \end{bmatrix} w(k) \tag{10.26}$$

where $X(k) = \begin{bmatrix} x & \dot{x} & y & \dot{y} \end{bmatrix}^T$ is the state vector, and $w(k)$ is the zero-mean white Gaussian noise with covariance matrix $Q = \begin{bmatrix} \sigma_w^2 & 0 \\ 0 & \sigma_w^2 \end{bmatrix}$. The measurement model is given as

$$z(k+1) = \begin{bmatrix} 1 & 0 & 0 & 0 \\ 0 & 0 & 1 & 0 \end{bmatrix} X(k+1) + v(k+1) \tag{10.27}$$

where $v(k)$ is the centroid measurement noise (zero mean/Gaussian) with covariance matrix

$$R = \begin{bmatrix} \sigma_x^2 & 0 \\ 0 & \sigma_y^2 \end{bmatrix} \tag{10.28}$$

These noise processes are assumed to be uncorrelated. A 2D array of 64×64 pixels is considered for the background image, which is modeled as a white Gaussian random field, $N(\mu_n, \sigma_n^2)$. A 2D array of pixels, modeled as a white Gaussian random field $N(\mu_t, \sigma_t^2)$, is used to generate a target of size (9×9). The total number of scans is 50 and the image frame rate (T) is 1 frame per second. The initial state vector of the target in the image frame is $X = \begin{bmatrix} x & \dot{x} & y & \dot{y} \end{bmatrix}^T = \begin{bmatrix} 10 & 1 & 10 & 1 \end{bmatrix}^T$. The synthetic image with these parameters is converted into a binary image using the upper $(I_U = 110)$ and lower $(I_L = 90)$ limits of a target layer and then grouped into clusters by the nearest neighbor data association method using the optimal proximity distance $d_p \, (d_p = 2)$. The centroids of the clusters are then computed. Since the background is very noisy, the cluster algorithm produces more clusters and more centroids. This requires a nearest neighbor Kalman filter (NNKF) or probabilistic data association filter (PDAF) to associate the true measurement with the target. The PFE, root mean square position (RMSP), and root mean square velocity (RMSV) metrics are given in Table 10.1. The parameters are within the acceptable limits, showing the tracker consistency.

10.2.5 Radar and Imaging Sensor Track Fusion

The CT algorithm is next used to provide input to state-vector fusion when the data on the position from ground-based radars are available in a Cartesian coordinate frame. In the fusion of data from imaging sensors

TABLE 10.1

PFE, Root Mean Square Percentage Error (RMSPE), and Root Mean Square Vector Error (RMSVE) Metrics

PFEx	PFEy	RMSPE	RMSVE
0.99	0.79	0.49	0.26

and ground-based radar, the data from the FLIR are first passed through the centroid detection algorithm, and then both types of data are used in the individual NNKF or PDAF tracker algorithms before carrying out the track-to-track fusion. The tracks (the state-vector estimates from the imaging sensor [track *i*] and the ground-based radar [track *j*]) and their covariance matrices at scan *k* are used for fusion as

$$\text{Track } i: \hat{X}_i(k), \hat{P}_i(k) \text{ and Track } j: \hat{X}_j(k), \hat{P}_j(k) \tag{10.29}$$

The fused state is given by

$$\hat{X}_c(k) = \hat{X}_i(k\,|\,k) + \hat{P}_i(k\,|\,k)\hat{P}_{ij}(k)^{-1}\left[\hat{X}_j(k\,|\,k) - \hat{X}_i(k\,|\,k)\right] \tag{10.30}$$

The combined covariance matrix associated with the state-vector fusion is given by

$$\hat{P}_c(k) = \hat{P}_i(k\backslash k) - \hat{P}_i(k\,|\,k)\hat{P}_{ij}(k)^{-1}\,\hat{P}_i(k\,|\,k) \tag{10.31}$$

Here, $\hat{P}_{ij}$ the cross-covariance between $\hat{X}_i(k\,|\,k)$ and $\hat{X}_j(k\,|\,k)$ is given by

$$\hat{P}_{ij}(k) = \hat{P}_i(k\,|\,k) + \hat{P}_j(k\,|\,k) \tag{10.32}$$

The PFE in *x* and *y* positions and RMSE in position and velocity metrics are given in Table 10.2. There was apparently a close match among the trajectories of radar, image centroid tracking algorithm (ICTA), true, and fused, when there was no data loss. Subsequently, the measurement data loss in the imaging sensor is simulated to occur from 15 to 25 seconds and in the ground-based radar from 30 to 45 seconds, during which time track extrapolation has been performed. Track deviation was observed during these periods. The PFE in *x* and *y* directions and RMSE in position and velocity metrics before and after fusion are shown in Table 10.3. Note that the fusion of data gives better results when there is a measurement loss in either of the sensors, thereby demonstrating robustness and better accuracy.

TABLE 10.2

The PFE, RMSPE, and RMSVE Metrics (without Measurement Loss)

	PFE*x*	PFE*y*	RMSPE	RMSVE
Imaging sensor	3.01	2.95	1.618	0.213
Radar	2.78	2.74	1.497	0.233
Fused	0.78	0.69	0.39	0.355

TABLE 10.3

PFE, RMSPE, and RMSVE (with Measurement Loss)

	PFEx	PFEy	RMSPE	RMSVE
Imaging sensor	4.29	2.86	1.984	0.374
Radar	2.66	3.59	1.688	0.51
Fused	1.31	1.46	0.736	0.478

10.3 Pixel-Level Fusion Algorithms

Multi-imaging sensor fusion (MSIF) should provide a better and enhanced image of an observed object. The fused image should have improved contrast, and it should be easy for the user to detect, recognize, and identify the targets and increase the user's situational awareness [10]. The fusion of images is of great value in microscopic imaging, medical imaging, remote sensing, robotics, and computer vision. Fused images, depending on the application, should preserve all the relevant information contained in the source images. Any irrelevant features and noise should be eliminated or reduced to the maximum extent [12]. When image fusion is carried out at the pixel level, the source images are combined without any preprocessing. The simplest MSIF is to take the average of the grey level images pixel by pixel, a kind of fusion process. This may produce undesired effects and reduced feature contrast. In some situations, various objects in the scene may be at different distances from the imaging sensor, so that if one object is in focus then another might be out of focus. In such situations the conventional fusion method does not work well.

We will consider two fusion architectures in this section: (1) the source images are considered a whole in the fusion process, and (2) the source images are decomposed into small blocks, and the blocks are then used in the fusion process. In the latter, the discrepancy would be reduced since local variations are considered in the fusion process. The block size and threshold are defined by the user, though it might be difficult to choose the threshold for optimal fusion. A modified algorithm that computes normalized spatial frequencies (SFs) can be used. Since the SFs of the source images are normalized, the user can choose the threshold as 0 to 0.5. A similar method can be used for principal component analysis (PCA)–based image fusion. The information in the source images should be registered prior to fusion.

10.3.1 Principal Component Analysis Method

PCA is a numerical procedure that transforms a number of correlated variables into a number of uncorrelated variables called principal components (PCs; see the Appendix) [9]: (1) the first PC accounts for much of the variance in the data, and each succeeding component accounts for much of the remaining variance; the first PC is along the direction with the maximum variance; (2) the second component is constrained to lie in the subspace perpendicular to the first component; within the subspace this component points to the direction of maximum variance; and (3) the third component is taken in the maximum variance direction in the subspace perpendicular to the first two, and so on. PC basis vectors depend on the data set. Let X be a d-dimensional random vector with zero mean, and orthonormal projection matrix V be such that $Y = V^T X$. The covariance of Y, $\mathrm{cov}(Y)$ is a diagonal matrix. Using simple matrix algebra we get [9]

$$\begin{aligned} \mathrm{cov}(Y) &= E\{YY^T\} \\ &= E\{(V^T X)(V^T X)^T\} \\ &= E\{(V^T X)(X^T V)\} \\ &= V^T E\{XX^T\}V \\ &= V^T \mathrm{cov}(X)V \end{aligned} \tag{10.33}$$

Multiplying both sides of Equation 10.33 by V, we get

$$\begin{aligned} V\,\mathrm{cov}(Y) &= VV^T \mathrm{cov}(X)V \\ &= \mathrm{cov}(X)V \end{aligned} \tag{10.34}$$

We can write V as $V = [V_1, V_2, \ldots, V_d]$ and $\mathrm{cov}(Y)$ in the diagonal form as

$$\begin{bmatrix} \lambda_1 & 0 & \cdots & 0 & 0 \\ 0 & \lambda_2 & \cdots & 0 & 0 \\ \vdots & \vdots & \ddots & \vdots & \vdots \\ 0 & 0 & \cdots & \lambda_{d-1} & 0 \\ 0 & 0 & \cdots & 0 & \lambda_d \end{bmatrix} \tag{10.35}$$

By substituting Equation 10.35 in Equation 10.34, we get

$$[\lambda_1 V_1, \lambda_2 V_2, \ldots, \lambda_d V_d] = [\mathrm{cov}(X)V_1, \mathrm{cov}(X)V_2, \ldots, \mathrm{cov}(X)V_d] \tag{10.36}$$

This is rewritten as

$$\lambda_i V_i = \mathrm{cov}(X)V_i \tag{10.37}$$

where $i = 1, 2, \ldots, d$ and V_i is an eigenvector of $\mathrm{cov}(X)$.

10.3.1.1 Principal Component Analysis Coefficients

The source images to be fused are arranged in two column vectors. The following steps are performed: (1) organize the data into column vectors yielding a matrix Z of $n \times 2$ dimensions; (2) compute the empirical mean for each column; the mean vector **M** has a dimension of 2×1; (3) subtract the mean vector from each column of the data matrix Z, yielding X of dimension $n \times 2$; (4) compute the covariance matrix C of Z as $C = X^T X$; (5) compute the eigenvectors V and eigenvalues D of C and sort V in decreasing order; both V and D are of dimension 2×2; and (6) consider the first column of V that corresponds to the largest eigenvalue to compute the PCs NPC_1 and NPC_2 as

$$\text{NPC}_1 = \frac{V(1)}{\sum V} \quad \text{and} \quad \text{NPC}_2 = \frac{V(2)}{\sum V} \tag{10.38}$$

10.3.1.2 Image Fusion

A flow diagram of PCA–based weighted-average image fusion process is depicted in Figure 10.3. PCs NPC_1 and NPC_2 (such that $\text{NPC}_1 + \text{NPC}_2 = 1$) are computed from the eigenvectors. The fused image is obtained by

$$I_f = \text{NPC}_1 I_1 + \text{NPC}_2 I_2 \tag{10.39}$$

This means that the PCs are used as weights for image fusion. The flow diagram of the PCA-based block-image fusion process is shown in Figure 10.4. The input images are decomposed into blocks (I_{1k} and I_{2k}) of size $m \times n$, where I_{1k} and I_{2k} denote the kth blocks of I_1 and I_2, respectively. If the PCs corresponding to the kth blocks are NPC_{1k} and NPC_{2k}, then the fusion of kth block of the image is

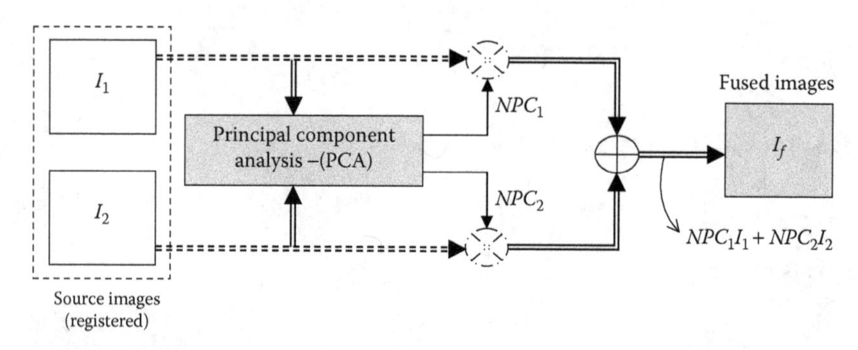

FIGURE 10.3
The principal component analysis–based image fusion.

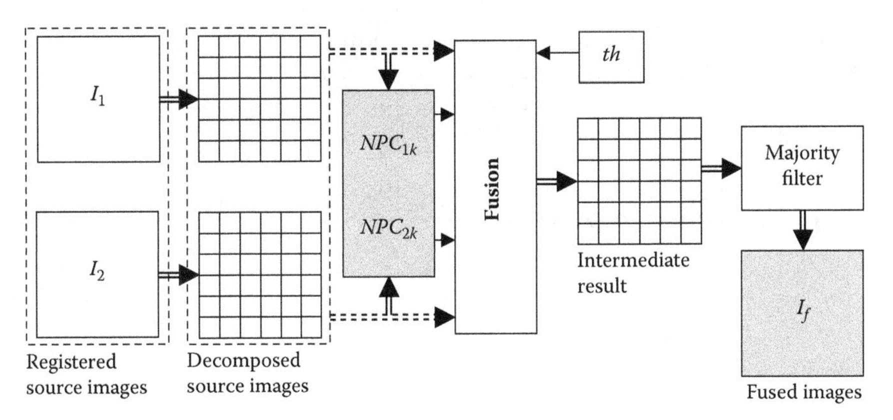

FIGURE 10.4
The principal component analysis–based block-image fusion process.

$$I_{fk} = \begin{cases} I_{1k} & \text{NPC}_{1k} > \text{NPC}_{2k} + th \\ I_{2k} & \text{NPC}_{1k} < \text{NPC}_{2k} - th \\ \dfrac{I_{1k} + I_{2k}}{2} & \text{otherwise} \end{cases} \tag{10.40}$$

where *th* is the user defined threshold, and $I_{1k} + I_{2k}/2$ is the gray level averaging of corresponding pixels.

10.3.2 Spatial Frequency

Spatial frequency (SF) measures the overall information level in an image [13,14], and for an image I of dimension $M \times N$, it is defined as follows:

1. Row frequency:

$$\text{RF} = \sqrt{\frac{1}{MN} \sum_{i=0}^{M-1} \sum_{j=1}^{N-1} \left[I(i,j) - I(i,j-1) \right]^2} \tag{10.41}$$

2. Column frequency:

$$\text{CF} = \sqrt{\frac{1}{MN} \sum_{j=0}^{N-1} \sum_{i=1}^{M-1} \left[I(i,j) - I(i-1,j) \right]^2} \tag{10.42}$$

3. SF:

$$\text{SF} = \sqrt{\text{RF}^2 + \text{CF}^2} \tag{10.43}$$

Here, M is the number of rows, N is the number of columns, (i, j) is the pixel index, and $I(i, j)$ is the gray value at pixel (i, j).

10.3.2.1 Image Fusion by Spatial Frequency

The SF–based weighted-image fusion process is shown in Figure 10.5. The computed SFs are normalized as

$$\text{NSF}_1 = \frac{\text{SF}_1}{\text{SF}_1 + \text{SF}_2} \quad \text{and} \quad \text{NSF}_2 = \frac{\text{SF}_2}{\text{SF}_1 + \text{SF}_2} \tag{10.44}$$

The fused image is obtained by

$$I_f = \text{NSF}_1 I_1 + \text{NSF}_2 I_2 \tag{10.45}$$

The SF–based block-image fusion process is shown in Figure 10.6. The images are decomposed into blocks I_{1k} and I_{2k}; then the normalized SFs for each block are computed. If the normalized SFs of I_{1k} and I_{2k} are NSF_{1k} and NSF_{2k}, respectively, then the fusion of the kth block of the (fused) image is given as

$$I_{fk} = \begin{cases} I_{1k} & \text{NSF}_{1k} > \text{NSF}_{2k} + th \\ I_{2k} & \text{NSF}_{1k} < \text{NSF}_{2k} - th \\ \dfrac{I_{1k} + I_{2k}}{2} & \text{otherwise} \end{cases} \tag{10.46}$$

10.3.2.2 Majority Filter

In the block-image fusion process, a majority filter is used to avoid the artifacts in the fused image. If the center block comes from I_1 and the

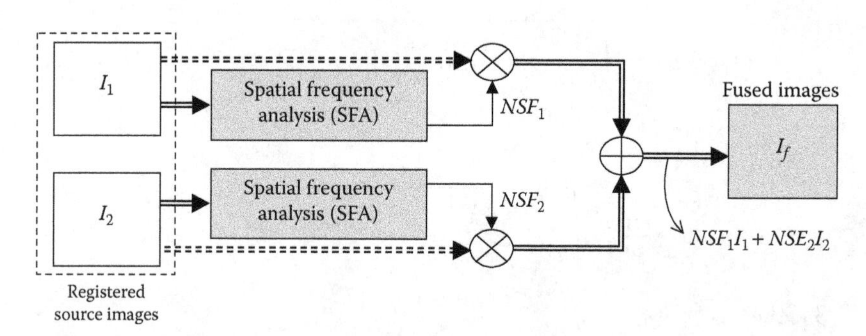

FIGURE 10.5
Spatial frequency–based image fusion process.

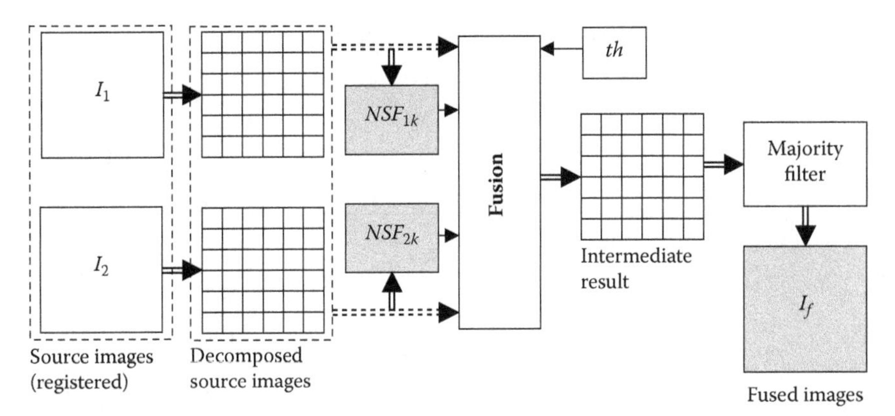

FIGURE 10.6
Spatial frequency–based block-image fusion process.

surroundings blocks are from I_2, then the center block will be replaced by the block from I_2 and vice versa [14]. Denote a and b as the block images coming from I_1 and I_2, respectively. If the blocks (a) from I_1 are six times and blocks (b) from I_2 are three times, then the majority filter replaces the center block with the block coming from I_1, since the majority of neighboring blocks are coming from the I_1.

10.3.3 Performance Evaluation

When the reference image is available, the performance of image fusion algorithms is evaluated using the following metrics:

1. RMSE is computed as the RMSE of the corresponding pixels in the reference I_r and the fused image I_f:

$$\text{RMSE} = \sqrt{\frac{1}{MN}\sum_{i=1}^{M}\sum_{j=1}^{N}\left(I_r(i,j) - I_f(i,j)\right)^2} \tag{10.47}$$

2. PFE is computed as the norm of the differences between the corresponding pixels of reference and fused images to the norm of the reference image:

$$\text{PFE} = \frac{\text{norm}(I_r - I_f)}{\text{norm}(I_r)} \times 100 \tag{10.48}$$

3. MAE is computed as the MAE of the corresponding pixels in reference and fused images:

$$\text{MAE} = \frac{1}{MN}\sum_{i=1}^{M}\sum_{j=1}^{N}\left|I_r(i,j) - I_f(i,j)\right| \tag{10.49}$$

4. SNR is computed as

$$\text{SNR} = 20\log_{10}\left(\frac{\sum_{i=1}^{M}\sum_{j=1}^{N}(I_r(i,j))^2}{\sum_{i=1}^{M}\sum_{j=1}^{N}(I_r(i,j) - I_f(i,j))^2}\right) \tag{10.50}$$

5. PSNR is computed as

$$\text{PSNR} = 20\log_{10}\left(\frac{L^2}{\frac{1}{MN}\sum_{i=1}^{M}\sum_{j=1}^{N}\left(I_r(i,j) - I_f(i,j)\right)^2}\right) \tag{10.51}$$

Here, L is the number of gray levels in the image.

6. *Correlation* shows the correlation between the reference and fused images:

$$\text{CORR} = \frac{2C_{rf}}{C_r + C_f} \tag{10.52}$$

Here, $C_r = \sum_{i=1}^{M}\sum_{j=1}^{N} I_r(i,j)^2$, $C_f = \sum_{i=1}^{M}\sum_{j=1}^{N} I_f(i,j)^2$, and

$$C_{rf} = \sum_{i=1}^{M}\sum_{j=1}^{N} I_r(i,j) I_f(i,j)$$

7. MI [16] is given as

$$\text{MI} = \sum_{i=1}^{M}\sum_{j=1}^{N} h_{I_r I_f}(i,j)\log_2\left(\frac{h_{I_r I_f}(i,j)}{h_{I_r}(i,j)h_{I_f}(i,j)}\right) \tag{10.53}$$

8. The *universal quality index* [16] measures how much of the salient information contained in the reference image has been transformed into the fused image. The metric range is –1 to 1 and the best value 1

would be achieved if and only if the reference and fused images are alike. The lowest value of –1 would occur when $I_f = 2\mu_{I_r} - I_r$:

$$\text{QI} = \frac{4\sigma_{I_r I_f}\left(\mu_{I_r} + \mu_{I_f}\right)}{\left(\sigma_{I_t}^2 + \sigma_{I_f}^2\right)\left(\mu_{I_r}^2 + \mu_{I_f}^2\right)} \tag{10.54}$$

Here,

$$\mu_{I_r} = \frac{1}{MN}\sum_{i=1}^{M}\sum_{j=1}^{N} I_r(i,j), \qquad \mu_{I_f} = \frac{1}{MN}\sum_{i=1}^{M}\sum_{j=1}^{N} I_f(i,j)$$

$$\sigma_{I_r}^2 = \frac{1}{MN-1}\sum_{i=1}^{M}\sum_{j=1}^{N}\left(I_r(i,j) - \mu_{I_r}\right)^2, \quad \sigma_{I_f}^2 = \frac{1}{MN-1}\sum_{i=1}^{M}\sum_{j=1}^{N}\left(I_f(i,j) - \mu_{I_f}\right)^2$$

$$\sigma_{I_r I_f}^2 = \frac{1}{MN-1}\sum_{i=1}^{M}\sum_{j=1}^{N}\left(I_r(i,j) - \mu_{I_r}\right)\left(I_f(i,j) - \mu_{I_f}\right)$$

9. *Measure of structural similarity* compares local patterns of pixel intensities that are normalized for luminance and contrast. Natural image signals are highly structured and their pixels reveal strong dependencies that carry information about the structure of the object. It is given as

$$\text{SSIM} = \frac{\left(2\mu_{I_r}\mu_{I_f} + C_1\right)\left(2\sigma_{I_r I_f} + C_2\right)}{\left(\mu_{I_r}^2 + \mu_{I_f}^2 + C_1\right)\left(\sigma_{I_r}^2 + \sigma_{I_f}^2 + C_2\right)} \tag{10.55}$$

Here, C_1 is a constant that is included to avoid instability when $\mu_{I_r}^2 + \mu_{I_f}^2$ is close to zero and C_2 is a constant that is included to avoid instability when $\sigma_{I_r}^2 + \sigma_{I_f}^2$ is close to zero.

10.3.3.1 Results and Discussion

The true image I_t is shown in Figure 10.7 and the source images to be fused, I_1 and I_2, are shown in Figure 10.8. The source images have been created by blurring a portion of the reference image with a Gaussian mask using a diameter of 12 pixels. The images fused by PCA and SF and the corresponding error images (with the procedure given for the first method) are shown in Figure 10.9. The performance metrics are shown in Table 10.4. From these results, note that image fusion by SF is marginally better or similar to PCA.

Figure 10.10a gives a 100×100 image block and Figure 10.10b through d shows the degraded images after blurring with a disk of radius 5, 9, and 21 pixels, respectively. Figure 10.10a is one of the source images I_1, and

FIGURE 10.7
True image (I_t).

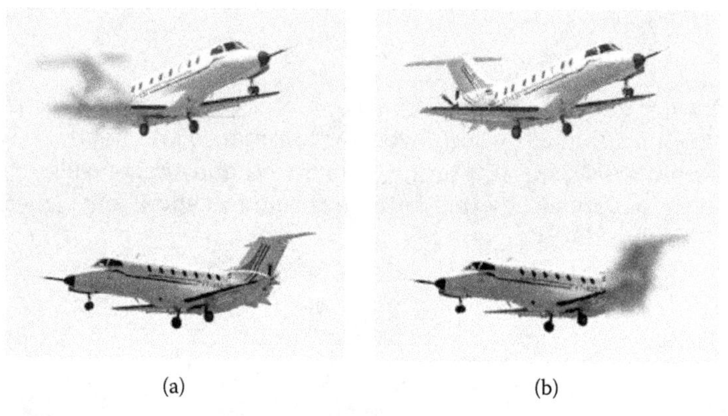

FIGURE 10.8
Source images for fusion: (a) image I_1, (b) image I_2.

any blurred image is taken as another source image I_2. The computed PCs and normalized SFs are given in Table 10.5. The performance metrics for different thresholds and block sizes for PCA and SF are given in Tables 10.6 and 10.7, respectively. Note that a threshold greater than 0.1 in case of PCA and 0.15 in case of SF show a degraded performance. When the chosen threshold is too high the fusion algorithms become a gray level averaging type of corresponding pixels, and block sizes of 4×4, 8×8, and 32×32 show degraded performance in both PCA and SF. The fused and error images by PCA and SF are shown in Figure 10.11 for block size 64×64 and $th = 0.025$. Table 10.8 shows performance metrics. Figure 10.12 shows the fused and error images by PCA and SF for block size 4×4 and $th = 0.2$. Table 10.9 shows the performance metrics.

Block-based image fusion scheme (the second approach) shows a somewhat enhanced performance, which may be due to the consideration

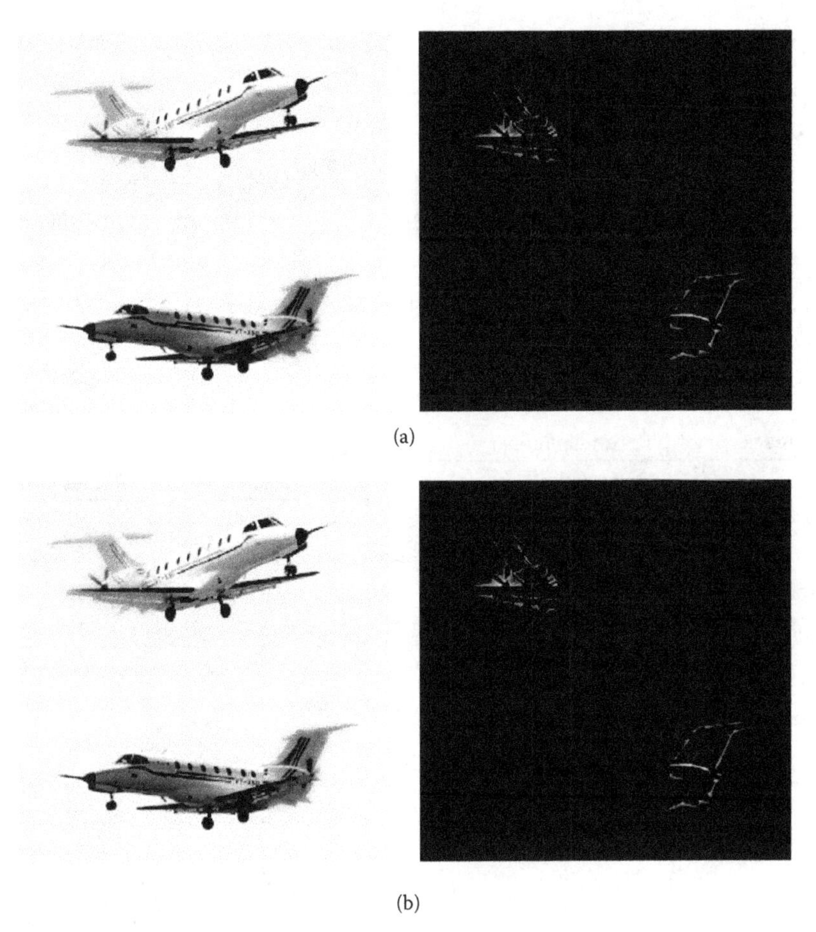

FIGURE 10.9
Fused and error images by (a) PCA and (b) SF.

TABLE 10.4
Performance Metrics (First Approach)

	RMSE	PFE	PSNR	SD	SF
PCA	5.8056	2.5388	40.5264	55.7286	16.7602
SF	5.7927	2.5332	40.5360	55.7302	16.7636

of local variations in the source images. The selection of block size and threshold is a relatively difficult job. One way to select block size and threshold is to compute the performance of the fused image for various block sizes and thresholds and then select the fused image with the best performance metric.

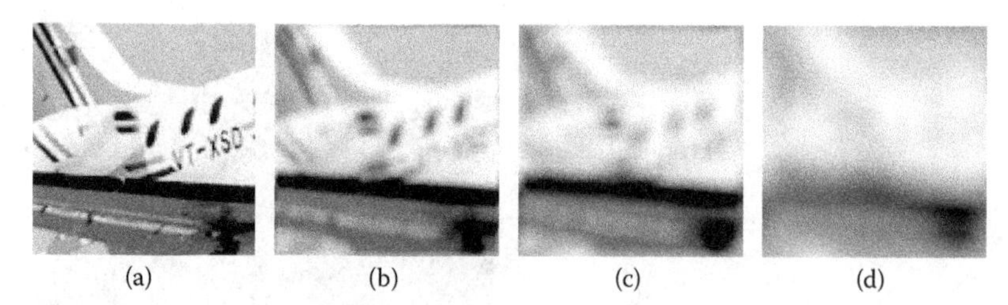

(a) (b) (c) (d)

FIGURE 10.10
(a) Original and its blurred versions; radius = 0 pixels. (b)–(d) with standard deviations of 10.

TABLE 10.5
PCA and SF of Blurred Images

	Radius = 0	Radius = 5	Radius = 9	Radius = 21
NPC1	0.5	0.5347	0.5611	0.6213
NPC2	0.5	0.4653	0.4389	0.3787
NSF1	0.5	0.7197	0.83	0.8936
NSF2	0.5	0.2803	0.17	0.1064

TABLE 10.6A
RMSE of the Fused Image by PCA (Various *th* and Block Sizes)

th	4 × 4	8 × 8	16 × 16	32 × 32	64 × 64	128 × 128	256 × 256
0	4.7355	3.5996	1.4115	2.5863	0.1665	0	0
0.025	4.7127	3.6013	1.4115	2.5863	0.1669	1.7458	0
0.05	4.6975	3.6150	1.4115	2.5863	0.1669	1.7458	3.7019
0.075	4.7195	3.6009	1.4115	2.5863	0.1610	3.3136	5.8080
0.1	4.7480	3.6127	1.4118	2.5863	0.1610	3.3136	5.8080
0.125	4.7693	3.6127	1.4118	2.5863	0.1677	4.0664	5.8080
0.15	4.8328	3.6081	2.1605	2.5867	0.1677	4.0664	5.8080
0.175	4.8103	3.6459	2.1605	2.5868	0.1677	4.7224	5.8080
0.2	4.8068	3.6459	2.1605	2.5863	0.1677	6.7926	5.8080
0.225	4.7936	3.6081	2.1605	2.5863	0.1677	6.7926	5.8080
0.25	4.8051	3.6081	2.1605	2.5863	0.1677	6.7926	5.8080
0.275	4.8215	3.5059	2.5705	2.5863	0.1677	6.7926	5.8080

10.3.3.2 Performance Metrics When No Reference Image Is Available

When the reference image is not available, the performance of image fusion methods can be evaluated using the following metrics:

1. *Standard deviation* (STD) measures the contrast in the fused image. An image with high contrast would have a high standard deviation:

$$SD = \sqrt{\sum_{i=0}^{L}(i-\bar{i})^2 h_{I_f}(i)}, \quad \bar{i} = \sum_{i=0}^{L} i h_{I_f} \tag{10.56}$$

TABLE 10.6B

PSNR of the Fused Image by PCA (Various *th* and Block Sizes)

th	4 × 4	8 × 8	16 × 16	32 × 32	64 × 64	128 × 128	256 × 256
0	41.4111	42.6023	46.6680	44.0381	55.9513	Inf	Inf
0.025	41.4321	42.6002	46.6680	44.0381	55.9395	45.7449	Inf
0.05	41.4461	42.5837	46.6680	44.0381	55.9395	45.7449	42.4806
0.075	41.4258	42.6007	46.6680	44.0381	56.0964	42.9617	40.5245
0.1	41.3997	42.5865	46.6672	44.0381	56.0964	42.9617	40.5245
0.125	41.3803	42.5865	46.6672	44.0381	55.9201	42.0727	40.5245
0.15	41.3228	42.5920	44.8192	44.0373	55.9201	42.0727	40.5245
0.175	41.3431	42.5468	44.8192	44.0372	55.9201	41.4232	40.5245
0.2	41.3462	42.5468	44.8192	44.0381	55.9201	39.8444	40.5245
0.225	41.3581	42.5920	44.8192	44.0381	55.9201	39.8444	40.5245
0.25	41.3478	42.5920	44.8192	44.0381	55.9201	39.8444	40.5245
0.275	41.3330	42.7168	44.0646	44.0381	55.9201	39.8444	40.5245

TABLE 10.7A

RMSE of the Fused Image by SF (Various *th* and Block Sizes)

th	4 × 4	8 × 8	16 × 16	32 × 32	64 × 64	128 × 128	256 × 256
0	3.9654	2.1281	1.4212	2.5942	0.1665	0	0
0.025	3.9016	1.8885	1.4212	2.5942	0.1610	0	0
0.05	3.9482	1.8885	1.4212	2.5942	0.1610	1.7458	0
0.075	3.9958	1.9016	1.4212	2.5942	0.1610	1.7458	0
0.1	4.0546	1.9226	1.4212	2.5942	0.1610	1.7458	0
0.125	4.0008	1.9226	1.4212	2.5942	0.1610	1.7458	0
0.15	4.0448	2.1034	1.4212	0.2027	0.1677	1.7458	0
0.175	3.9632	2.2860	1.4212	0.2027	0.1677	2.9331	3.7019
0.2	4.0151	2.2860	1.4212	0.2027	0.1677	2.9331	3.7019
0.225	3.9613	2.1687	1.4212	0.2027	0.1677	2.9331	5.8080
0.25	3.9624	2.2751	1.4212	0.2027	0.1677	2.9331	5.8080
0.275	3.7989	2.2751	1.4212	0.2027	0.1677	2.9331	5.8080

TABLE 10.7B

PSNR of the Fused Image by SF (Various *th* and Block Sizes)

th	4 × 4	8 × 8	16 × 16	32 × 32	64 × 64	128 × 128	256 × 256
0	42.1820	44.8848	46.6382	44.0248	55.9513	Inf	Inf
0.025	42.2524	45.4037	46.6382	44.0248	56.0964	Inf	Inf
0.05	42.2009	45.4037	46.6382	44.0248	56.0964	45.7449	Inf
0.075	42.1488	45.3736	46.6382	44.0248	56.0964	45.7449	Inf
0.1	42.0853	45.3260	46.6382	44.0248	56.0964	45.7449	Inf
0.125	42.1433	45.3260	46.6382	44.0248	56.0964	45.7449	Inf
0.15	42.0959	44.9356	46.6382	55.0964	55.9201	45.7449	Inf
0.175	42.1843	44.5741	46.6382	55.0964	55.9201	43.4915	42.4806
0.2	42.1279	44.5741	46.6382	55.0964	55.9201	43.4915	42.4806
0.225	42.1864	44.8028	46.6382	55.0964	55.9201	43.4915	40.5245
0.25	42.1852	44.5947	46.6382	55.0964	55.9201	43.4915	40.5245
0.275	42.3682	44.5947	46.6382	55.0964	55.9201	43.4915	40.5245

TABLE 10.8

Performance Metrics for a Block Size of 64 × 64 with $th = 0.025$

	RMSE	PFE	PSNR	SD	SF
PCA	0.1669	0.073	55.9395	57.0859	18.8963
SF	0.161	0.0704	56.0964	57.086	18.8962

(a) (b)

FIGURE 10.11
Fused and error images by the principal component analysis and spatial frequency block methods (a) 0PCA for $th = 0.025$ and block size 64 × 64 and (b) by SF for $th = 0.025$ and block size 64 × 64.

Here, $h_{I_f}(i)$ is the normalized histogram of the fused image $I_f(x,y)$ and L is the number of frequency bins in the histogram.

2. *Entropy* [15] is sensitive to noise and other unwanted rapid fluctuations. The information content of a fused image is

$$He = -\sum_{i=0}^{L} h_{I_f}(i) \log_2 h_{I_f}(i) \tag{10.57}$$

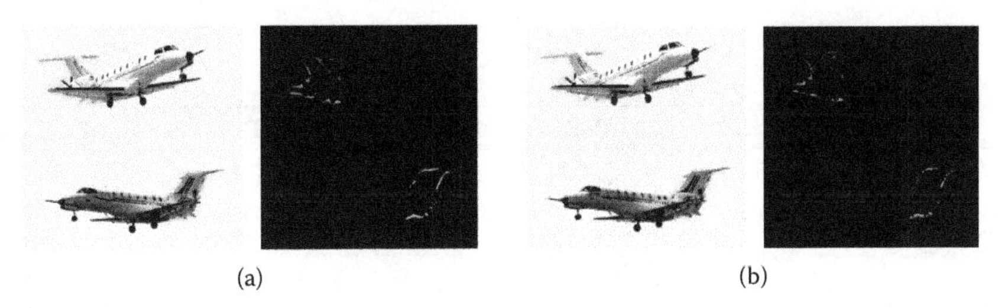

(a) (b)

FIGURE 10.12
Fused and error images (a) by PCA for $th = 0.2$ and block size 4 × 4 and (b) by SF for $th = 0.2$ and block size 4 × 4.

3. *Cross-entropy* [17] evaluates similarities in information content between input images and fused images. Overall cross-entropy of the source images I_1 and I_2 and the fused image I_f is

$$\mathrm{CE}(I_1, I_2; I_f) = \frac{\mathrm{CE}(I_1; I_f) + \mathrm{CE}(I_2; I_f)}{2} \tag{10.58}$$

Here,

$$\mathrm{CE}(I_1; I_f) = \sum_{i=0}^{L} h_{I_1}(i) \log\left(\frac{h_{I_1}(i)}{h_{I_f}(i)}\right) \text{ and } \mathrm{CE}(I_2; I_f) = \sum_{i=0}^{L} h_{I_2}(i) \log\left(\frac{h_{I_2}(i)}{h_{I_f}(i)}\right)$$

4. *Fusion MI* [17] measures the degree of dependence of the two images. A larger measure signifies a better quality. If the joint histogram between $I_1(x, y)$ and $I_f(x, y)$ is defined as $h_{I_1 I_f}(i, j)$ and $I_2(x, y)$ and $I_f(x, y)$ as $h_{I_2 I_f}(i, j)$, then the MI between the source and fused images is given as

$$\mathrm{FMI} = MI_{I_1 I_f} + MI_{I_2 I_f} \tag{10.59}$$

Here,

$$MI_{I_1 I_f} = \sum_{i=1}^{M} \sum_{j=1}^{N} h_{I_1 I_f}(i, j) \log_2\left(\frac{h_{I_1 I_f}(i, j)}{h_{I_1}(i, j) h_{I_f}(i, j)}\right);$$

$$MI_{I_2 I_f} = \sum_{i=1}^{M} \sum_{j=1}^{N} h_{I_2 I_f}(i, j) \log_2\left(\frac{h_{I_2 I_f}(i, j)}{h_{I_2}(i, j) h_{I_f}(i, j)}\right)$$

5. The *fusion quality index* (FQI) [18] with a range of 0 to 1 (1 indicating that the fused image contains all the information from the source images) is given as follows:

$$\mathrm{FQI} = \sum_{w \in W} c(w)\left(\lambda(w) QI(I_1, I_f \mid w) + \left(1 - \lambda(w)\right) QI(I_2, I_f \mid w)\right) \tag{10.60}$$

TABLE 10.9

Performance Metrics for a Block Size of 4×4 with $th = 0.2$

	RMSE	PFE	PSNR	SD	SF
PCA	4.8068	2.102	41.3462	56.7722	18.7141
SF	4.0151	1.7558	42.1279	56.8962	18.8518

Here, $\lambda(w) = \sigma_{I_1}^2 / \sigma_{I_1}^2 + \sigma_{I_2}^2$ and $C(w) = \max\left(\sigma_{I_1}^2, \sigma_{I_2}^2\right)$ are computed over window, $c(w)$ is a normalized version of $C(w)$, and $QI(I_1, I_f \mid w)$ is the quality index over a window for a given source image and fused image.

6. The *fusion similarity metric* (FSM) [17] takes into account the similarities between the source and fused image block within the same spatial position. The range is 0 to 1, 1 indicating that the fused image contains all the information from the source images. It is given as

$$\mathrm{FSM} = \sum_{w \in W} \mathrm{sim}(I_1, I_2, I_f \mid w)\left(QI(I_1, I_f \mid w) - QI(I_2, I_f \mid w)\right) + QI(I_2, I_f \mid w) \tag{10.61}$$

Here, $\mathrm{sim}(I_1, I_2, I_f \mid w) = \begin{cases} 0 & \text{if } \dfrac{\sigma_{I_1 I_f}}{\sigma_{I_1 I_f} + \sigma_{I_2 I_f}} < 0 \\ \dfrac{\sigma_{I_1 I_f}}{\sigma_{I_1 I_f} + \sigma_{I_2 I_f}} & \text{if } 0 \le \dfrac{\sigma_{I_1 I_f}}{\sigma_{I_1 I_f} + \sigma_{I_2 I_f}} \le 1 \\ 1 & \text{if } \dfrac{\sigma_{I_1 I_f}}{\sigma_{I_1 I_f} + \sigma_{I_2 I_f}} > 1 \end{cases}$

7. In SF [13,14] the frequency in the spatial domain indicates the overall activity level in the fused image.

10.3.4 Wavelet Transform

The theory of wavelets for signal processing is an extension of Fourier transform and short-time Fourier transform (STFT) theory [9,19,20]. In wavelets, a signal is projected on a set of wavelet functions (see the Appendix). The wavelet provides a good resolution in both the time and frequency domains. It is used extensively in image processing and provides a multiresolution decomposition of an image on a biorthogonal basis. The bases are wavelets and are functions generated by the translation and dilation of a mother wavelet. In WT analysis, a signal is decomposed into scaled (dilated or expanded) and shifted (translated) versions of the chosen mother wavelet or function $\psi(t)$. The wavelet is a small wave that grows and decays within a limited time period. It should satisfy the following two properties:

1. The time integral property:

$$\int_{-\infty}^{\infty} \psi(t)dt = 0 \tag{10.62}$$

2. The square of wavelet integrated over time property:

$$\int_{-\infty}^{\infty} \psi^2(t)dt = 1 \tag{10.63}$$

The WT of the 1D signal $f(x)$ onto a basis of wavelet functions is defined as

$$W_{a,b}\left(f(x)\right) = \int_{x=-\infty}^{\infty} f(x)\psi_{a,b}(x)dx \tag{10.64}$$

The basis is obtained by the translation and dilation operations of the mother wavelet as

$$\psi_{a,b}(x) = \frac{1}{\sqrt{a}}\psi\left(\frac{x-b}{a}\right) \tag{10.65}$$

The mother wavelet localizes in the spatial and frequency domains and should satisfy the zero mean constraint. For a discrete WT (DWT), the dilation factor is $a = 2^m$ and the translation factor is $b = n2^m$. Here, m and n are integers.

The process in one level of the 2D image decomposition is shown in Figure 10.13. The WT separately filters and downsamples the 2D data image in the vertical and horizontal directions using a separable filter bank. The input image $I(x,y)$ is filtered by a low-pass filter L and a high-pass filter H in the horizontal direction. Then, it is downsampled by a factor of two to create the coefficient matrices $I_L(x,y)$ and $I_H(x,y)$ while keeping the alternative samples. The coefficient matrices $I_L(x,y)$ and $I_H(x,y)$ are both low-pass filtered and high-pass filtered in the vertical direction and downsampled by a factor of two to create subbands (subimages): $I_{LL}(x,y), I_{LH}(x,y), I_{HL}(x,y)$, and $I_{HH}(x,y)$ [19]. The $I_{LL}(x,y)$ subband contains the average image information corresponding to the low-frequency band of multiscale decomposition. It could be considered a smoothed and subsampled version of the source image $I(x,y)$, and represents the approximation of the source image $I(x,y)$. $I_{LH}(x,y)$, $I_{HL}(x,y)$, and $I_{HH}(x,y)$ are detailed subimages and contain directional information (horizontal, vertical, and diagonal) of the source image $I(x,y)$ due to spatial orientation. The multiresolution process can be performed by recursively applying the same algorithm to low-pass coefficients from the previous decomposition. The labeled subbands (subimages) can be seen in Figure 10.14 [19,20].

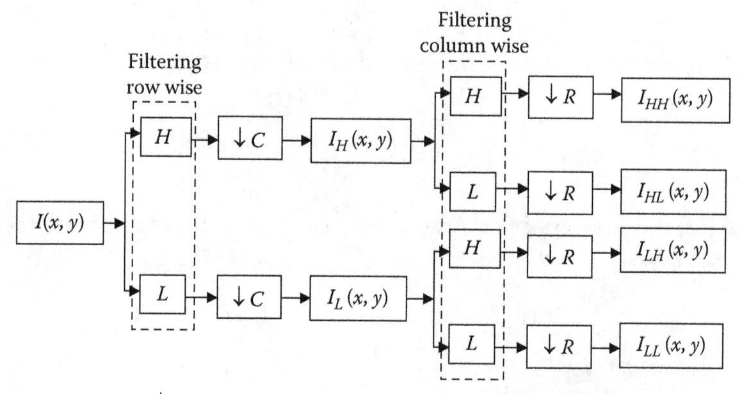

FIGURE 10.13
One level of 2D image decomposition. (From Naidu, V. P. S., and J. R. Raol. 2008. *Def Sci J*, 58:338–52. With permission.)

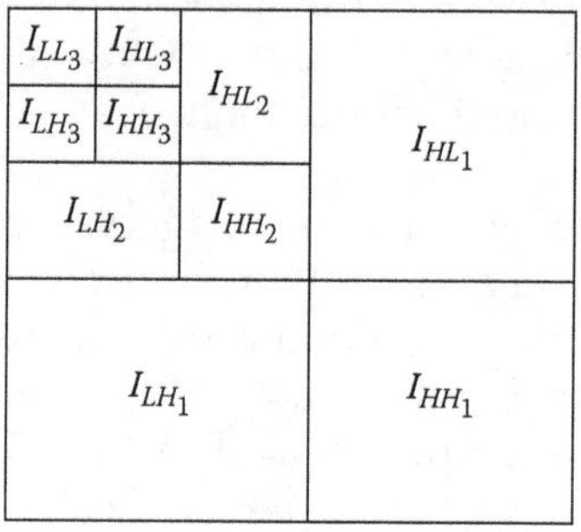

I_{LL_3} I_{HL_3} I_{LH_3} I_{HH_3} I_{HL_2} I_{LH_2} I_{HH_2} I_{HL_1} I_{LH_1} I_{HH_1}

FIGURE 10.14
Subbands.

An inverse 2D wavelet transform (IWT) process is used to restore the image $I(x,y)$ from subimages $I_{LL}(x,y)$, $I_{LH}(x,y)$, $I_{HL}(x,y)$, and $I_{HH}(x,y)$ by column-up sampling and filtering using a low-pass filter $\tilde{L}$ and a high-pass filter $\tilde{H}$ for each of the subimages, as shown in Figure 10.15. Row-up sampling and filtering of the resulting images with low-pass filter $\tilde{L}$ and high-pass filter $\tilde{H}$, and summation of all matrices, is used to restore the image $I(x,y)$. The finite impulse response filter coefficient of the low-pass and high-pass filters in both images decomposition/analysis and restoration/synthesis should fulfill the following conditions [19,20]:

$$\begin{aligned}
&\sum_{n=1}^{m} \mathbf{H}(n) = \sum_{n=1}^{m} \tilde{\mathbf{H}}(n) = 0 \\
&\sum_{n=1}^{m} \mathbf{L}(n) = \sum_{n=1}^{m} \tilde{\mathbf{L}}(n) = \sqrt{2} \\
&\tilde{\mathbf{H}}(n) = (-1)^{n+1}\mathbf{L}(n) \\
&\tilde{\mathbf{L}}(n) = (-1)^{n}\mathbf{H}(n) \\
&\mathbf{H}(n) = (-1)^{n}\mathbf{L}(m-n+1)
\end{aligned} \tag{10.66}$$

Here, m is the number of coefficients in the filter and n is the index of the filter coefficient, **L** and **H** are the vectors of numerator coefficients of the low- and high-pass filters, respectively, used in image decomposition, and $\tilde{\mathbf{L}}$ and $\tilde{\mathbf{H}}$ are the vectors of numerator coefficients of low- and high-pass filters, respectively, used in image reconstruction.

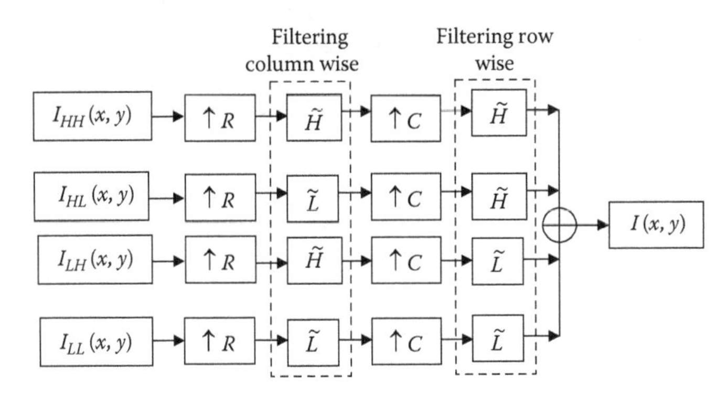

FIGURE 10.15
One level of 2D image restoration. (From Naidu, V. P. S., and J. R. Raol. 2008. *Def Sci J*, 58:338–52. With permission.)

10.3.4.1 Fusion by Wavelet Transform

The scheme for WT-image fusion is shown in Figure 10.16. The source images, $I_1(x,y)$ and $I_2(x,y)$ are decomposed into approximation and detailed coefficients at the required level using DWT. The coefficients of both images are subsequently combined using a fusion rule. The fused image $I_f(x,y)$ is then obtained by the inverse DWT as follows [9]:

$$I_f(x,y) = \text{IDWT}\left[\phi\left\{\text{DWT}\left(I_1(x,y)\right), \text{DWT}\left(I_2(x,y)\right)\right\}\right] \tag{10.67}$$

The fusion rule used is to average the approximation coefficients and pick the detailed coefficient in each subband with the largest magnitude.

10.3.4.2 Wavelet Transforms for Similar Sensor Data Fusion

A WT-based image fusion algorithm is used to fuse out-of-focus images obtained from similar sensors. To evaluate the fusion algorithm, the out-of-focus or complimentary pair input images I_1 and I_2 are taken, as shown in Figure 10.17. These images are created by blurring the size 512×512 reference image with a Gaussian mask using a diameter of 12 pixels. The blurring occurs at the left half and the right half, respectively. Figure 10.18 shows the fused and error images given by the WT image fusion algorithm. The error difference image is computed by taking the corresponding pixel difference of reference image and fused image, that is $I_e(x,y) = I_r(x,y) - I_f(x,y)$. The performance metrics with and without the reference image are given in Tables 10.10 and 10.11. Note that when the reference image is not available, the metrics SD, SF, and FMI are well-suited to evaluate the fusion results.

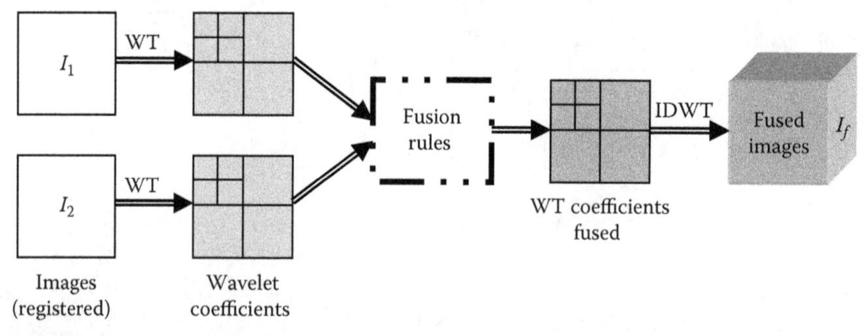

FIGURE 10.16
Image fusion process with multiscale decomposition.

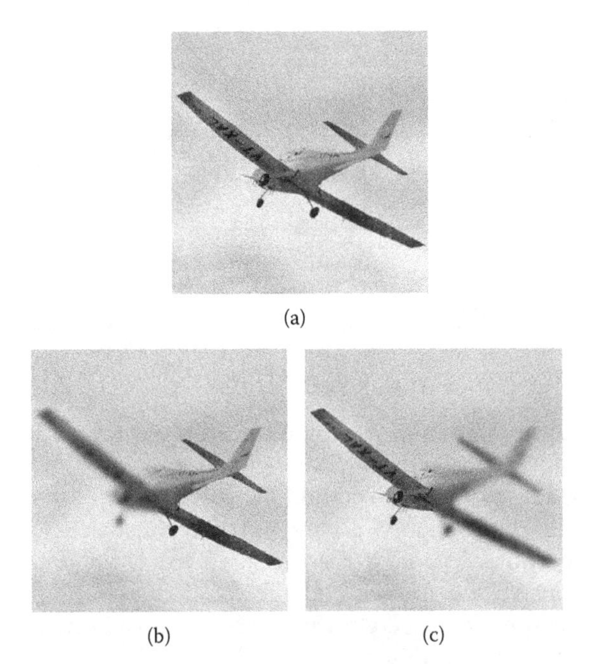

FIGURE 10.17
The images for fusion (a) truth image I, (b) image I_1, (c) image I_2.

FIGURE 10.18
Fused images with wavelet transform and the error image.

TABLE 10.10

Performance Metrics (with Reference Image)

	RMSE	PFE	MAE	CORR	SNR	PSNR	MI	QI	SSIM
PCA	2.01	1.05	0.13	1	39.6	45.1	1.92	0.983	0.997
SFA	**0.49**	**0.25**	**0.01**	1	**51.95**	**51.28**	**2**	**1**	**1**
WT	2.54	1.32	1.16	1	37.61	44.11	1.511	0.91	0.99

TABLE 10.11

Performance Metrics (without Reference Image)

	He	SD	CE	SF	FMI	FQI	FSM
PCA	6.36	44.12	0.02	12.31	2.99	0.78	**0.75**
SFA	6.26	**44.25**	0.02	**12.4**	**3**	0.78	0.74
WT	6.2	43.81	0.4	12.34	2.73	0.78	0.68

10.4 Fusion of Laser and Visual Data

Laser scanners are capable of providing accurate range measurements of the environment in large angular fields and at very fast rates. Laser scanned data and visual information can be fused to infer accurate 3D information. Simple 3D models are initially constructed according to 2D laser range data. Vision is then utilized to (1) validate the correctness of the constructed model; and (2) qualitatively and quantitatively characterize inconsistencies between the laser and visual data wherever such inconsistencies are detected [21]. The visual depth information is extracted only where laser range information is incomplete. Figure 10.19 depicts the fusion scheme [21].

10.4.1 3D Model Generation

To generate a local 3D model of the environment, an infinite horizontal plane (floor) is assumed right below the robot, at a known distance from the robot's coordinate system, which is the position of the range finding device. The defined line segments are extended to form rectangular vertical surfaces of infinite height. For each line segment, the plane that is perpendicular to the floor and contains the line segment is inserted into the 3D model. The coordinate system of the generated 3D model presumably coincides with the coordinated system of the robot. A local 3D model of the robot's environment is constructed based on a single 2D range scan. The environment presumably consists of a flat horizontal floor surrounded by piecewise vertical planar walls. The range measurements are initially grouped into clusters of connected points according to their sphere-of-influence graph. Clusters are then further grouped into line segments by utilizing an iterative-end-point-fit (IEPF) algorithm. Each of the resulting line segments corresponds to a vertical planar surface in the resulting model. These line segments extend to form rectangular vertical surfaces. For each line segment, the plane that is perpendicular to the floor and

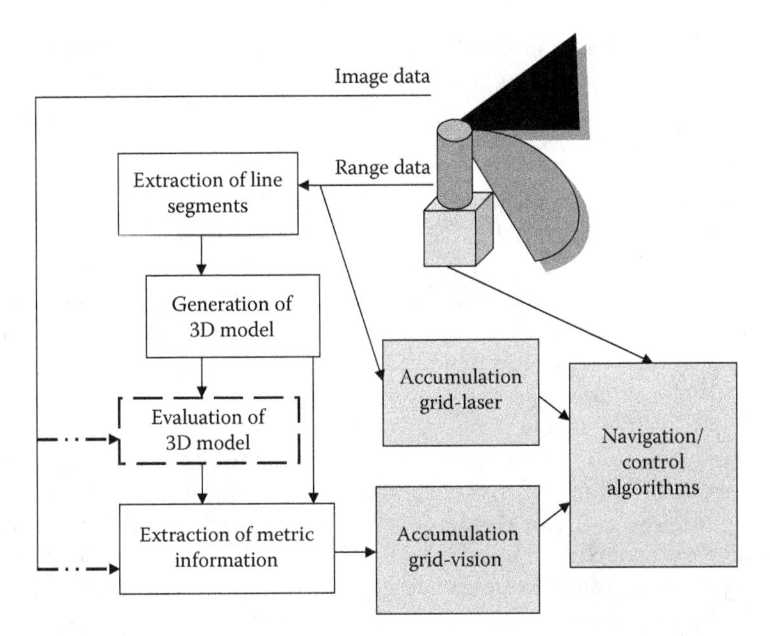

FIGURE 10.19
Laser and vision data fusion scheme.

contains the line segment is inserted into 3D model. A split-and-merge algorithm (SAMA) is given as follows [21]:

1. Initial: Set S1 consists of *N* points. Put S1 in a list *L*.
2. Fit a line to the next set Si in *L*.
3. Detect point *P* with maximum distance *Dp* to the line.
4. If *Dp* is less than a threshold, continue (if not, go to 2).
5. Otherwise, split Si at *P* into Si1 and Si2, replace Si in the list *L* by Si1 and Si2, and continue (if not, go to 2).
6. When all sets (segments) in *L* have been checked, merge the collinear segments.

A pair of images acquired by a calibrated stereo vision rig is used for validation of the 3D model. The points of the first image are ray-traced to the 3D model, and 3D coordinates are estimated. Based on this information, the image points are reprojected onto the frame of the second camera. If the assumed 3D model is correct, then the image constructed by reprojection should be identical to the image actually acquired by the second camera. Wherever the model is not correct, the images differ. A local correlation of image intensity values reveals regions with such inconsistencies. Vision is then further utilized to provide additional depth information in regions where the 2D range data proved to be insufficient.

10.4.2 Model Evaluation

Model evaluation procedure is described next [21]. Let M be the 3D model built according to range data acquired at time $t1$, and let the I1 be an image acquired by a camera C1 at the same time, $t1$. For each image point P1 = ($x1$, $y1$) of I1, the coordinates (x, y, and z) of the corresponding 3D point P can be computed by ray-tracing it to the model M. If the assumptions made for constructing the 3D model are correct, the coordinates (x, y, and z) found by the above procedure correspond to a real-world point on M. Let I2 be a second image acquired by the second camera of the stereoscopic system. Since the coordinate system of C2 with respect to the coordinate system of M is also known, the projection P2 = ($x2$, $y2$) of P = (x, y, and z) on C2 can also be computed. Ray-tracing the points of I1 to find 3D world coordinates and then back-projecting them to I2 leads to an analytical computation of point correspondences between I1 and I2.

If the assumptions made to form model M are correct, corresponding image points would actually be projections of the same world points and thus would share the same attributes, such as color, intensity, values, and intensity gradients. If the image points have different attributes, then there is a strong indication that the model is locally invalid. The normalized cross-correlation metric can be used to evaluate the correctness of the calculated point correspondences. Low values indicate the regions within the images that depict parts of the environment that do not conform to the 3D model.

10.5 Feature-Level Fusion Methods

2D and 3D intensity and color images have been extensively researched for object recognition problems [22]. A 3D representation provides additional information. A 3D picture lacks texture information. A 2D image can complement 3D information. 2D images are localized in many details, such as eyebrows, eyes, nose, mouth, and facial hair, whereas 3D images are difficult to localize and are not accurate for such details. A robust system may require the fusion of 2D and 3D images. Then, nonclarity in one aspect, such as lighting problem, may be compensated by other aspects, such as depth features. Image fusion can be performed at the feature level, matching score level, or decision level, utilizing different fusion models, for example, sum, product, maximum, minimum, and major voting to combine the individual scores (normalized to range [0,1]). A multivariate polynomial (MP) technique provides an effective way to describe complex nonlinear input/output (I/O) relationships. MP is also tractable for optimization, sensitivity analysis, and predication of confidence intervals. With the incorporation of a decision criterion into the model output, the MP can be utilized for pattern analysis. It can then serve as a fusion model to overcome the limitations of the existing decision fusion models.

We can use an extended reduced MP model (RMPM) [22] to fuse appearance and depth information for object or face recognition. RMPM is useful in problems with a small number of features. To apply RMPM to a recognition problem, PCA is used for dimension reduction and feature extraction and a two-stage PCA + RMPM can be used for the recognition problem. A recursive formulation for online learning of new-user parameters is also possible. There are three main techniques for 3D capture: (1) passive stereo, using at least two cameras to capture an image and using a computational matching method; (2) structured lighting, by projecting a pattern on a face; and (3) the use of laser range-finding device.

10.5.1 Fusion of Appearance and Depth Information

A new feature is formed by linking together the appearance of an image and the depth or disparity of the image. A RMPM can be trained using the combined 2D and 3D features. The general MP model is given as [22]

$$g(\mathbf{a},\mathbf{x}) = \sum_{i=1}^{K} \mathbf{a}_i \mathbf{x}_1^{n_1} \ldots \mathbf{x}_l^{n_l} \tag{10.68}$$

Here, the summation is over all nonnegative integers ($\leq r$). r is the order of approximation, $\mathbf{a}$ is the parameter vector estimated, and $\mathbf{x}$ is a regression vector containing inputs as l. A second-order bivariate polynomial model ($r = 2$ and $l = 2$) is given by

$$g(\mathbf{a},\mathbf{x}) = \mathbf{a}^T p(\mathbf{x}) \tag{10.69}$$

in which $\mathbf{a}$ has six elements and $p(\mathbf{x}) = [1 \;\; \mathbf{x}_1 \;\; \mathbf{x}_2 \;\; \mathbf{x}_1^2 \;\; \mathrm{x}_1 \;\; \mathrm{x}_2 \;\; \mathbf{x}_2^2]^T$. Let m data points be given with $m > K$ ($K = 6$). Using the least-squares error minimization, the objective is given by

$$s(\mathbf{a},\mathrm{x}) = \sum_{i}^{m} \left[\mathrm{y}_i - g(\mathbf{a},\mathbf{x}_i) \right]^2 = [\mathbf{y} - P\mathbf{a}]^T [\mathrm{y} - P\mathbf{a}] \tag{10.70}$$

Vector $\mathbf{a}$ can be estimated from

$$\mathbf{a} = (P^T P)^{-1} P_T \mathbf{y} \tag{10.71}$$

Here, matrix P is the classical Jacobian matrix, and vector $\mathbf{y}$ is the known inference vector from the training data. It is also possible to use a reduced MP with only a few terms. RMPM is useful for problems with a small number of features and a large number of examples. Since the face space of the image is large, dimension reduction might be necessary, for which PCA can be used. PCA is used for appearance and depth images. The fusion of appearance and depth information is accomplished at the feature level by linking the eigenface features of the appearance and depth images. The learning algorithm of an RMPM is expressed as

$$P = RM\left(r, \left\{W_{\text{eigenappearance}} \; W_{\text{eigendepth}}\right\}\right) \tag{10.72}$$

Here, r is the order of the RMPM, and Ws are the eigenappearance and eigendepth of the eigenface features. The parameters of the RMPM can be determined from the training samples using the LS method explained earlier in this section. In the testing phase, a probe face is identified as a face of the gallery if the output element of the reduced model classifier (appearance and depth) is the maximum (and = 0.5) among all the faces in the training gallery.

10.5.2 Stereo Face Recognition System

Figure 10.20 shows the schematic of this process [22]. The output of a stereo vision system is a set containing three images: left image, right image, and disparity image. Facial features can be detected from either the left image or the disparity image because they are assumed to be fully registered. In addition, nose tip can be detected from the disparity image and eye corners from the left image. By combining the disparity and intensity images, the 3D pose of either the head or the face can be tracked. The head is tracked if the facial features are not available, for example, when the person is far away from the stereo head or when the face is viewed in profile. The face is tracked once the facial features, such as nostrils and eyebrows, are found. Disparity maps of the face can be obtained at the frame rate using commercially available stereo software (such as the SRI International Small Vision System). Assuming that the person of interest is the nearest object to the camera, the range data are extracted from the disparity map. The head contour can be modeled as an ellipse, which can then be least-squares fitted to points in the watershed segmentation. The eye and mouth corners can be extracted using a corner detector and the nose tip in the disparity image by a template matching. The head pose can be estimated using the calibrated parameters of the vision system.

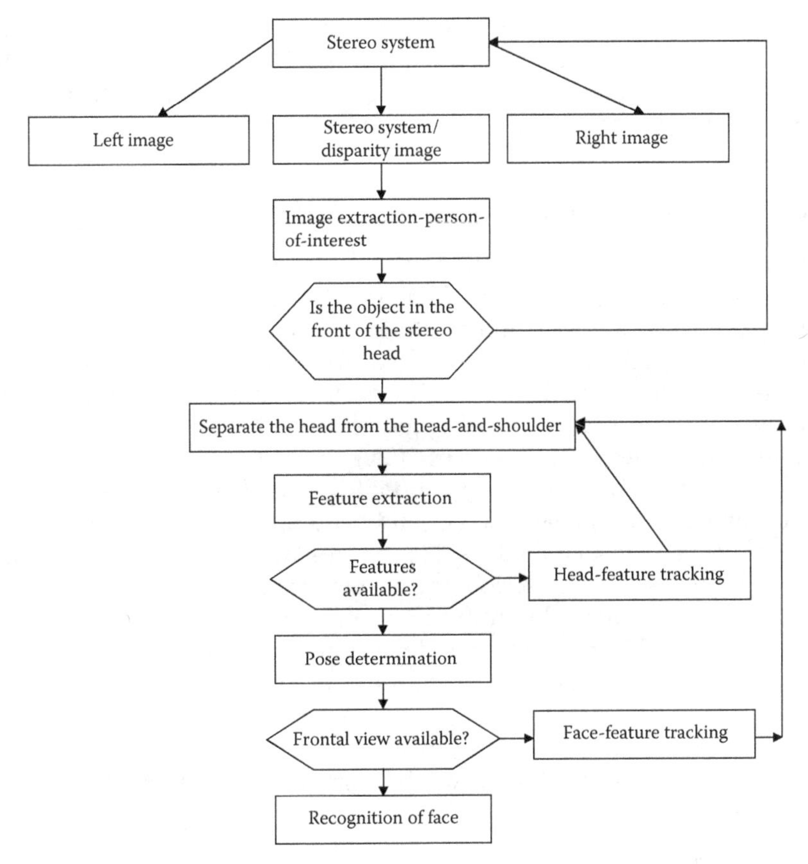

FIGURE 10.20
Head–face tracking scheme.

10.5.2.1 Detection and Feature Extraction

Face detection includes locating the face, extracting the facial features, and normalizing the image. Face detection is easier with available 3D information. An object-oriented segmentation can be applied to ascertain the person of interest: the one closest to the camera. The nearer the object to the camera, the brighter the pixel in the disparity image. Hence, a histogram-based segmentation can be applied. The subject of interest can be segmented using thresholds that are selected based on the peak analysis of the disparity histogram. This would help in tracking the objects. Two persons at different distances in front of the camera are separated using a disparity map. Some feature extraction methods are based on artificial template matching, which is a small rectangular intensity image that contains, for example, an eye corner. The corner is located in the center of the template. The image region that best matches the artificial template is extracted from current image. To apply the corner detector, we need to establish a rectangular search region for the mouth and two rectangular regions for the left and right eyes, respectively.

10.5.2.2 Feature-Level Fusion Using Hand and Face Biometrics

Multibiometric systems use information from multiple biometric sources to verify the identity of an individual, for example, face and fingerprints, multiple fingers of a user, and multiple matchers [23]. Information from these multiple sources are consolidated into three distinct levels: (1) the feature extraction level, (2) the match score level, and (3) the decision level. Fusion at the feature level is a relatively less attended problem and can be studied in three ways: (1) fusion of PCA and LDA coefficients of the face; (2) fusion of LDA coefficients corresponding to the R, G, and B channels of a face image; and (3) fusion of face and hand modalities. Thus, multibiometric systems consolidate the evidence presented by multiple biometric traits or sources. Such systems improve the recognition performance of a biometric system in addition to improving population coverage, deterring spoof attacks, increasing the degrees of freedom, and reducing the failure-to-enroll rate. The storage requirements, processing time, and computational demands of a multibiometric system are much higher than a monobiometric system.

The information in a multibiometric system can be integrated at the following levels:

1. Sensor level: Data acquired from multiple sensors are processed and integrated in order to generate new data from which features can be extracted, for example, in the case of face biometrics, both 2D texture information and 3D depth/range information, obtained using two different sensors, can be fused to generate a 3D texture image of the face that is then subjected to feature extraction and matching.
2. Feature level: Feature sets extracted from multiple data sources can be fused to create a new feature set that represents the individual, for example, the geometric features of the hand may be augmented with the eigencoefficients of the face to construct a new high-dimensional feature vector. The feature selection and transformation method may be used to elicit a minimal feature set from the high-dimensional feature vector.
3. Match-score level: Multiple classifiers output a set of match scores that are fused to generate a single scalar score. The match scores generated by the face and hand modalities of a user may be

combined via the simple sum to obtain a new match score, which is then used to make the final decision.

4. Rank level: This fusion is used in identification systems in which each classifier associates a rank with every enrolled identity, and a higher rank indicates a good match. The multiple ranks associated with an identity are consolidated and a new rank that would aid in establishing the final decision is determined.
5. Decision level: When each matcher outputs its own class label (i.e., the accept or reject decision in a verification system or the identity of a user in an identification system), a single class label can be obtained by using techniques such as majority voting and behavior knowledge space.

10.5.3 Feature-Level Fusion

Feature-level fusion can be accomplished by linking the feature sets obtained from multiple information sources. Let $\mathbf{X} = [\mathbf{x}_1, \mathbf{x}_2, \ldots, \mathbf{x}_m]$ and $\mathbf{Y} = [\mathbf{y}_1, \mathbf{y}_2, \ldots, \mathbf{y}_m]$ be the feature vectors that represent the information extracted using two different sources. The fusion of these two sets should give a new feature vector that is generated by first augmenting vectors **X** and **Y**, and then performing feature selection on the resultant vector. The process is described in Sections 10.5.3.1 through 10.5.3.3.

10.5.3.1 Feature Normalization

The location (mean) and the scale (variance) of the feature values are modified due to a possible large variation in these values. The minimum–maximum method can give the new value = $(x - \min \mathrm{F}(x)/(\max \mathrm{F}(x) - \min \mathrm{F}(x))$. This requires that the minimum and maximum values be computed beforehand. In the median normalization method, the new value = $(x - \text{median } F(x)/(\text{median } (|x - \text{median } F(x)|)$. The denominator is an estimate of the scale parameter of the feature value. The modified feature vectors are subsequently used.

10.5.3.2 Feature Selection

The two normalized feature vectors are put in $\mathrm{Z} = \{\mathbf{X}, \mathbf{Y}\}$ form. A minimal feature set of size k $[k < (m + n)]$ is chosen that enhances the classification performance on a training set of feature vectors. The new feature vector is selected based on the genuine accept rate (GAR) at four levels of false accept rate (FAR) values: 0.05, 0.1, 1, and 10%, in the receiver operating characteristics (ROC) curve related to the training data set.

10.5.3.3 Match Score Generation

Let there be two composite feature vectors, one at each time instant $Z(t)$ and $Z(t+d)$, for example, $\mathbf{Z} = (\mathbf{X}, \mathbf{Y})$. Let Sx and Sy be the normalized match (i.e., Euclidean distance) generated by matching each **X** and **Y** from the two sets. Then $S = (Sx + Sy)/2$ is the fused match score. A distance measure or metric must be computed here.

Once we have the match score level and the feature-level information, we combine them using the simple sum rule to obtain the final score. In the genuine pairs, the high match score would be the effect of a few features' values that constitute the fused vector. The feature selection process eliminates the redundant features and the features that are correlated with the other features. A detailed treatment with results is given in [23].

10.6 Illustrative Examples

The following examples will help in image analysis and image-related fusion exercises, although some of them may not be directly related to image fusion.

EXAMPLE 10.1

Create a current frame and reference object with the following MATLAB® code.

```
% Current frame
Ic = zeros(32,32);
Ic(14:21,14:21) = 1;
% Reference target/object
Ir = ones(7,7);
%END of the program
```

Determine the centroid of the reference object in the current frame using the correlation, sum absolute difference (SAD), and FFT methods.

SOLUTION 10.1

Run the MATLAB code *ch10_ex1*. It contains the codes for the correlation, SAD, and FFT methods. The centroid based on (1) correlation is 17.0, 17.0, (2) SAD is 17.0, 17.0, and (3) FFT is 16.0, 17.0.

EXAMPLE 10.2

Compute the coefficients (for the transformation function in Equation 10.3) to register two images using the LS method and CPs from the images to be registered. Assume that the CPs from first image are (52.64, 201.66) (261.53, 206.21)

(143.46, 246.17) (158.91, 197.13) and (5.4, 201.67) and the CPs for the second image are (86.51, 189.87) (218.48, 190.78) (144.76, 242.65) (152.95 186.22) and (40.09, 191.89).

SOLUTION 10.2

Run the MATLAB code *ch10_ex2*. It contains the code to perform estimation using the LS method. The results are as follows:

a0 a1 a2:	27.79	0.68	0.08
b0 b1 b2:	−47.56	−0.02	1.19

EXAMPLE 10.3

Compute the performance evaluation metrics for the given true and filtered images. Assume that the true and filtered images are generated using the following MATLAB code:

```
% image generation
rand('seed,'9119);randn('seed,'7117);
s = floor(128*rand(4,4)); % true or ground truth
y = floor(s + randn(4,4)); % filtered image
% END of the program
```

SOLUTION 10.3

Run the MATLAB code *ch10_ex3*. It contains the codes for the RMSE, MAE, PFE, SNR, PSNR performance metrics. RMSE is 1.03, MAE is 0.81, PFE is 1.73, SNR is 35.24, and PSNR is 48.00.

EXAMPLE 10.4

Find the centroid of the following object using the nonconvolution method.

$$I = \begin{bmatrix} 2 & 2 & 2 & 2 \\ 2 & 3 & 3 & 2 \\ 2 & 2 & 2 & 2 \end{bmatrix}$$

SOLUTION 10.4

Run the MATLAB code *ch10_ex4*. The estimated centroid is at [2.0, 2.50].

EXAMPLE 10.5

Find the match between the current frame and reference signal based on SAD. Assume that the current frame is $Ic = [1,5,5,4,5,3,2,2,6,7,1,2,9,5,6,4,5,6,7,8,0,0]$ and the reference signal is $Ir = [2,2,6]$.

Solution 10.5

Run the MATLAB code *ch10_ex5*.

Match =
Columns 1 through 13
5 8 6 8 8 5 0 5 14 10 4 8 10
Columns 14 through 20
9 7 5 8 11 17 14

The reference signal is matched with current signal at index number 7, since the value of SAD is zero at the seventh position.

EXAMPLE 10.6

Find the match between the current frame and the first and second reference signals. Assume that the current signal is Ic = [8, 9, 7, 5, 3, 4, 1], the first reference signal is Ir1 = [8, 9, 7], and the second reference signal is Ir2 = [3, 4, 1].

Solution 10.6

Run the MATLAB code *ch10_ex6*. The current frame is matched with the first reference at 3 and also with the second reference at 3.

EXAMPLE 10.7

Simulate a target of size (9,9) with intensity 64 and standard deviation 2. Place this target on a black background at a location of (19,19). Find the match between the simulated frame and the target based on SAD. Show the results in terms of contour and mesh plots.

Solution 10.7

Run the MATLAB code *ch10_ex7*. The resulting plots are shown in Figures 10.21 to 10.23.

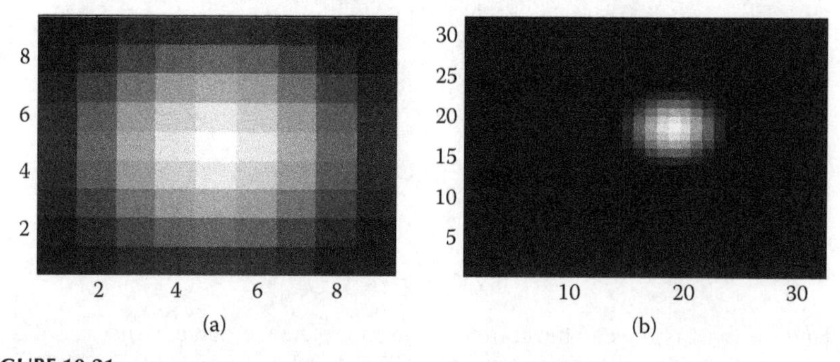

FIGURE 10.21
(a) The simulated target and (b) the target in the background for Example 10.7.

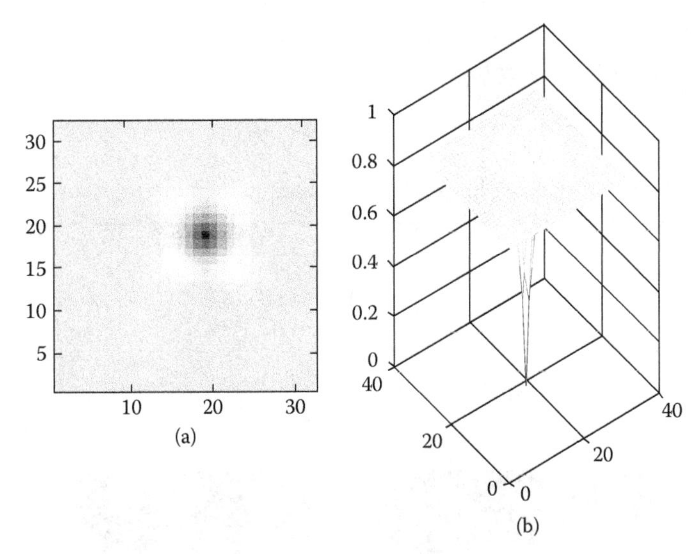

FIGURE 10.22
(a) The matched image and (b) the mesh for Example 10.7.

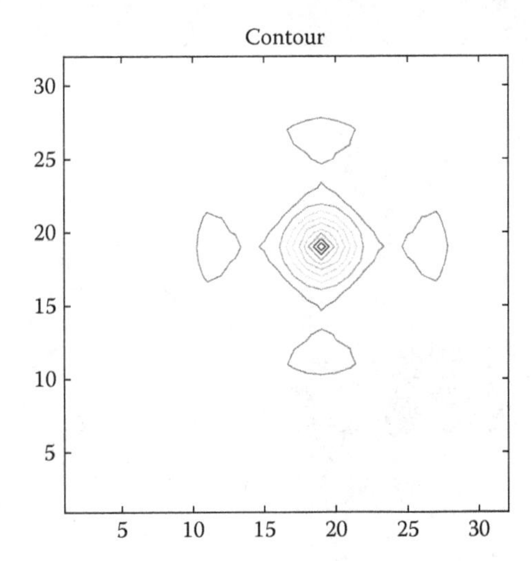

FIGURE 10.23
The contour for Example 10.7.

EXAMPLE 10.8

Use the data from Example 10.7 and find the match between the simulated frame and the target based on NCC. Show the results in terms of contour and mesh plots.

SOLUTION 10.8

Run the MATLAB code *ch10_ex8*. The resulting plots are shown in Figures 10.24 to 10.26.

EXAMPLE 10.9

Compute the PCs of two images. Assume that the two images are generated with following MATLAB code:

```
%
rand('seed,'3443);
I1 = floor(128*rand(4,4));
rand('seed,'1991);
I2 = floor(128*rand(4,4));
%
```

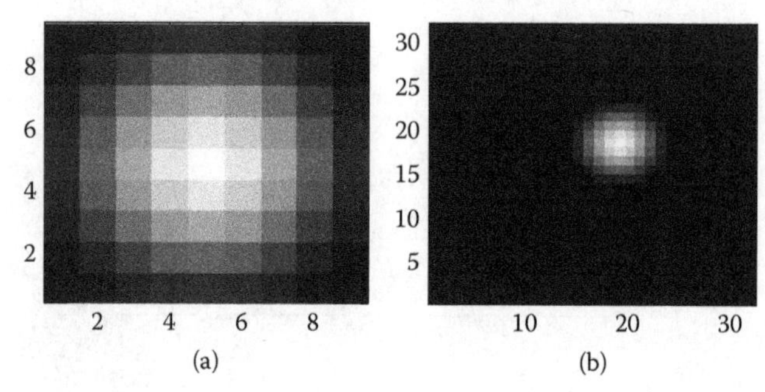

FIGURE 10.24
(a) The simulated target and (b) the target in the background for Example 10.8.

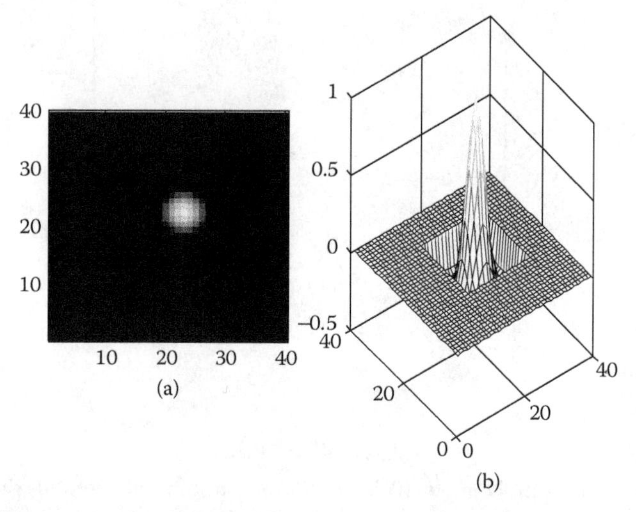

FIGURE 10.25
(a) The matched image and (b) the mesh for Example 10.8.

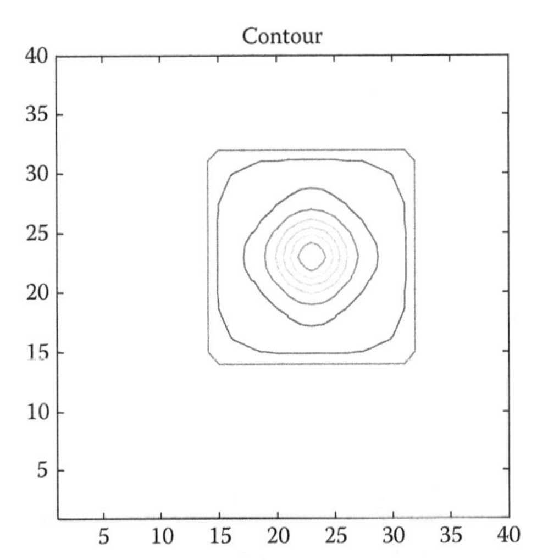

FIGURE 10.26
The contour for Example 10.8.

SOLUTION 10.9

Run the MATLAB code *ch10_ex9*. The first PC is 0.93, and the second is 0.07.

EXAMPLE 10.10

Compute the SF of an image. Assume that the image is generated with the following MATLAB code:

```
%
rand('seed,'3443);
I = floor(128*rand(4,4));
%
```

SOLUTION 10.10

Run the MATLAB code *ch10_ex10*. RF is 44.82, CF is 51.64, and SF is 68.38.

EXAMPLE 10.11

Find the target location by measuring the azimuth angles to the target from the sensors. Assume that sensor 1 is at (0,0) and sensor 2 is at (50,0). The target is seen by sensor 1 at an angle of 0.2013 radian and by sensor 2 at an angle of 1.6705 radian. Use the triangulation method.

SOLUTION 10.11

Run the MATLAB code *ch10_ex11*. The target location is $x = 49$ and $y = 10$.

11

Performance Evaluation of Image-Based Data Fusion Systems

An evaluation of the performance of a data fusion (DF) system, algorithms, and software is very important in order to establish confidence in the DF system. Various performance evaluation methods for tracking filters have already been discussed in Chapter 4, and Sections 10.2.3, 10.3.3, and 10.3.3.2 discussed methods for tracking filter/image fusion when the reference image is either available or not available. In this chapter, the performance of various tracking filters and related algorithms for data analysis supported by image and/or acoustic data is evaluated.

11.1 Image Registration and Target Tracking

In a tracking procedure that uses data from imaging sensors: (1) the target is specified, and then an image-registration algorithm searches for the target in each subsequent image obtained by the imaging sensor; (2) the measurements resulting from the image-registration algorithm are passed on to a target-state estimator; and (3) the estimator continually estimates the position of the target. The image-registration algorithm's sum of the absolute difference (SAD), normalized cross-correlation (NCC), and Kalman filter (KF) tracking algorithm can be used for image tracking.

11.1.1 Image-Registration Algorithms

An image-registration algorithm finds the centroid of the target in the current frame by registering the target's reference image (I) within the current image frame.

11.1.1.1 Sum of Absolute Differences

The SAD of two 1D discrete signals, $I_c(x)$ of length M and $I_r(x)$ of length P, is computed using the following expression:

$$\text{SAD}(x) = \sum_{i=0}^{P-1} | I_c(x+i) - I_r(i) |, \quad x = 0,1,2,\ldots,M-1 \tag{11.1}$$

The reference signal is aligned with each pixel in the search/current frame and then is subtracted from it. This process yields another signal in which each pixel contains the SAD values. The SAD will be minimal at the position where similarity is maximal. The SAD of two 2D images, $I_c(x,y)$ of length $M \times N$ and $I_r(x,y)$ of length $P \times Q$, is computed using the following expression [24]:

$$\mathrm{SAD}(x,y)=\sum_{i=0}^{P-1}\sum_{j=0}^{Q-1}|I_c(x+i,y+j)-I_r(i,j)|, \quad \begin{array}{l} x=0,1,2,\ldots,M-1 \\ y=0,1,2,\ldots,N-1 \end{array} \tag{11.2}$$

In a 2D view, if the reference image and the current frame are aligned as in Figure 11.1, then the image formed by the SAD is as shown in Figure 11.2.

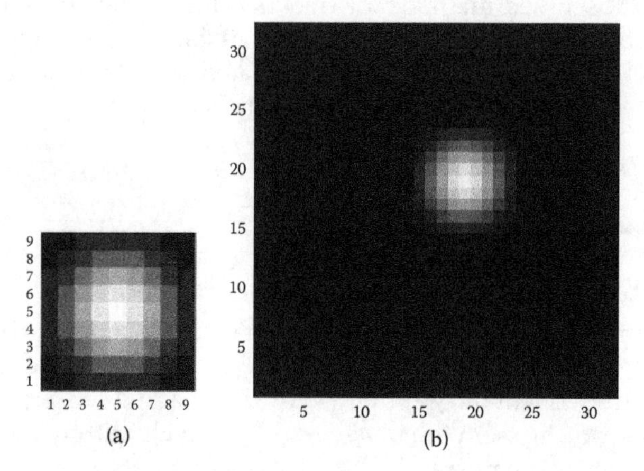

FIGURE 11.1
(a) Reference image, and (b) current frame.

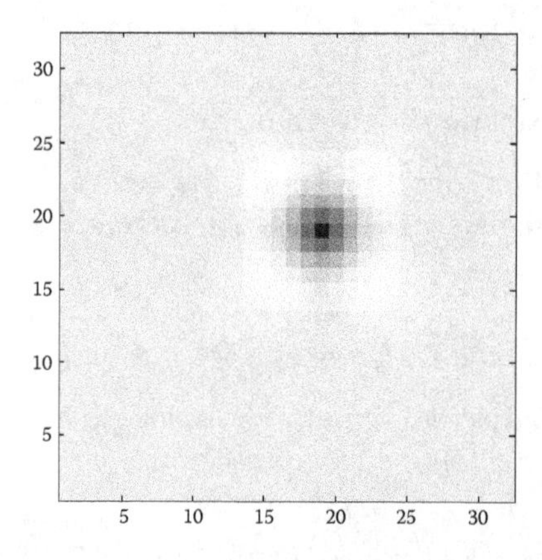

FIGURE 11.2
The SAD image.

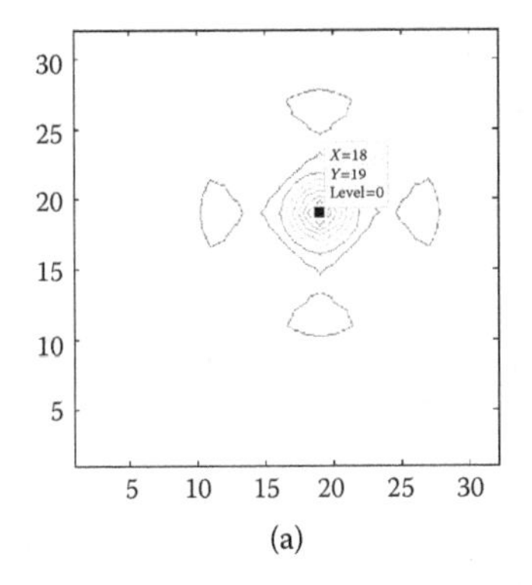

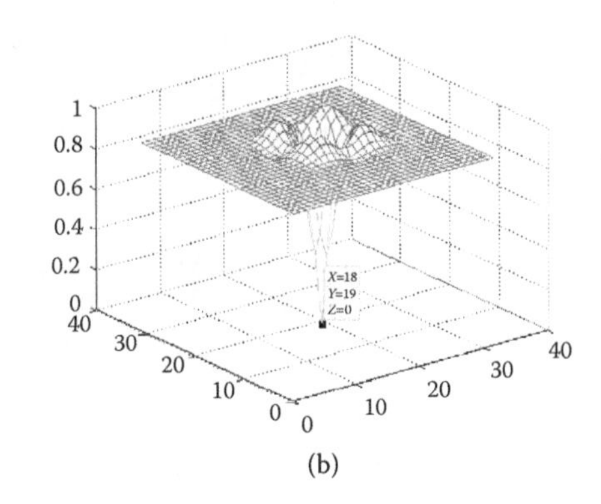

(a) (b)

FIGURE 11.3
(a) 2D and (b) 3D representations from the SAD algorithm.

The 2D and 3D representations of the results obtained using the SAD algorithm are illustrated in Figure 11.3, which shows that the minimum value is obtained at pixel (19, 19), the centroid of the reference image in the current frame. This means that the reference image and the current frame are matched/registered at location (19, 19).

11.1.1.2 Normalized Cross Correlation

The cross-correlation of two 1D discrete signals $I_c(x)$ and $I_r(x)$ of lengths M and N, respectively, will be a sequence of length $M+N-1$ and is computed using the equation

$$\mathrm{CC}(x)=\sum_{i=0}^{N-1} I_c(x+i)I_r(i) \quad x=0,1,2,\ldots,M+N-1 \tag{11.3}$$

The NCC is then given as follows [25–27]:

$$\mathrm{NCC}(x)=\sum_{i=0}^{N-1}\frac{I_c(x+i)I_r(i)}{\sqrt{\sum_{i=0}^{N-1} I_c^2(x+i)}\sqrt{\sum_{i=0}^{N-1} I_r^2(i)}} \quad x=0,1,2,\ldots,M+N-1 \tag{11.4}$$

The cross-correlation of two 2D discrete signals, $I_c(x,y)$ and $I_r(x,y)$, of dimensions $M\times N$ and $P \times Q$ (a 2D correlation sequence of dimensions $(M+P-1)\times(N+Q-1)$, is given as

$$\mathrm{CC}(x,y)=\sum_{i=0}^{N-1}\sum_{j=0}^{M-1} I_c(x+i,y+j)I_r(i,j) \quad \begin{array}{l} x=0,1,2,\ldots,M+P-1 \\ y=0,1,2,\ldots,N+Q-1 \end{array} \tag{11.5}$$

The NCC of the two 2D discrete signals, $I_c(x,y)$ and $I_r(x,y)$ of dimensions $M \times N$ and $P \times Q$, results in a 2D correlation sequence of dimensions $(M+P-1)\times(N+Q-1)$, as depicted below:

$$\mathrm{NCC}(x,y)=\sum_{i=0}^{N-1}\sum_{j=0}^{M-1}\frac{I_c(x+i,y+j)I_r(i,j)}{\sqrt{\sum_{i=0}^{N-1}\sum_{j=0}^{M-1}I_c^2(x+i,y+j)}\sqrt{\sum_{i=0}^{N-1}\sum_{j=0}^{M-1}I_r^2(i,j)}} \tag{11.6}$$

Figure 11.4 shows the image obtained as a result of the NCC of the two images. The dimensions of the correlated image are $(32+9-1)\times(32+9-1)$, i.e., 40 × 40. Figure 11.5 shows the 2D and 3D images of the NCC. The maximum value is obtained at point (23, 23). This implies that the maximum correlation is achieved when the bottom-left corner of the reference image is at (−9 + 1 + 23, −9 + 1 + 23) or (15, 15). Thus, the centroid will be at (15 + (9 − 1)/2, 15 + (9 − 1)/2), i.e., (19, 19). The above algorithms provide the point in the current frame around which the similarity to the reference image is maximized; in effect, these algorithms yield the position of the centroid of the reference image within the current frame.

11.1.2 Interpolation

An interpolation may be required when an image corresponds to a large physical area. There are many types of interpolations, such as linear,

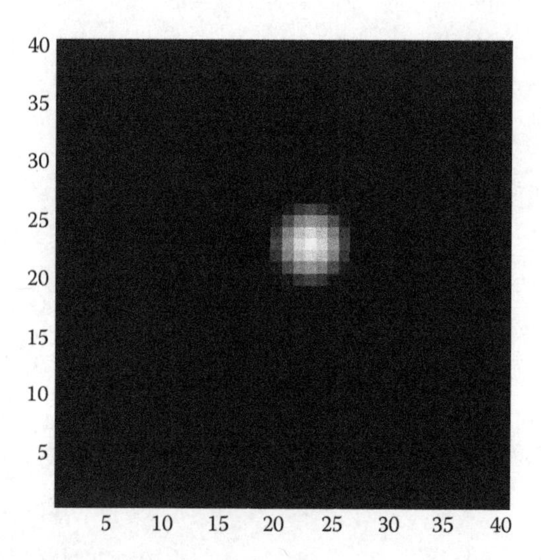

FIGURE 11.4
A 2D NCC image.

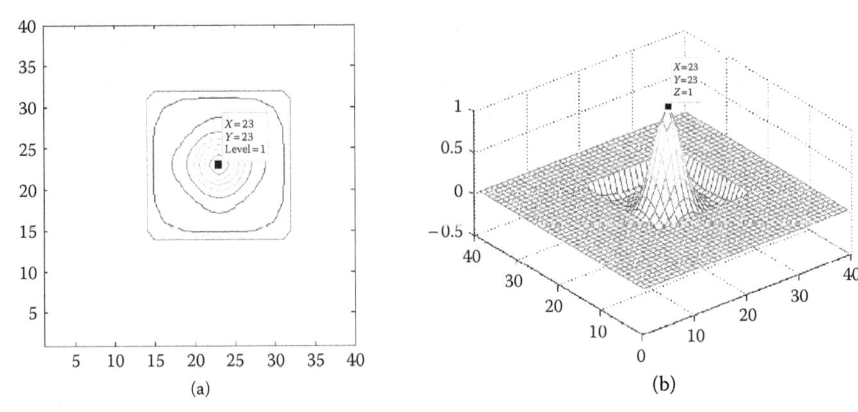

FIGURE 11.5
(a) 2D and (b) 3D representations from an NCC algorithm.

polynomial, and spline. The centroid determined by the centroid-search algorithm has integer or half-integer values, which are interpolated using the function for a paraboloid based on the following formulae [24]:

$$\begin{aligned}
a &= q(x_0, y_0) \\
b &= \frac{1}{2}(q(x_1, y_0) - q(x_{-1}, y_0)) \\
c &= \frac{1}{2}(q(x_0, y_1) - q(x_0, y_{-1})) \\
d &= -q(x_0, y_0) + \frac{1}{2}q(x_1, y_0) + \frac{1}{2}q(x_{-1}, y_0) \\
e &= -q(x_0, y_0) + \frac{1}{2}q(x_0, y_1) + \frac{1}{2}q(x_0, y_{-1})
\end{aligned} \tag{11.7}$$

Here, the actual centroid of the target is $(x_c, y_c) = \left(\frac{b}{2d}, \frac{c}{2e}\right)$; q is a 3 × 3 matrix, with the peak at the center and the immediate neighbors of the peak at their corresponding positions.

EXAMPLE 11.1

If the values corresponding to the peak are

$$q = \begin{bmatrix} (x_{-1}, y_1) & (x_0, y_1) & (x_1, y_1) \\ (x_{-1}, y_0) & (x_0, y_0) & (x_1, y_0) \\ (x_{-1}, y_{-1}) & (x_0, y_{-1}) & (x_1, y_{-1}) \end{bmatrix} = \begin{bmatrix} 3 & 3 & 3 \\ 4 & 5 & 3 \\ 3 & 4 & 3 \end{bmatrix}$$

then the peak may be at the position (2, 2); but upon interpolation, the peak is said to be at (2.1667, 1.8333).

11.1.3 Data Simulation and Results

Data simulation (DS) is carried out using PC-MATLAB® with a graphical user interface (GUI), as shown in Figure 11.6.

1. Target: The DS allows a choice of parameters, such as size, intensity, and standard deviation (sigma), for generating a target with Gaussian intensities, or alternatively, for importing an image file to act as the target.
2. Compression: The data may be stored as desired in an audio/video interleaved (AVI) file using any type of compression.
3. Movie parameters: The size of the background on which the target is moved may be changed according to the needs of the simulation. The number of frames captured in the simulation may also be customized. The sampling interval and the initial state vector may be defined. An image file may also be used as a background for the simulated data.
4. Noise parameters: The data may be corrupted with two types of noise—salt and pepper, with selectable noise density, or Gaussian noise, with selectable variance. The target may be simulated as a

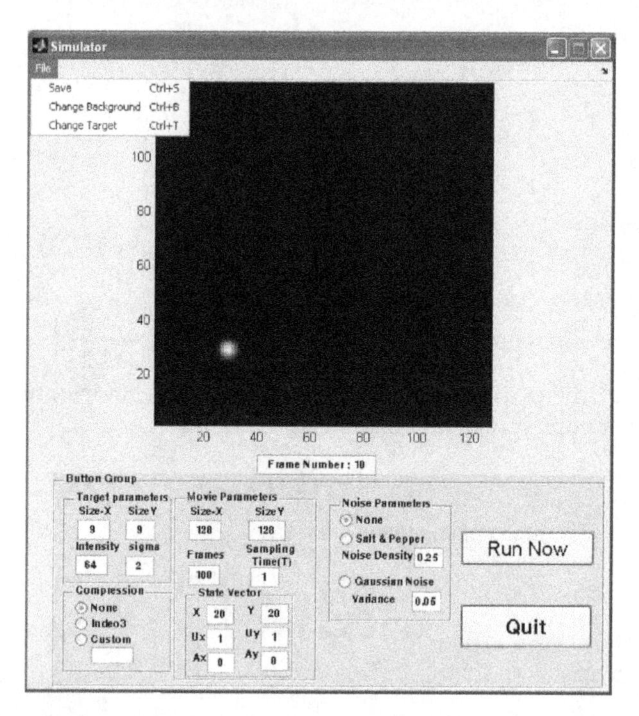

FIGURE 11.6
Data simulator.

rectangular block, having a Gaussian distribution of intensities around its center. A Gaussian distribution is given as follows:

$$f_g(x) = \frac{1}{\sqrt{2\Pi\alpha^2}} e^{-\frac{(x-a)^2}{2\sigma^2}} \tag{11.8}$$

Here, σ is the standard deviation, and a is the location parameter or mean, which corresponds to the center of the rectangle in this case.

The size, intensity, and standard deviation for computing the probability distribution function (pdf) can be chosen, leading to different appearances of the target's image. The target is placed in consecutive video frames at a specific position, depending on the state vector and the sampling time. The state vector has six fields: position, velocity, and acceleration, along both the x- and y-axes, i.e., $[x \quad \dot{x} \quad \ddot{x} \quad y \quad \dot{y} \quad \ddot{y}]$. The simulation also allows the addition of two types of noise: salt and pepper and Gaussian noise, with noise density or standard deviation as needed in the simulation. One of the simulated frames that contains both the target and the image is shown in Figure 11.7.

A data set for testing is created that contains 100 frames of video corrupted by salt and pepper noise of noise density 0.05, in which the target moves with constant acceleration (CA) in a parabolic trajectory from the bottom-left corner to the bottom-right corner.

This data set is used to test the image-registration algorithms for target tracking with KF. Figure 11.8 shows the GUI for target tracking. The tracking performance is evaluated using performance-check metrics and the following process:

1. The tracker allows AVI files containing the video data to be opened.
2. In the first frame, the user is prompted to select the target that will act as the reference image.

FIGURE 11.7
Noisy image frame: salt and pepper (density = 0.05).

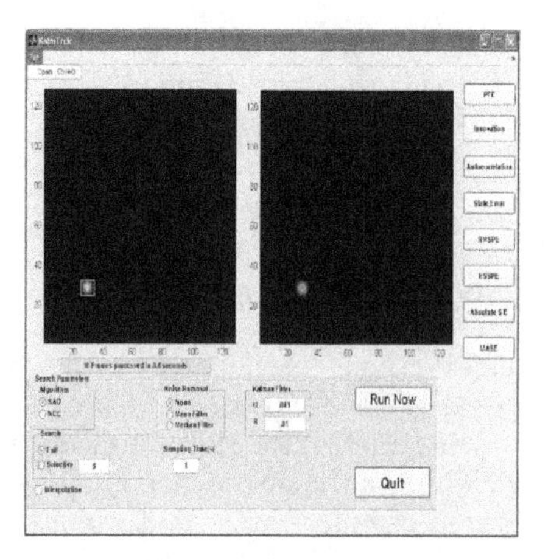

FIGURE 11.8
Target tracker.

3. Consequently, one of the two image-registration algorithms (SAD or NCC) is used to obtain the centroid of the target. The left axes show the output of the image-registration algorithm, and the right axes show the output of the KF. During tracking, a count of the number of frames processed and the time spent in doing so are displayed.
4. The search for the reference image may be conducted across the entire image or only in the vicinity of the estimated position of the target. The size of the vicinity is defined by the selective parameter.
5. Interpolation may be used to obtain subpixel-level tracking accuracy.
6. One of two filters, mean or median, may be used to remove or reduce the noise in the video data.
7. The sampling time is the time between two successive frames and corresponds to the sampling time of the video-capture device.
8. KF-tuning parameters Q (process-noise covariance) and R (measurement-noise covariance) may also be set.
9. The buttons show the filter performance-evaluation parameters.

The percentage fit error (PFE), root mean square error in position (RMSPE) and mean absolute error (MAE) metrics are presented in Table 11.1 and depicted in Figure 11.9. Interpolation reduces the state-estimation error in situations where the images are corrupted with salt and pepper noise.

The innovation sequence with theoretical bounds, state-position errors with theoretical bounds, root-sum-square (RSS) position errors, and absolute errors (AE) in x and y positions when the prefiltering and interpolation

TABLE 11.1

Performance Metrics: PFE, RMSPE, and MASE

Metric	Without Median Filter, Without Interpolation		With Median Filter, Without Interpolation		Without Median Filter, With Interpolation		With Median Filter, With Interpolation	
	SAD	NCC	SAD	NCC	SAD	NCC	SAD	NCC
PFEx	0	0.39	0	0	0.04	0.36	0.07	0.12
PFEy	0.51	0.55	0.51	0.51	0.51	0.53	0.46	0.48
RMSPE	0.249	0.339	0.249	0.249	0.249	0.317	0.229	0.24
MAEx	0	0.124	0	0	0.024	0.180	0.039	0.071
MAEy	0.303	0.325	0.303	0.303	0.302	0.314	0.275	0.282

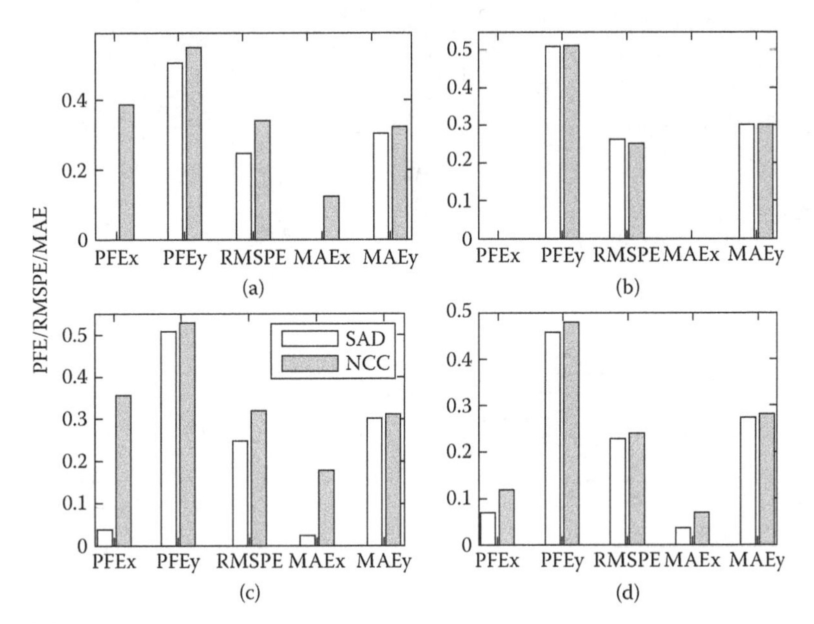

FIGURE 11.9
Performance metrics (a) without both filter and interpolation; (b) with mean filter, but without interpolation; (c) without filter but with interpolation; and (d) with both mean filter and interpolation.

are not considered are shown in Figures 11.10 through 11.13, respectively. SAD performed marginally better than NCC.

Figures 11.14 and 11.15 show the RSS-position and the AEs in the x and y positions, where prefiltering is considered and interpolation is not considered. Here, both SAD and NCC performed almost equally well. Figures 11.16 and 11.17 show the RSS-position and AEs in the x and y positions, where prefiltering is not considered and interpolation is considered. SAD

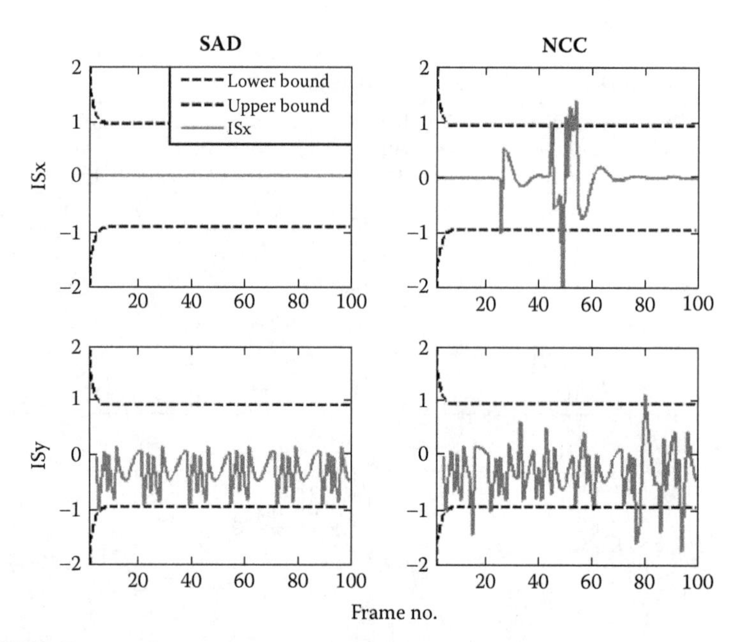

FIGURE 11.10
Innovations (without filtering or interpolation).

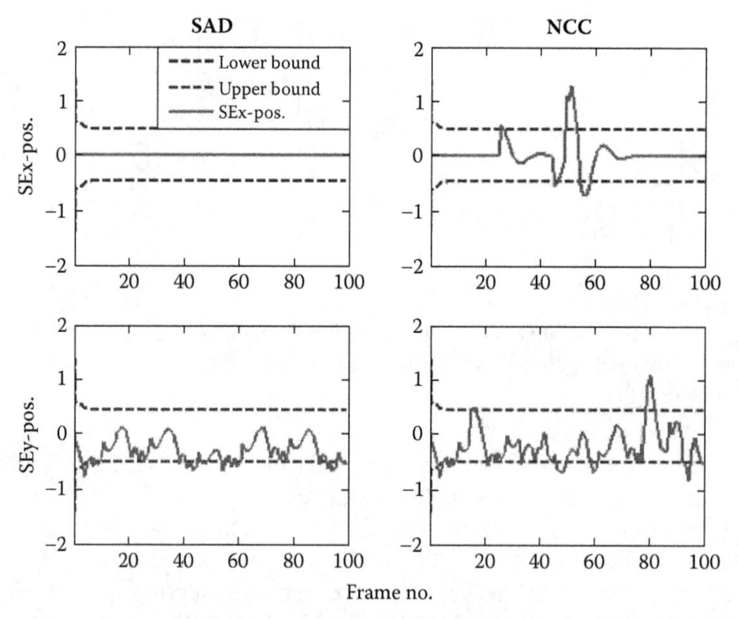

FIGURE 11.11
State errors in the x and y positions, processed without filtering or interpolation.

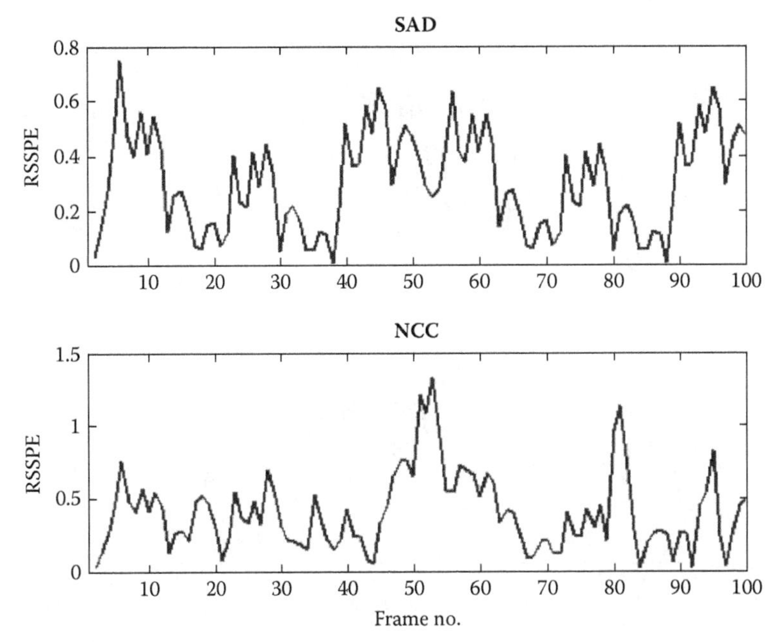

FIGURE 11.12
RSSPE without filtering or interpolation.

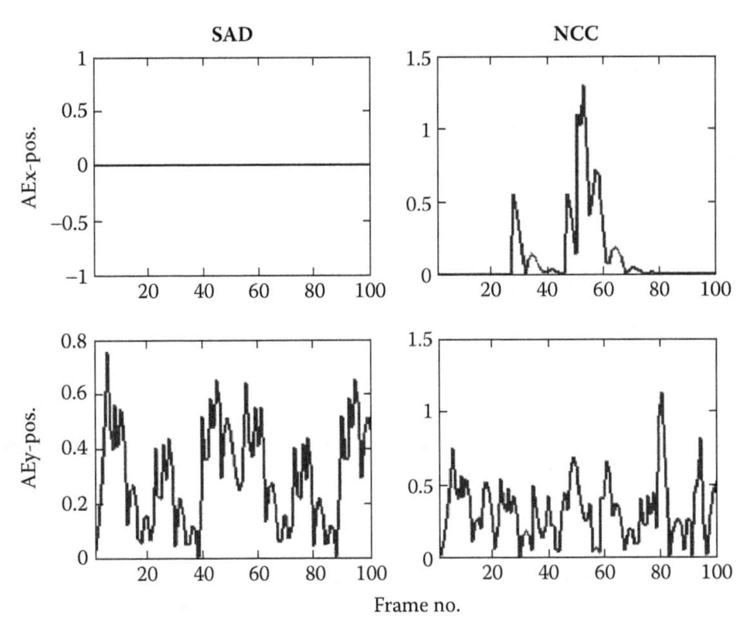

FIGURE 11.13
Absolute errors in the x and y positions without filtering or interpolation.

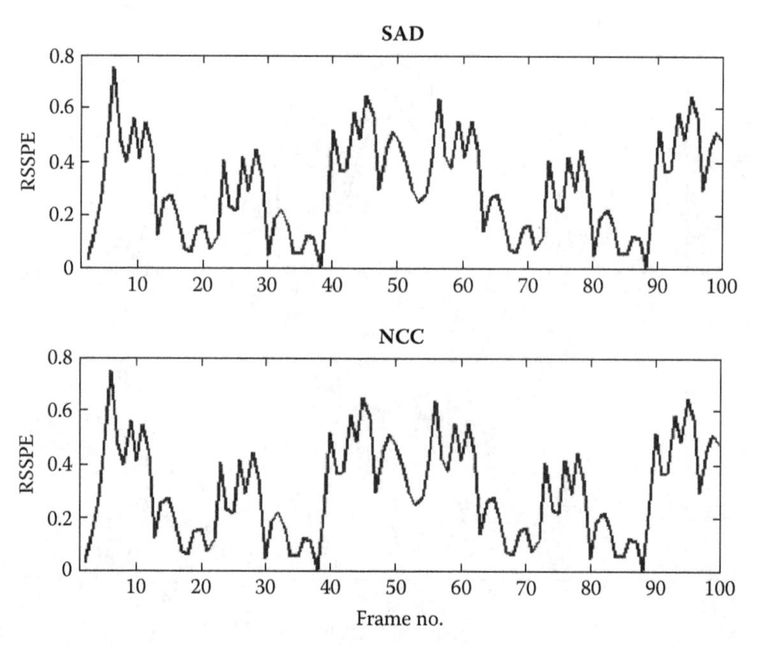

FIGURE 11.14
RSSPE with median filter, without interpolation.

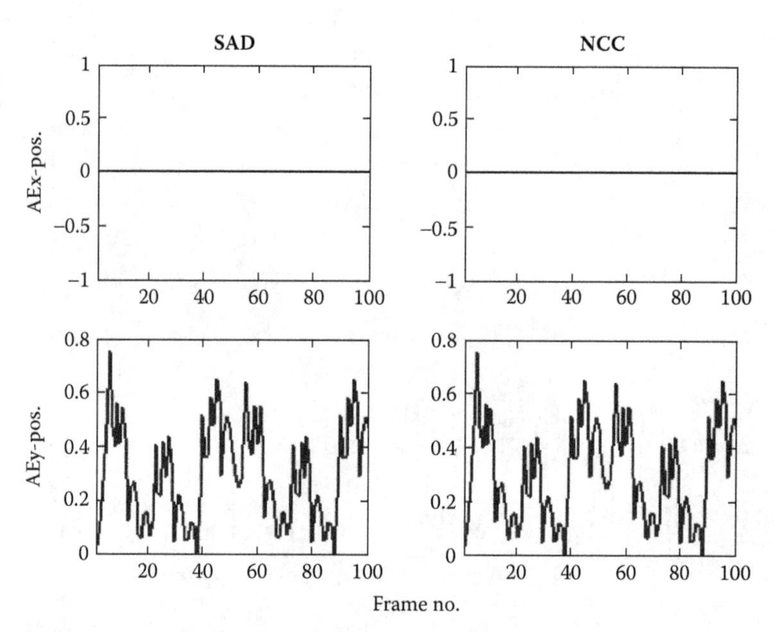

FIGURE 11.15
Absolute errors with median filter, without interpolation.

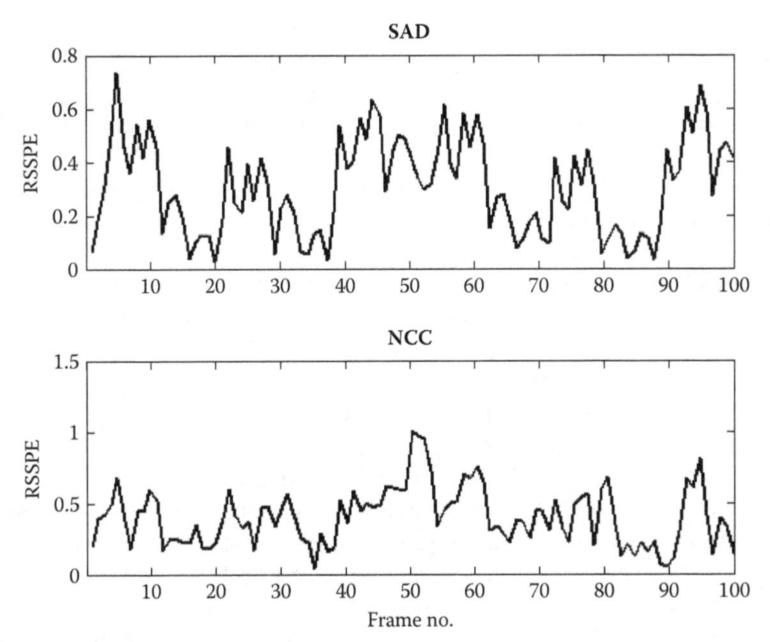

FIGURE 11.16
RSSPE with interpolation, without filtering.

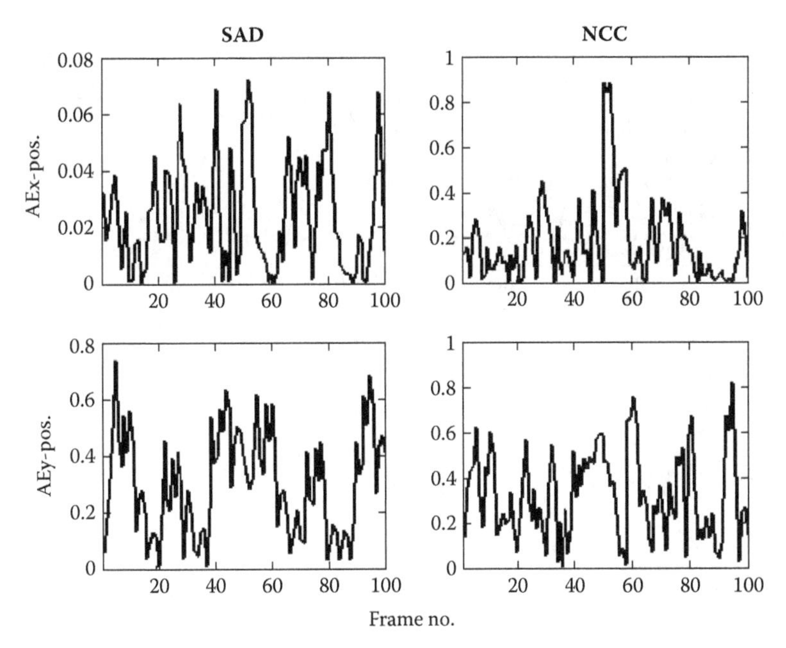

FIGURE 11.17
Absolute errors with interpolation, without filtering.

performed marginally better than NCC. Figures 11.18 and 11.19 show the RSS-position and AEs in the x and y positions, when both prefiltering and interpolation are considered. Both SAD and NCC performed almost equally well here. A spatial filter (a median filter in this case) therefore improves the results, as does filtering with interpolation. Moreover, SAD fares better than NCC in the estimation of states using this data.

In addition, a spatial filter (a mean filter in this case) improves tracking performance, as does interpolation, although only marginally. NCC also fares better than SAD in the estimation of the state vector using this data. When comparing SAD and NCC as image-registration techniques, the performance of SAD is better than that of NCC when the input-image data are corrupted by salt and pepper noise.

Image-registration algorithms, namely SAD and NCC, spatial filtering algorithms as a preprocessing step, and an interpolation algorithm to achieve subpixel-level accuracy were implemented using PC-MATLAB. A KF was used to track the centroid of the target obtained using the image-registration algorithm. Pertaining to the comparison of SAD and NCC as image-registration techniques, the following points should be noted: (1) in the absence of noise, both image-registration techniques proved to be equally accurate; (2) in the presence of salt and pepper noise, SAD proved to be more accurate than NCC; and (3) in the presence of Gaussian noise,

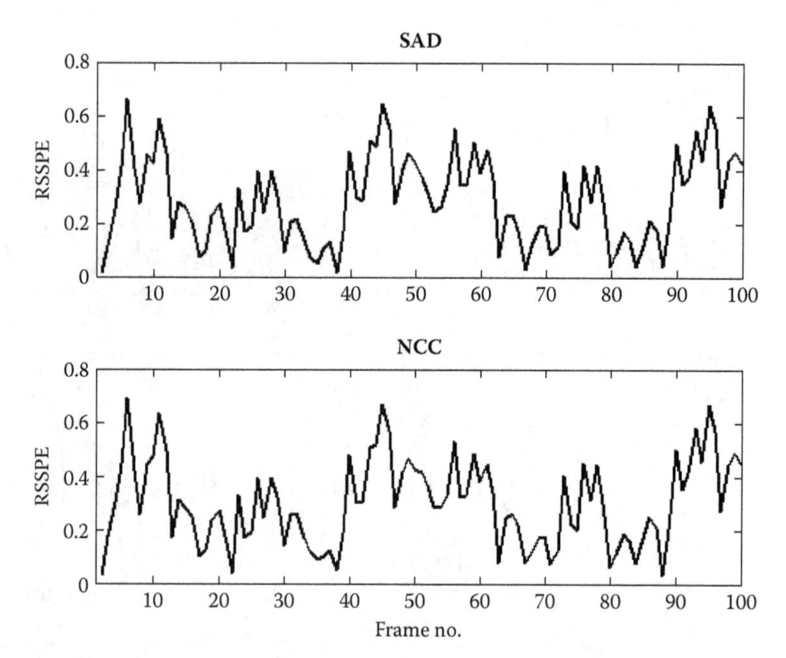

FIGURE 11.18
RSSPE with median filter and interpolation.

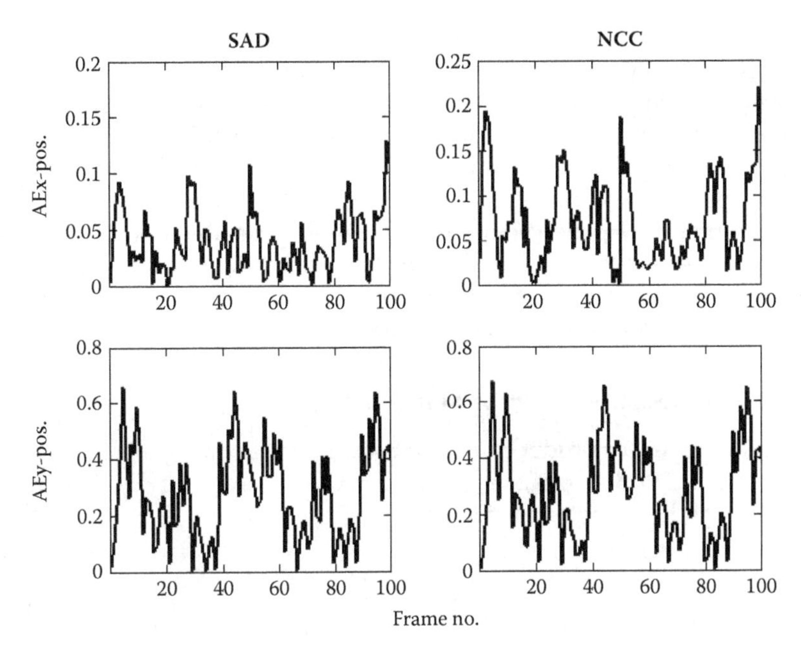

FIGURE 11.19
Absolute errors in the x and y positions (with median filter and interpolation).

NCC proved to be more accurate than SAD. The main points pertaining to the effect of spatial filtering are as follows: (1) in the presence of salt and pepper noise, the median filter drastically reduced the state-estimation error when either of the image-registration techniques was used; and (2) in the presence of Gaussian noise, the mean filter drastically reduced the error in state estimation when either of the image-registration techniques were used (the results are not shown here). Finally, interpolation reduces the error in state estimation.

11.2 3D Target Tracking with Imaging and Radar Sensors

Multiple sensors are often used to enhance target-tracking capabilities. Radar can measure range with good resolution, but angular measurements have poor resolution. Radar provides sufficient information to track the target because it measures both the angles and ranges of a target. The uncertainty associated with radar might be represented as a volume whose dimensions are relatively large perpendicular to and small along the measured line of sight. An infrared search-and-track sensor (IRST) can measure the azimuth and elevation of a target with good resolution. It can provide only the direction of a target but not its location because it does not provide the range. The uncertainty associated with IRST might be represented as a square whose dimensions are comparatively small perpendicular to the measured line of sight. The fusion of measurements from radar and IRST results in less uncertainty of the estimated position of the target. The state-vector fusion of tracks obtained from imaging and radar sensors is discussed here. An interacting multiple-model KF (IMMKF) is used to track a maneuvering target using the measurements obtained from both imaging and radar sensors.

11.2.1 Passive Optical Sensor Mathematical Model

The imaging sensor produces a 2D data array representing the intensity of the pixels [28,29]. The image at time k is $I(k)$, with size N by N. The intensity, $I_{ij}(k)$, of the ijth pixel is given by

$$I_{ij}(k) = \begin{cases} s_{ij}(k) + n_{ij}(k) \\ n_{ij}(k) \end{cases} \tag{11.9}$$

Here, $s_{ij}(k)$ is the target intensity of pixel (i, j) at time k, and $n_{ij}(k)$ is the background intensity of pixel (i, j) at time k. Generally, an imaging sensor provides angular measurements. The centroid of the target in the pixel coordinates is determined; one can then transform the pixel coordinates into equivalent angular coordinates. Figure 11.20 shows the relation between the angular and pixel coordinates [28]. The centroid Z_c can be computed as

$$Z_c(k) = \begin{bmatrix} c_h(k) \\ c_v(k) \end{bmatrix} = \frac{\sum_{i=-(N-1)/2}^{(N-1)/2} \sum_{j=-(N-1)/2}^{(N-1)/2} I_{ij}(k) \begin{bmatrix} i \\ j \end{bmatrix}}{\sum_{i=-(N-1)/2}^{(N-1)/2} \sum_{j=-(N-1)/2}^{(N-1)/2} I_{ij}(k)} \tag{11.10}$$

Here, c_h is the centroid of the target in the i coordinate, and c_v is the centroid of the target in the j coordinate. The distance spanned by each pixel in the i and j coordinates is given as [28]

$$u_h = f \frac{\tan\left(\frac{\psi_h}{2}\right)}{\frac{(N-1)}{2}} \quad \text{and} \quad u_v = f \frac{\tan\left(\frac{\psi_v}{2}\right)}{\frac{(N-1)}{2}} \tag{11.11}$$

Here, ψ_h is the number of degrees spanned in i coordinates, ψ_v is the number of degrees spanned in j coordinates, u_h is the distance spanned in i

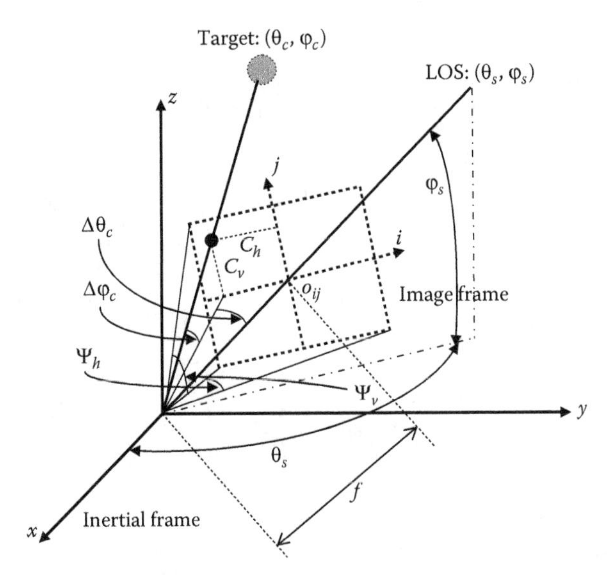

FIGURE 11.20
Angular and pixel coordinates.

coordinates, u_v is the distance spanned in j coordinates, and f is the focal length. The angular coordinates $(\Delta\theta_c(k), \Delta\phi_c(k))$ of the centroid $Z_c(k)$ are computed as

$$\Delta\theta_c(k) = \tan^{-1}\left(\frac{c_h(k)u_h}{f}\right) \quad \text{and} \quad \Delta\phi_c(k) = \tan^{-1}\left(\frac{c_v(k)u_v}{f}\right) \tag{11.12}$$

The imaging-sensor measurements in the inertial frame are modeled as

$$\begin{aligned} \theta(k) &= \theta_s(k) + \Delta\theta_c(k) + v_\theta(k) \text{ (rad)} \\ \phi(k) &= \phi_s(k) + \Delta\phi_c(k) + v_\phi(k) \text{ (rad)} \end{aligned} \tag{11.13}$$

Here, $\theta_s(k)$ and $\phi_s(k)$ are the line of sight of the sensor, and $v_\theta(k)$ and $v_\phi(k)$ are zero-mean white-noise sequences with standard deviations σ_θ and σ_ϕ, respectively.

11.2.2 State-Vector Fusion for Fusing IRST and Radar Data

Two tracking algorithms can be used to track the target independently; the final estimated target states can then be fused using state-vector fusion to obtain the improved target states [28,29]. The scheme is shown in Figure 11.21.

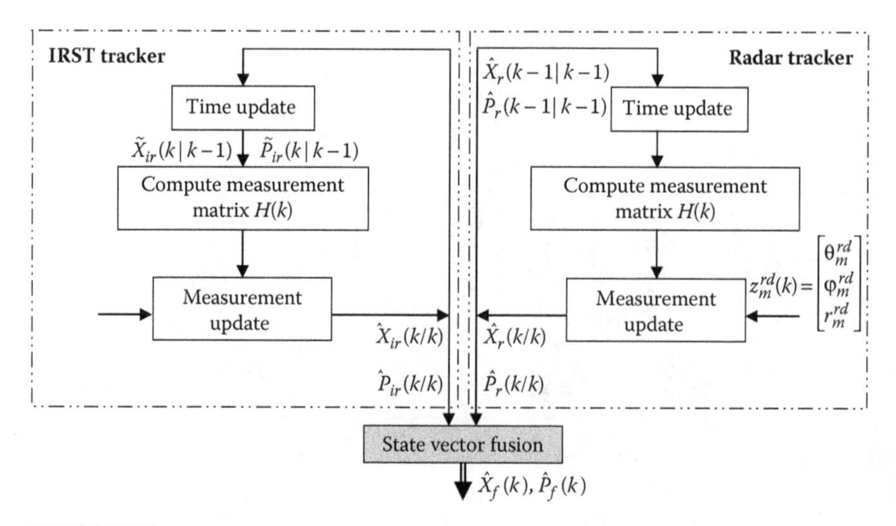

FIGURE 11.21
Fusion of tracks estimated from IRST and radar data.

11.2.2.1 Application of Extended KF

A motion model for target tracking is given as

$$X(k) = FX(k-1) + Gw(k-1) \tag{11.14}$$

$$z(k) = h(X(k)) + v(k) \tag{11.15}$$

Radar and IRST provide the measurements in a spherical coordinate system. In most cases, the state vector can be estimated in the Cartesian coordinate system. Consider a state vector consisting of position, velocity, and acceleration components in the x, y, and z directions as

$$\begin{bmatrix} x & \dot{x} & \ddot{x} & y & \dot{y} & \ddot{y} & z & \dot{z} & \ddot{z} \end{bmatrix} \tag{11.16}$$

The IRST sensor-measurement vector consists of azimuth θ and elevation φ and is given in vector form as follows:

$$z(k) = [\theta \quad \varphi]^T \tag{11.17}$$

The radar-measurement vector consists of azimuth, elevation, and range r and is given in the vector form as

$$z(k) = [\theta \quad \varphi \quad r]^T \tag{11.18}$$

The predicted state is in the following form:

$$\begin{bmatrix} \tilde{x} & \dot{\tilde{x}} & \ddot{\tilde{x}} & \tilde{y} & \dot{\tilde{y}} & \ddot{\tilde{y}} & \tilde{z} & \dot{\tilde{z}} & \ddot{\tilde{z}} \end{bmatrix} = \tilde{X}(k|k-1) \tag{11.19}$$

The predicted measurement of the IRST sensor is

$$\tilde{z}(k\,|\,k-1) = h\left[\tilde{X}(k\,|\,k-1)\right] = \left[\tilde{\theta} \quad \tilde{\varphi}\right]^T \tag{11.20}$$

The predicted measurement of the radar is

$$\tilde{z}(k\,|\,k-1) = h\left[\tilde{X}(k\,|\,k-1)\right] = \left[\tilde{\theta} \quad \tilde{\varphi} \quad \tilde{r}\right]^T \tag{11.21}$$

Components in the predicted measurement are computed from the predicted state vector in Equation 11.19.

$$\tilde{\theta} = \tan^{-1}\left(\frac{\tilde{x}}{\tilde{y}}\right), \quad \tilde{\varphi} = \tan^{-1}\left(\frac{\tilde{z}}{\sqrt{\tilde{x}^2 + \tilde{y}^2}}\right), \quad \tilde{r} = \sqrt{\tilde{x}^2 + \tilde{y}^2 + \tilde{z}^2} \tag{11.22}$$

11.2.2.2 State-Vector Fusion

Consider the tracks from IRST and radar, whose state estimations and covariance matrices during the scan k are as follows:

$$\begin{aligned} &\text{Track from IRST sensor: } \hat{X}_1(k\,|\,k),\ \hat{P}_1(k\,|\,k) \\ &\text{Track from Radar: } \hat{X}_2(k\,|\,k),\ \hat{P}_2(k\,|\,k) \end{aligned} \tag{11.23}$$

These two tracks can be fused or combined using any one of the following schemes.

11.2.2.2.1 State-Vector Fusion 1 or SVF1

Without considering the cross-covariance matrix, the fused state vector and covariance matrix can be given as

$$X_f(k) = \hat{X}_1(k\,|\,k) + \hat{P}_1(k\,|\,k)\left[\hat{P}_1(k\,|\,k) + \hat{P}_2(k\,|\,k)\right]^{-1}\left[\hat{X}_2(k\,|\,k) - \hat{X}_1(k\,|\,k)\right] \tag{11.24}$$

or

$$\begin{aligned} X_f(k) = {} & P_2(k\,|\,k)\left[\hat{P}_1(k\,|\,k) + \hat{P}_2(k\,|\,k)\right]^{-1}\hat{X}_1(k\,|\,k) \\ & + P_1(k\,|\,k)\left[\hat{P}_1(k\,|\,k) + \hat{P}_2(k\,|\,k)\right]^{-1}\hat{X}_2(k\,|\,k) \end{aligned} \tag{11.25}$$

$$\hat{P}_f(k) = \hat{P}_1(k\,|\,k) - \hat{P}_1(k\,|\,k)\left[\hat{P}_1(k\,|\,k) + \hat{P}_2(k\,|\,k)\right]^{-1}\hat{P}_1^T(k\,|\,k) \tag{11.26}$$

11.2.2.2.2 State-Vector Fusion 2

The following computation is carried out during the time-updating step, concurrent with the consideration of the cross-covariance matrix. The predicted covariance matrices of the individual sensors and the cross-covariance matrix between two sensors are computed as

$$\tilde{P}_{11}(k|k-1) = F\hat{P}_{11}(k-1|k-1)F^T + GQG^T \tag{11.27}$$

$$\tilde{P}_{22}(k|k-1) = F\hat{P}_{22}(k-1|k-1)F^T + GQG^T \tag{11.28}$$

$$\tilde{P}_{12}(k|k-1) = F\hat{P}_{12}(k-1|k-1)F^T + GQG^T \tag{11.29}$$

$$\tilde{P}_{21}(k|k-1) = F\hat{P}_{21}(k-1|k-1)F^T + GQG^T \tag{11.30}$$

These covariance and cross-covariance matrices are updated during measurement updates. Covariance and cross-covariance matrices are updated according to the following equations:

$$\hat{P}_{11}(k|k) = (I - K_1(k)H_1(k))\tilde{P}_{11}(k|k-1) \tag{11.31}$$

$$\hat{P}_{22}(k|k) = (I - K_2(k)H_2(k))\tilde{P}_{22}(k|k-1) \tag{11.32}$$

$$\hat{P}_{12}(k|k) = \left[I - K_1(k)H_1(k)\right]\tilde{P}_{12}(k|k-1)\left[I - H_2^T(k)K_2^T(k)\right] \tag{11.33}$$

$$\hat{P}_{21}(k|k) = \left[I - K_2(k)H_2(k)\right]\tilde{P}_{21}(k|k-1)\left[I - H_1^T(k)K_1^T(k)\right] \tag{11.34}$$

Here, K_1 and K_2 are the Kalman gains from the IRST and radar trackers, respectively. H_1 and H_2 are the measurement matrices, which are nonlinear functions of state vectors from the IRST and radar trackers, respectively. Next, the fused state vector and covariance matrix are computed as follows:

Let

$$\begin{aligned} A_1 &= \hat{P}_{11}(k|k) - \hat{P}_{12}(k|k) \\ A_2 &= \hat{P}_{22}(k|k) - \hat{P}_{21}(k|k) \end{aligned} \tag{11.35}$$

The fusion covariance matrix can be computed as follows:

$$P_f(k) = P_{11}(k|k) - [P_{11}(k|k) - P_c(k|k)]P_e^{-1}(k)[P_{11}(k|k) - P_c(k|k)]^T \tag{11.36}$$

Here,

$$P_e(k) = P_{11}(k\,|\,k) + P_{22}(k\,|\,k) - P_c(k\,|\,k) - P_c(k\,|\,k)^T \tag{11.37}$$

$$\begin{aligned} P_c(k\,|\,k) &= (I - K_1(k)H_1(k))FP_c(k-1\,|\,k-1)F^T(I - K_2(k)H_2(k)) \\ &\quad + (I - K_1(k)H_1(k))GQG^T(I - K_2(k)H_2(k))^T \end{aligned} \tag{11.38}$$

with $P_c(0\,|\,0) = 0$.

Next, we obtain

$$\begin{aligned} P_c(k\,|\,k) &= [I - K_1(k)H_1(k)]F\tilde{P}_{12}(k-1\,|\,k-1)F^T[I - H_2(k)K_2(k)] \\ &= \hat{P}_{12}(k\,|\,k) \end{aligned} \tag{11.39}$$

$$\begin{aligned} P_c(k\,|\,k)^T &= [I - K_2(k)H_2(k)]F\tilde{P}_{21}(k-1\,|\,k-1)F^T[I - H_1(k)K_1(k)] \\ &= \hat{P}_{21}(k\,|\,k) \end{aligned} \tag{11.40}$$

$$P_e(k) = \hat{P}_{11}(k\,|\,k) + \hat{P}_{22}(k\,|\,k) - \hat{P}_{12}(k\,|\,k) - \hat{P}_{21}(k\,|\,k) = A_1 + A_2 \tag{11.41}$$

Finally, the fused covariance matrix can be written as follows:

$$P_f(k) = \hat{P}_{11}(k\,|\,k) - (\hat{P}_{11}(k\,|\,k) - \hat{P}_{12}(k\,|\,k))P_e^{-1}(\hat{P}_{11}(k\,|\,k) - \hat{P}_{12}(k\,|\,k))^T \tag{11.42}$$

$$P_f(k) = \hat{P}_{11}(k\,|\,k) - A_1[A_1 + A_2]^{-1}A_1^T$$

The fused state vector can be computed as

$$X_f(k) = \hat{X}_1(k\,|\,k) + A_1\left[A_1 + A_2\right]^{-1}\left[\hat{X}_2(k\,|\,k) - \hat{X}_1(k\,|\,k)\right] \tag{11.43}$$

or equivalently

$$X_f(k) = A_2[A_1 + A_2]^{-1}\hat{X}_1(k\,|\,k) + A_1[A_1 + A_2]^{-1}\hat{X}_2(k\,|\,k) \tag{11.44}$$

11.2.3 Numerical Simulation

The 3-degrees-of-freedom (DOF) kinematic model, with position, velocity, and acceleration components in each of the three Cartesian coordinates x, y, and z, has the following transition and process-noise gain matrices:

$$F = \text{diag}[\Phi \quad \Phi \quad \Phi] \quad G = \text{diag}[\zeta \quad \zeta \quad \zeta] \tag{11.45}$$

Here, $\Phi = \begin{bmatrix} 1 & T & T^2/2 \\ 0 & 1 & T \\ 0 & 0 & 1 \end{bmatrix}$ $\zeta = \begin{bmatrix} T^3/6 \\ T^2/2 \\ T \end{bmatrix}$ T is a sampling interval, F is a state-transition matrix, and G is a process noise-gain matrix. The algorithm is validated using simulated data, which utilizes the following parameters: (1) sampling interval = 1 second, (2) process-noise variance = 0.00001^2, (3) measurement-noise variance, per Table 11.2, and (4) duration of simulation = 500 seconds. Figure 11.22 shows the simulated IRST and radar measurements. The initial-state vector is

$$[x \ \dot{x} \ \ddot{x} \ y \ \dot{y} \ \ddot{y} \ z \ \dot{z} \ \ddot{z}] = [100 \ -0.2 \ -0.05 \ 100 \ -2 \ 0.01 \ 100 \ -0.5 \ 0.1]$$

TABLE 11.2

Measurement Variances

Sensor	Azimuth (rad)	Elevation (rad)	Range (m)
IRST	10^{-6}	10^{-6}	–
Radar	10^{-2}	10^{-2}	50

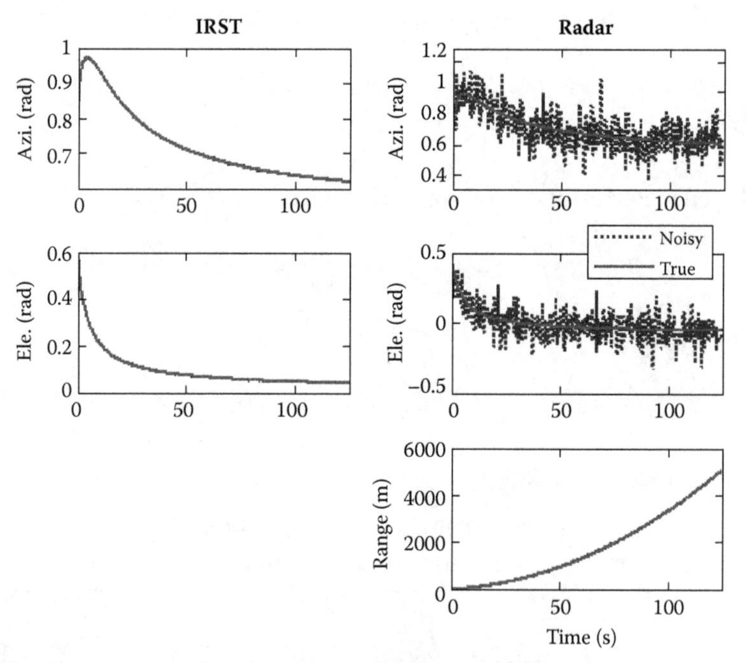

FIGURE 11.22
Measurements from IRST and radar.

TABLE 11.3

PFE and MAE Metrics

	PFEx	PFEy	PFEz	MAEx	MAEy	MAEz
IRST	9.07	9.07	9.07	130	100	8.94
Radar	0.36	0.6	4.21	5.29	6.36	4.09
SVF1	0.05	0.08	0.33	0.67	0.76	0.26

TABLE 11.4

RMSPE, RMSVE, and RMSAE Metrics

	RMSPE	RMSVE	RMSAE
IRST	123.99	2.415	0.031
Radar	7.0	0.247	0.108
SVF1	0.853	0.108	0.009

The tracking algorithm is evaluated using 75 Monte Carlo runs. The PFE and MAE metrics for the x, y, and z positions are given in Table 11.3; the RMSPE and the root–mean–square errors in velocity (RMSVE) and acceleration (RMSAE) metrics are presented in Table 11.4. After fusion, the PFE is comparatively less, which indicates that the true and estimated positions are well matched. Figure 11.23 shows the RSS acceleration errors; after fusion, these errors are comparatively much less and the estimated trajectory follows the true trajectory. After fusion, the uncertainty in estimation is also comparatively reduced. Figure 11.24 shows the innovation sequences and their bounds for both IRST and radar trackers.

11.2.4 Measurement Fusion

This section describes two measurement vector fusion schemes. In the first method (MF1), the measurements from IRST and radar are merged into an augmented measurement vector; and in the second method (MF2), the measurements from IRST and radar are combined using covariance matrices as weights. The information-flow diagram is shown in Figure 11.25.

11.2.4.1 Measurement Fusion 1 Scheme

The measurement vectors $z_m^{\text{ir}}(k)$ and $z_m^{\text{rd}}(k)$ obtained from the IRST and radar are merged into an augmented measurement vector as shown below:

$$z_m(k) = \begin{bmatrix} z_m^{\text{ir}}(k) \\ z_m^{\text{rd}}(k) \end{bmatrix} \tag{11.46}$$

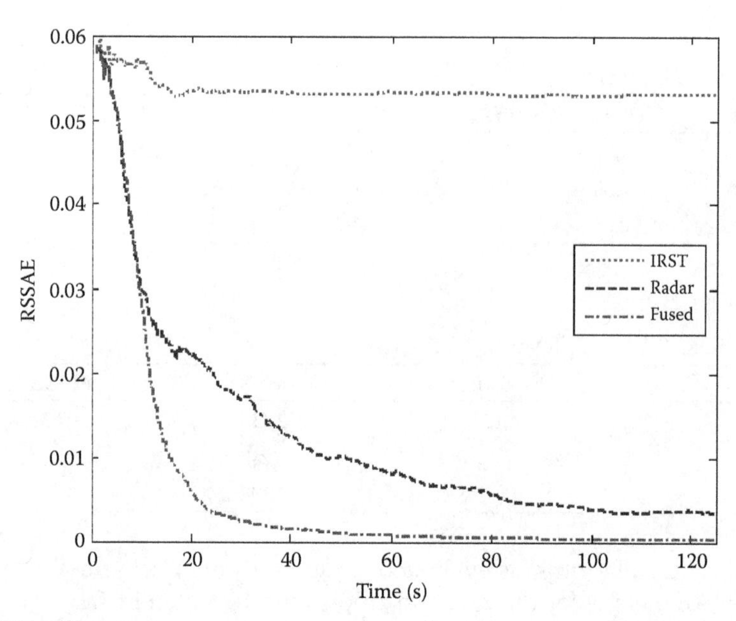

FIGURE 11.23
RSS acceleration errors.

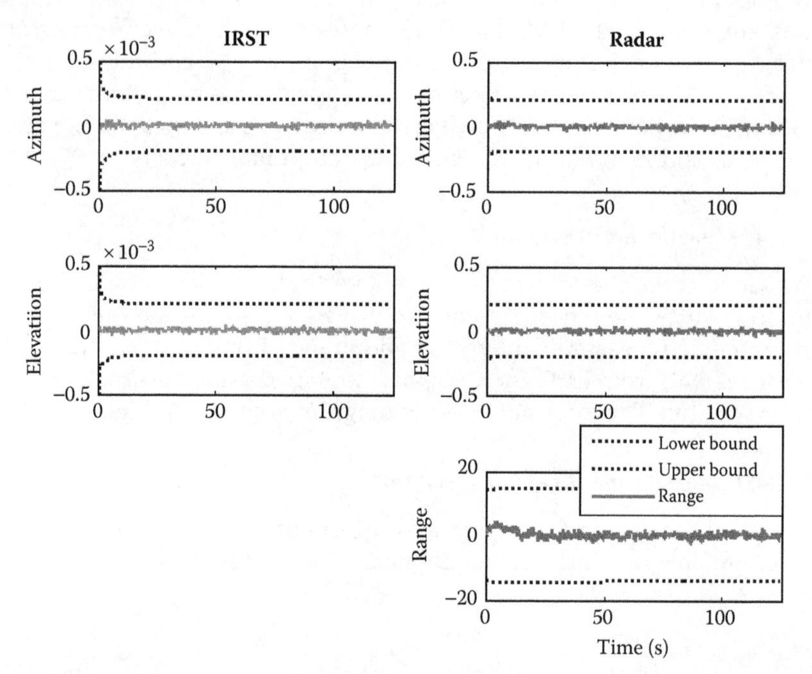

FIGURE 11.24
Innovation sequences.

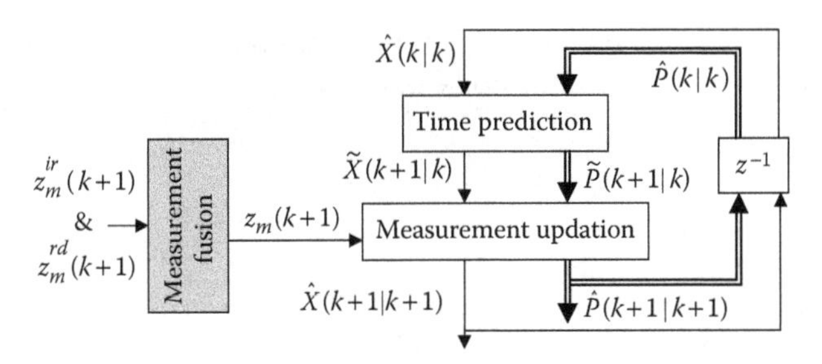

FIGURE 11.25
Flow diagram for MF.

Here, $z_m^{\text{ir}}(k) = \left[\theta_m^{\text{ir}}(k) \quad \phi_m^{\text{ir}}(k)\right]^T$, and $z_m^{\text{rd}}(k) = \left[\theta_m^{\text{rd}}(k) \quad \phi_m^{\text{rd}}(k) \quad r_m^{\text{rd}}(k)\right]^T$.

The observation matrices of IRST and radar are merged into the observation matrix as follows:

$$h(X(k)) = \begin{bmatrix} h^{\text{ir}}(X(k)) \\ h^{\text{rd}}(X(k)) \end{bmatrix} \tag{11.47}$$

The measurement-noise covariance matrices of IRST and radar are merged to yield the following:

$$R = \begin{bmatrix} R^{\text{ir}} & 0 \\ 0 & R^{\text{rd}} \end{bmatrix} \tag{11.48}$$

Here, $R^{\text{ir}} = \begin{bmatrix} \left(\sigma_\theta^{\text{ir}}\right)^2 & 0 \\ 0 & \left(\sigma_\phi^{\text{ir}}\right)^2 \end{bmatrix}$, and $R^{\text{rd}} = \begin{bmatrix} \left(\sigma_\theta^{\text{rd}}\right)^2 & 0 & 0 \\ 0 & \left(\sigma_\phi^{\text{rd}}\right)^2 & 0 \\ 0 & 0 & \left(\sigma_r^{\text{rd}}\right)^2 \end{bmatrix}$

11.2.4.2 Measurement Fusion 2 Scheme

In this scheme, the weighted combination of measurements based on covariance matrices is considered to fuse measurement vectors [29], as shown below:

$$z_m(k) = \frac{c_1 z_m^{\text{ir}}(k) + c_2 z_m^{\text{rd}}(k)}{c_1 + c_2} \tag{11.49}$$

where c_1 and c_2 are the weights. The weights in Equation 11.49 are computed from the measurement-noise covariance matrix as follows:

$$c_1 = \frac{1}{R^{\mathrm{ir}}} \quad \text{and} \quad c_2 = \frac{1}{R^{\mathrm{rd}}}$$

$$z_m(k) = \frac{\frac{1}{R^{\mathrm{ir}}} z_m^{\mathrm{ir}}(k) + \frac{1}{R^{\mathrm{rd}}} z_m^{\mathrm{rd}}(k)}{\frac{1}{R^{\mathrm{ir}}} + \frac{1}{R^{\mathrm{rd}}}} \tag{11.50}$$

The MF2 scheme is represented as follows:

$$z_m(k) = \frac{R^{\mathrm{rd}} z_m^{\mathrm{ir}}(k) + R^{\mathrm{ir}} z_m^{\mathrm{rd}}(k)}{R^{\mathrm{ir}} + R^{\mathrm{rd}}} \tag{11.51}$$

or

$$z_m(k) = \frac{R^{\mathrm{rd}} z_m^{\mathrm{ir}}(k)}{R^{\mathrm{ir}} + R^{\mathrm{rd}}} + \frac{R^{\mathrm{ir}} z_m^{\mathrm{rd}}(k)}{R^{\mathrm{ir}} + R^{\mathrm{rd}}} + \frac{R^{\mathrm{ir}} z_m^{\mathrm{ir}}(k)}{R^{\mathrm{ir}} + R^{\mathrm{rd}}} - \frac{R^{\mathrm{ir}} z_m^{\mathrm{ir}}(k)}{R^{\mathrm{ir}} + R^{\mathrm{rd}}}$$

$$z_m(k) = z_m^{\mathrm{ir}}(k) + R^{\mathrm{ir}}(R^{\mathrm{ir}} + R^{\mathrm{rd}})^{-1}\left(z_m^{\mathrm{rd}}(k) - z_m^{\mathrm{ir}}(k)\right) \tag{11.52}$$

The associated measurement-noise covariance matrix (R) of the fused measurement vector (Equation 11.52) is computed as follows:

$$\frac{1}{R} = \frac{1}{R^{\mathrm{ir}}} + \frac{1}{R^{\mathrm{rd}}} \tag{11.53}$$

$$R = ((R^{\mathrm{ir}})^{-1} + (R^{\mathrm{rd}})^{-1})^{-1} \tag{11.54}$$

or

$$R = \frac{R^{\mathrm{ir}} R^{\mathrm{rd}}}{R^{\mathrm{ir}} + R^{\mathrm{rd}}} = \frac{R^{\mathrm{ir}} R^{\mathrm{rd}}}{R^{\mathrm{ir}} + R^{\mathrm{rd}}} - R^{\mathrm{ir}} + R^{\mathrm{ir}}$$

$$R = R^{\mathrm{ir}} + \frac{R^{\mathrm{ir}} R^{\mathrm{rd}} - R^{\mathrm{ir}}(R^{\mathrm{ir}} + R^{\mathrm{rd}})}{R^{\mathrm{ir}} + R^{\mathrm{rd}}} \tag{11.55}$$

$$R = R^{\mathrm{ir}} - R^{\mathrm{ir}}(R^{\mathrm{ir}} + R^{\mathrm{rd}})^{-1}(R^{\mathrm{ir}})^T$$

11.2.5 Maneuvering Target Tracking

In tracking applications for a target moving with constant velocity (CV), the state model includes the first derivative of the position, and for a target moving with CA, the state model includes the second derivative of position. Models with second-order derivatives are preferred for tracking maneuvering targets and are referred to as acceleration models. The IMMKF (see Section 3.5) is an adaptive state estimator that is based on the assumption that a finite number of models are required to characterize the motion of the target at all times.

11.2.5.1 Motion Models

The following two types of models in the Cartesian frame are considered for tracking a maneuvering targets:

1. The CV model (2-DOF kinematic model, with position and velocity) with state-transition and process-noise gain matrices given as

$$F_1 = F_{CV} = \begin{bmatrix} \Phi_V & 0_{3\times3} & 0_{3\times3} \\ 0_{3\times3} & \Phi_V & 0_{3\times3} \\ 0_{3\times3} & 0_{3\times3} & \Phi_V \end{bmatrix} \quad G_1 = G_{CV} = \begin{bmatrix} \varsigma_V & 0_{3\times1} & 0_{3\times1} \\ 0_{3\times1} & \varsigma_V & 0_{3\times1} \\ 0_{3\times1} & 0_{3\times1} & \varsigma_V \end{bmatrix} \tag{11.56}$$

   Here,

$$\Phi_V = \begin{bmatrix} 1 & T & 0 \\ 0 & 1 & 0 \\ 0 & 0 & 0 \end{bmatrix} \varsigma_V = \begin{bmatrix} T^2/2 \\ T \\ 0 \end{bmatrix} \quad 0_{3\times3} = \begin{bmatrix} 0 & 0 & 0 \\ 0 & 0 & 0 \\ 0 & 0 & 0 \end{bmatrix} \quad 0_{3\times1} = \begin{bmatrix} 0 \\ 0 \\ 0 \end{bmatrix}$$

   The state vector is given as $X = [x \;\; \dot{x} \;\; 0 \;\; y \;\; \dot{y} \;\; 0 \;\; z \;\; \dot{z} \;\; 0]^T$. The process-noise intensity (air turbulence, slow turns, and small linear acceleration) in each axis is generally assumed to be small and equal in all three axes $\left(\sigma_x^2 = \sigma_y^2 = \sigma_z^2\right)$. Note that the acceleration components in the model, although identically equal to zero, have been retained for dimensional compatibility with the third-order acceleration model.

2. The CA model (3-DOF kinematic model, with position, velocity, and acceleration) has the following state-transition and process-noise gain matrices:

$$F_2 = F_{CA} = \begin{bmatrix} \Phi_A & 0_{3\times3} & 0_{3\times3} \\ 0_{3\times3} & \Phi_A & 0_{3\times3} \\ 0_{3\times3} & 0_{3\times3} & \Phi_A \end{bmatrix} \quad G_2 = G_{CA} = \begin{bmatrix} \varsigma_A & 0_{3\times1} & 0_{3\times1} \\ 0_{3\times1} & \varsigma_A & 0_{3\times1} \\ 0_{3\times1} & 0_{3\times1} & \varsigma_A \end{bmatrix} \tag{11.57}$$

Here,

$$\Phi_A = \begin{bmatrix} 1 & T & T^2/2 \\ 0 & 1 & T \\ 0 & 0 & 1 \end{bmatrix} \quad \varsigma_A = \begin{bmatrix} T^3/6 \\ T^2/2 \\ T \end{bmatrix} \quad 0_{3\times3} = \begin{bmatrix} 0 & 0 & 0 \\ 0 & 0 & 0 \\ 0 & 0 & 0 \end{bmatrix} \quad 0_{3\times1} = \begin{bmatrix} 0 \\ 0 \\ 0 \end{bmatrix}$$

The state vector in this model is $X = [x \quad \dot{x} \quad \ddot{x} \quad y \quad \dot{y} \quad \ddot{y} \quad z \quad \dot{z} \quad \ddot{z}]^T$. A moderate value of the process-noise variance, Q_{CA} (relatively higher than Q_{CV}), will yield a nearly CA motion. The noise variances are assumed as $\left(\sigma_x^2 = \sigma_y^2 = \sigma_z^2\right)$.

11.2.5.2 Measurement Model

The measurement vector of the IRST is azimuth (θ) and elevation (ϕ), as represented in the expression

$$z_m^{\mathrm{ir}}(k) = \left[\theta_m^{\mathrm{ir}}(k) \quad \phi_m^{\mathrm{ir}}(k)\right]^T \tag{11.58}$$

The measurement vector of the radar is range (r), azimuth, and elevation and is given by the following equation:

$$z_m^{\mathrm{rd}}(k) = \left[\theta_m^{\mathrm{rd}}(k) \quad \phi_m^{\mathrm{rd}}(k) \quad r_m^{\mathrm{rd}}(k)\right]^T \tag{11.59}$$

These are related to the various states in a nonlinear fashion as follows:

$$\begin{aligned} x(k) &= r(k)\cos\theta(k)\cos\phi(k) \\ y(k) &= r(k)\sin\theta(k)\cos\phi(k) \\ z(k) &= r(k)\sin\phi(k) \end{aligned} \tag{11.60}$$

The measurement equation is given by

$$z_m(k+1) = h(X(k+1) + v(k+1)) \tag{11.61}$$

11.2.5.3 Numerical Simulation

The target trajectory is simulated in Cartesian coordinates, with a sampling interval of 0.25 seconds using the CV model for the first 50 seconds, followed by the CA model for 25 seconds, and then the CV model for 50 seconds. Acceleration magnitudes of 3 m/s^2 and –3 m/s^2 are injected at the time of maneuver along the x and y axes, respectively. The initial state vector is $X = [x \ \dot{x} \ \ddot{x} \ y \ \dot{y} \ \ddot{y} \ z \ \dot{z} \ \ddot{z}]^T = [100 \ 50 \ 0 \ 100 \ 50 \ 0 \ 10 \ 1 \ 0]^T$. The range, azimuth, and elevation data of the radar are obtained from the following transformations:

$$\begin{aligned} r_m^{\mathrm{rd}}(k) &= \sqrt{x^2(k) + y^2(k) + z^2(k)} + v_r^{\mathrm{rd}}(k) \\ \theta_m^{\mathrm{rd}}(k) &= \tan^{-1}\left(\frac{y(k)}{x(k)}\right) + v_\theta^{\mathrm{rd}}(k) \\ \phi_m^{\mathrm{rd}}(k) &= \tan^{-1}\left(\frac{z(k)}{\sqrt{x^2(k) + y^2(k)}}\right) + v_\phi^{\mathrm{rd}}(k) \end{aligned} \quad (11.62)$$

The azimuth and elevation data of IRST are obtained from the following transformations:

$$\begin{aligned} \theta_m^{\mathrm{ir}}(k) &= \tan^{-1}\left(\frac{y(k)}{x(k)}\right) + v_\theta^{\mathrm{ir}}(k) \\ \phi_m^{\mathrm{ir}}(k) &= \tan^{-1}\left(\frac{z(k)}{\sqrt{x^2(k) + y^2(k)}}\right) + v_\phi^{\mathrm{ir}}(k) \end{aligned} \quad (11.63)$$

Measurement-noise variances are given in Table 11.5. The duration of the simulation is 125 seconds. Figure 11.26 shows the resultant noisy IRST and radar measurements. Process-noise variance of $\sigma_x^2 = \sigma_y^2 = \sigma_z^2 = 0.00001$ is considered for the CV model, and $\sigma_x^2 = \sigma_y^2 = \sigma_z^2 = 3$ is considered for the CA model. The Markov chain transition matrix is $p = \begin{bmatrix} 0.99 & 0.01 \\ 0.01 & 0.99 \end{bmatrix}$, and the initial mode probabilities $\mu = [\mu_1 \ \mu_2]^T$, corresponding to the respective nonmaneuver and maneuver modes, can be considered 0.9 and 0.1, respectively.

Figures 11.27 and 11.28 show the true and estimated velocities and accelerations along all the axes. The estimated positions, velocities, and accelerations match with the true trajectories. There is little delay in the acceleration estimation, because only the position measurements

TABLE 11.5

Measurement Variances

Sensor	Azimuth (rad)	Elevation (rad)	Range (m)
IRST	1E-6	1E-6	–
Radar	1E-2	1E-2	100

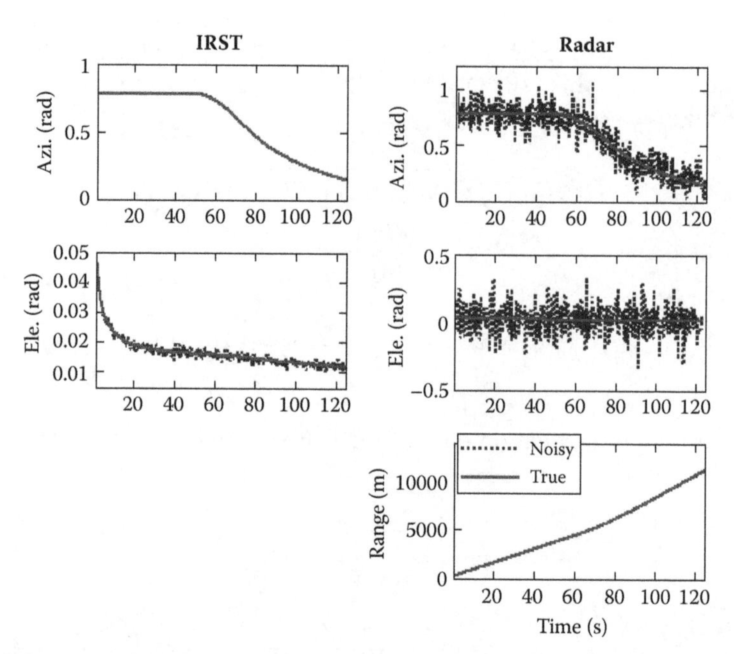

FIGURE 11.26
Simulated measurements from IRST and radar.

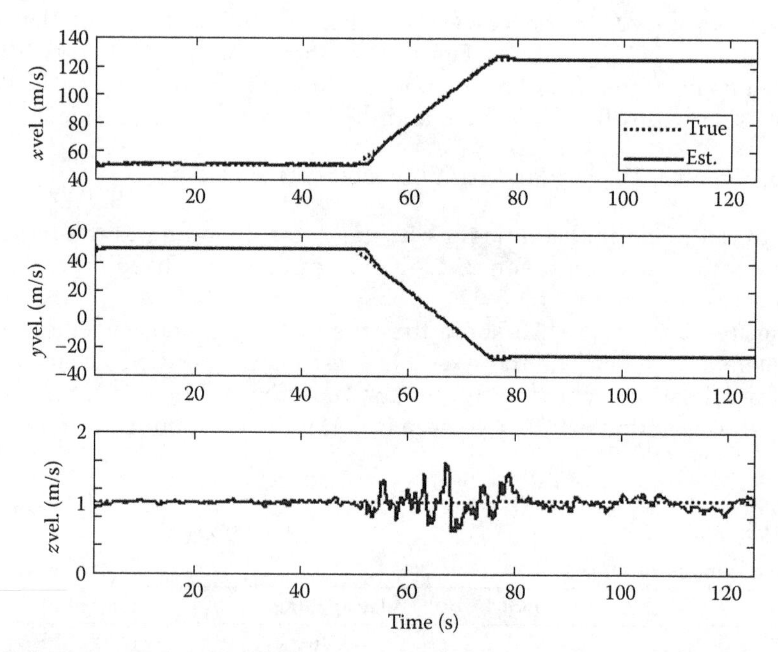

FIGURE 11.27
True and estimated velocities.

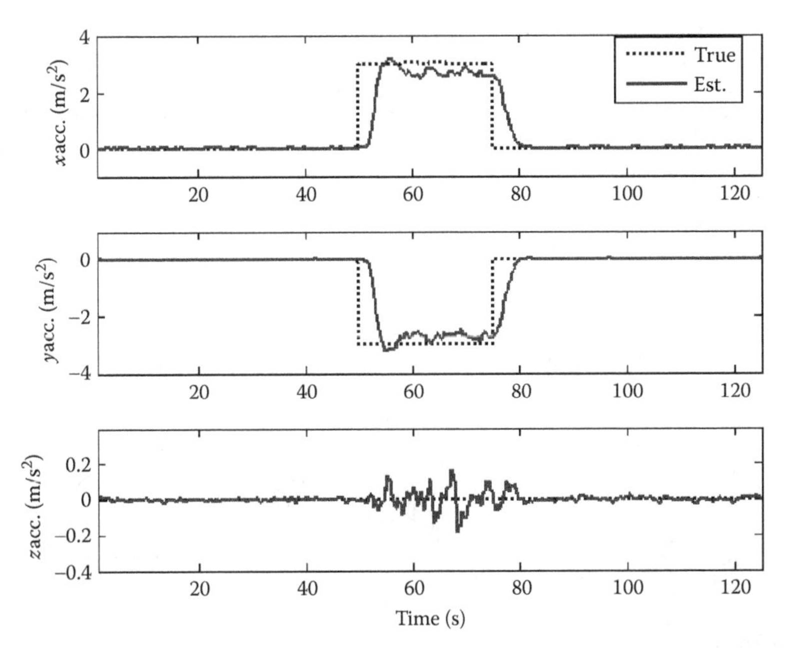

FIGURE 11.28
True and estimated accelerations.

are utilized in the state estimation. The velocity and acceleration estimates along the z axis are a little degraded during the maneuvering portion. This could be due to the high process noise or due to the active CA model, although there is no maneuver in the z axis. The problem can be solved by providing less process noise to this axis. Figure 11.29 shows the RSS errors in position, velocity, and acceleration. These errors are high during the maneuvering period. The state errors in the x, y, and z positions, with their theoretical bounds, are shown in Figure 11.30. These errors are within the bounds. Similar observations can be made for the velocities and acceleration states. The bounds are high because the CA model is active during this portion. The root-sum variances in position, velocity, and acceleration, denoted as RSvarP, RSvarV, and RSvarA, respectively, are shown in Figure 11.31. As expected, the variances are very low in the nonmaneuvering portion and high in the maneuvering portion. This could be due to the fact that the CV model is active during the nonmaneuvering portion, whereas the CA model is active during the maneuvering portion. This information could be very useful in the fusion process because it is a realistic representation of the maneuver. The mode probabilities are shown in Figure 11.32, showing the alternation between the models. This information is useful for knowing which model is active in a given time and useful for confidence in the state and

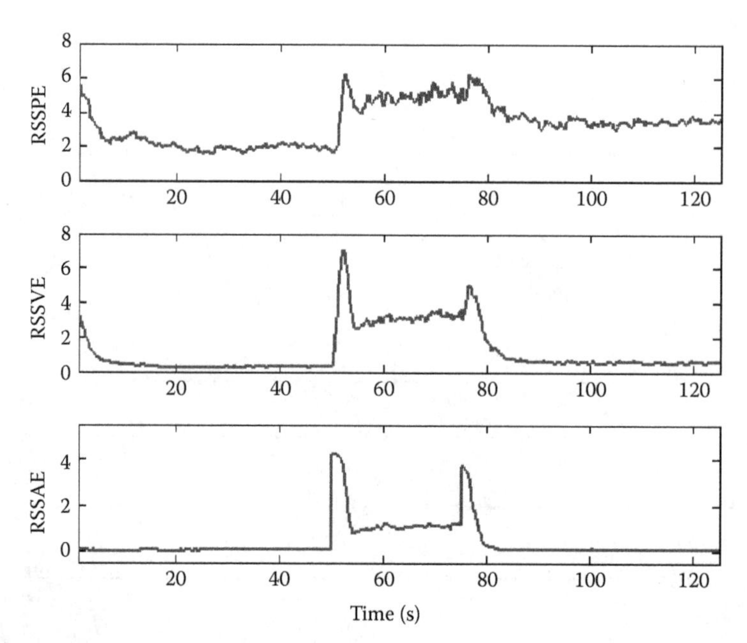

FIGURE 11.29
RSS errors in position, velocity, and acceleration.

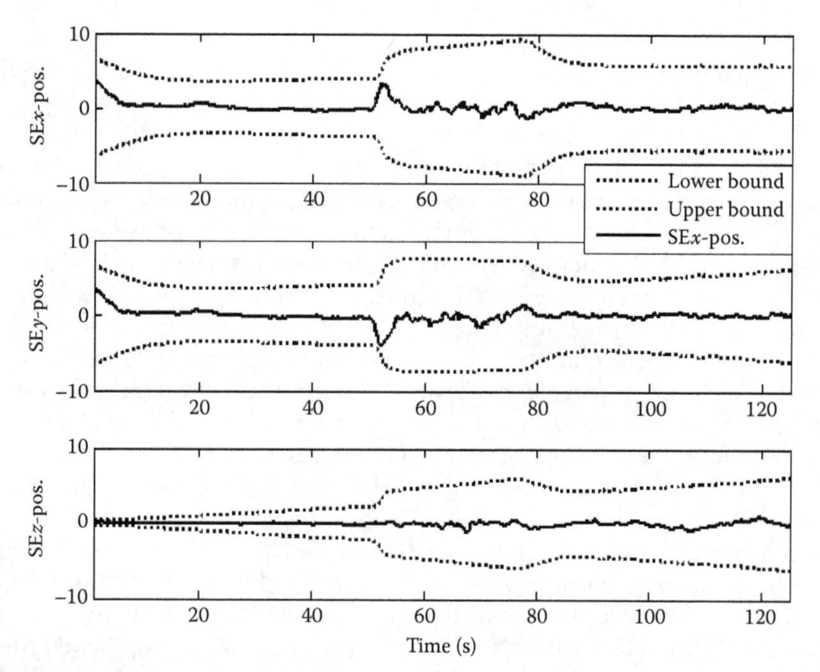

FIGURE 11.30
State errors in positions with bounds.

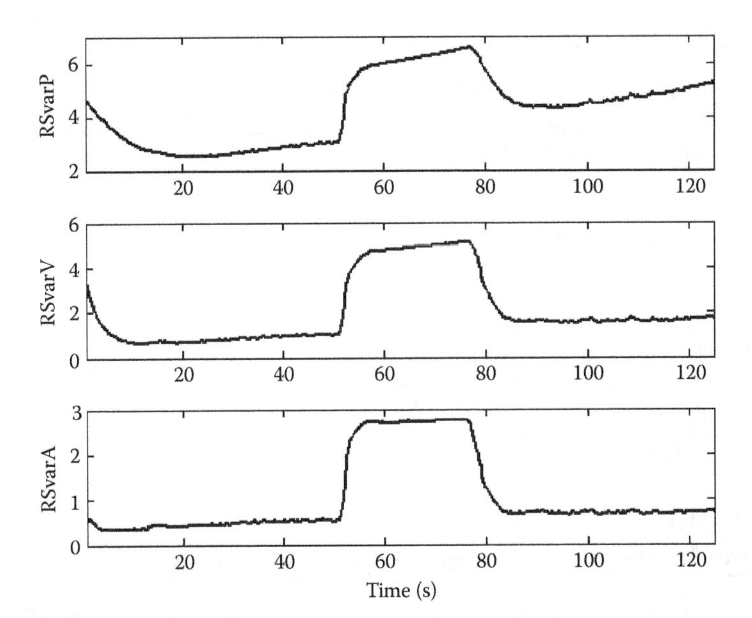

FIGURE 11.31
The root-sum variances in position (RSvarP), velocity (RSvarV), and acceleration (RSvarA).

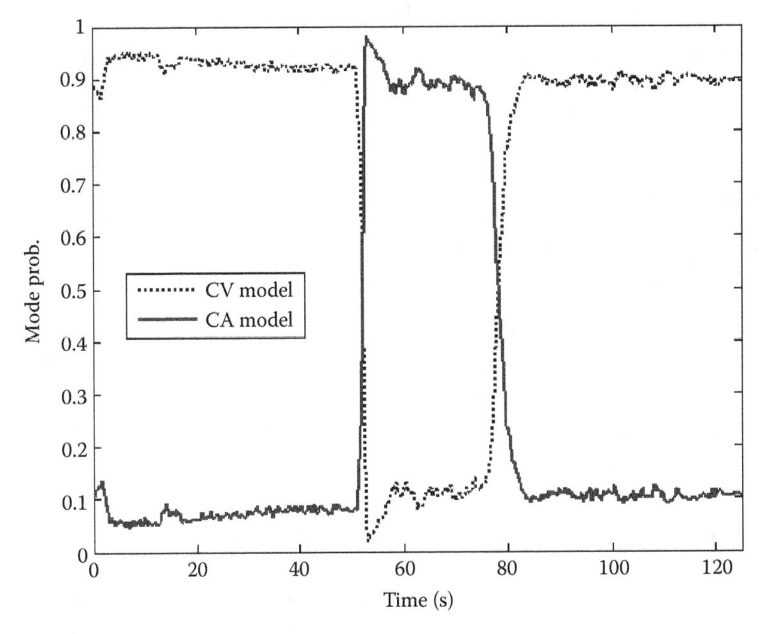

FIGURE 11.32
Mode probabilities.

TABLE 11.6A

PFE Metrics along the x, y, and z axes for the Positions, Velocities, and Accelerations

PFEx	PFEy	PFEz	PFE$\dot{x}$	PFE$\dot{y}$	PFE$\dot{z}$	PFE$\ddot{x}$	PFE$\ddot{y}$	PFE$\ddot{z}$
0.049	0.113	2.175	1.43	3.56	78.11	46.947	46.987	61212.7

TABLE 11.6B

MAE Metrics along the x, y, and z axes for the Positions, Velocities, and Accelerations

MAEx	MAEy	MAEz	MAE$\dot{x}$	MAE$\dot{y}$	MAE$\dot{z}$	MAE$\ddot{x}$	MAE$\ddot{y}$	MAE$\ddot{z}$
1.1995	1.864	1.23	0.734	0.714	0.428	0.245	0.245	0.11

TABLE 11.6C

RMS Error Metrics for Position, Velocity, and Acceleration

RMSPE	RMSVE	RMSAE
2.339	1.185	0.536

state-error covariance matrix combination. The PFE metrics for positions, velocities, and accelerations along all three axes are depicted in Table 11.6a. The MAEs in positions, velocities, and accelerations of all the axes are shown in Table 11.6b. The RMSPE, RMSVE, and RMSAE are listed in Table 11.6c.

11.3 Target Tracking with Acoustic Sensor Arrays and Imaging Sensor Data

This section discusses an approach to joint audio-video tracking based on acoustic sensor array (ASA) modeling, uniform linear array sensor data generation, and tracking a target in Cartesian coordinates. It also discusses the joint audio-video tracking of a target in 3-DOF Cartesian coordinates using decentralized KF.

11.3.1 Tracking with Multiple Acoustic Sensor Arrays

Target tracking using ASA is very useful in air-traffic control, air defense, mobile-user location in cellular communications, military, underwater tracking, and acoustic source localization [30]. In ASA tracking, a sensor in the array receives an acoustic signal emitted by the moving target. An ASA-based target-tracking procedure is inexpensive compared to other modalities and requires less power. The system may include magnetic sensors and imaging sensors. When the target is near the ASA, acoustic sounds can be used to determine the location, range, and angle of the target in polar coordinates, and when the target is in the far field of the ASA, only the bearing (angle) information can be extracted.

Hence, multiple acoustic-sensor arrays (MASAs) are needed to determine the target location. ASA output data are analyzed by the well-known multiple signal classification algorithm MUSIC to estimate the direction of arrival (DoA) of the emitted acoustic signal [31,32]. The DoA estimation is based on batch processing, like with MUSIC and minimum norm, and does not reuse the information from previous batches to help refine the estimates of the current batch. Because ASA provides only the angle (azimuth), it would be difficult to track the target in Cartesian coordinates with single-angle information. Hence, MASAs are required to provide multiple-angle measurements of the target at one point in time; the target location could then be computed by the triangulation method [33,34]. It is assumed that the target is in the far field of the array. Two such arrays are used to get the angle bearing of the target. Three tracking schemes for an acoustic target using Cartesian coordinates are studied here.

1. A digital filter with least-squares (LS) estimation is used for the estimation of the target position in Cartesian coordinates.
2. A KF with LS estimation is used to track the target in Cartesian coordinates. This could be a straight, forward-oriented array, because the time propagation and measurement updating are in polar coordinates.
3. An extended KF (EKF) is used, where the state estimation is in Cartesian coordinates and measurement updating is in polar coordinates.

11.3.2 Modeling of Acoustic Sensors

The ASA contains m sensors uniformly placed (with distance d) along a line, as shown in Figure 11.33, with $d = \lambda/2$ preventing spatial aliasing (λ is the carrier wavelength) [30–33,35]. It is assumed that the sound waves reaching each sensor are arriving in parallel. The direction perpendicular to the ASA is the broadside direction (BD). When a target is moving, its DoA would be measured with respect to the BD. The target is assumed to be moving with CV with minor random perturbations and emitting narrow band signals of wavelength λ, which impinge on the ASA (which are

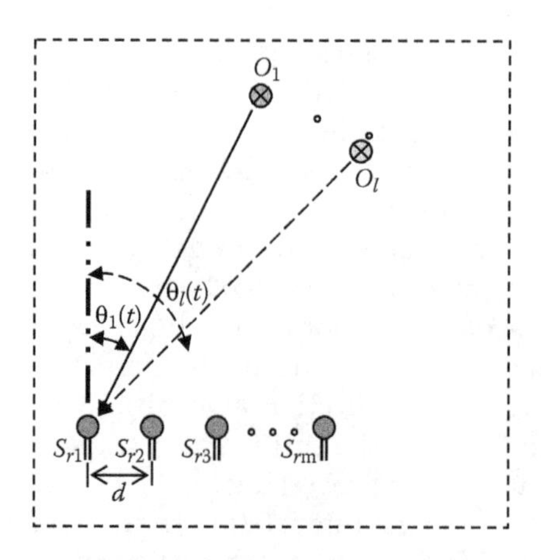

FIGURE 11.33
Geometry of the ASA.

passive) at an angle of θ. The signal from the target reaches the sensors at different times because each sound wave must travel different distances to reach the different sensors. For example, the signal that depends on sensor S_{r1} has to travel an extra distance of $d\sin(\theta_1)$ compared to the signal that depends on sensor S_{r2}. The delay is $-d\sin(\theta_1)/c$, where c is the rate of propagation of the signal. Let $l(l < m)$ be the number of targets in the test scenario; then, one can express the jth sensor as the sum of the shifted versions of the source signals represented below [29]:

$$\begin{aligned} X_{a_j}(t) &= \sum_{i=1}^{l} S_{a_i}\left(t+\tau_{ij}(t)\right)+e_{a_j}(t) \\ \tau_{ij}(t) &= \frac{1}{c}\underline{z}_j\underline{k}_i(t) \\ \text{with} \quad & i=1,2,\ldots,l \quad j=1,2,\ldots,m \end{aligned} \tag{11.64}$$

where $X_{a_j}(t)$ is the jth sensor at time t; $S_{a_i}(t)$ is the ith target signal waveform; $\tau_{ij}(t)$ is the relative time delay induced by the ith target signal in the jth sensor; $e_{a_j}(t)$ is the additive noise at the jth sensor; $\underline{k}_i(t)=[\sin\theta_i(t)]$ for $\theta \in = [-90^\circ,\ldots,0,\ldots,90^\circ]$ is the ith direction vector; $\underline{z}_j$ is the jth sensor; c is the velocity of the acoustic signal in the medium; and τ is the time delay between any two neighboring sensors in the array. Thus, the array outputs are written as shown below:

$$X_a(t) = A\left[\theta(t)\right]S_a(t)+e_a(t) = Y \tag{11.65}$$

where $X_a(t) = X_{a_1}(t) \cdots X_{a_m}(t)^T$ is an m vector of the complex envelopes of the sensor outputs; $S_a(t)=\left[S_{a_1}(t) \cdots S_{a_l}(t)\right]^T$ is an l vector of the complex envelopes of the target signals; $e_a(t)=\left[e_{a_1}(t) \cdots e_{a_m}(t)\right]^T$ denotes an m vector of the complex envelopes of the additive measurement noises (with the usual assumptions); $A\left[\theta(t)\right]=\left[a(\theta_1) \cdots a(\theta_l)\right]$ is the $m \times l$ Vandermode matrix (steering matrix); $a(\theta_i)=\left[1e^{-j2\pi\frac{d}{\lambda}\sin\theta_i} \cdots e^{-j(m-1)2\pi\frac{d}{\lambda}\sin\theta_i}\right]^T$ is the steering vector for the ith source/target signal with unit-magnitude elements; and $\theta(t)=\left[\theta_1(t) \cdots \theta_l(t)\right]^T$ is the unknown DoA vector (of the l targets at time t) to be estimated. The term $\theta(t)$ is a slowly varying function. Because the change in $\theta(t)$ is either zero or negligible in each interval of $\left[nT,(n+1)T\right]$, $n=0,1,2,\ldots$, the term $\theta(t)$ can be further modified as

$$\theta(t) \approx \theta(nT) \quad \text{for} \quad t \in [nT,(n+1)T], \quad n=0,1,2,\ldots, \tag{11.66}$$

In an interval N, snapshots of the sensor data are available for processing. Based on the assumptions made in Equation 11.66, these N snapshots of sensor data are expressed as shown below:

$$X_a(k) \approx A[\theta(n)]S_a(k)+e_a(k), \quad k=n,\, n+1,\, n+2,\ldots,\, n+N-1 \tag{11.67}$$

Equation 11.67 is the discrete form of Equation 11.65 with a sampling interval of T/N.

11.3.3 DoA Estimation

The MUSIC algorithm is one of the most popular subspace methods used in spectral estimation [31]. Let us consider the following ASA output data from Equation 11.65:

$$Y = AS_a + e_a \tag{11.68}$$

If the signals from different targets/sources are uncorrelated, the correlation matrix of Y is written as

$$\begin{aligned} R_a &= E\{YY^T\} = E\left\{AS_aS_a^TA^T\right\}+E\left\{e_ae_a^T\right\} \\ R_a &= AR_sA^T+\sigma^2 I = Z+\sigma^2 I \end{aligned} \tag{11.69}$$

Here, $R_s = E\{S_a S_a^T\}$ is the $l \times l$ autocorrelation matrix with rank l, and I is the $l \times l$ identity matrix. The term σ^2 is the noise variance, and Z is the $l \times l$ signal covariance matrix. The singular value decomposition (SVD) is represented as [29]

$$UDU^T = svd(R_a) \tag{11.70}$$

Here, $U = [u_1 \ u_2 \ \cdots \ u_m]$, $D = \begin{bmatrix} \lambda_1+\sigma^2 & 0 & \cdots & 0 & 0 & \cdots & 0 \\ 0 & \vdots & \cdots & \vdots & \vdots & \cdots & 0 \\ \vdots & \vdots & \ddots & \vdots & \vdots & \cdots & \vdots \\ 0 & 0 & \cdots & \lambda_l+\sigma^2 & 0 & \cdots & 0 \\ 0 & 0 & \cdots & 0 & \sigma^2 & \cdots & 0 \\ \vdots & \vdots & \cdots & \vdots & \vdots & \ddots & \vdots \\ 0 & 0 & \cdots & 0 & 0 & \cdots & \sigma^2 \end{bmatrix}$

The matrix U could be partitioned into a matrix U_s with l columns, corresponding to the l signal values, and a matrix U_e with $m - l$ columns, corresponding to the singular values of noise, where U_s is the signal subspace and U_e is the noise subspace, both orthogonal to each other. All noise eigenvectors are orthogonal to the signal steering vectors and the pseudospectrum is computed as

$$P_M(\theta) = \frac{1}{A^T U_e U_e^T A} \tag{11.71}$$

If θ = DoA, the denominator becomes zero, causing peaks in function $P_M(\theta)$, the accuracy of which is limited by the discretization (to which P is computed). Because the search algorithm is computationally demanding, the estimation of DoA from the pseudospectrum is not very practical. These limitations are overcome by the root-MUSIC (RM) algorithm, a model-based parameter estimation method. The RM algorithm uses a steering vector model of the received signal (Equation 11.72) as a function of the DoA, which is the parameter in the model. DoA, or θ, is estimated based on the model and the received signal. The RM algorithm is described as follows [29]:

Let $z = e^{jkd\cos\theta}$ and $A[\theta] = [1 \ \ z \ \ z^2 \ \dots \ z^{m-1}]^T$; then, we have

$$u_p^T A = \sum_{q=0}^{m-1} u_{pq}^* z^q = u_p(z) \tag{11.72}$$

The DoA is thus obtained from Equation 11.72, where $u_p \perp A[\theta]$, $p = l+1$, $l+2,\ldots, m$. Furthermore, we have

$$P_M^{-1} = A^T(\theta)U_e U_e^T A(\theta) = A^T(\theta)CA(\theta) \tag{11.73}$$

$$P_M^{-1} = \sum_{p=0}^{m-1}\sum_{q=0}^{m-1} z^q C_{pq} z^{-p} = \sum_{p=0}^{m-1}\sum_{q=0}^{m-1} z^{(q-p)} C_{pq} \tag{11.74}$$

The double summation can be avoided by setting $r = q - p$, i.e., Equation 11.74 can be written as

$$P_M^{-1}(\theta) = \sum_{r=-(m-1)}^{m+1} C_r z^r \tag{11.75}$$

with

$$C_r = \sum_{q-p=r} C_{pq} \tag{11.76}$$

Here, C_r is the sum of the elements on the qth diagonal. Moreover, because z and $1/z^*$ have the same phase and reciprocal magnitude, one zero is within the unit circle and the other is outside the circle. The phase carries the desired information. Without noise, the roots fall on the unit circle and are used to estimate the DoA. The steps involved in estimating the DoA using RM are (1) estimate the correlation matrix R_a; (2) carry out the SVD, i.e., $[UDU^T] = svd(R_a)$; (3) partition U to obtain U_e, corresponding to $m-l$ smallest singular values; (4) compute C_r; (5) find the zeros of the polynomial; (6) from the $m-1$ roots within the unit circle, choose the l roots closest to the unit circle $(z_q, z = 1,2,\ldots,l)$; and (7) finally, obtain the DoA as

$$\theta_q = \sin^{-1}\left(\frac{2z_q}{d\pi}\right)\frac{180}{\pi} \tag{11.77}$$

11.3.4 Target-Tracking Algorithms

Three target-tracking schemes will be evaluated here [35]:

1. A second-order digital filter is used to remove the noise, with the measurements in polar coordinates, and the LS-estimation method is used to compute the target positions in Cartesian coordinates.
2. KF is used to estimate the states of the target in the polar frame, where the ASA provides measurements in the same plane, and the LS method is used to compute the position of the target.
3. EKF is used for estimating the target states in the Cartesian coordinates, with the measurements in polar coordinates. These schemes are shown in Figures 11.34 and 11.35 [35].

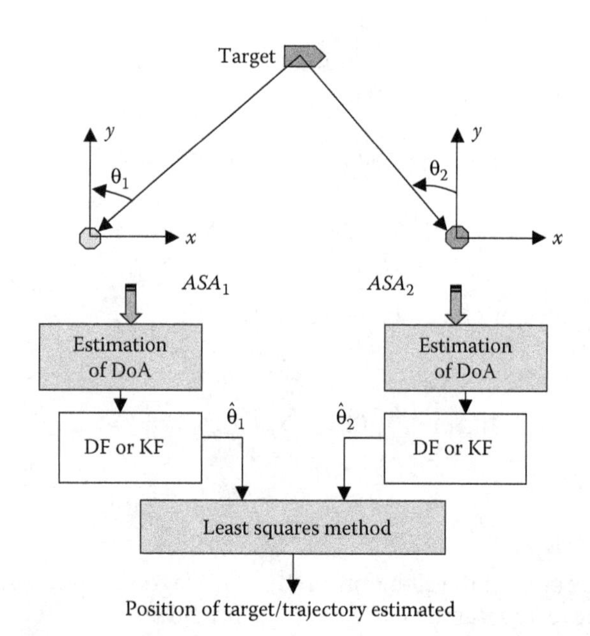

FIGURE 11.34
Target tracker for schemes 1 and 2.

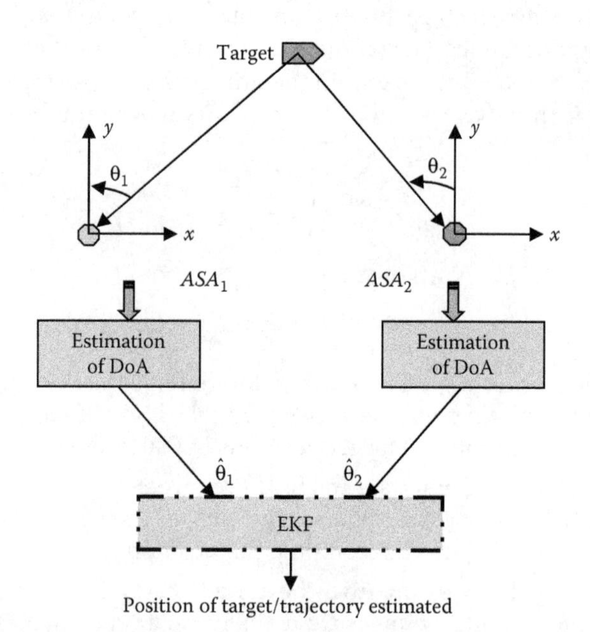

FIGURE 11.35
Target tracker for scheme 3.

11.3.4.1 Digital Filter

The ASA provides noisy measurements in polar coordinates. A second-order low-pass digital Butterworth filter with a normalized cutoff frequency of 0.5 is used to remove the noise. The transfer function, $H(z)$, of the filter is

$$H(z) = \frac{b_0 + b_1 z^{-1} + b_2 z^{-2}}{1 + a_1 z^{-1} + a_2 z^{-2}} \tag{11.78}$$

Here, n is the order of the filter; b_0, b_i, and b_2 are numerator coefficients, and a_i and a_2 are the denominator coefficients.

11.3.4.2 Triangulation

For scheme 1, the measurements are in polar coordinates. To obtain the target position in Cartesian coordinates, LS estimation is used [33]. If the true target positions in Cartesian coordinates are (x_{tr}, y_{tr}), the locations of the two sensors S_{r1} and S_{r2} are (S_{r1x}, S_{r1y}) and (S_{r2x}, S_{r2y}), and the target is at θ_1 and θ_2 with respect to S_{r1} and S_{r2}, respectively, then

$$\tan(\theta_1) = \frac{y_{tr} - S_{r1y}}{x_{tr} - S_{r1x}} \quad \text{and} \quad \tan(\theta_2) = \frac{y_{tr} - S_{r2y}}{x_{tr} - S_{r2x}} \tag{11.79}$$

After rearrangement, we obtain

$$\begin{bmatrix} S_{r1y} - S_{r1x}\tan(\theta_1) \\ S_{r2y} - S_{r2x}\tan(\theta_2) \end{bmatrix} = \begin{bmatrix} -\tan(\theta_1) & 1 \\ -\tan(\theta_2) & 1 \end{bmatrix} \begin{bmatrix} x_{tr} \\ y_{tr} \end{bmatrix} \tag{11.80}$$

which is represented as

$$f = Bg \tag{11.81}$$

Here, B and f are known and g is unknown; then, the target position g is computed from $g = B^{-1} f$.

11.3.4.3 Results and Discussion

For the performance evaluation of the DoA algorithm, the parameters given in Table 11.7 and 50 Monte Carlo simulations are used to obtain the results. Figure 11.36a is obtained by varying the variance level from 0 to 0.1 and keeping the other parameters constant, as shown in Table 11.7. The AE in DoA increases with an increase in the noise level. Keeping the other parameters constant, the effect of the snapshots is shown in Figure 11.36b. The AE decreases with an increase in the number of snapshots.

TABLE 11.7

Parameters for Evaluation of DoA Estimation

Parameter	Value
No. of sensors (m)	5
No. of snapshots (N)	50
Noise variance (σ^2)	0.00001
No. of targets (l)	1
Theta (θ)	30°
Spacing between the sensors (d)	0.5λ

Source: Naidu, V. P. S., and J. R. Raol. 2007. *Def Sci J* 57(3):289–303. With permission.

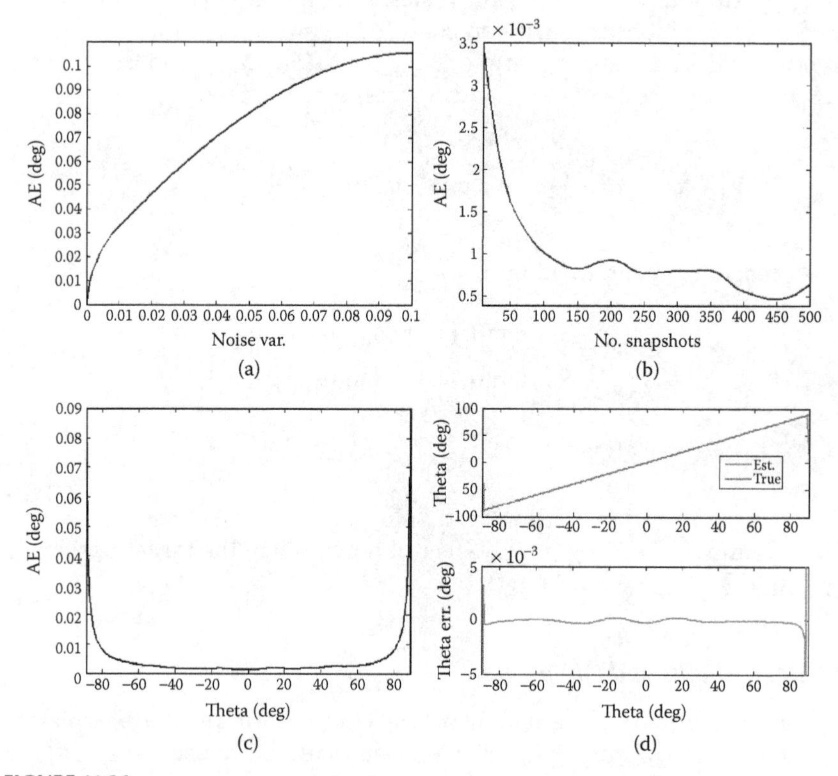

FIGURE 11.36

Effect of various parameters on the DoA estimation: (a) Effect of noise variance on AE in DoA estimation; (b) effect of snapshots on AE in DoA estimation; (c) effect of theta on DoA estimation; and (d) effect of theta on DoA estimation-error. (From Naidu, V. P. S., and J. R. Raol. 2007. *Def Sci J* 57(3):289–303. With permission.)

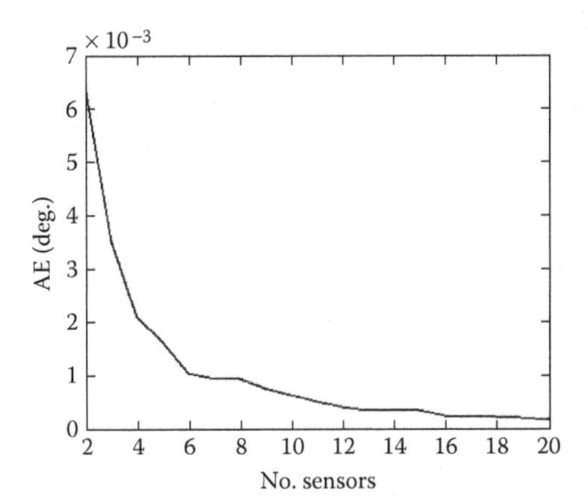

FIGURE 11.37
Effect of the number of sensors on the DoA estimation. (From Naidu, V. P. S., and J. R. Raol. 2007. *Def Sci J* 57(3):289–303. With permission.)

Figure 11.36c and d show the effect of the direction of the signal incoming to the ASA on the DoA estimation. The error is not much, except at 90°. From Figure 11.37, note that the AE decreases with an increase in the number of sensors; a greater number of signals provide an average effect that would reduce the effect of noise.

11.3.5 Target Tracking

A 2-DOF kinematic model in the Cartesian coordinates x and y is given as

$$F = \begin{bmatrix} \Phi & 0 \\ 0 & \Phi \end{bmatrix} \quad \text{and} \quad G = \begin{bmatrix} \varsigma & 0 \\ 0 & \varsigma \end{bmatrix} \tag{11.82}$$

Here, $\Phi = \begin{bmatrix} 1 & T \\ 0 & 1 \end{bmatrix}$ and $\varsigma = \begin{bmatrix} T^2/2 \\ T \end{bmatrix}$

The simulation utilizes the following parameters:

(1) $T = 0.1$ second; (2) process-noise variance is 0.001; (3) the simulation is run for 100 seconds; (4) the initial values are $[x \quad \dot{x} \quad y \quad \dot{y}] = [0 \quad 1 \quad 10 \quad 0]$; and (5) the sensor locations are $S_{1r} = [S_{r1x} \quad S_{r1y}] = [0 \quad 0]$ and $S_{r2} = [S_{r2x} \quad S_{r2y}] = [50 \quad 0]$s.

Figure 11.38a shows the target trajectory. The trajectory of the target in Cartesian coordinates can be converted into polar coordinates using

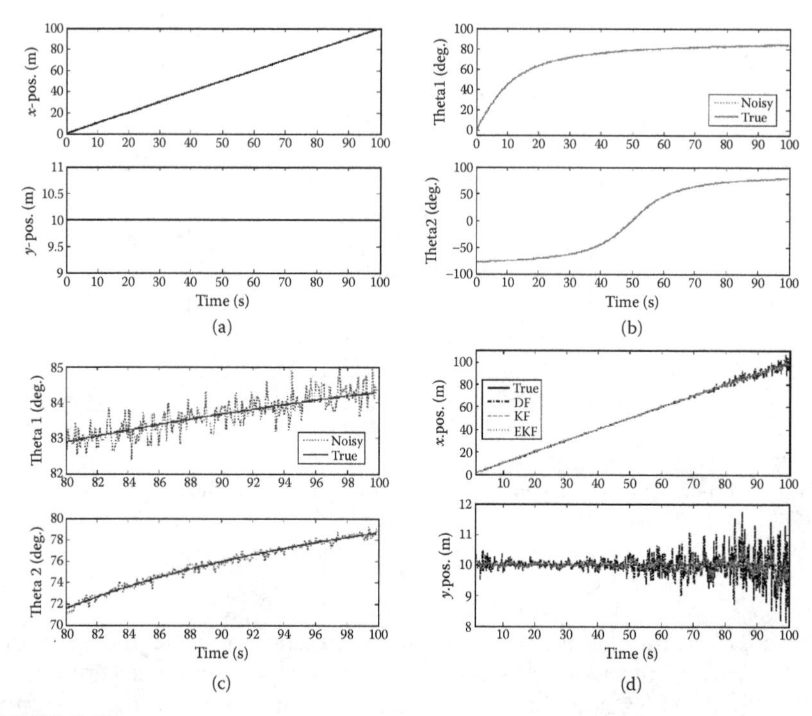

FIGURE 11.38
Target trajectory in Cartesian coordinates and results: (a) Target trajectory; (b) true and noisy measurements (DoA); (c) enlarged view of a portion from part (b); and (d) true and estimated x- and y-positions. (From Naidu, V. P. S., and J. R. Raol. 2007. *Def Sci J* 57(3):289–303. With permission.)

Equation 11.79, and ASA output generated for each sample. The ASA contains five sensors. It produces 100 snapshots, with a noise variance of 0.00001. The RM algorithm is used to estimate DoA. Measurement noise, with a variance of 0.001, is added to the estimated DoAs. Figure 11.38b shows the simulated DoA data, and Figure 11.38c shows the enlarged view of a small portion from Figure 11.38b. The initial states for both the linear and EKF filters are computed with the first two measurements. Identity matrices, of the same order of the state vector, are multiplied by 0.01 and are taken as the initial state-covariance matrices for both the filters. Figure 11.38d shows the true and estimated positions, and Figure 11.39 shows the enlarged view of a portion form Figure 11.38d. The performance metrics are given in Table 11.8. Figure 11.40a shows the AE in relation to the positions, and Figure 11.40b shows an enlarged view of a portion from Figure 11.40a. Figure 11.40c shows the RSS position errors; Figure 11.40d shows an enlarged view of a small portion. Note that the errors increase when the target moves away from the sensors. From Table 11.8, note that

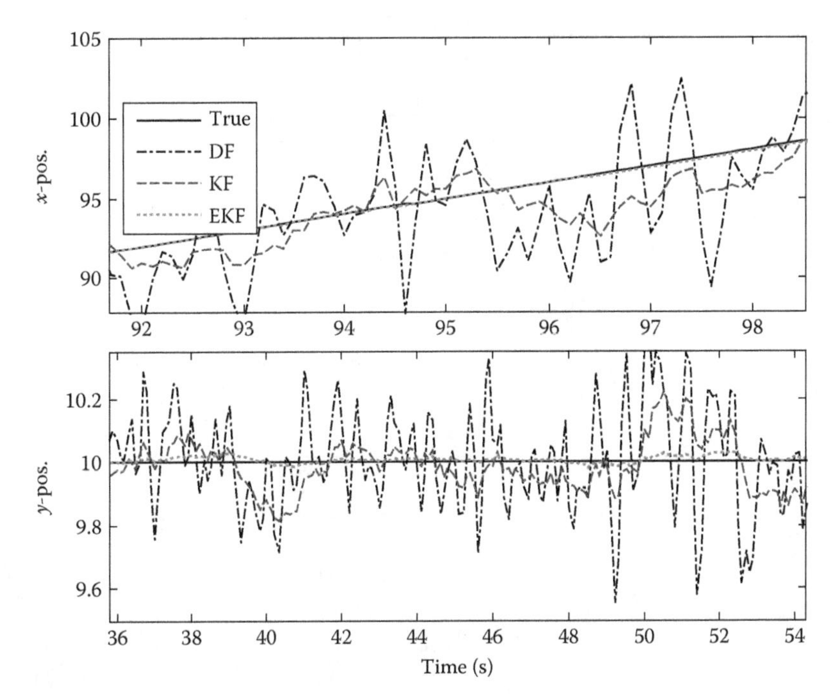

FIGURE 11.39
Enlarged view of a portion of Figure 11.38d. (From Naidu, V. P. S., and J. R. Raol. 2007. *Def Sci J* 57(3):289–303. With permission.)

TABLE 11.8

Performance Metrics for the Different Schemes

	PFEx	PFEy	MAEx	MAEy	RMSPE	Execution Time (s)
DF	2.43	5.349	0.67	0.251	1.06	0.04
KF	0.98	1.503	0.29	0.101	0.41	0.15
EKF	0.06	0.173	0.03	0.013	0.03	0.18

Source: Naidu, V. P. S., and J. R. Raol. 2007. *Def Sci J* 57(3):289–303. With permission.

the execution time for the first scheme is the least among the three schemes, however, this scheme does not perform very well.

11.3.5.1 Joint Acoustic-Image Target Tracking

The target-localization and tracking procedure based on acoustic and visual measurements is important in various applications, such as

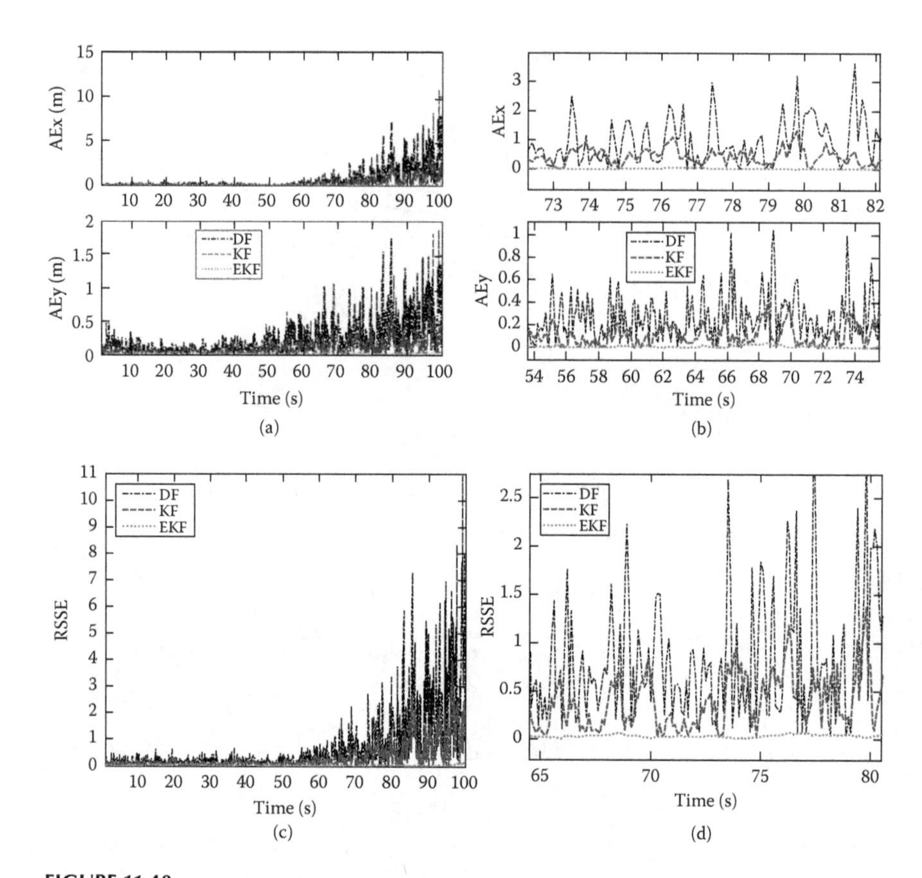

FIGURE 11.40
AE and RSS-position errors: (a) AE in x- and y-positions; (b) enlarged view of a portion from part (a); (c) RSS position errors; and (d) enlarged view of a portion from part (c). (From Naidu, V. P. S., and J. R. Raol. 2007. *Def Sci J* 57(3):289–303. With permission.)

analysis of dynamic scenes, video conferences, and analysis of traffic situations [36]. This section discusses a scheme for target tracking using joint acoustic and imaging sensor data, for which a decentralized KF is used. Since both the sensors provide angle measurements, a triangulation method is used to get the 3D position of the target. Figure 11.41 shows the flow diagram of the scheme. To get the 3D position, at least two sensors are required, because these sensors provide only 2D measurements.

11.3.5.2 Decentralized KF

The two local KFs (LKFs) produce track state estimates based on their respective local sensors. A fusion center, namely, a global KF (GKF),

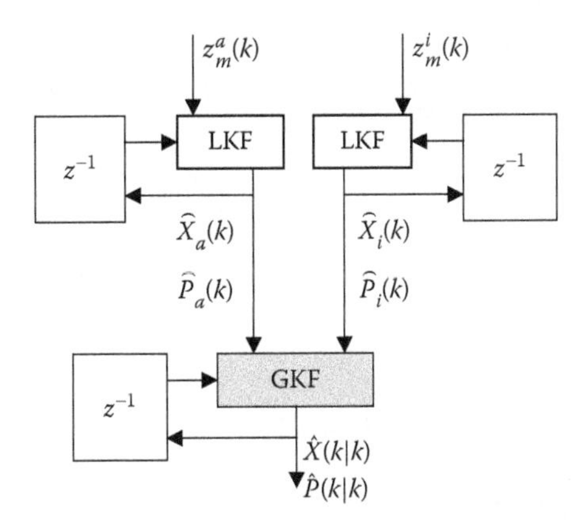

FIGURE 11.41
A decentralized KF scheme.

combines these local target state estimates into a global target state estimate. The target dynamics are given as follows:

$$X(k+1) = FX(k) + Gw(k) \tag{11.83}$$

Here, $X(k)$ is the state vector of the target at instant k, F is the time-invariant state-transition matrix, G is the process noise-gain matrix, and $w(k)$ is the process-noise sequence with covariance matrix Q. The linear measurement models are given by

$$z_m^a(k+1) = HX(k+1) + v_a(k+1) \tag{11.84}$$

$$z_m^i(k+1) = HX(k+1) + v_i(k+1) \tag{11.85}$$

Here, $z_m^a(k)$ is the measurement vector of the acoustic sensor, $z_m^i(k)$ is the measurement vector of the imaging sensor, H is the observation matrix, $v_a(k)$ is the measurement noise sequence with covariance matrix R_a, and $v_i(k)$ is the measurement noise sequence with covariance matrix R_i. In the first step, when the LKF is active, it takes the measurements from the sensors and estimates the target states. In the second step, the GKF takes the local estimates with their associated covariances and fuses them to obtain the global target state estimate.

The first step at the LKF is time prediction. It is assumed that there are two sensors, the acoustic and imaging sensors, and that these sensors provide measurements at the same time. The processing steps followed at the LKFs, where the standard KF functions, are time prediction and measurement updation.

Time prediction is carried out as follows:

(a) At the acoustic sensor:

$$\tilde{X}_a(k|k-1) = F\hat{X}_a(k-1|k-1) \tag{11.86}$$

$$\tilde{P}_a(k|k-1) = F\hat{P}_a(k-1|k-1)F^T + GQG^T \tag{11.87}$$

(b) At the imaging sensor:

$$\tilde{X}_i(k|k-1) = F\hat{X}_i(k-1|k-1) \tag{11.88}$$

$$\tilde{P}_i(k|k-1) = F\hat{P}_i(k-1|k-1)F^T + GQG^T \tag{11.89}$$

Measurement update is carried out as follows:

(a) At the acoustic sensor:

$$\vartheta = z_m^a(k) - H\tilde{X}_a(k|k-1) \tag{11.90}$$

$$S = H\tilde{P}_a(k|k-1)H^T + R_a \tag{11.91}$$

$$K = \tilde{P}_a(k|k-1)H^T S^{-1} \tag{11.92}$$

$$\hat{X}_a(k|k) = \tilde{X}_a(k|k-1) + K\vartheta \tag{11.93}$$

$$\hat{P}_a(k|k) = [I - KH]\tilde{P}_a(k|k-1) \tag{11.94}$$

$$\widehat{X}_a(k) = \hat{P}_a^{-1}(k|k)\hat{X}_a(k|k) - \tilde{P}_a^{-1}(k|k-1)\tilde{X}_a(k|k-1) \tag{11.95}$$

$$\widehat{P}_a(k) = \widehat{P}_a^{-1}(k|k) - \widehat{P}_a^{-1}(k|k-1) \tag{11.96}$$

(b) At the imaging sensor:

$$\vartheta = z_m^i(k) - H\tilde{X}_i(k|k-1) \tag{11.97}$$

$$S = H\tilde{P}_i(k|k-1)H^T + R_i \tag{11.98}$$

$$K = \tilde{P}_i(k|k-1)H^T S^{-1} \tag{11.99}$$

$$\hat{X}_i(k|k) = \tilde{X}_i(k|k-1) + K\vartheta \tag{11.100}$$

$$\hat{P}_i(k|k) = [I - KH]\tilde{P}_i(k|k-1) \tag{11.101}$$

$$\hat{X}_i(k) = \hat{P}_i^{-1}(k \mid k)\hat{X}_i(k \mid k) - \tilde{P}_i^{-1}(k \mid k-1)\tilde{X}_i(k \mid k-1) \tag{11.102}$$

$$\hat{P}_i(k) = \hat{P}_i^{-1}(k \mid k) - \hat{P}_i^{-1}(k \mid k-1) \tag{11.103}$$

The processing steps followed at the fusion center, where the GKF is functional, are as follows:

(a) Time prediction:

$$\tilde{X}(k \mid k-1) = F\hat{X}(k-1 \mid k-1) \tag{11.104}$$

$$\tilde{P}(k \mid k-1) = F\hat{P}(k-1 \mid k-1)F^T + GQG^T \tag{11.105}$$

(b) Estimate correction:

$$\hat{P}^{-1}(k \mid k) = \tilde{P}^{-1}(k \mid k-1) + \hat{P}_a(k) + \hat{P}_i(k) \tag{11.106}$$

$$\hat{X}(k \mid k) = \hat{P}(k \mid k)\left[\tilde{P}^{-1}(k \mid k-1)\tilde{X}(k \mid k-1) + \hat{X}_a(k) + \hat{X}_i(k)\right] \tag{11.107}$$

11.3.5.3 3D Target Tracking

Because the ASA provides only angle measurements, triangulation is used to obtain the target's 3D position. The two sensors are placed at two different locations. Each set contains both acoustic and imaging sensors. Figure 11.42 shows the general scheme. The estimates from the two fusion centers are used to determine the 3D target position using triangulation. Let the target state estimate from the first and the second fusion centers be

$$\hat{X}_1(k \mid k) = \left[\theta_1 \;\; \dot{\theta}_1 \;\; \ddot{\theta}_1 \;\; \phi_1 \;\; \dot{\phi}_1 \;\; \ddot{\phi}_1\right]^T \tag{11.108}$$

$$\hat{X}_2(k \mid k) = \left[\theta_2 \;\; \dot{\theta}_2 \;\; \ddot{\theta}_2 \;\; \phi_2 \;\; \dot{\phi}_2 \;\; \ddot{\phi}_2\right]^T \tag{11.109}$$

Similarly, let the locations of the two sensor sets, S_{r1} and S_{r1}, be $(S_{r1x}, S_{r1y}, S_{r1z})$ and $(S_{r2x}, S_{r2y}, S_{r2z})$, respectively. Let the target be at an angle of (θ_1, ϕ_1) and (θ_2, ϕ_2) with respect to S_{r1} and S_{r2}. Let the target position in Cartesian coordinates be (x_{tr}, y_{tr}, z_{tr}). Then

$$\tan(\theta_1) = \frac{y_{tr} - S_{r1y}}{x_{tr} - S_{r1x}} \tag{11.110}$$

$$\tan(\theta_2) = \frac{y_{tr} - S_{r2y}}{x_{tr} - S_{r2x}} \tag{11.111}$$

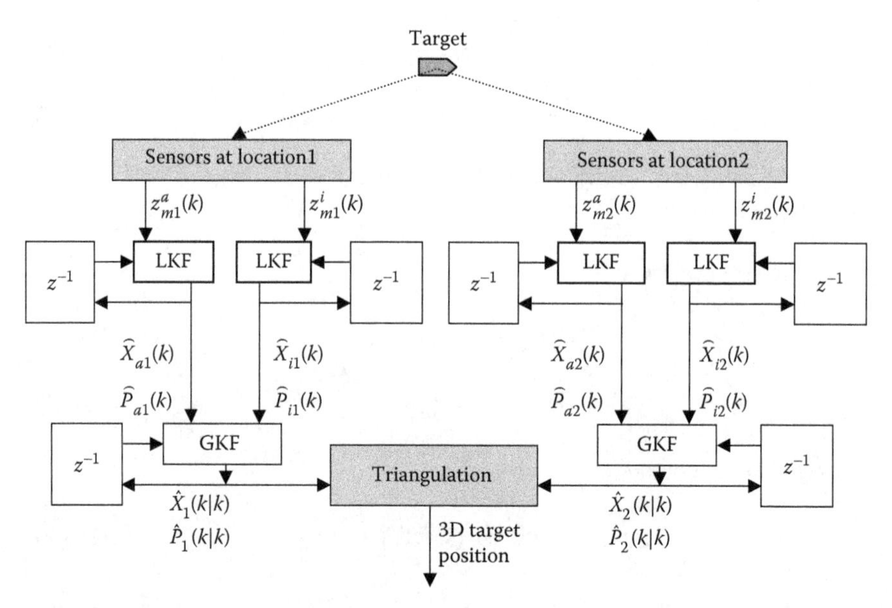

FIGURE 11.42
A decentralized KF with triangulation.

$$\tan(\phi_1) = \frac{z_{\text{tr}} - S_{r1z}}{\sqrt{(x_{\text{tr}} - S_{r1x})^2 + (y_{\text{tr}} - S_{r1y})^2}} \quad \text{and} \tag{11.112}$$

$$\tan(\phi_2) = \frac{z_{\text{tr}} - S_{r2z}}{\sqrt{(x_{\text{tr}} - S_{r2x})^2 + (y_{\text{tr}} - S_{r2y})^2}} \tag{11.113}$$

By arranging the above equations, the following expression is obtained:

$$\begin{bmatrix} S_{r1y} - S_{r1x}\tan(\theta_1) \\ S_{r2y} - S_{r2x}\tan(\theta_2) \\ S_{r1z} - S_{r1x}\tan(\phi_1)\sqrt{1+\tan^2(\theta_1)} \end{bmatrix} = \begin{bmatrix} -\tan(\theta_1) & 1 & 0 \\ -\tan(\theta_1) & 1 & 0 \\ -\tan(\phi_1)\sqrt{1+\tan^2(\theta_1)} & 0 & 1 \end{bmatrix} \begin{bmatrix} x_{\text{tr}} \\ y_{\text{tr}} \\ z_{\text{tr}} \end{bmatrix} \tag{11.114}$$

The above equation is represented as $f = Bg$, where B and f are known and g is unknown. The vector g of the target position is computed as $g = B^{-1}f$.

11.3.6 Numerical Simulation

The 3-DOF kinematic model, with position, velocity, and acceleration components in Cartesian coordinates of the x, y, and z axes, is represented as follows:

$$F = \text{diag}[\Phi \ \ \Phi \ \ \Phi] \quad \text{and} \quad G = \text{diag}[\varsigma \ \ \varsigma \ \ \varsigma] \tag{11.115}$$

where $\Phi = \begin{bmatrix} 1 & T & T^2/2 \\ 0 & 1 & T \\ 0 & 0 & 1 \end{bmatrix}$ and $\varsigma = \begin{bmatrix} T^3/6 \\ T^2/2 \\ T \end{bmatrix}$.

The simulation utilizes the following parameters: sampling interval $T = 0.25$, process noise variance along the three axes $\sigma_x^2 = \sigma_y^2 = \sigma_z^2 = 10^{-6}$, duration of simulation = 125 s, and initial state vector $X_i = [100 \ \ 10 \ \ 0.01 \ \ 100 \ \ -10 \ \ 0.01 \ \ 125 \ \ -1 \ \ 0]^T$. The trajectory of the target in Cartesian coordinates is converted into polar coordinates using Equation 11.114. The following measurement noises are added to obtain realistic measurements. The two sensor locations are $S_{r1} = (0,0,0)$ and $S_{r1} = (1000, -1000, 0)$. Noise with a variance of $\left(\sigma_\theta^a\right)^2 = \left(\sigma_\phi^a\right)^2 = 10^{-4}$ is added to obtain the acoustic measurements, and noise with a variance of $\left(\sigma_\theta^i\right)^2 = \left(\sigma_\phi^i\right)^2 = 10^{-6}$ is added to obtain imaging sensor measurements. Figures 11.43 and 11.44 show the acoustic and imaging sensor data at locations 1 and 2, respectively. For each LKF and GKF at each sensor location with the angular position, velocity, and acceleration components in each of the two polar coordinates, the 3-DOF kinematic model has the following transition (F) and process-noise gain (G) matrices:

$$F = \begin{bmatrix} \Phi & 0 \\ 0 & \Phi \end{bmatrix} \quad \text{and} \quad G = \begin{bmatrix} \varsigma & 0 \\ 0 & \varsigma \end{bmatrix} \tag{11.116}$$

Here, $\Phi = \begin{bmatrix} 1 & T & T^2/2 \\ 0 & 1 & T \\ 0 & 0 & 1 \end{bmatrix}$, and $\varsigma = \begin{bmatrix} T^3/6 \\ T^2/2 \\ T \end{bmatrix}$

The initial true state vector is $X_t = \left[\theta \ \ \dot{\theta} \ \ \ddot{\theta} \ \ \phi \ \ \dot{\phi} \ \ \ddot{\phi}\right]^T$, computed using the first two true measurements. The state errors for each sensor location were found to be within the bounds. Figures 11.45 and 11.46 show the innovation sequences. Figure 11.47 shows the RSS errors in the angular position. Table 11.9 shows the PFEs and AEs in angular positions, while Table 11.10 shows the RMS errors in angular positions. Figure 11.48 shows the

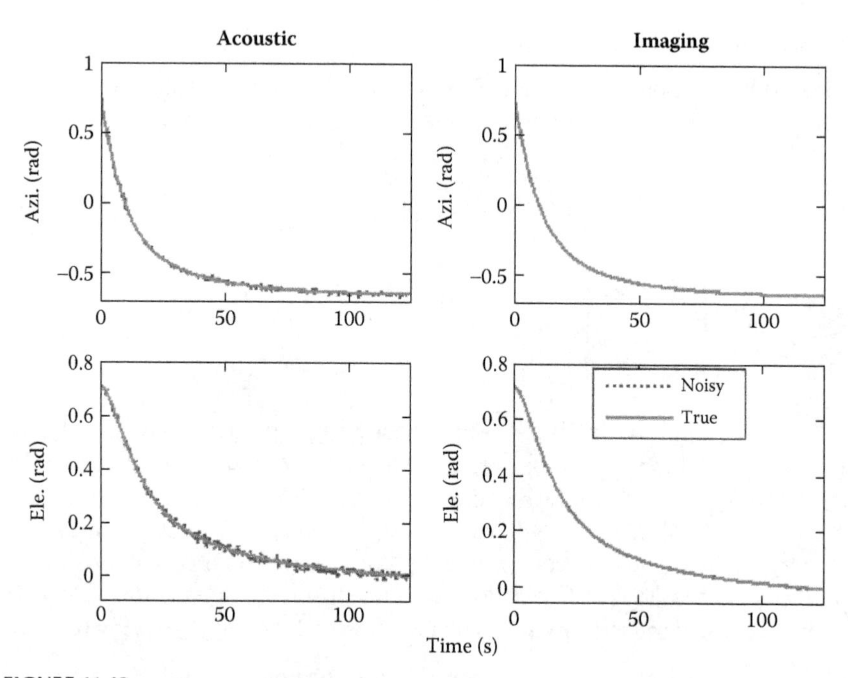

FIGURE 11.43
True/noisy measurements from the acoustic and imaging sensors at sensor location 1.

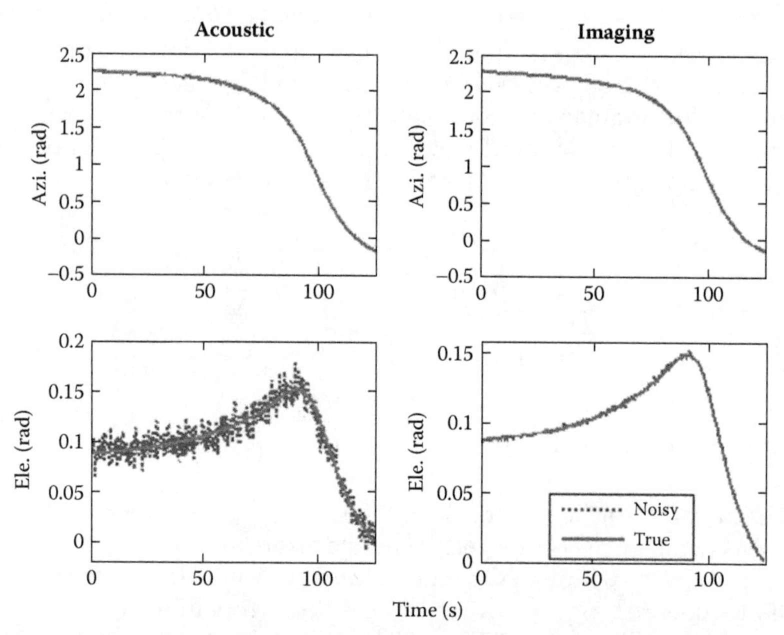

FIGURE 11.44
True/noisy measurements from the acoustic and imaging sensors at sensor location 2.

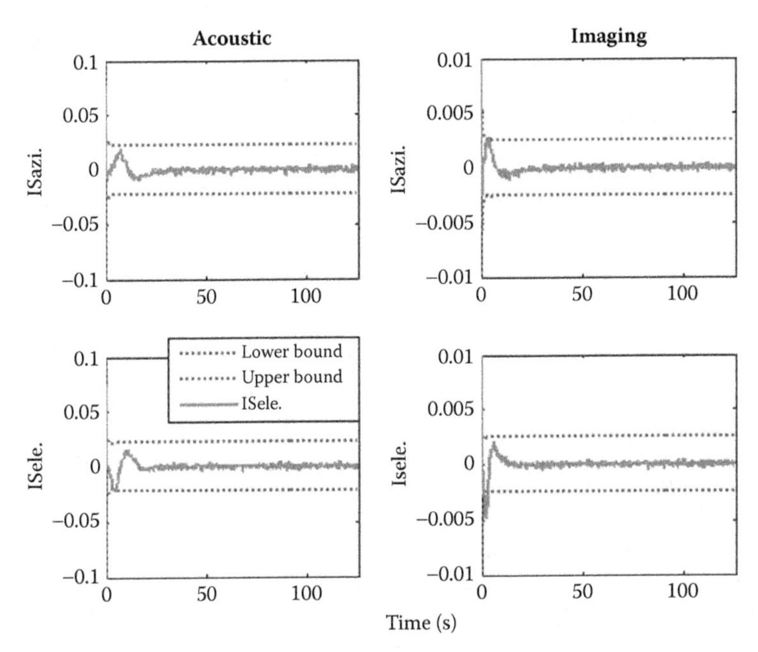

FIGURE 11.45
Innovation sequences for acoustic and imaging sensors, obtained from sensor location 1.

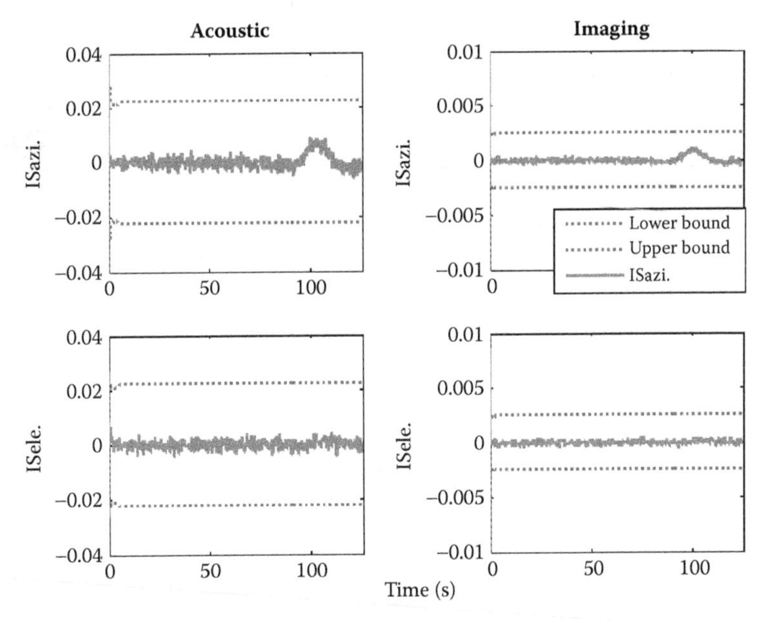

FIGURE 11.46
Innovation sequences for acoustic and imaging sensors, obtained from sensor location 2.

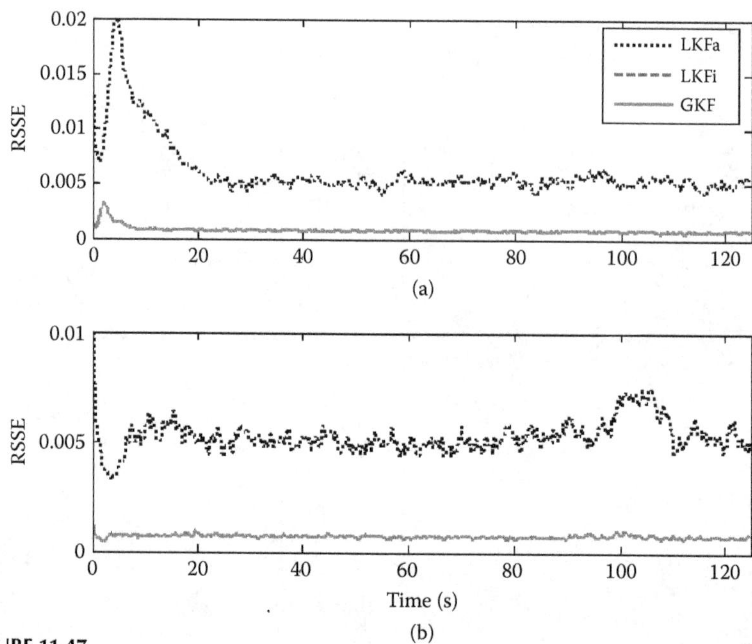

FIGURE 11.47
RSSE in angle for (a) location 1 and (b) location 2.

TABLE 11.9
PFE and MAE in Angular Positions

	Sensor Location 1				Sensor Location 2			
	PFEθ_1	**PFEϕ_1**	**MAEθ_1**	**MAEϕ_1**	**PFEθ_1**	**PFEϕ_1**	**MAEθ_1**	**MAEϕ_1**
LKFa	0.8824	2.2399	0.0038	0.0041	0.242	3.8994	0.0035	0.0033
LKFi	0.1135	0.2887	0.0005	0.0005	0.0328	0.5538	0.0005	0.0005
DKF	0.1129	0.2875	0.0005	0.0005	0.0326	0.5514	0.0005	0.0005

TABLE 11.10
RMSE in Angular Position

	Sensor Location 1	Sensor Location 2
LKFa	0.0052	0.0042
LKFi	0.0007	0.0006
GKF	0.0007	0.0006

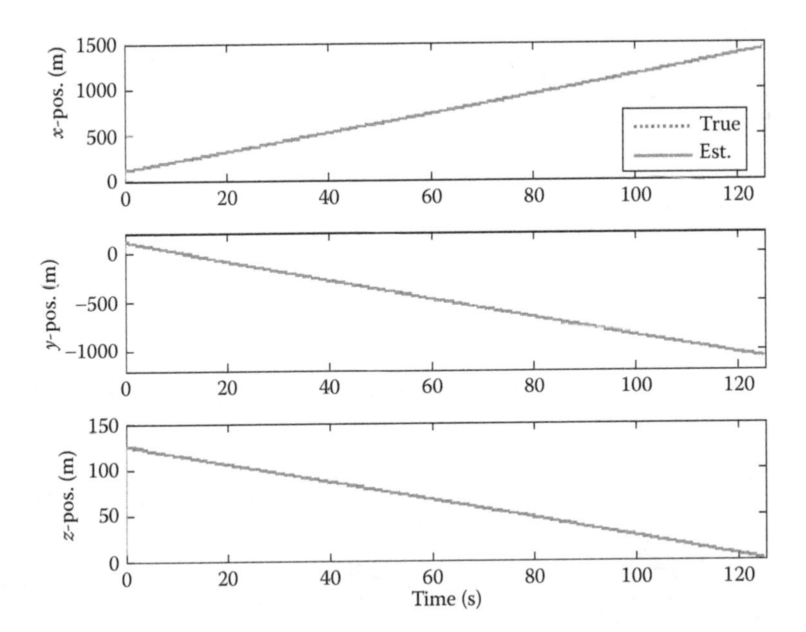

FIGURE 11.48
True/estimated target positions.

TABLE 11.11
PFE and MAE in the x, y, and z Positions

PFEx	PFEy	PFEz	MAEx	MAEy	MAEz
0.1194	0.1231	0.9217	0.7175	0.5427	0.5082

true and estimated target positions from triangulation. Table 11.11 shows the percentage-fit error and mean AE in all three x, y, and z positions. We can infer that EKF shows better performance than other filters, but with a longer execution time.

EXAMPLE 11.2

This demo example, Demo for 3D Tracking with IRST and Radar Data, demonstrates the 3D target-tracking procedure using both IRST and radar measurements. The sampling time is 0.25 seconds and the total flight is 120 seconds. The measurement noise variances are given in Table 11.12. Process noise is taken as 1E-6. The CA model is used. The initial state vector is [10,2,0.5,10,5,0.3,10,1,0.01].

TABLE 11.12

Measurement Variances (Example 11.2)

Sensor	Azimuth (rad)	Elevation (rad)	Range (m)
IRST	1E-6	1E-6	–
Radar	1E-4	1E-4	10

Solution 11.2

Run the MATLAB code ch11_ex2. The results are as follows:

PFEP : [0.0264 0.0386 0.3316]
PFEV : [0.1679 0.3524 2.0282]
PFEA : [1.7784 2.6751 18.1935]
MAEP : [0.4040 0.4129 0.2778]
MAEV : [0.0429 0.0501 0.0256]
MAEA : [0.0037 0.0036 0.0014]
RMSPE RMSVE RMSAE : [0.4867 0.0675 0.0070]
mean NIS : 4.9767
mean NEES : 7.7232

The various plots generated by this example are not included here in the text; however, when the program is run using MATLAB (in the command window), all of the plots are displayed.

Epilogue

There is a scope to adopt the unscented Kalman filter (derivative-free KF), or particle filter, with appropriate modifications, to track the centroid of the extended target, instead of the probabilistic data association filter (to replace the KF part). Different kinds of image-registration techniques are used in medical image processing. In some situations, the objects in the scene would be at different distances from the imaging sensor. Inexpensive sensors would not properly focus everywhere. If one object is in focus, then another will be out of focus. This would yield a slightly out-of-focus fused image [13,37]. Multisensor image-fusion applications are treated in a study by Blum and Liu [38]. Although estimation fusion rules and algorithms (see Section 2.2) are very useful for target-tracking and fusion applications, their use in image-tracking applications (centroid tracking, etc.) is very limited. Further details on the development of image fusion and related tracking algorithms and applications can be found in [39–41].

Exercises

III.1. In what ways can the fusion of 3D vision data and active tactile sensing be useful?

III.2. In what ways can the fusion of vision information and thermal sensing be useful in object recognition and, hence, target tracking?

III.3. What are the merits of fusing information from laser radar (LADAR) and forward-looking infrared (FLIR) images?

III.4. Explain the difference between decision fusion and data fusion in the context of image fusion.

III.5. Consider an image fusion example and prove that decision fusion is a special case of data fusion.

III.6. How would you use the processes of data fusion and decision fusion in the context of acoustic- and vision-based information (say, for speech recognition)?

III.7. In light of the previous exercise, can you construct and show the scheme for the fusion of acoustic and visual information using artificial neural networks?

III.8. What is the distinction between active and passive sensors in the context of visual sensors [40]?

III.9. What is an "active vision"?

III.10. What are *passive* imaging sensors/systems?

III.11. What are *active* sensing systems/techniques?

III.12. What are the intrinsic and extrinsic parameters of a camera?

III.13. Why is calibration of a camera necessary?

III.14. Explain the procedure of fusion of a stereo vision system.

III.15. Give examples of some fusion configurations (sensor network configurations) in the context of image/vision systems.

III.16. Why are image synchronization and registration important for the image-fusion process?

III.17. An RBF can be used in the transform model of the ABM for image registration (see Section 10.1.3.2). What then is the difference between this method and the RBF neural network?

III.18. Which fusion rule (as discussed in Chapter 9) is implied by Equation 10.67 (see Section 10.3.4.1)?

III.19. Which rule for image fusion is implied by the algorithm given in Chapter 9?

III.20. Is it possible to model image noise as something other than linear additive noise?

III.21. Why is segmentation required in the process of image analysis/fusion?

III.22. What is actually the function of principal-component analysis or PCA?

III.23. What is implied by the formula of Equation 10.39? What are the weights?

III.24. What is implied by the formula of Equation 10.45? What are the weights?

III.25. If sensor 1 has an uncertainty ellipse with its major axis in the x direction and sensor 2 has an uncertainty ellipse with its major axis in the y direction, and if these 2D images are fused, what is the likely shape of the uncertainty ellipse?

III.26. What it the significance of Exercise III.25?

III.27. What is the useful feature available in a thermal imager?

III.28. Do we get a 3D view of an image from a conventional television?

III.29. What type of sensory information do we obtain or use in pixel-level and feature-level fusion processes?

III.30. What type of sensory information do we use for pixel-level and feature-level image-fusion processes?

III.31. What are the registration aspects for pixel- and feature-level fusion algorithms or processes?

III.32. What is the fundamental fusion process for pixel- and feature-level image fusion?

III.33. What are the common noise sources that can contaminate an image?

III.34. What are possible methods for reducing the effect of noise processes in acquired images?

III.35. What are the spatial filter types and their fundamental characteristics?

III.36. What is the condition on the convolution mask in the case of the linear mean filters?

III.37. What are the main characteristics of nonlinear spatial filters?

III.38. What is "noise" in the context of an image?

III.39. What is "uncorrelated noise" in the context of an image?

III.40. What are the distinct noise processes in the context of an image?

III.41. What are the possible causes of noise in an image?

III.42. How is salt and pepper noise generated for an image?

III.43. What are the major types of image segmentation?

III.44. How is the image-registration algorithm used to help in target tracking?

III.45. When comparing two images for image registration, when is the sum of the absolute difference minimal?

III.46. While scanning an image, what is the difference between intrascan and interscan filters?

III.47. What are the fundamental requirements for fusion results in the case of image fusion?

III.48. What is the major significance of image fusion from the viewpoint of a fused image?

III.49. What does spatial frequency do in the context of an image?

III.50. What is blocks-based image fusion?

III.51. What are multiwavelets, and how can they be used for image fusion?

References

1. Hong, G. 2007. Image fusion, image registration, and radiometric normalization for high resolution image processing. PhD Thesis, TR No. 247, Department of Geodesy and Geomatics Engineering, University of New Brunswick, Fredericton, New Brunswick, Canada, p. 198.
2. Goshtasby, A. A. and S. Nikolov, eds. 2007. Image fusion: Advances in the state of the art. *Inf Fusion* 8:114–8.
3. Zitova, B., and J. Flusser. 2003. Image registration methods: A survey. *Image Vis Comput* 21:977–1000.
4. Gonzalez, R. C., and R. E. Woods. 1993. *Digital image Processing*. New York: Addison-Wesley Inc.
5. Hill, P., N. Canagarajah, and D. Bull. 2002. Image fusion using complex wavelets. BMVC. http://www.bmva.ac.uk/bmvc/2002/papers/88/full_88.pdf (accessed December 2008).
6. Qu, J., and C. Wang. 2001. A wavelet package-based data fusion method for multi-temporal remote sensing image processing. 22nd Asian conference on remote sensing, November 2001, Singapore.
7. Kumar, A., Y. Bar Shalom, and E. Oron. 1995. Precision tracking based on segmentation with optimal layering for imaging sensor. *IEEE Trans Pattern Anal Mach Intell* 17:182–8.
8. Oron, E., A. Kumar, and Y. Bar Shalom. 1993. Precision tracking with segmentation for imaging sensor. *IEEE Trans Aerosp Electron Syst* 29:977–87.
9. Waltz, E. and J. L. Llinas. 1990. Multi-sensor data fusion. Boston: Artech House.
10. Naidu, V. P. S., and J. R. Raol. 2008. Pixel-level image fusion using wavelets and principal component analysis—A comparative analysis. *Def Sci J* 58:338–52.
11. Leung, L. W., B. King, and V. Nohora. 2001. Comparison of image fusion techniques using entropy and INI. 22nd Asian conference on remote sensing, November 2001, Singapore.
12. Naidu, V. P. S., G. Girija, and J. R. Raol. 2007. Data fusion for identity estimation and tracking of centroid using imaging sensor data. *Def Sci J* 57:639–52.
13. Pajares, G., and J. M. de la Cruz. 2004. A wavelet-based image fusion tutorial. *Pattern Recognit* 37:1855–72.
14. Li, S., J. T. Kwok, and Y. Wang. 2001. Combination of images with diverse focuses using the spatial frequency. *Inf Fusion* 2:169–76.
15. Eskicioglu, A. M., and P. S. Fisher. 1995. Image quantity measures and their performance. *IEEE Trans Commun* 43:2959–65.
16. Cover, T. M., and J. A. Thomas. 1991. *Elements of information theory*. New York: Wiley.
17. Wang, Z., and A. C. Bovik. 2002. A universal image quality index. *IEEE Signal Process Lett* 9:81–4.
18. Cvejic, N., A. Loza, D. Bull, and N. Cangarajah. 2005. A similarity metric for assessment of image fusion algorithms. *Int J Signal Process* 2:178–82.

19. Piella, G., and H. Heijmans. 2003. A new quality metric for image fusion. Proceedings of the IEEE International Conference on Image Processing, Barcelona, Spain, pp. 173–6.
20. Jalili-Moghaddam, M. 2005. Real-time multi focus image fusion using discrete wavelet transform and Laplasican pyramid transform. Master's thesis, Chalmess University of Technology, Goteborg, Sweden.
21. Daubechies, I. 1992. *Ten lectures on wavelets. Regional conference series in applied maths*, Volume 91. Philadelphia: SIAM.
22. Baltzakis, H., A. Argyros, and P. Trahanias. 2003. Fusion of laser and visual data for robot motion planning and collision avoidance. *Mach Vis Appl* 15:92–100.
23. Wang, J. G., K. A. Toh, E. Sung, and W. Y. Yau. 2007. A feature-level fusion of appearance and passive depth information for face recognition. In *Face recognition*, ed. K. Delac, and M. Grgic. Vienna, pp. 537–58, Austria: I-Tech.
24. Ross, A., and G. Rohin. 2005. Feature level fusion using hand and face biometircs. In *Proceedings of SPIE conference on biometric technology for human identification II*, Vol. 5779, 196–204, Orlando, FL.
25. Chang, K. C., R. K. Saha, and Y. Bar-Shalom. 1997. On optimal track-to-track fusion. *IEEE Trans Aerosp Electron Syst* 33:1271–6.
26. Cheezum, M. K., W. F. Walker, and W. H. Guilford. 2001. Quantitative comparison of algorithms for tracking single fluorescent particles. *Biophys J* 81:2378–88.
27. Pratt, W. K. 1974. Correlation techniques of image registration. *IEEE Trans Aerosp Electron Syst* 10:353–8.
28. Chang, S., and C. P. Grover. 2002. Centroid detection based on optical correlation. *Opt Eng* 41:2479–86.
29. Romine, J. B., and E. W. Kamen. 1996. Modeling and fusion of radar and imaging sensor data for target tracking. *Opt Eng* 35:659–72.
30. Naidu, V. P. S., G. Girija, and J. R. Raol. 2005. Target tracking and fusion using imaging sensor and ground based radar data, AIAA guidance, navigation and control conference and exhibit, Paper No. AIAA-2005-5842, San Francisco, CA.
31. Dogancay, K., and A. Hashemi-Sakhtsari. 2005. Target tracking by time difference of arrival using recursive smoothing. *Signal Process* 85:667–79.
32. Stoica, P., and A. Nehorai. 1989. MUSIC, maximum likelihood, and Cramer-Rao bound. *IEEE Trans Acoust Speech Signal Process* 37:720–41.
33. Johnson, D. H., and D. E. Dudgeon. 1993. *Array signal processing: Concepts and techniques. Prentice Hall Signal Processing Series.* Englewood Cliffs, NJ: Prentice Hall.
34. Wijk, O., and H. I. Christensen. 2000. Triangulation-based fusion of sonar data with application in robot pose tracking. *IEEE Trans Rob Autom* 16:740–52.
35. Dufour, F., and M. Mariton. 1991. Tracking a 3D maneuvering target with passive sensors. *IEEE Trans Aerosp Electron Syst* 27:725–38.
36. Naidu, V. P. S., and J. R. Raol. 2007. Target tracking with multiacoustic array sensors data. *Def Sci J* 57(3):289–303.
37. Strobel, N., S. Spors, and R. Rabenstein. 2001. Joint audio-video object localization and tracking. *IEEE Signal Process Mag* 18:22–31.

38. Naidu, V. P. S., and J. R. Raol. 2008. Fusion of out of focus images using principal component analysis and spatial frequency. *J Aerosp Sci Technol* 60:216–225.
39. Blum, R. S., and Z. Liu. 2006. *Multi-sensor image fusion and its applications.* Boca Raton, FL: CRC Press.
40. Harris, C., A. Bailey, and T. Dodd. 1998. Multisensor data fusion in defense and aerospace. *J Aerosp Soc* 162:229–44.
41. Naidu, V. P. S., M. Mudassar, G. Girija, and J. R. Raol. 2005. Target location and identity estimation and fusion using disparate sensor data, AIAA guidance, navigation and control conference and exhibit, Paper No. AIAA-2005-5841, San Francisco, CA.
42. Sam Ge, S., and F. L. Lewis, eds. 2006. *Autonomous mobile robots—Sensing, control, decision making and applications.* Control engineering series. Boca Raton, FL: CRC Press.

Part IV

A Brief on Data Fusion in Other Systems

A. Gopal and S. Utete

12

Introduction: Overview of Data Fusion in Mobile Intelligent Autonomous Systems

12.1 Mobile Intelligent Autonomous Systems

Intelligent field robotic systems are mobile robots that operate in an unconstrained and dynamic environment. Unlike their conventional manufacturing counterparts that operate in controlled environments, field robotic systems are more challenging due to the significantly more complex levels of perception, navigation, planning, and learning required to achieve their missions [1].

Perception addresses the challenges associated with understanding the environment in which the robotic system is operating. Environmental understanding is achieved through the creation of models that are truthful representations of the robotic system's operating environment. Many types of sensors can be used for this world-modeling task, including sonar, laser scanners, and radar (see Section 2.5); however, the primary focus of a mobile intelligent autonomous system (MIAS) within the perception research domain is to acquire and interpret visual data. Navigation uses the knowledge of the perceived environment as well as additional information from other on-board sensors to address the challenges of localizing the system within the environment and controlling the platform actuators to maintain a given trajectory. Mapping an unknown environment while simultaneously localizing within that environment is also part of the navigation challenge. Planning addresses the challenge of integrating the knowledge of the high-level goals with the current system perception and localization to generate appropriate behaviors that will take the system closer to successfully achieving these goals. The objective of planning is to generate behavior options and then decide which behavior will be executed. Finally, learning closes the loop by incorporating the knowledge of previous states, behaviors, and results in order to improve the system's reasoning capability and increase the probability of choosing the optimal behavior.

These four MIAS areas overlap, to some degree, with the common thread among them being suitable knowledge representation. A unified framework definition for interfaces and data flow (system architecture) is also required to facilitate the integration of all MIAS subsystems into a single demonstrator system. It is at this system-level integration that the benefits of data fusion will be most noticeable. The distinguishing characteristic of MIAS is that these systems operate in an environment that changes, often unpredictably, and as a result they have to first detect the change and then assess the impact of the change on their task or objective. Depending on the result of the assessment, the system must make appropriate planning and actuation decisions to compensate for the change while still achieving its objective.

Examples of MIAS global research include the following (Figure 12.1): (1) the driverless car and underground mine safety initiatives at the Council

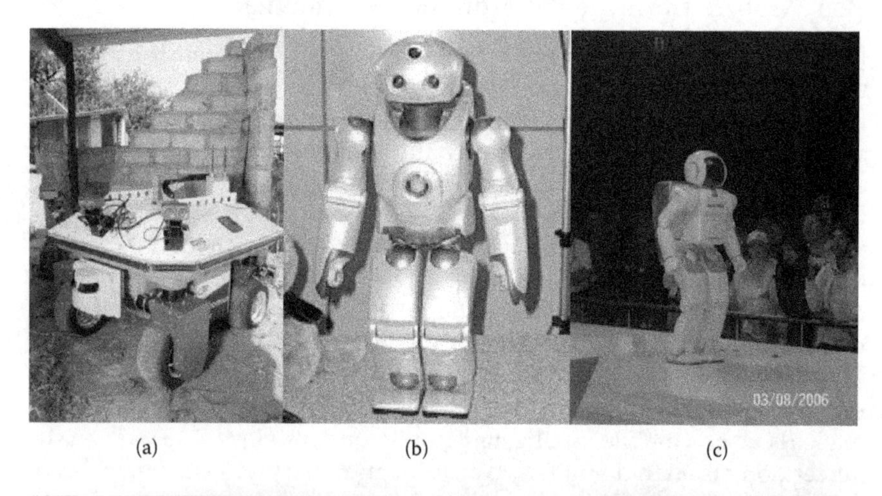

(a) (b) (c)

(d) (e)

FIGURE 12.1
Some systems where data fusion and AI could play major role: (a) Driverless vehicle research platforms (from www.csir.co.za/mias. With permission.); (b) Sony's QRIO humanoid (from en.wikipedia.org/wiki/Qrio. With permission.); (c) Honda's Asimo humanoid (photo taken by the author at the Auto Africa motor show in South Africa); (d) Israel Aircraft Industries' RQ-2 Pioneer UAV (from en.wikipedia.org/wiki/RQ-2_Pioneer. With permission.); and (e) Bluefin-12 AUV/Bluefin Robotics in the USA. (from en.wikipedia.org/wiki/AUV. With permission.)

for Scientific and Industrial Research (CSIR), South Africa, (2) QRIO of Sony, (3) Honda's Asimo humanoid programs, (4) unmanned military ground and air vehicles, and (5) the Bluefin-12 autonomous underwater vehicle (AUV) from Bluefin Robotics in the United States.

12.2 Need for Data Fusion in MIAS

There are several reasons for data fusion in MIAS: (1) the environment in which the system is operating is not always fully observable; (2) the interpretation of the environment is done via sensors (and only one sensor will not provide complete information about the scene); (3) in many instances, some information is uncertain; (4) the real world is spatially and temporally linked, and data fusion provides a useful methodology for realizing this relationship [2]. A good example of an environment that is not fully observable is a cluttered environment with many obstacles to obscure sensors, in which a mobile robot needs to navigate, or when a mobile robot must navigate to a point that is beyond the range of its exterioceptive sensors. In this case, fusing the system localization with an *a priori* map of the environment will facilitate better path-planning strategies in order to reach the destination point.

Mobile robots rely on sensors to generate a description of internal states and the external environment. Internal state sensors are referred to as *proprioceptive sensors,* and sensors for measuring the external environment are referred to as *exterioceptive sensors* [3]. Proprioceptive sensors are typically used to measure robot velocity, acceleration, attitude, current, voltage, temperature, and so on, and are used for condition monitoring (fault detection), maintaining dynamic system stability, and controlling force/contact interaction with the external environment. Exterioceptive sensors are typically used to measure range, color, shape, light intensity, and so on, and are used to construct a world model in order to interact with the environment through navigation or direct manipulation. These sensors measure very specific information such as range and color and thus provide only partial views of the real world, which need to be fused together in order to provide a complete picture. Also, the information that needs to be fused does not need to be limited to a numerical format as is typical of measurement sensors, but can be linguistic or in other formats, which allows the system to build a more descriptive world model that can be used to determine if the end objective has been achieved.

In addition, data fusion increases prediction accuracy and the robustness of information and reduces noise and uncertainty. Dead-reckoning errors for mobile robot localization are cumulative, and fusion with exterioceptive sensor information is required to determine the robot's location more accurately. Data fusion also facilitates the detection and extraction of patterns that are not present in the unfused data and is therefore very useful for system condition monitoring and target tracking. Lastly, robotic systems operating under a decentralized processing architecture require data fusion to incorporate data communicated from neighboring nodes with local observations [4].

12.3 Data Fusion Approaches in MIAS

Data fusion in mobile robotics occurs at different levels: (1) time series data fusion of a single sensor, (2) fusion of data from redundant sensors, (3) fusion of data from multiple sensors, and (4) fusion of sensor and goal information. Each level of fusion has challenges associated with it that depend on the type of data being fused, the requirements of the application, and the reliability of the outputs.

Approaches to data fusion in MIAS applications generally include (1) statistical analysis, such as the Kalman filter, (2) heuristic methods, such as Bayesian networks, (3) possibility models, such as fuzzy logic and Dempster–Shafer, (4) mathematical models, (5) learning algorithms based on neural networks, (6) genetic algorithm-type evolutionary algorithms, and (7) hybrid systems [5].

A brief overview of some of the research being conducted internationally is presented here to give the reader a broad perspective of the field. It is not intended to be a comprehensive mathematical description of all available data fusion methods, many of which have been covered in the preceding sections of the book.

The Kalman filter–based approach to data fusion is probably the most significant and widely studied method to date. It depends on linear state-space models and Gaussian probability density functions; the extended Kalman filter is based on a Jacobian linearization and can be applied to nonlinear models. However, errors introduced in the linearization process in real applications such as navigation, where the underlying process is typically nonlinear, can cause inaccuracies in the state-vector fusion approach. Reference 3 asserts that in these cases the measurement fusion approach is preferred to state-vector fusion. Measurement fusion directly fuses sensor data to obtain a weighted or augmented output, which can then be input into a Kalman filter. The data from multiple sensors is combined using weights to reduce the mean-square-error estimate, and the dimension of the observation vector remains unchanged. In the state-vector approach, data from multiple sensors are added to the observation vector and result in an increased dimension. A separate Kalman filter is required for each sensor observation, and the results of the individual estimates are then fused to give an improved estimate. According to [3], the state-vector fusion method is computationally more efficient, but the measurement fusion method has more information and should be more accurate; the net result is that both methods are theoretically functionally equivalent, if the sensors have identical measurement matrices (see Section 3.1).

An example of Kalman filter–based fusion is described in [6], where the issue of cumulative errors in odometry for localization is addressed through the fusion of laser range data and image intensity data. An extended Kalman filter is used to determine the location of vertical edges, corners, and door and window frames, which is then compared to an *a priori* map to obtain a better estimation of the robot's location. The Kalman filter is also used to fuse laser and stereo range data to model the environment using 3D planes [7].

Bayesian estimators were very prominent in earlier works on sensor fusion for mobile robot applications, but research with this method had been stagnant for years before again increasing in popularity recently. Some researchers use Bayesian estimators to fuse sonar and stereovision range data for marking cells in an occupancy grid as occupied, empty, or unknown.

Examples of mathematics-based data fusion methods include exponential mixture density (EMD) models and vector field histograms. In [8], the theoretical properties of exponential mixture density models are investigated to determine their applicability to robust data fusion in distributed networks. These developments take advantage of the previous use of EMDs in expert data fusion and the adaptation of nonlinear filtering methods, such as particle filters, to suit the EMD's computational requirements and develop upper and lower bounds on the probabilities of the result. The authors claim a consistent and conservative data fusion method for information with unknown correlations. In [9], evidence grid methodology, also referred to as occupancy grids, certainty grids, or histogram grids, is used for navigation in conjunction with vector field histograms (VFHs) to fuse CCD camera data and laser rangefinder data. Data fusion was successfully demonstrated on the navigator wheeled platform. Evidence grid methodology can also implement probabilistic measurements as demonstrated by Martin and Moravec from Carnegie Mellon University (cited in [9]). Approaches based on possibility theory data fusion center around the application of fuzzy logic theory (see Chapter 6). The use of neural networks and genetic algorithms for data fusion are also possible applications in MIAS.

Research in data fusion is also being conducted from the system architecture perspective; in [5], a general data fusion architecture based on the unified modeling language (UML) is developed that allows for dynamic changes in the framework in order to respond to changes in a dynamic environment. This approach introduces taxonomy for data fusion and facilitates various levels of fusion.

The current challenges in data fusion for MIAS are in the characterization of uncertainty as well as vagueness in sensor measurements [3]. The problem of quantitative gain from data fusion to justify the cost implications of additional sensors must also be addressed. A good challenge in MIAS is that of the SLAM (simultaneous localization and mapping), which uses extended Kalman filter and data association methods and applications to multirobot systems, mining robot arms that can take advantage of the data fusion approaches discussed in the previous sections of the book. The real challenge is in building MIAS so that the system not only carries out its various tasks rationally, but also thinks rationally, like a human being!

13

Intelligent Monitoring and Fusion

Data fusion has an extensive range of applications in aeronautics, space systems, and robotics. It also has applications in monitoring the health or condition of systems [10], where fusion can be used to improve the observations made about a system or environment for event detection. This section considers various aspects that contribute to the development of intelligent monitoring and fusion systems—systems capable of making decisions about the existence of precursors to events through the acquisition and combination of data about an environment or system.

Condition monitoring involves the acquisition and analysis of data to determine changes in an observed system or environment. Examples include the monitoring of systems, such as jet engines, for preventive maintenance purposes [11,12] and the monitoring of environments, such as mine work areas, as described in [13]. Condition monitoring can involve the processing of data for fault characterization, change detection, or novelty or anomaly detection [12]. In novelty detection, a model or description of the normal state of the observed system or environment is developed. Changes from this model or description are then detected. This approach is important for high-integrity systems, such as aircraft engines, where faults by definition are few.

Other applications of condition monitoring include multiple-server monitoring, which is relevant to companies such as Web search entities. In robotics, condition-monitoring applications can be used for the management of fleets of robots or in the use of mobile robots as sensor platforms to make observations in mining environments (Candy, L., June 2008, pers. comm).

13.1 The Monitoring Decision Problem

Intelligent monitoring and fusion involve the combination of data acquired by intelligent agents, such as intelligent sensors, which make observations of a system or environment in order to detect precursors to events or anomalies. Monitoring involves a decision problem about the presence or absence of events through the detection of changes in the state, which are

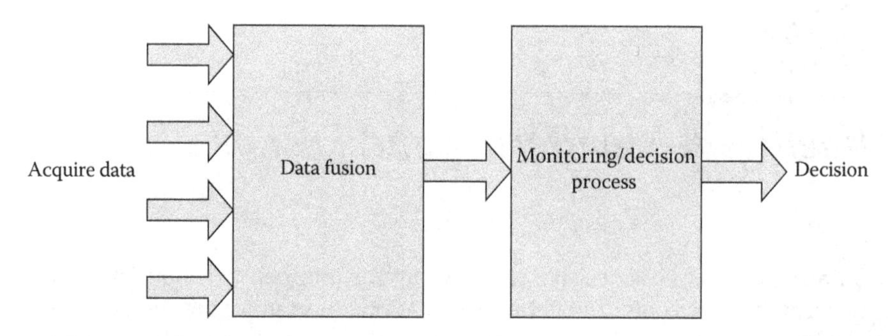

FIGURE 13.1
Batched processing method for offline monitoring.

indicative of an underlying change in the system. Detection of anomalies is required at an early stage, before events occur, so that action can be taken, if necessary.

In offline condition monitoring, data can be batched, and decision making can take place once a predetermined set of data has been acquired (Figure 13.1); this is essentially a centralized data and decision fusion scheme. In online condition monitoring, data must be examined and classified as it is being acquired, and a delayed decision might be of little value. For large volumes of data, online monitoring bears some resemblance to stream data-processing problems, which arise, for example, in the management and analysis of network traffic [14].

Given the time limits of online monitoring, suboptimal strategies for data analysis and decision making are relevant to event detection. One method with a potential application is ordinal optimization [15]. Ordinal optimization can be employed to reduce the size of the data set that must be examined to achieve a goal [15]. Given the restricted time available to examine the data in online monitoring, one approach is to develop robust decision methods that utilize subsets of representative data or can provide a decision about the state of a data stream, rather than computing estimates explicitly. Abstraction of data, as in voting [16], is also potentially useful.

Table 13.1 compares a single sonar sensor's beliefs about the target type with a fusion of the beliefs of 15 such sensors using Dempster's rule [14]. For a description of Dempster–Shafer reasoning, see [17] and Section 2.3. The sensors seek to differentiate between planes and corners. At both of the tabulated angular positions, the multisensor system outperforms the single sensor in terms of classification accuracy. The performances of the sensor groups using Dempster's fusion rule and voting are compared in Table 13.2. In each case, five sensors are grouped, and their beliefs are fused. For these groups of distributed sensors, voting gives a higher percentage of correct decisions. This can be explained by voting's relative insensitivity to the conflicting decisions of the sensors in the selected groups [16]. For some monitoring tasks, the number of sensors supporting a view might be more meaningful than the intensity with which the view is held. In such cases, the abstraction of data through voting can prove useful. The judicious selection of sensor groups in multisensor systems is also important: in one of the cases reported in [16], a group of five sensors gives a higher correct-classification percentage than does the set of all 15 sensors. This is one motivation to use adaptive sampling in intelligent monitoring, as discussed below.

The decision making stage of monitoring can lead to the following: (1) detection of a precursor; (2) failure to detect a precursor; and (3) a state where there is insufficient information to determine whether a significant change has occurred. The third state can arise, for example, in online voting-based distributed sensor systems, when a predetermined quorum has not yet been achieved.

TABLE 13.1

Comparison of Single-Sensor Beliefs with Fusion of Beliefs from 15 Sensors

		Angular Position 1	Angular Position 2
Sensor 1	m(plane)	0.260	0
	m(corner)	0	0.509
	m(unknown)	0.740	0.491
Dempster's rule (all 15 sensors)	m(plane)	0.9988	0.0002
	m(corner)	0.0007	0.9992
	m(unknown)	0.0005	0.0005

Source: Utete, S. W., B. Barshan, and B. Ayrulu. 1999. *Int J Robot Res*, 18:401–13.

TABLE 13.2

Comparison of Percentages of Correct Decisions Regarding Target Discrimination by Groups of Distributed Sonar Sensors

Sensor Grouping	Dempster's Fusion Rule (%)	Voting (%)
Group 1	63.25	76.07
Group 2	81.20	89.74
Group 3	87.18	91.88
Group 4	58.97	82.48
Group 5	71.37	76.92
Group 6	63.68	84.62
Group 7	69.66	80.34
Group 8	76.07	85.04

Source: Utete, S. W., B. Barshan, and B. Ayrulu. 1999. *Int J Robot Res*, 18:401–13.

13.2 Command, Control, Communications, and Configuration

Athans poses control problems arising in C2 and C3 systems and makes a case for the adaptation of methods from the command, control, and communication theory for civilian applications. Note that issues in condition-monitoring problems resemble the problems in situational assessment. Incorporating situational assessment into intelligent monitoring would be an example of the type of crossover proposed by Athans [18].

C3 systems are constrained by a fourth "C," configuration. In situations where a system of intelligent sensors or agents exchange information on a network, the network configuration becomes a part of the information-processing paradigm, with potential implications for decision making, as in the case reported in [19]. The area of what might be termed "4C" systems is relevant to the development of robust methods for updating information, decision, and position in highly autonomous networks of sensor nodes or agents that collaborate to collect information about an observed environment.

Some of the issues that arise in this scenario are (1) the structure of the system through which decisions are communicated [20,21], (2) suboptimal strategies for decision making in highly autonomous networks [19], and (3) novelty detection by distributed agents.

Methods from graph theory can be applied to address network configuration issues. For example, detecting a change of state using physical agents such as sensors is similar to early bird synchronization problems in automata theory [22]; this can be considered a physical novelty-detection problem. Results from extremal graph theory [23] and automata theory [22] can assist in understanding the questions that can be posed on particular network configurations [19]. It is also possible to draw on these areas for the analysis and development of sensor systems for intelligent monitoring.

13.3 Proximity- and Condition-Monitoring Systems

Awareness of configuration is relevant to civilian situation-assessment applications, such as the proximity-detection systems of AcuMine products [24]. A proximity-monitoring network is an example of a 4C system. In AcuMine products [24], vehicle updates can be coordinated by a base station through interaction with other mobile units. The mobile units and the base station form a distributed system of monitoring agents (in this case, making observations about proximity and threats during task performance). The system can also operate without the base unit, a situation that creates a higher degree of autonomy for individual units and an accompanying need for greater levels of situational awareness and coordination. Position updates in such systems are relevant in terms of the decisions that can be made as the autonomy of mobile units increases. Wider applications can be made in other problems of multiple-robot or multiple-agent coordination.

Adaptive sampling systems are reactive systems [25] that respond strategically to the availability of relevant data; sampling occurs in response to changes in the information content. Spatial adaptive sampling systems have been developed for applications such as ocean monitoring [25].

Network-monitoring sensor systems have been developed for mine environment monitoring in the Council for Scientific and Industrial Research (CSIR) Mining Group's AziSA project (CSIR NRE Mining, 2008, pers. comm). Both the AziSA system and AcuMine's proximity-management systems [24] incorporate instances of adaptive sampling as follows: (1) Considering the difficult environmental conditions in underground mines and limitations to access, low-power sensors are required [CSIR NRE Mining, 2008, pers. comm.; Vogt, D., May 2008, pers. comm.]. In the case of the CSIR Mining Group's AziSA network, the possibility of sleep–wake states for sensors has been considered [12]. Such states are an explicit instance of adaptive sampling. (2) Vehicles in the proximity-management system of AcuMine products [24] use an adaptive sampling time to inform other units of their most recent data acquisition. The demands of environmental monitoring necessitate that sensor nodes be active only in the presence of relevant new information. Adaptive sampling by sensor networks is an explicit form of novelty detection.

In an intelligent monitoring system, the processes of fusion and decision making can be augmented with position or sampling rate updates, as in Figure 13.2. Data is fused when it is acquired, and the processed values are used to generate a decision about the presence or absence of precursors. This information can be used to determine whether to increase or maintain the active rate of data sampling. In such dynamic monitoring systems, this decision could be about the position updates required from sensing platforms. Intelligent monitoring requires the development of techniques for adaptively sampling a monitored environment in order to focus attention and energy for data acquisition on the most information-rich spatial regions and time slots. This problem is fundamentally linked to condition monitoring in a wide range of contexts. Problems in sensor-network organizations, such as those discussed by Utete et al. [16], can also be related to this framework.

Thus, some aspects contributing to intelligent monitoring and fusion systems have been highlighted, with a special application to robotics. The

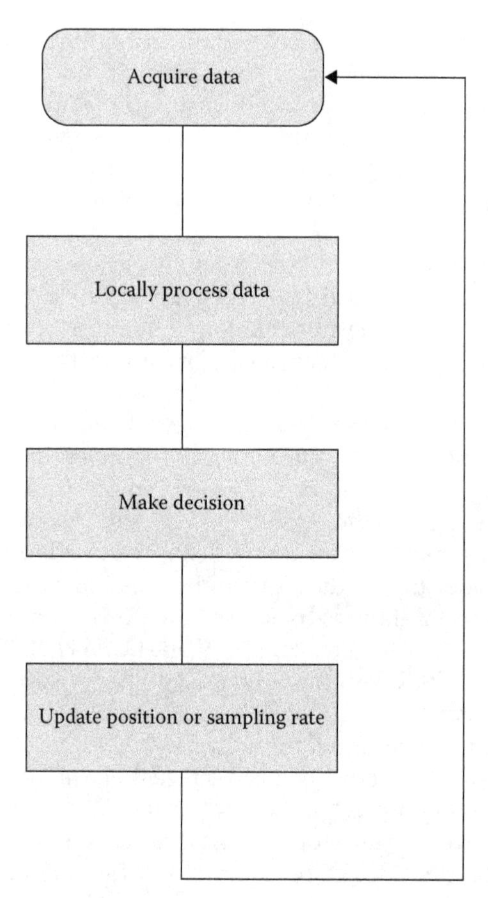

FIGURE 13.2
Data flow in a monitoring scheme with information-acquisition updates.

problem of detecting precursors compels the consideration of suboptimal or abstraction methods, such as voting, for data fusion. The configuration of mobile platforms is considered an input to the monitoring and fusion problem; configuration is a part of the information available to a data fusion system and plays a role in determining what information can be combined in the system. This is relevant to monitoring by mobile robots, as well as to proximity-monitoring applications. Monitoring methods can also be applied to the adaptive sampling of data, enabling a monitoring system to acquire data selectively. Intelligent monitoring and fusion are of great importance for a potentially large range of applications in robotics, where intelligent sensors are deployed to monitor environments, particularly in remote operations, such as those in mining [13] and space exploration.

Epilogue

Mobile intelligent autonomous systems (MIAS), a more general subject than robotics, is an emerging research area in many developing countries. The main research and development (R&D) issues involved are the following: (1) perception and reasoning, (2) mobility and navigation, (3) haptics and teleoperation, (4) image fusion and computer vision, (5) mathematical modeling of manipulators, (6) hardware and software architectures for planning and behavior learning, (7) path- and motion-planning of a robot or vehicle, and (8) the application results of artificial neural networks, fuzzy systems, probabilistic reasoning, Bayesian networks, and genetic algorithms to the above problems. Multisensor data fusion can play a very crucial role at many levels of the data fusion process, such as kinematic fusion (position/bearing tracking), image fusion (for scene recognition, building a world model), and decision fusion (for tracking and control actions). Many of the concepts, techniques, and algorithms discussed in Parts I, II, and III of this book are very much applicable to robotics and to MIAS, an emerging technology. Such systems are useful for the automation of complex tasks, surveillance in hazardous and hostile environments, human-assistance systems for very difficult manual activities, medical robotics, hospital systems, autodiagnostic systems, and many other related civil and military systems. It is important to understand the environment in which robotics and other intelligent systems operate, as they make decisions in the context of this environment in order to achieve their goals. *Perception* occurs when a robotic system or intelligent agent uses its sensors to acquire sensory information about its surroundings and then interprets this information to create models of its environment, which provide a truthful representation of that environment. *Reasoning* is the process by which a robotic system or intelligent agent incorporates new perceptions about its environment into its current knowledge and belief representations and uses its total knowledge to decide which actions to carry out in order to achieve its goals. Research areas include sensors and sensor modeling, computer vision, scene understanding, tactile sensing, human–machine interfaces, knowledge representation, learning, decision making, planning, and logic. Knowledge in the disciplines of statistical modeling, artificial intelligence, machine learning, pattern recognition, statistical learning methods, probabilistic reasoning, algorithmic design, and computer science are useful tools for researching MIAS technology.

Mobility addresses the physical issues in moving the system between points A and B. *Navigation* addresses issues related to planning the motion of the system to achieve mobility, as well as accomplishing mobility in the most efficient way as measured against performance criteria such as energy or time. Localization, path planning, low-level obstacle avoidance, and map building are also very important issues. *Haptics* and *teleoperation* address issues in human–machine interfaces, with a focus on teleoperation using haptic feedback. Research on reality augmentation, human response to tactile and visual stimuli, real-time feedback control, and 3D visualization are also very important research areas. The dynamic systems considered could be autonomous systems, field robots, mobile/fixed systems, mini/microrobots, mining robots, surveillance systems, networked/multirobot systems, and so on. The importance of this field calls for a separate volume on data fusion for autonomous intelligent systems.

Exercises

IV.1. In what ways are inertial navigation systems, global positioning systems integration, and data fusion (Section 4.5) useful to a mobile robot?

IV.2. What are the three most important and major aspects of robotic architecture?

IV.3. What is the importance of each of the tasks in Exercise IV.2?

IV.4. What is map building for an autonomous robot? How would sensor data fusion be useful in this process?

IV.5. What roles can neural networks, fuzzy logic, and genetic algorithms play in an autonomous robotic system?

References

1. Sam Ge, S., and F. L. Lewis, eds. 2006. *Autonomous mobile robots—Sensing, control, decision making and applications* (Control Engineering Series). Boca Raton, FL: CRC Press.
2. Vitko, A. 2008. Data fusion, context awareness and fault detection in autonomous robotics, Faculty of Electrical Engineering and Information Technology, Institute of Control and Industrial Informatics, Slovakia.
3. Hu, H., and J. Q. Gan. 2005. Sensors and data fusion algorithms in mobile robotics, Technical report: CSM-422, Department of Computer Science, University of Essex, United Kingdom.

4. Durrant-Whyte, H. 2000. *A beginners guide to decentralised data fusion,* Australian Centre for Field Robotics, The University of Sydney, Australia.
5. Carvalho, H. S., W. B. Heinzelman, A. L. Murphy, and C. J. N. Coelho. 2003. A general data fusion architecture. In *Proceedings of the Sixth International Conference of Information Fusion*, 2:1465–72.
6. Neira, J., J. D. Tardos, J. Horn, and G. Schmidt. 1999. Fusing range and intensity images for mobile robot localisation. *IEEE Trans Rob Autom* 15:76–83.
7. Grandjean, P., and A. R. Robert de Saint Vincent. 1989. 3-D modeling of indoor scenes by fusion of noisy range and stereo data. In *Proceedings of the IEEE International Conference in Robotics and Automation*, 681–687, Scottsdale, AZ.
8. Julier, S. J., T. Bailey, and J. K. Uhlmann. 2006. Using exponential mixture models for suboptimal distributed data fusion. In *Proceedings of the Nonlinear Statistical Signal Processing Workshop*, 160–163. UK: IEEE.
9. Conner, D.C., P. R. Kedrowski, and C. F. Reinholtz. 2001. Multiple camera, laser rangefinder, and encoder data fusion for navigation of a differentially steered 3-wheeled autonomous vehicle. In *Proceedings of SPIE*, 4195:76–83, Bellingham, WA.
10. Raheja, D., J. Llinas, R. Nagi, and C. Romanowski. 2006. Data fusion/data mining-based architecture for condition based maintenance. *Int J Prod Res* 44:2869–87.
11. Nairac, A., N. Townsend, R. Carr, S. King, P. Cowley, and L. Tarassenko. 1999. A system for the analysis of jet engine vibration data. *Integr Comput Aided Eng* 6:53–66.
12. Hayton, P., S. Utete, D. King, S. King, P. Anuzis, and L. Tarassenko. 2007. Static and dynamic novelty detection methods for jet engine condition monitoring. In *Philosophical Transactions of the Royal Society Part A*. Special Issue on Structural Health Monitoring, ed. K. Worden and C. Farrar, 365:493–514. London: The Royal Society.
13. Vogt, D., and V. Z. Brink. 2007. The Azisa standard for mine sensing and control. Presentation at Automation in Mining Conference.
14. Hayes, B. 2008. The Britney Spears problem. *Am Sci* 96:274–9.
15. Ho, Y.-C. 1994. Heuristics, rules of thumb, and the 80/20 proposition. *IEEE Trans Automat Contr* 39:1025–7.
16. Utete, S. W., B. Barshan, and B. Ayrulu. 1999. Voting as validation in robot programming. *Int J Robot Res* 18:401–13.
17. Krause, P., and D. Clark. 1993. *Representing uncertain knowledge: An artificial intelligence approach*. Oxford: Intellect Books.
18. Athans, M. 1987. Command and control (C2) theory: A challenge to control science. *IEEE Trans Automat Control* 32:286–93.
19. Utete, S. 1994. Network Management in Decentralised Sensing Systems, PhD Thesis, Department of Engineering Science, University of Oxford, UK.
20. Grime, S. 1992. Communication in Decentralised Sensing Architectures. PhD Thesis, Department of Engineering Science, University of Oxford, UK.
21. Grime, S., and H. F. Durrant-Whyte. 1994. Data fusion in decentralised sensor networks. *Control Eng Pract* 2:849–63.
22. P. Rosenstiehl, J. R. Fiksel, and A. Holliger. 1972. Intelligent graphs: networks of finite automata capable of solving graph problems. In *Graph theory and computing*, ed. R. Read. London: Academic Press.

23. Bollobás, B. 1978. *Extremal graph theory*. London: Academic Press.
24. AcuMine products. 2008. http://www.acumine.com/_Products/products.php (accessed January 2009).
25. Ehrich Leonard, N., D. A. Paley, F. Lekien, R. Sepulchre, D. M. Fratantoni, and R. E. Davis. 2007. Collective motion, sensor networks, and ocean sampling. *Proc IEEE* 95.

Appendix: Numerical, Statistical, and Estimation Methods

Data fusion (DF) could become a core technology for current and future onboard data-processing schemes in many aerospace and ground-based systems. DF is based on many theories, concepts, and methods, such as signal processing, control and estimation theory, artificial intelligence, pattern recognition, spectral analysis, reliability theory, statistical tools, wavelet analysis, artificial neural networks (ANNs), fuzzy logic (FL), genetic algorithms (GAs), knowledge-based expert systems, and decision theory [1–12]. In essence, we can obtain combined information with higher-quality and refined data and thus can eliminate redundancy. Statistical matching is often used as a core fusion algorithm. The approach can be compared to the *k*–NN (nearest neighbor) prediction with two sets of data: the training set and the test set. The Euclidian distance is used as a measure for matching the distance. A prediction of the fusion variables is made using the set of nearest neighbors found. Nearest neighbors are found by computing averages, modes, or distributions. The training set data could be a very large sample of the general population; the training set variables should be good predictors for the fusion variables. For numerical data, the methods of estimation theory are directly adaptable to the DF problem, as discussed in Section 2.2.

A.1 Some Definitions and Concepts

Several definitions and concepts that are useful in data processing and DF are compiled here [1–3].

A.1.1 Autocorrelation Function

For $x(t)$, a random signal

$$R_{xx}(\tau) = E\{x(t)x(t+\tau)\} \tag{A.1}$$

Here, τ is the time lag and E is the mathematical expectation operator. If R_{xx} decreases with an increase in time, this means that the nearby values of the signal x are not correlated. The autocorrelation of the white noise process is an impulse function, and R for discrete-time signals or residuals is given by

$$R_{rr}(\tau)=\frac{1}{N-\tau}\sum_{k=1}^{N-\tau} r(k)r(k+\tau); \quad \tau=0,\ldots,\tau_{\max} \text{(discrete time lags)} \qquad \text{(A.2)}$$

A.1.2 Bias in Estimate

The bias $\beta = \beta - E(\hat{\beta})$ is the difference between the true value of the parameter β and the expected value of its estimate. For the large amount of data used to estimate a parameter, the estimate is expected to center closely around the true value. The estimate is unbiased if $E\{\hat{\beta}-\beta\}=0$.

A.1.3 Bayes' Theorem

This is based on the conditional probability—the probability of event A, given the fact that event B has occurred: P(A/B)P(B) = P(B/A)P(A). P(A) and P(B) are the unconditional probabilities of events A and B. Bayes' theorem allows the use of new information to update the conditional probability of an event given the *a priori* information. P(A|B) is the degree of belief in hypothesis A, after the experiment that produced data B. P(A) is the prior probability of A being true, P(B|A) is the ordinary likelihood function (LF), and P(B) is the prior probability of obtaining data B.

A.1.4 Chi-Square Test

The variable χ^2 given as $\chi^2 = x_1^2 + x_2^2 + \cdots + x_n^2$ (x_i is the normally distributed variable with a zero mean and unit variance) has a probability density function (pdf) with n DOF: $p(\chi^2) = 2^{-n/2}\Gamma(n/2)^{-1}(\chi^2)^{\frac{n}{2}-1}\exp(-\chi^2/2)$, where $\Gamma(n/2)$ is the Euler's gamma function, and $E(\chi^2)=n$; $\sigma^2(\chi^2)=2n$. In the limit, the χ^2 distribution tends to be the Gaussian distribution with mean n and variance $2n$. When the pdf is numerically computed from the random data, the χ^2 test is used to determine whether the computed pdf is Gaussian. In the case of normally distributed and mutually uncorrelated variable x_i with mean m_i and variance σ_i, the normalized sum of squares is computed as $s=\sum_{i=1}^{n}\frac{(x_i-m_i)^2}{\sigma_i^2}$, where s obeys the χ^2 distribution with n degrees of freedom (DOF). This χ^2 test is used for hypothesis testing.

A.1.5 Consistency of Estimates Obtained from Data

As the number of data increases, the estimate should tend to the true value—this is the property of "consistency." Consistent estimates are asymptotically unbiased. The convergence is assumed to happen with probability 1.

$$\lim_{N\to\infty} P\left\{|\hat{\beta}(z_1, z_2, \ldots, z_n) - \beta| < \delta\right\} = 1 \quad \forall\delta > 0 \tag{A.3}$$

This means that the probability that the error in the estimates, with respect to the true values, is less than a certain small positive value is 1.

A.1.6 Correlation Coefficients and Covariance

The correlation coefficient is defined as $\rho_{ij} = \dfrac{\text{cov}(x_i, x_j)}{\sigma_{x_i}\sigma_{x_j}}$; $-1 \le \rho_{ij} \le 1$. It is zero for independent variables x_i and x_j and for definitely correlated processes $\rho = 1_{ij}$. If a variable d is dependent on several x_i, then the correlation coefficient for each of the x_i can be used to determine the degree of this correlation with d as

$$\rho(d, x_i) = \frac{\sum_{k=1}^{N}\left(d(k) - \underline{d}\right)\left(x_i(k) - \underline{x}_i\right)}{\sqrt{\sum_{k=1}^{N}\left(d(k) - \underline{d}\right)^2}\sqrt{\sum_{k=1}^{N}\left(x_i(k) - \underline{x}_i\right)^2}} \tag{A.4}$$

If $|\rho(d, x_i)|$ tends to 1 then d is considered linearly related to that particular x_i. The covariance is defined as $\text{Cov}(x_i, x_j) = E\left\{\left[x_i - E(x_i)\right]\left[x_j - E(x_j)\right]\right\}$. If x is the parameter vector, then we get the parameter estimation error covariance matrix. The square roots of the diagonal elements of this matrix give the standard deviations of the errors in the estimation of states or parameters. The inverse of the covariance matrix is the indication of information in the signals about the parameters or states—a large covariance means a higher uncertainty and not much information or confidence in the estimation results.

A.1.7 Mathematical Expectations

A mathematical expectation is defined as $E(x) = \sum_{i=1}^{n} x_i P(x = x_i)$ and $E(x) = \int_{-\infty}^{\infty} xp(x)dx$, with P as the probability distribution of variables x and p the pdf of variable x. It is a weighted mean, and the expected value of the sum of two variables is the sum of their expected values $E(X+Y) = E(X) + E(Y)$; similarly $E(a \times X) = a \times E(X)$.

A.1.8 Efficient Estimators

For $\hat{\beta}_1$ and $\hat{\beta}_2$ (the unbiased estimates of the parameter vector $\boldsymbol{\beta}$) the estimates are compared in terms of error covariance: $E\left\{(\beta - \hat{\beta}_1)(\beta - \hat{\beta}_1)^T\right\} \leq E\left\{(\beta - \hat{\beta}_2)(\beta - \hat{\beta}_2)^T\right\}$. If the inequality is satisfied, $\hat{\beta}_1$ is said to be superior to $\hat{\beta}_2$ and if the inequality is satisfied for any other unbiased estimator, then we obtain the efficient estimator. The efficiency of an estimator is defined by the Cramer–Rao (CR) inequality, and gives a theoretical limit to the achievable accuracy: $E\left\{\left[\hat{\beta}(z) - \beta\right]\left[\hat{\beta}(z) - \beta\right]^T\right\} \geq I_m^{-1}(\beta)$. The matrix I_m is the Fisher information matrix. Its inverse is a theoretical covariance limit. An unbiased estimator with valid equality is called an efficient estimator. The CR inequality means that for an unbiased estimator, the variance of parameter estimates cannot be lower than its theoretical bound $I_m^{-1}(\beta)$. Thus, it bounds uncertainty levels around the estimates obtained using the maximum likelihood (ML) or the output error method (OEM).

A.1.9 Mean-Squared Error (MSE)

The mean-squared error (MSE) is given as $\text{MSE}(\hat{\beta}) = E\left\{(\hat{\beta} - \beta)^2\right\}$ and measures on average how far away the estimator is from the true value in repeated experiments.

A.1.10 Mode and Median

The median is the smallest number of the middle value, such that at least half the numbers are no greater than it; that is, if the values have an odd number of entries, then the median is the middle entry after sorting the values in an increasing order. The mode is the most common or frequently occurring value, and there can be more than one mode. In estimation, the mode defines the value of x for which the probability of observing the random variable is the maximum.

A.1.11 Monte Carlo Data Simulation

In a system simulation, we can study the effect of random noise on parameter and state estimates and evaluate the performance of these estimators.

First, obtain one set of estimated states, then change the seed number (for the random number generator), and add these random numbers to the measurements as noise to get the estimates of the states with the new data. Formulate a number of such data sets with different seeds, and obtain states to establish the variability of these estimates across different realizations of the random data. Then, obtain the mean value and the variance of the estimates using all of the individual estimates from all of these realizations. The mean of the estimates should converge to the true values. Often, as many as 100 runs or as few as 20 runs are used to generate the average results. It is ideal to use 500 runs for data simulation.

A.1.12 Probability

For x, a continuous random variable, there is a pdf $p(x)$, such that for every pair of numbers $a \le b$, $P(a \le X \le b)$ = (area under p between a and b). The Gaussian pdf is given as $p(x) = \frac{1}{\sqrt{2\pi\sigma}} \exp\left(-\frac{(x-m)^2}{2\sigma^2}\right)$, with mean m and variance σ^2 of the distribution. For the recorded random variables, given the state x (or parameters), the pdf is given as

$$p(z \mid x) = \frac{1}{(2\pi)^{n/2} \mid R \mid^{1/2}} \exp\left(-\frac{1}{2}(z - Hx)^T R^{-1}(z - Hx)\right) \tag{A.5}$$

Here, R is the covariance matrix of the measurement noise.

A.2 Decision Fusion Approaches

Decision fusion performs reduces data from multiple inputs to a smaller number of outputs. The inputs could be raw data, pixel values, some extracted features, or control signals [9]. The outputs could be target types, recognized scenes or events, or enhanced features. As opposed to estimation fusion, decision fusion does not usually assume any parametric statistical model (see Section 2.2). The decision fusion approach aims to combine the individual beliefs (see Section 2.3) of the used set of models into a single consensus measure of belief. There are three popular decision fusion approaches: (1) the linear opinion pool (see Section 2.5.3), (2) the logarithmic opinion pool, and (3) the voting or ranking approach.The individual (sensor data or processed) inputs to these decision-making expressions can be based on *possibility* theory, which is based on FL and not necessarily on the *probability* theory.

The linear opinion pool is a commonly used decision fusion method and is very convenient, since the fusion output is obtained as a weighted sum of the probabilities from each model. The probability distribution of the combined output might be multimodal. If the weights are constrained in the logarithm opinion pool, then the pooling process yields a probability distribution (see Section 2.5.3). The output distribution of the log opinion pool is typically unimodal. If any model assigns a probability of zero, then the combined probability is also zero. Hence, an individual model can "veto" the decision. In the linear opinion pool, the zero probability is averaged with other probabilities. In the voting method, a voting procedure is used such that each model generates a decision instead of a score. The popular decision is the majority vote; other voting techniques are the maximum, minimum, and median votes. Ranking methods are appropriate for problems that involve numerous classes; they do not use class labels or numerical scores, rather, they utilize the order of classes as estimated by the model(s). Ranking methods employ the class set reduction method to reduce the number of class candidates without losing the true class. By reducing the number of classes and reordering the remaining classes, the true class moves to the top of the ranking.

Other computational methods for decision fusion are the Dempter–Shafer method (see Section 2.3) and the fuzzy integral methods (see Section 6.7). These methods combine the beliefs of various models into an overall consensus.

A.3 Classifier Fusion

Pattern recognition and classification are important applications for decision fusion. Multiple classifier systems are efficient solutions for difficult pattern recognition tasks such as the following [9]: (1) combining multiple classifiers, (2) decision combination, (3) the mixture of experts, (4) classifier ensembles, (5) classifier fusion, (6) consensus aggregation, (7) dynamic classifier selection, and (8) hybrid methods. Specialized classifiers are superior in different cases. The traditional approach to classifier selection, where the available classifiers are compared to a set of representative sample data and the best one is chosen, is in contrast to classifier combining.

We may simply regard a classifier as a black box that receives an input sample x and outputs label y_j, or in short, $C(x) = y_j$ [9]. A Bayes' classifier may supply N values of posterior probabilities $P(y_j|x)$, $j = 1,\ldots, N$ for each possible label. The final label y_j is the result of a maximum value selected from the N values. This selection discards some information that may be useful for a multiclassifier combination. The output information that various classification algorithms supply can be divided into the following three levels: (1) the abstract level, where a classifier C outputs a unique label y_j or subset Y_J from Δ; (2) the rank level, in which C ranks all the labels in Δ (or a subset $J \subset \Delta$) in a queue, with the label at the top being the first choice; and (3) the measurement level, where C attributes a value to each label in Δ for the degree that x has the label.

The measurement level contains the most information, and the abstract level contains the least. From the measurements attributed to each label, we can rank all the labels in Δ according to a ranking rule. By choosing the label at the top rank or by choosing the label with the maximal value at the measurement level, we can assign a unique label to x. From the measurement level to the abstract level, there is an information reduction process or abstraction process.

A.3.1 Classifier Ensemble Combining Methods

The combination of a set of imperfect estimators may be used to manage the limitations of the individual estimators. The combination of classifiers can minimize the overall effect of these errors.

A.3.1.1 Methods for Creating Ensemble Members

Identical classifiers are not useful in gaining an advantage. A set of classifiers can vary in terms of the following parameters [9]: (1) their weight, (2) the time they take to converge, and (3) their architecture. Yet they may provide the same solution, since they may result in the same pattern of errors when tested. The aim is to find classifiers that generalize differently. There are many training parameters that can be manipulated: (1) initial conditions, (2) the training data, (3) the typology of the classifiers, and (4) the training algorithm.

The commonly used methods are as follows [9]: (1) varying the set of initial random weights—a set of classifiers is created by varying the initial random weights from which each classifier is trained, while maintaining the same training data; (2) varying the topology—a set of classifiers can be created by varying the topology or architecture, while maintaining constant training data; (3) varying the algorithm employed—the algorithm used to train the classifiers can be varied; and (4) varying the data.

A.3.1.2 Methods for Combining Classifiers in Ensembles

The voting schemes include the minimum, maximum, median, average, and product schemes. The weighted average approach attempts to evaluate the optimal weights for the various classifiers used. The behavior-knowledge space selects the best classifier in a region of the input space and bases a decision on its output. Other approaches to combining classifiers are the Bayes' approach, Dempster–Shafer theory, fuzzy theory, probabilistic schemes, and combination by ANNs. The combiner is a scheme to assign weights of value to classifiers. The weights can be independent of or dependent on the data.

Methods for combining the classifiers are as follows: (1) averaging and weighted averaging—the linear opinion pools refer to the linear combination of outputs of the ensemble members' distributions, so a single output can be created from a set of classifier outputs; a weighted average that takes into account the relative accuracies of the classifiers; (2) nonlinear combining methods—Dempster–Shafer belief-based methods, the use of rank-based information, voting, and order statistics; (3) supra-Bayesian—the opinions of the experts are themselves data, therefore the probability distribution of the experts can be combined with its own prior distribution; and (4) stacked generalization—a nonlinear estimator learns how to combine the classifiers with weights that vary over the feature space. The outputs from a set of level 0 generalizers are used as the input to a level 1 generalizer that is trained to produce the appropriate output.

A.4 Wavelet Transforms

In the wavelet transform, mathematical transformations are applied to a raw or original signal to obtain further information from that signal that is not readily available [10]. Most time-domain signals contain information hidden in their frequency content. The frequency content is made up of the frequency components (spectral components) of that signal, and is revealed by Fourier transform (FT), whereas the wavelet transform (WT) provides the time-frequency representation. The WT is capable of simultaneously providing the time and frequency information. The WT was developed as an alternative to the short time FT (STFT) to overcome some resolution-related problems of the STFT.

In STFT, the time-domain original signal passes through various high-pass and low-pass filters, which filter out either the high-frequency or the low-frequency portions of the signal. This procedure is repeated every time a portion of the signal, corresponding to some frequencies, is removed. A signal that has frequencies up to 1000 Hz is split up into two parts by passing the signal through a high-pass and a low-pass filter, which results in two different versions of the same signal: the parts of the signal corresponding to 0–500 Hz and 500–1000 Hz. Then, the decomposition procedure is repeated with either portion. We now have three sets of data, each corresponding to the same signal at frequencies 0–250 Hz, 250–500 Hz, and 500–1000 Hz. Then we take the low-pass portion again and pass it through low-pass and high-pass filters. We now have four sets of signals corresponding to 0–125 Hz, 125–250 Hz, 250–500 Hz, and 500–1000 Hz. This leads to a group of signals that actually represents the same signal, but corresponds to different frequency bands. We can get a 3D plot with time on one axis, frequency on the second axis, and amplitude on the third axis. This will show us which frequencies exist at which times. Thus, we can know only what frequency bands exist at what time intervals. From there, we can investigate what spectral components exist at any given interval of time. This is a problem of resolution. STFT gives a fixed resolution at all times.

The WT gives a variable resolution. Higher frequencies are better resolved in time, and lower frequencies are better resolved in frequency. This means that a high-frequency component can be better located in time, and with less relative error, than a low-frequency component. On the contrary, a low-frequency component can be better located in frequency when compared to a high-frequency component. Wavelet analysis is performed in a similar way to the STFT analysis in the sense that the signal is multiplied with a function (the wavelet), similar to the window function in the STFT, and the transform is computed separately for different segments of the time-domain signal. There are two main differences between STFT and the continuous WT (CWT): (1) the FTs of the windowed signals are not taken, and therefore a single peak will correspond to a sinusoid; that is, negative frequencies are not computed; and (2) the width of the window is changed as the transform is computed for each spectral component; this is the most significant characteristic of the WT. The CWT is defined as follows [10]:

$$CWT_x^{\psi}(\tau, s) = \Psi_x^{\psi}(\tau, s) = \frac{1}{\sqrt{|s|}} \int x(t)\psi^*\left(\frac{t-T}{s}\right)dt \qquad \text{(A.6)}$$

The transformed signal is a function of two variables: the translation parameter and the scale parameter s. The "psi" is the transforming function called the mother wavelet. Wavelet means a small wave and refers to the condition that this function is of finite length. The term "mother" signifies the functions with different regions of support that are used in the transformation process and are derived from one main function. The term "translation" is related to the location of the window as the window is shifted through the signal. This term corresponds to the time information in the transform domain. We have a scale parameter, which is defined as 1/frequency.

A.5 Type-2 Fuzzy Logic

Type-2 FL handles uncertainties by modeling them and minimizing their effects [11]. If all uncertainties disappear, Type-2 FL is reduced to Type-1 FL. A rule-based FL system (FLS) consists of a fuzzifier, an inference mechanism or engine (fuzzy inference system [FIS] associated with the specified rules), an output processor ([o/p]; fuzzy implication method [FIM] for operations on fuzzy sets that are characterized by membership functions), and the defuzzification process. The o/p for a Type-1 FLS is a defuzzifier, and it transforms a Type-1 fuzzy set into a Type-0 fuzzy set. The o/p for a Type-2 FLS has two components: (1) Type-2 fuzzy sets are transformed into Type-1 fuzzy sets by means of type reduction; and (2) the type-reduced set is then transformed into a number by means of defuzzifcation. Type-1 FLS does not directly handle rule uncertainties because it uses Type-1 fuzzy sets that are certain. Type-2 FLS is very useful in circumstances in which it is difficult to determine an exact membership function for a fuzzy set. They can be used to handle rule uncertainties and even measurement uncertainties.

The output of a Type-1 FLS (the defuzzified output) is analogous to the mean of a pdf. In probability theory, the variance provides a measure of dispersion about the mean and captures more information about the probabilistic uncertainty. Type-2 FL provides this measure of dispersion, which is fundamental to the design of systems that include linguistic or numerical uncertainties that translate into rule or input uncertainties. Linguistic and random uncertainties "flow" through a Type-2 FLS. These effects can be evaluated using the defuzzified output and the type-reduced output of that system. The type-reduced output provides a measure of dispersion about the defuzzified output and can be thought of as related to a linguistic confidence interval. The type-reduced set increases as linguistic or random uncertainties increase; hence a Type-2 FLS is analogous to a probabilistic system (via first and second moments), whereas a Type-1 FLS is analogous to a probabilistic system only through the first moment, the mean.

A Type-2 FLS has more design DOF than a Type-1 FLS. This is because Type-2 fuzzy sets are described by more parameters than are Type-1 fuzzy sets; this signifies that a Type-2 FLS can have better performance than a Type-1 FLS. Although there is no mathematical proof that this will always be the case, in every application of Type-2 FLS, a better performance has been obtained using a Type-2 FLS than is obtained using a Type-1 FLS [11]. Thus, Type-2 FLS is required to directly model uncertainties and minimize their effects within the framework of rule-based FLSs. This can also be extended to MSDF and other decision fusion approaches based on FL.

A.6 Neural Networks

Biological neural networks (BNNs) and ANNs have the ability to continue functioning with noisy and/or incomplete information, have robustness or fault tolerance, and can adapt to changing environments by learning [1]. ANNs attempt to mimic and exploit the parallel-processing capability of the human brain. ANNs are modeled based on the parallel biological structures found in the brain and thus simulate a highly interconnected parallel computational structure with relatively simple individual processing elements called neurons. The human brain has between 10 and 500 billion neurons and approximately 1000 main modules with 500 neural networks (NWs). Each neural NW has 100,000 neurons [5–7]. The neuron is the fundamental building block of the nervous system. A comparison of BNNs and ANNs is given in Table A.1.

Conventional computers cannot match the human brain's capacity for several tasks: (1) speech processing, (2) image processing, (3) pattern recognition, and (4) heuristic reasoning and universal problem solving. Each biological neuron is connected to approximately 10,000 other neurons, giving it a massively parallel computing capability. The brain solves certain problems with two main characteristics: (1) problems that are ill-defined; and (2) problems that require a very large amount of processing. Hence, the similarity between BNNs and ANNs is that each system typically consists of a large number of simple elements that learn and are collectively able to solve complicated problems. ANNs have been successfully applied in image processing, pattern recognition, nonlinear curve fitting or mapping, aerodynamic modeling, flight data analysis, adaptive control, system

TABLE A.1

A Comparison of Biological and Artificial Neuronal Systems

ANN System	BNN System
The signals or data enter the input layer	Dendrites are the input branches and connect to a set of other neurons
The hidden layer is between the input and output layers	The soma is the cell body wherein all the logical functions of the neurons are performed
The output layer produces the NW's output responses, linear or nonlinear	The axon nerve fiber is the output channel; the signals are converted into nerve pulses (spikes) to target cells
Weights and coefficients are the strength of the connection between nodes	Synapses are the contacts on a neuron; they interface some axons to the spines of the input dendrites, and can enhance or dampen the neuronal excitation

identification and parameter estimation, robot path planning, and DF. ANNs are used for prediction of the phenomena they have previously learned through known samples of data. The features that motivate a strong interest in ANNs are as follows: (1) NW expansion is a basis of weights (recall Fourier expansion and orthogonal functions are the bases for expression of periodic signals); (2) NW structure performs extrapolation in adaptively chosen directions; (3) NW structure uses adaptive base functions, for example the shape and location of these functions can be adjusted by the measured data; (4) the NW's approximation capability is generally good; and (5) NW structure is repetitive, which provides resilience to faults. The two types of neurons used are sigmoid and binary. In a sigmoid neuron, the output is a real number, and the sigmoid function normally used is given by

$$z = 1/\left(1+\exp(-y)\right) = \left(1+\tan h(y)\right)/2$$

The y could be a simple linear combination of the input values, or it could be from complicated operations. y = summation of (weights *x) bias. Thus, y = weighted linear sum of the components of the input vector and the threshold. In a binary neuron, the nonlinear function is

$$\begin{aligned} z = \text{saturation}(y) = \{&1 \quad \text{if} \quad y \geq 0 \\ \{&0 \quad \text{if} \quad y < 0 \end{aligned}$$

The binary neuron has saturation or hard nonlinear functions. It is called a perception; however, one can have a higher-order perceptron with y being a second-order polynomial as a function of input data. Weight adaptation in the ANN is performed using the LMS algorithm (Widrow–Hoff delta rule), which minimizes the sum of squares of the linear errors over the training set. ANNs called multilayer perceptrons (MLPNs) are more useful in practical applications with an input layer, a hidden layer, and an output layer and a sigmoid function in each layer. In this NW, the input and output are real-valued signals or data. The NW is capable of approximating arbitrary nonlinear mapping using a given set of data. The generalization property is an important property of the NW; it is a measure of how well the NW performs on an actual problem once the training is accomplished.

A.6.1 Feed-Forward Neural Networks

Feed-forward neural networks (FFNNs) are an information processing system consisting of a large number of simple processing elements without any feedback. FFNNs can be thought of as a nonlinear black box, the parameters (weights) of which can be estimated by any conventional optimization method. The postulated FFNN is first trained using training data, and then is used for prediction using another input set of data that belongs to the same class of data. The FFNN training algorithm gives a procedure for adjusting the initial set of synaptic weights in order to minimize the difference between the NW's output of each input and the output with which the given input is known or desired to be associated. The FFNN training algorithm can be described using matrix or vector notation. The FFNN has the following variables: (1) u_0 as input to (input layer of) the network; (2) n_i as the number of input neurons (of the input layer) equal to the number of inputs u_0; (3) n_h as the number of neurons of the hidden layer; (4) n_0 as number of output neurons (of the output layer) equal to the number of outputs z; (5) $W_1 = n_h \times n_i$ as the weight matrix between the input and hidden layers; (6) $W_{10} = n_h \times 1$ as the bias weight vector; (7) $W_2 = n_o \times n_h$ as the weight matrix between the hidden and output layers; (8) $W_{20} = n_o \times 1$ as the bias weight vector; and (9) μ as the learning rate or step size. The steps in the training algorithm are given as follows:

The initial computation is

$$y_1 = W_1 u_0 + W_{10} \tag{A.7}$$

$$u_1 = f(y_1) \tag{A.8}$$

The $f(y_1)$ is a sigmoid activation function given by

$$f(y_i) = \frac{1 - e^{-\lambda y_i}}{1 + e^{-\lambda y_i}} \tag{A.9}$$

The λ is a scaling factor. The signal between the hidden and output layer is computed as

$$y_2 = W_2 u_1 + W_{20} \tag{A.10}$$

$$u_2 = f(y_2) \tag{A.11}$$

$$f'(y_i) = \frac{2\lambda_i e^{-\lambda y_i}}{(1 + e^{-\lambda y_i})^2} \tag{A.12}$$

The error of the output layer is expressed as

$$e_{2b} = f'(y_2)(z - u_2) \tag{A.13}$$

The recursive weight update formula for the output layer is given as

$$W_2(i+1) = W_2(i) + \mu e_{2b} u_1^T \tag{A.14}$$

The back propagation of the error and the update rule for the weights W_1 are given as

$$e_{1b} = f'(y_1)W_2^T e_{2b} \tag{A.15}$$

$$W_1(i+1) = W_1(i) + \mu e_{1b}u_0^T + \Omega\left[W_1(i) - W_1(i-1)\right] \tag{A.16}$$

The measured signals are presented to the FFNN in a sequential manner, but with the initial weights as the outputs from the previous cycle until the convergence is reached.

A.6.2 Recurrent Neural Networks

Recurrent neural networks (RNNs) are NWs with feedback of output to the internal states. RNNs are dynamic NWs that are amenable to explicit parameter estimation in state-space models and can also be used for trajectory matching. RNNs are a type of Hopfield NW (HNN). One type of RNN is shown in Figure 2.18, and its dynamics are given by

$$\dot{x}_i(t) = -x_i(t)R^{-1} + \sum_{j=1}^{n} w_{ij}\beta_j(t) + b_i; \quad j = 1,\ldots,n \tag{A.17}$$

Here, (1) x is the internal state of the neurons; (2) β is the output state; (3) w is a neuron weight; (4) b is the bias input to neurons; (5) f is the sigmoid nonlinearity, $\beta = f(x)$; (6) R is the neuron impedance; and (7) n is the dimension of the neuron state. Several variants of RNNs are possible.

A.7 Genetic Algorithm

Natural biological systems are more robust, efficient, and flexible compared to many sophisticated artificial systems. Genetic algorithms (GAs) are heuristic search methods based on the evolution of biological species [8]. GAs mimic the genetic dynamics of natural evolution and search for optimal and global solutions for general optimization problems. GAs work with the coding of parameter sets. The parameters must be coded as a finite string. A simple code can be generated by a string of binary numbers. GAs require only objective function values associated with each point.

A.7.1 Chromosomes, Populations, and Fitness

A chromosome is a string constructed by the encoding of information. The GA operates on a number of such chromosomes. A collection of chromosomes is a population. Each chromosome consists of a string of bits. Each bit is a binary 0 or 1, or it may be a symbol from a set of more than two elements. Each member is called an individual and is assigned a fitness value based on the objective function.

A.7.2 Reproduction, Crossover, Mutation, and Generation

In reproduction, chromosomes are copied so that the individuals with a higher fitness have a higher number of copies compared with other individuals. In crossover, members of the newly reproduced strings in the pool are mated at random. Each chromosome is split at a crossover site, selected randomly, and then new chromosomes are formed by joining the top piece of one chromosome with the tail of another. Mutation involves a random change of the value of a string position. The GA consists of a number of such generations of the above operations. In each generation, pairs of individual chromosomes cross over, mutation is randomly applied during the crossover operation, and the new chromosomes that evolve from these operations are placed at a relevant ranked place in the population according to their fitness. The fittest individuals replace the weakest individuals from the previous generation. Thus, a fitter population evolves. For real numbers, the crossover can be performed using the average of the two chromosomes.

A.8 System Identification and Parameter Estimation

Dynamic systems can generally be described by a set of difference or differential equations. Such equations are formulated in the state space, which has a nice matrix structure. Therefore, the dynamics of the systems can be represented fairly well by linear or nonlinear state-space equations. Then, the problem of parameter estimation pertains to the determination of the numerical values of these matrices' elements, which form the structure of the state-space equations [1]. In the system identification problem, the coefficients of transfer function (numerator polynomial or denominator polynomial) are determined from the input–output data of the system using optimization procedures [2–4]. Parameter estimation is an important aspect in model building using the empirical data of the system.

Figure A.1 shows an approach to system identification and parameter estimation problems [1]. The parameters of the postulated mathematical model are adjusted iteratively, using a parameter estimation rule, until the responses of the model closely match the measured outputs of the system in the sense specified by the minimization criterion.

The minimization of an error criterion will lead to a set of equations which are a function of the data and contain unknown parameters. When solved, this set will give estimates of the parameters of the dynamic system. The error can be defined in any of the three ways: (1) output error—the differences between the output of the model to be estimated and the input and output data, the input to the model being the same as the system input; (2) equation error—if accurate measurements of $\dot{x}$, x (state of the system) and u (control input) are available, then equation error is defined as $(\dot{x}_m - Ax_m - Bu_m)$; and (3) parameter error—the difference between the estimated value of a parameter and its true value.

A.8.1 Least-Squares Method

Usually the data are contaminated by measurement noise. These noisy data are used in an identification or estimation procedure or algorithm to arrive at the optimal estimates of the unknown parameters [2–4]. We need to define a mathematical model of the dynamic algebraic system

$$\mathbf{z} = H\boldsymbol{\beta} + v,\ y = H\boldsymbol{\beta} \tag{A.18}$$

Here, y is the true output and z is the $(m \times 1)$ vector of the measurements of the unknown parameters (through H) that are affected by noise; $\boldsymbol{\beta}$ is the

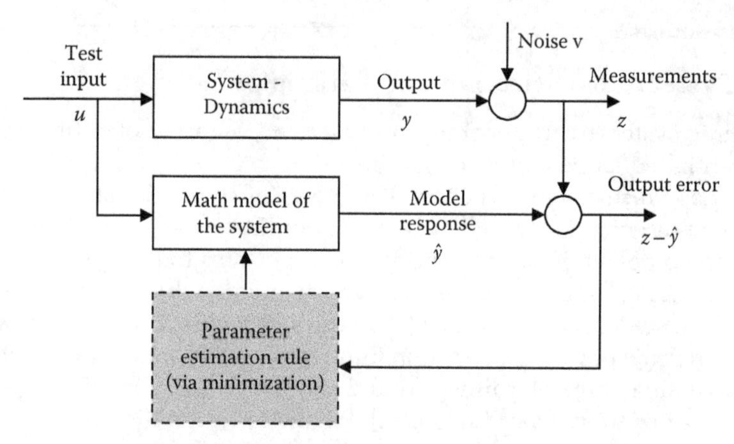

FIGURE A.1
System identification (parameter estimation process).

$(n \times 1)$ vector of the unknown parameters; and v is the measurement noise or errors. The noise is assumed to be zero mean, white, and Gaussian. In general, least-squares (LS) method is applicable to linear and nonlinear problems. The LS estimate of $\boldsymbol{\beta}$ is given as

$$\hat{\beta}_{LS} = (H^T H)^{-1} H^T \mathbf{z} \tag{A.19}$$

Some properties of the LS estimates are as follows: (1) $\hat{\beta}_{LS}$ is a linear function of the data vector $\mathbf{z}$. H could contain input or output data of the system; (2) the error in the estimator is a linear function of the measurement errors (v_k); (3) if the measurement errors are large, then the error in estimation is large; (4) if $E\{v\}$ is zero, then the LS estimate is unbiased; and (5) the covariance of the estimation error is given as

$$E\left\{\tilde{\beta}_{LS}\tilde{\beta}_{LS}^T\right\} \cong P = (H^T H)^{-1} H^T R H (H^T H)^{-1} \tag{A.20}$$

Here, R is the covariance matrix of v. If v is uncorrelated and its components have identical variances, then $R = \sigma^2 I$, where I is an identity matrix. Thus, we have

$$\text{cov}\left(\tilde{\beta}_{LS}\right) = P = \sigma^2 (H^T H)^{-1} \tag{A.21}$$

Hence, the standard deviation of the parameter estimates can be obtained as $\sqrt{P_{ii}}$, ignoring the effect of cross terms of the matrix P. The use of a weighting matrix W in the LS criterion function gives the cost function for the Gaussian LS (GLS) method

$$J = (z - H\beta)^T W (z - H\beta) \tag{A.22}$$

W is the symmetric and positive definite and is used to control the influence of specific measurements upon the estimates of β. The solution exists if the weighting matrix is positive definite.

A.8.2 Maximum Likelihood and Output Error Methods

The ML method invokes the probabilistic aspect of random variables (e.g., measurements or errors) and uses a process which obtains estimates of the parameters [1]. These parameters are most likely to produce model responses that closely match the measurements. An LF, akin to the pdf, is defined when the measurements are collected and used in the pdf. The LF is maximized to obtain the ML estimates of the parameters of the dynamic system. The OEM is an ML estimator (MLE) for data containing only measurement noise. The Gaussian LS (GLS) differential correction method is also an OEM, but it is not based on the ML principle. In any estimator, the error is made in the estimates relative to the true values of the parameters. The true parameters are unknown in any real case. We get only a few statistical indicators for the errors made. The Cramer–Rao (CR) lower bound is one such useful measure for these errors. For unbiased and efficient estimators, the CR lower bound is given as

$$\sigma^2_{\beta e} = I_m^{-1}(\beta)$$

The inverse of the information matrix is the covariance matrix. The information matrix can be computed from the LF or related data. The CR lower bound signifies that the variance in the estimator for an efficient estimator will be at least equal to the predicted variance, whereas for other cases, it could be greater but not less than the predicted value. Thus, the predicted value provides the lower bound. Let a linear dynamical system be described as

$$\dot{x}(t) = Ax(t) + Bu(t) \tag{A.23}$$

$$y(t) = Hx(t) \tag{A.24}$$

$$z(k) = y(k) + v(k) \tag{A.25}$$

The following assumptions are made regarding the measurement noise $v(k)$:

$$E\{v(k)\} = 0; \quad E\{v(k)v^T(l)\} = R\delta_{kl} \tag{A.26}$$

In Equation A.26, we assume that the measurement noise is zero mean and white Gaussian with R as the covariance matrix of this noise. This assumption allows us to use the Gaussian probability concept for deriving the MLE. The parameter vector $\boldsymbol{\beta}$ is obtained by maximizing the LF with respect to β by minimizing the negative (log) LF:

$$L = -\log p(\mathbf{z} \mid \beta, R) = \frac{1}{2}\sum_{k=1}^{N}\left[\mathbf{z}(k) - y(k)\right]^T R^{-1}\left[\mathbf{z}(k) - y(k)\right] + \frac{N}{2}\log|R| + \text{const} \tag{A.27}$$

Minimization of L w.r.t. β (under not too restrictive conditions) results in

$$\frac{\partial L}{\partial \beta} = -\sum_k \left(\frac{\partial y(\beta)}{\partial \beta}\right)^T R^{-1}\left(\mathbf{z} - y(\beta)\right) = 0 \tag{A.28}$$

This yields a system of nonlinear equations. In the present case, we obtain an iterative solution by using quasi-linearization method (also known as the *modified Newton–Raphson* or *Gauss Newton method*), that is, we expand

$$y(\beta) = y(\beta_0 + \Delta\beta) \tag{A.29}$$

as

$$y(\beta) = y(\beta_0) + \frac{\partial y(\beta)}{\partial \beta}\Delta\beta \tag{A.30}$$

Finally we get

$$-\sum_k \left[\frac{\partial y(\beta)}{\partial \beta}\right]^T R^{-1}\left[(\mathbf{z} - y(\beta_0)) - \frac{\partial y(\beta)}{\partial \beta_0}\Delta\beta\right] = 0 \tag{A.31}$$

$$\left[\sum_k \left[\frac{\partial y(\beta)}{\partial \beta}\right]^T R^{-1}\frac{\partial y(\beta)}{\partial \beta}\right]\Delta\beta = \sum_k \left[\frac{\partial y(\beta)}{\partial \beta}\right]^T R^{-1}(\mathbf{z} - y) \tag{A.32}$$

Next we have

$$\Delta\beta = \left\{\sum_k \left[\frac{\partial y(\beta)}{\partial \beta}\right]^T R^{-1}\frac{\partial y(\beta)}{\partial \beta}\right\}^{-1}\left\{\sum_k \left[\frac{\partial y(\beta)}{\partial \beta}\right]^T R^{-1}(\mathbf{z} - y)\right\} \tag{A.33}$$

The ML estimate is obtained as

$$\hat{\beta}_{\text{new}} = \hat{\beta}_{\text{old}} + \Delta\beta \tag{A.34}$$

The absence of true parameter values (in the case of real systems) makes the task of determining the accuracy very difficult. CR bound (CRB) is one good criterion for evaluating accuracy of the estimated parameters. The MLE gives the measure of parameter accuracy without any extra computation. For a single parameter case, for unbiased estimate $\hat{\beta}(\mathbf{z})$ of β, we have

$$\sigma^2_{\hat{\beta}} \geq I_m^{-1}(\beta)$$

Here, the information matrix is given as

$$I_m(\beta) = E\left\{-\frac{\partial^2 \log p(\mathbf{z} \mid \beta)}{\partial \beta^2}\right\} = E\left\{\left(\frac{\partial \log p(\mathbf{z} \mid \beta)}{\partial \beta}\right)^2\right\} \tag{A.35}$$

For several parameters, CR inequality is given by

$$\sigma^2_{\hat{\beta}_i} \geq (I_m^{-1})_{ii}$$

Here, the information matrix is

$$(I_m)_{ij} = E\left\{-\frac{\partial^2 \log p(\mathbf{z}\,|\,\beta)}{\partial\beta_i \partial\beta_j}\right\} = \left\{\left(\frac{\partial \log p(\mathbf{z}\,|\,\beta)}{\partial\beta_i}\right)\cdot\left(\frac{\partial \log p(\mathbf{z}\,|\,\beta)}{\partial\beta_j}\right)\right\} \quad \text{(A.36)}$$

For an efficient estimation, the equality holds and the covariance matrix of the estimation errors is given by

$$P = I_m^{-1}$$

The standard deviation of the parameters is given by

$$\sigma_{\hat{\beta}_i} = \sqrt{P_{ii}} = \sqrt{P(i,i)}$$

and the correlation coefficients are given as

$$\rho_{\hat{\beta}_i,\hat{\beta}_j} = \frac{P_{ij}}{\sqrt{P_{ii}P_{jj}}} \quad \text{(A.37)}$$

For MLE we have

$$\log p(z\,|\,\beta) = -\frac{1}{2}\sum_{k=1}^{N}\left[\mathbf{z}(k) - y(k)\right]^T R^{-1}\left[\mathbf{z}(k) - y(k)\right] + \text{const} \quad \text{(A.38)}$$

The information matrix is obtained as follows: differentiate both sides with respect to β_i to get

$$\left(\frac{\partial \log p(\mathbf{z}\,|\,\beta)}{\partial\beta_i}\right) = \sum_k\left(\frac{\partial y}{\partial\beta_i}\right)^T R^{-1}(\mathbf{z} - y) \quad \text{(A.39)}$$

Differentiate both sides again with respect to β_j to get

$$\left(\frac{\partial^2 \log p(\mathbf{z}\,|\,\beta)}{\partial\beta_i \partial\beta_j}\right) = \sum_k\left(\frac{\partial^2 y}{\partial\beta_i \partial\beta_j}\right)^T R^{-1}(\mathbf{z} - y) - \sum_k\left(\frac{\partial y}{\partial\beta_i}\right)^T R^{-1}\frac{\partial y}{\partial\beta_j} \quad \text{(A.40)}$$

If we take the mathematical expectation of the terms in Equation A.40 we obtain

$$(I_m)_{ij} = E\left\{-\frac{\partial^2 \log p(z\,|\,\beta)}{\partial\beta_i \partial\beta_j}\right\} = \sum_{k=1}^{N}\left(\frac{\partial y(k)}{\partial\beta_i}\right)^T R^{-1}\frac{\partial y(k)}{\partial\beta_j} \tag{A.41}$$

We recall here that the increment in the parameter estimate $\Delta\beta$ is given by

$$\Delta\beta = \left[\sum_k\left(\frac{\partial y}{\partial\beta}\right)^T R^{-1}\frac{\partial y}{\partial\beta}\right]^{-1}\sum_k\left(\frac{\partial y}{\partial\beta}\right)^T R^{-1}(\mathbf{z}-y) \tag{A.42}$$

The output error approach is based on the assumption that only the observations contain measurement noise and that there is no noise in the state equations. The mathematical model of a linear system consists of vector $\mathbf{x}$ representing the system states, vector $\mathbf{y}$ representing the computed system response (model output), vector $\mathbf{z}$ as the measured variables, and $\mathbf{u}$ as the control input vector. A, B, and H contain the parameters to be estimated. The OEM assumes that the measurement vector is corrupted with noise that is zero mean and has a Gaussian distribution with covariance R. The aim is to minimize the error (criterion) between the measured and model outputs by adjusting the unknown parameters in matrices A, B, and H. Let the parameter vector to be estimated be represented by Θ where Θ = [elements of $A, B,$ and $H,$ initial condition of x].

The estimate of Θ is obtained by minimizing the cost function

$$J = \frac{1}{2}\sum_{k=1}^{N}\left[\mathbf{z}(k)-y(k)\right]^T R^{-1}\left[\mathbf{z}(k)-y(k)\right] + \frac{N}{2}\ln|R| \tag{A.43}$$

The estimate of R is obtained from

$$\hat{R} = \frac{1}{N}\sum_{k=1}^{N}\left[\mathbf{z}(k)-\hat{y}(k)\right]\left[\mathbf{z}(k)-\hat{y}(k)\right]^T \tag{A.44}$$

The estimate of Θ at the $(i+1)^{th}$ iteration is obtained as

$$\Theta(i+1) = \Theta(i) + \left[\nabla^2_{\Theta}J(\Theta)\right]^{-1}\left[\nabla_{\Theta}J(\Theta)\right] \tag{A.45}$$

Here, the first and the second gradients are defined as

$$\nabla_{\Theta}J(\Theta) = \sum_{k=1}^{N}\left[\frac{\partial y}{\partial\Theta}(k)\right]^T R^{-1}\left[\mathbf{z}(k)-y(k)\right] \tag{A.46}$$

$$\nabla^2_\Theta J(\Theta) = \sum_{k=1}^{N} \left[\frac{\partial y}{\partial \Theta}(k) \right]^T R^{-1} \left[\frac{\partial y}{\partial \Theta}(k) \right] \tag{A.47}$$

Starting with suitable initial parameter values, the model response is computed with the input used to obtain measurement data. The estimated responses and the measured responses are compared, and the response errors are used to compute the cost function. Using Equation A.45, the updated parameter values are once again applied in the mathematical model in order to compute the new estimated response and the new response error. This updating procedure continues until convergence is achieved. The diagonal elements of the inverse of the information matrix give the individual covariance values, and the square root of these elements is a measure of the standard deviations called the CRB.

$$\text{Fisher information matrix} = \nabla^2_\Theta J(\Theta) \tag{A.48}$$

$$\begin{aligned}\text{Standard deviation of estimated parameters} &= \text{CRB}(\Theta) \\ &= \text{diag}\left[\sqrt{[\nabla^2_\Theta J(\Theta)]^{-1}} \right]\end{aligned} \tag{A.49}$$

ML estimates are consistent, asymptotically unbiased, and efficient. Computation of the coefficients of parameter vector $\boldsymbol{\Theta}$ requires: (1) the initial values of the coefficients in $\boldsymbol{\Theta}$; (2) the current values of variables y at each discrete time point k; and (3) the sensitivity matrix $(\partial y/\partial \Theta)_{ij} = \partial y_i/\partial \Theta_j$. The current state values are computed by the numerical integration of the system state equations. The sensitivity coefficients are approximately computed using a numerical difference method.

A.9 Reliability in Information Fusion

The real test of information fusion (IF) depends on how well the knowledge displayed or generated by the fusion process represents reality. This in turn depends on (1) the adequacy of the data, (2) the adequacy of the uncertainty model used, and (3) the accuracy, appropriateness, or applicability of the prior knowledge [12]. It is important to consider the reliability of these models and the fusion results. Different models may have different reliability, and it is necessary to account for this fact.

The difficulty of modeling uncertainty stems from the difficulty of finding an adequate belief model. An inadequate choice of metrics or poor estimation of the LFs can result in an inadequate belief model and can lead to a combination of conflicting and unreliable beliefs. Moreover, beliefs can be models within different uncertainty frameworks, and dealing with different sources may also mean dealing with different framework theories.

Modeling beliefs has always some limitations, and models are only valid within a certain range. Therefore, we have to take into account the range and the limitations of the belief model used for each source when combining information provided by many sources. The most natural way to deal with this problem is to establish the reliability of the beliefs computed within the framework of the model selected. This may be achieved using reliability coefficients, which introduce the second level of uncertainty (uncertainty of evaluation of uncertainty) and represent a measure of the adequacy of the model used and the state of the environment observed. There are at least two approaches used for defining reliability as a higher-order uncertainty: (1) reliability is understood as the relative stability of the first order uncertainty—reliability is often measured by the performance of each source (e.g., by recognition or by false alarm rates); and (2) to measure the accuracy of predicted beliefs—reliability coefficients represent the adequacy with which each belief model represents the reality. In general, fusion operators are based on the assumption that the sources are reliable. If the sources are not reliable, the fusion operators must account for their reliability. This can be represented as F = function of (degree of beliefs, reliability, and coefficients). Reliability coefficients control how respective sources influence fusion results.

The reliability coefficient is close to 0 if the source is unreliable and close to 1 if it is more reliable. Reliability coefficients depend not only on the model selected but also on the characteristics of the environment and the particular domain of the input. The problem of source reliability is related to the problem of conflict. Indeed, the existence of conflict indicates the existence of at least one unreliable source. Global knowledge about the sources, environment, and the nature and properties of the particular credibility model can provide different information about reliability. Several situations that differ by the level of knowledge about source reliability can be considered: (1) It is possible to assign a numerical degree of reliability to each source; each value of reliability may be "relative" or "absolute" and may or may not be linked by an equation, such as the sum of reliability coefficients being equal to 1; (2) only an order of the reliabilities of the sources is known; their precise values are unknown; and (3) a subset of sources is reliable, it is unclear which one in particular is reliable. Dealing with these situations calls for one or the combination of the following two strategies: (1) explicitly utilizing reliability of the sources; in this case two substrategies may be conceivable: (a) including reliability before fusion in the modeling belief for each source, to compensate for their different reliability and make them totally, or at least equally, reliable before fusion; then fuse the transformed beliefs (separable case), and (b) using reliability coefficients to modify the fusion operator considered (nonseparable case) because each source cannot be transformed independently; and (2) identifying the quality of the data input to the fusion processes and eliminating data of poor reliability.

A.9.1 Bayesian Method

In the Bayesian method, degrees of belief are represented by *a priori*, conditional, and posterior probabilities. The decision is made on the basis of the computed posterior probabilities. Fusion is performed by the Bayesian rule. The fusion operator assumes the total reliability of the sources.

A.9.1.1 Weighted Average Methods

The linear opinion pool can be used:

$$P(x \mid Z^k; R_j) = \sum_i R_i P\left(x \mid \mathbf{z}_i^k\right) \tag{A.50}$$

Here, the reliability coefficients, associated with the sources, are used as the weights (see Equation 2.55). x is a hypothesis and z is the measurement or feature vector.

The other methods are logarithmic opinion pools:

$$p(x \mid Z^k; R_i) = \sum_i R_i \log\left\{P\left(x \mid \mathbf{z}_i^k\right)\right\} \tag{A.51}$$

and

$$P(x \mid Z^k; R_i) = P(x) \prod_i \left\{P\left(x \mid \mathbf{z}_i^k\right)/P(x)\right\}^{R_i} \tag{A.52}$$

A.9.2 Evidential Methods

In these methods (Dempster–Shafer), the decision (discussed in Section 2.3.2) is based on the following formula:

$$m^{1,2}(C) \propto \sum_{A \cap B = C} m^1(A) m^2(B) \tag{A.53}$$

Here, instead of probabilities, "mass" is assigned. The mass can be perceived as a probability, but it is not a probability. In terms of use of reliability, we have the following formula with R as the reliability coefficient.

$$m(A) = \sum_i R_i m_i(A) \tag{A.54}$$

A.9.3 Fuzzy Logic–Based Possibility Approach

In the fuzzy logic–based possibility approach, the information obtained from a sensor is represented by a possibility distribution [12]:

$$\Pi : \Theta -> [0,1] : \max_{\theta \in \Theta} \Pi(\theta) = 1 \tag{A.55}$$

Most of the combination rules in fuzzy logic theory are based on *t*-norms and *s*-norms (*t*-conorms). The disjunctive rule is used when at least one source of data is reliable, but the reliable source is not known. This is the "OR" or the "max" (union) rule in fuzzy operator theory (see Section 6.1). When equally reliable sources are available, a conjunctive operation is used. This is the "AND," "min," or "inf" (infimum, intersection) operator. The decision fusion rule, based on the possibility theory, is given as

$$\Pi_p(\theta) = \sum_i R_i \Pi_i(\theta) \tag{A.56}$$

Interestingly, the operator in Equation A.56 is not linked to the "min" and "max" operators.

The reliability coefficients can be obtained or determined (1) by utilizing domain knowledge and contextual information; (2) from the training data; (3) by using a degree of consensus among various sources; or (4) by experts' subjective probabilities and judgements. As a result of the incorporation of the reliability into the fusion process, the fusion system performance would most likely increase. However, the reliability of the fusion results should be studied.

A.10 Principal Component Analysis

Principal component analysis (PCA) involves an eigenanalysis of the correlation or covariance matrix of the given data set. It reduces the data dimensionality by performing an analysis of covariance. PCA is used to uncover hidden trends in the data, such as image intensities, and will reveal the relevant components, patterns in an image scene, and so on. Once the pattern is found, the data can be compressed by removing the unimportant aspects. This method is also used in image compression. At least some of the variables in a given data set might be correlated with each other, implying some redundancy in the information provided by these variables. PCA exploits the redundancy in such multivariate data and helps detemine the patterns or relationships in the variables. It also reduces the dimensionality of the data set without much loss of information.

A.11 Reliability

Reliability is formally defined as the probability that an item will perform a required function without failure under the stated conditions for a stated period of time. Several other definitions of reliability are available in literature on the subject: (1) the ability of equipment, machines, or systems to consistently perform their intended or required function or mission on demand and without degradation or failure; (2) the probability of a failure-free performance over an item's useful life or a specified timeframe, under specified environmental and duty-cycle conditions, and (3) the consistency and validity of test results determined through statistical methods after repeated trials. Reliability in the context of sensor or DF pertains to reliability of the sensors or data. If the sensors or data are not reliable, then there is no need to fuse these data.

References

1. Raol, J. R., G. Girija, and J. Singh. 2004. *Modelling and parameter estimation of dynamic systems, IEE Control Engineering Series Book*. Vol. 65. London: IEE (With associated MODEST software in MATLAB via IEE book. NAL web sites).
2. Sorenson, H. W. 1980. *Parameter estimation—Principles and problems*. New York: Marcel Dekker.
3. Eykhoff, P. 1972. *System identification: Parameter and state estimation*. London: John Wiley.
4. Ljung, L. 1987. *System identification: Theory for the user*. Englewood Cliffs, NJ: Prentice Hall.
5. Eerhart, R. C., and R. W. Dobbins. 1993. *Neural network PC tools—A practical guide*. New York: Academic Press.
6. Hush, D. R., and B. G. Horne. 1993. Progress in supervised neural networks—What is new since Lippmann. *IEEE Signal Process Mag* 10(1):8–39.
7. King, R. E. 1999. *Computational intelligence in control engineering*. New York: Marcel Dekker.

8. Goldberg, D. E. 1989. *Genetic algorithms in search, optimization and machine learning*. Boston: Addison-Wesley.
9. Sinha, A., H. Chen, D. G. Danu, T. Kirubarajan, and M. Farooq. 2008. Estimation and decision fusion: A survey. *Neurocomputing* 71(13–15):2650–6.
10. Polikar, R. From the Fourier transform to the wavelet transform. rowan.du/~polikar/WAVELETS/WTutorial.html – 92k (accessed January 2009).
11. Mendel, J. M. 2001. Why we need Type-2 fuzzy logic system. http://www.informit.com/articles/article.aspx?p=21312-32k. May 2001.
12. Rogova, G. L., and V. Nimier. 2004. Reliability in information fusion. In *Proceedings of the 7th International Conference on Information Fusion*, eds. P. Svensson and J. Schubert.

Index

F

J

K

L

M

N

CPSIA information can be obtained
at www.ICGtesting.com
Printed in the USA
LVHW04*2102120918
589940LV00003B/3/P